D0203400

DISCARD

BRYANT & STRATTON
SYRACUSE NORTH CAMPUS
RESOURCE CENTER

Online around the World

Online around the World

A Geographic Encyclopedia of the Internet, Social Media, and Mobile Apps

LAURA M. STECKMAN AND
MARILYN J. ANDREWS, EDITORS

ABC-CLIO™

An Imprint of ABC-CLIO, LLC
Santa Barbara, California • Denver, Colorado

Copyright © 2017 by ABC-CLIO, LLC

All rights reserved. No part of this publication may be reproduced, stored in a retrieval system, or transmitted, in any form or by any means, electronic, mechanical, photocopying, recording, or otherwise, except for the inclusion of brief quotations in a review, without prior permission in writing from the publisher.

Library of Congress Cataloging-in-Publication Data

Names: Steckman, Laura M., editor. | Andrews, Marilyn J., editor.
Title: Online around the world : a geographic encyclopedia of the Internet,
    social media, and mobile apps / Laura M. Steckman and Marilyn J. Andrews,
    editors.
Description: Santa Barbara, California : ABC-CLIO, [2017] | Includes
    bibliographical references and index.
Identifiers: LCCN 2016048706 | ISBN 9781610697750 (hbk : alk. paper) |
    ISBN 9781610697767 (ebook)
Subjects: LCSH: Online social networks—Encyclopedias. | Social
    media—Encyclopedias. | Mobile apps—Encyclopedias.
Classification: LCC HM742 .O552 2017 | DDC 302.30285—dc23
LC record available at https://lccn.loc.gov/2016048706

ISBN: 978-1-61069-775-0
EISBN: 978-1-61069-776-7

21 20 19 18 17   1 2 3 4 5

This book is also available as an eBook.

ABC-CLIO
An Imprint of ABC-CLIO, LLC

ABC-CLIO, LLC
130 Cremona Drive, P.O. Box 1911
Santa Barbara, CA 93116-1911
www.abc-clio.com

This book is printed on acid-free paper ∞

Manufactured in the United States of America

Contents

Preface | ix

Introduction | xi

Overview of Popular International Social Media Sites | xxiii

Chronology of Significant Events in Cyber History | xxxvii

Afghanistan | 1

Algeria | 5

Angola | 8

Argentina | 11

Australia | 16

Bangladesh | 21

Botswana | 24

Brazil | 27

Cameroon | 33

Canada | 37

Chile | 41

China | 45

China: Hong Kong | 51

China: Macau | 54

Colombia | 55

Cuba | 59

Denmark | 65

Ecuador | 71

Egypt | 74

El Salvador | 79

Estonia	82
Ethiopia	87
Finland	93
France	96
Georgia	101
Germany	103
Ghana	107
Greece	110
Honduras	115
Hungary	118
Iceland	123
India	126
Indonesia	130
Iran	134
Iraq	139
Ireland	142
Israel	144
Italy	148
Jamaica	153
Japan	155
Kazakhstan	161
Kenya	164
Kyrgyzstan	168
Libya	173
Malaysia	177
Mexico	181
Mongolia	184
Morocco	188
Mozambique	191

Myanmar	194
Namibia	199
Nepal	204
Nigeria	208
North Korea	214
Pakistan	221
Paraguay	224
Peru	227
Philippines	230
Poland	234
Qatar	239
Romania	243
Russia	247
Saudi Arabia	255
Senegal	258
Singapore	262
Somalia	266
South Africa	270
South Korea	274
Spain	281
Suriname	284
Syria	288
Taiwan	293
Tanzania	297
Thailand	301
Timor-Leste (East Timor)	305
Tunisia	308
Turkey	313
Ukraine	317

United Arab Emirates   320

United Kingdom   324

United States   328

Uzbekistan   333

Venezuela   337

Vietnam   342

Yemen   347

Zimbabwe   351

Bibliography   355

About the Editors and Contributors   367

Index   373

Preface

Online around the World aims to give readers a glimpse of how people use the internet, social media, and social apps differently around the globe. There is a tendency to think that because they are popular forms of communication, everyone uses them the same way and for the same purpose. The reality, however, is far more complex. For example, there is no singular version of Facebook or Instagram, the two most popular social media platforms worldwide at the time of this writing. How they are used in the United States varies—at times greatly—from how people use them in Nigeria, Colombia, or other areas. In some parts of the world, they aren't used at all, either by choice or through government restrictions.

This encyclopedia focuses on explaining some of the major differences between the online experience in multiple countries, often in countries occupying the same geographic regions. It includes information on the history of the internet, gives social media and app profiles of countries that include user demographics when available, and examines some of the freedoms, restrictions, and activism that the web has enabled. In some countries, the internet is not always used for progress or positive social change; when this is the case, the corresponding entries include information on the darker side of the internet and social media. The entries can be read either individually or used to compare and contrast across hemispheres, continents, regions, and next-door neighbors.

Compiling a book on the internet and social media has its challenges. Because the environment is robust, expanding, and moving at an increasingly rapid pace, it can be difficult to capture the essence of what it looks like at any given moment for an entire country. In some cases, how it functions in the national capital and urban centers is vastly different from rural or more remote areas. In composing this volume, the most reliable, up-to-date sources were available online rather than being based primarily in academic journals and books—a contrast to other encyclopedias. Each entry, as it was written, encompassed the general online preferences and realities of the country at that time; inevitably, as is the nature of the online world, some of that information has already become outdated as people change how they access the internet, where they go online, and with whom they choose to exchange information.

This volume is a culmination of the research and hard work of various talented researchers. The editors were already familiar with a community of researchers working on different online environments in various regions and multiple languages. Starting with that original group, the author pool expanded to include referrals from the community and, on occasion, the identification of researchers posting exceptional papers and scholarship on sites such as Academia.edu and Research-Gate. Our sincerest thanks go to all our contributors, as well as the publishing team at ABC-CLIO, without whose hard work this book would not have been possible.

Introduction

Worldwide, the internet and social media constitute phenomena that have become integral to people's lives. They permit real-time, rapid communication, including information and content sharing with people all over the world, at the touch of a few buttons. Less than twenty years ago, while the internet was spreading across the world, it was almost unforeseeable that it would lead to the invention and mass adoption of social media, and later applications or *apps,* that would revolutionize how people consume and create online content. The online information environment started very small—mostly at the behest of universities and government-sponsored institutions—and has grown to the extent that today, it reaches almost every corner of the globe. The early internet grew from a few providers and several budding social media platforms into a dynamic, high-velocity, ubiquitous, and interactive environment that reaches almost every enabled device. It stands to reason that this environment will continue to improve faster and grow ever more complex as new technologies are introduced and people keep up their demand for and interest in new uses for them. At the same time, these technologies are ultimately used and produced by people from various cultures and online needs. These differences between peoples have led to innovations in emerging communications technology, as well as global entrepreneurship that develops new ideas to meet mainstream and niche cultural needs.

Internet

The term *internet* refers to the hardware and software infrastructure that connects multiple networks located all over the globe. It is, in a nutshell, the overarching framework that connects and enables the sharing of information among these networks. The U.S. government laid the initial groundwork for the internet in the late 1950s. In the midst of the space race between the United States and the Soviet Union that began after World War II (1939–1945), the United States created the Advanced Research Project Agency (ARPA) to invigorate research and development of the country's science and technology in 1958. ARPA's portfolio contained some of the early technologies that paved the way for the internet, although it did not emerge in its current form until after 1993, when the European Organization for Nuclear Research, known colloquially as *CERN,* released the software required to run the World Wide Web over the internet. After 1993, the internet grew exponentially, allowing freer and faster exchanges of information.

Many people envisioned the internet as a tool to promote freedom of speech and open access to information. The freedom of speech issue usually resonates in local politics and is often attached to internet speech and communications; arguably, while the internet has supplied the means to communicate with others quickly from anywhere, it has not been entirely successful at promoting freedom of speech throughout the world. In terms of access, there are infrastructural and financial considerations that limit mass access to the internet—namely, speed and cost. *Internet speed* is a measure of how fast data travels from a computer, over the internet via the World Wide Web, to another wired device. The measure is often different for *download speed*, or how quickly information goes from the web to a device, and *upload speed*, the speed of transfer from a device to a computer. The unit used for this speed is megabits per second (Mbps).

Average internet speed is one of the most reported measurements. It refers to the mean number derived from tests conducted at multiple places throughout a twenty-four-hour period. In early 2016, Akamai's State of the Internet (2016) report found the world's average connection speed to be 6.3 Mbps. In a comparison of average connection speeds worldwide, the company designated South Korea as the overall speed winner, at 29 Mbps. Rounding out the top ten were Norway, Sweden, Hong Kong, Switzerland, Latvia, Japan, the Netherlands, the Czech Republic, and Finland. The United States did not make the list because its overall average was too low; in a closer study of the country, the District of Columbia had the fastest internet speed, at 24 Mbps, with the runners-up being Delaware, Rhode Island, Massachusetts, Utah, New Jersey, Maryland, New York, Virginia, and Washington State.

Sometimes average peak internet speed is reported. Peak speed is similar to "internet rush hour." It is the mean derived from testing internet speeds in multiple locations during the times when the most people are online. In the United States, the internet rush hour period is 7–10 p.m., when the majority of people are home from work. Rush hours can change by country, though typically they are in the evenings. When the internet is being accessed by a lot of people, there can be congestion that slows connection speeds. In some countries, average peak connection speeds can be higher than the average speed over a day. For example, Mongolia ranks in the top ten for average peak connection speed worldwide, but performs much worse in terms of average speed overall (Akamai 2016).

Internet costs around the world are difficult to measure because telecommunications companies set the price relative to download and upload speed. In addition, to keep prices relatively affordable for the middle and upper classes, many of whom rely on mobile phones for internet access, companies often consider where to place their cellular towers and other equipment to increase access and therefore increase their customer base. Prices can fluctuate based on competition among providers and local consumer demand. There are also cultural factors that determine what a typical data plan is; it could be unlimited or set to a certain data amount, such as 1 gigabyte (GB) or 4 GB. Data plans could also be offered as prepaid, pay-as-you-go, or monthly packages. To provide some comparisons, a Pakistani researcher priced country-specific broadband packages offering unlimited data plans through cable or digital subscriber line (DSL) at 10 Mbps. He determined that the cost in Pakistan averaged $35.81. In contrast, the same plan costs $84.24 in the

United Arab Emirates (UAE), $55.81 in Australia, $48.41 in the United States, $18.28 in China, and $6.80 in Russia (Sheikh 2015). This example demonstrates that multiple factors contribute to price differences among countries, or essentially, that internet and data pricing is not universal.

The percentage of people in a country who have access to the internet is expressed as a country's internet penetration. Many people and organizations believe that everyone should have internet access. These people and groups are experimenting with different ways for populations with little or no access to gain it. Examples of efforts to achieve global connectivity include Facebook's work with drones and Google's Project Loon, an initiative that involves using inflated balloons that fly over unconnected areas and provide internet access. These examples represent two novel ways to provide internet access to areas without stable internet connections, as companies seek innovative avenues to connect people with internet service. In countries where expensive technologies are prohibitive, such as Bangladesh, a group of women called the Info Ladies bicycle through the countryside, carrying computers and other equipment to bring the internet to isolated villages and other remote areas.

Speed, cost, and access barriers, in addition to a country's internal politics, laws, and sociocultural norms, affect which members of the population have internet access. The International Telecommunications Union (ITU), an organ of the United Nations, released a report in December 2015 describing discrepancies in global connectivity (ITU 2015). It discussed the fact that approximately 3.2 billion people—slightly less than half of the world's population—had access to the internet. However, that access was not equally available across the world; the least-developed countries accounted for only 9.5 percent of the 3.2 billion. The report also stated that there is no gender equality for internet users, and in some cases, there are wide regional variations as well. For example, the report concluded that 18.4 percent of women had internet access on the African continent, as opposed to men, at 23.1 percent. In contrast, in Europe, 74.3 percent of women had internet access, as compared to 81 percent of men. For men, however, there was little difference between American users and users from the former Soviet Union, with 65.8 percent and 62.2 percent online, respectively (Truong 2015). There is no standard internet user demographic that can be applied globally. Infrastructure, costs, politics, and sociocultural factors play a role in who has internet access where.

Finally, the internet is a dynamic system that is always in flux. The internet of the future is already here in some ways. It is poised to change how people connect, in person and through their devices, with the world around them. On top of the internet itself, which is essentially a network of networks, the number of enabled objects, sensors, and devices that can connect and communicate over the internet is rapidly accelerating. This phenomenon is often referred to as the *Internet of Things* (*IoT*). IoT is not really a new idea; in fact, Mark Weiser (1952–1999), the chief scientist at Xerox PARC Labs, offered a vision of IoT in 1991 under the name "ubiquitous computing" (Weiser 1991). Weiser's conceptualization differs slightly from how IoT has developed technologically, but there is little doubt that his work laid the foundation for the current number of devices that can transmit information via the internet.

Social Media and Social Networking Sites

The term *social media* currently has no accepted single definition; there are dozens of proposed definitions, yet the term either means different things to different people or gets boiled down to specific platforms such as Facebook or Twitter. In fact, in spoken American English, *Facebook* and *Twitter* are sometimes used synonymously with *social media*. In contrast, many definitions are too vague, in that they categorize anything posted to the internet as social media. Other definitions explain social media by the back-end technology used to create specific platforms. Andreas Kaplan (1977–) and Michael Haenlein (n.d.–) proposed a definition of *social media* that bridges the two viewpoints, stating that social media encompasses "a group of Internet-based applications that build on the ideological and technological foundations of Web 2.0 and that allow the creation and exchange of User Generated Content" (Kaplan and Haenlein 2010, 60). In other words, social media is characterized by the tools, such as the platforms and/or apps that allow users to create and transmit content. It is akin to broadcasting information to large groups of people. In a technical sense, social media is like a broadcasting channel rather than a specific site or location.

Occasionally, the terms *social media* and *social networking site* are used interchangeably. However, there are distinct differences between the two. Social networking sites are online spaces, which could be platforms, websites, or apps, where users go to join and communicate with a like-minded community of interest. Social networking sites allow people to converse and share information about specific topics, such as a hobby, sports team, or other common interest. Whereas the emphasis on social media falls on the apps that allow content creation and exchange, social networking sites enable communication and unite people of similar interests. Sometimes there are even instances where a platform has the capacity to do both. Facebook, which began as a social media platform in 2004, later added Facebook Messenger, to enhance the platform with a social networking component. Thus, Facebook could be considered as either social media or a social networking site, depending on the elements of the site being discussed.

People have several primary motivations for using social media and social networking sites. According to the Global Web Index (2015), there are ten main reasons that people use social media: (1) staying in touch with friends and family, (2) keeping up with the news and current events, (3) passing time, (4) searching

---

Social media data, along with other information posted on the internet, comprise what analysts now refer to as "big data." If all the big data produced online in one day were burned to recordable DVDs, the resulting stack of DVDs would be tall enough to reach the moon twice (Nabila 2014). DVDs are 1.2 millimeters thick, meaning that the stacks would total more than 640 billion discs. Every day, as the amount of data posted online increases, those stacks would continue to reach even farther into the solar system.

for content, (5) sharing opinions and reviews, (6) sharing photos and video, (7) maintaining a presence in social circles, (8) networking, (9) meeting new people and dating, and (10) sharing personal details. These are the general categories that came from the report, though they do not represent all the motivations that people have when they access social media. For instance, social media can become a way for social activists to engage a larger audience, help a small business attract new consumers and launch new products, foster romantic relationships, and so much more. Often, people will choose a social media or social networking site that corresponds to the community that they want to reach and is considered culturally acceptable by that community.

Social media and communications preferences across the world are not the same. People may choose certain social media apps over others, based on how they transmit information or whom the information is most likely to reach. The same is true of social networking sites. Preferences change based on language and script compatibility, existing interpersonal social relations and means of communications, privacy policies, and the purpose of the site. For example, Facebook attracts "friends," as opposed to Twitter, which attracts "followers." The types of information shared with friends can be different than with people who declare themselves followers. Likewise, different genders and age groups may prefer one site over another. In the United States, for instance, Pinterest claims that women comprise its majority user demographic; Snapchat declares itself the clear winner for the eighteen- to twenty-four-year-old user bracket (Becker 2015). The diversity of social media allows groups of almost any age, gender, and culture to find a suitable space to discuss common interests. It just may be, however, that young male bicyclists from South Korea would use an entirely different platform than young men with the same interests in Peru. Sometimes a social media site will bring global enthusiasts or hobbyists together and create a new space for multicultural dialogue and exchange.

User demographics and preferences can fluctuate widely by country and region. Vincenzo Cosenza (1973–), an Italian blogger and researcher, created the World Map of Social Networks to map major social networking sites. It is updated every six months to track the popularity and changes in global demand for the major platforms. The current global winner is Facebook. As of January 2016, it was the most popular site in 127 countries and boasted over 1.6 billion unique monthly users. Twitter, commonly believed to be a popular competitor to Facebook, has struggled in many markets. In 2016, it held the most popular spot only in Japan. Apart from Facebook and, in Japan, Twitter, other notable popular sites that held the top spot in at least one country were Qzone (China), VKontakte (Russia), Facenama (Iran), and Odnoklassniki (Moldova, Uzbekistan, and Kyrgyzstan). The second place winner worldwide was Instagram. Some notable deviations for the second place rank included Facebook (Japan), LinkedIn (India), Reddit (Norway), Odnoklassniki (Russia, Germany, Israel, and Ukraine), and VKontakte (Czech Republic). The map is updated every January and August. Twitter struggled in third place overall, though it maintained second place in the United Kingdom, Spain, France, Egypt, Saudi Arabia, and Pakistan.

Applications or Apps

Applications, commonly referred to as *apps*, are software programs that appear as colorful icons on a tablet or mobile phone. The earliest apps were arcade games, calculators, calendars, and other utility programs. In fact, one of the earliest apps to reach mass popularity was Nokia's Snake game in 1997, which sold an estimated 400 million copies. The Snake game captivated many people, despite the fact that its design was pretty simple, containing only small black squares that moved around a solid, dark-colored background.

The Snake game would not have been called an app during its heyday. The term *app* has become prevalent with the popularity of smartphones and tablets in the 2000s, though technically they have existed for a much longer time. As consumers became accustomed to the internet in the late 1990s, they wanted access to the same information on their phones. Developers had to contend with the fact that webpages were already colorful, used dynamic scripts (such as Java), and contained too much information to be rendered as simply as the earlier apps had. Through a series of advances in technology, mobile phones started evolving into smartphones, which enabled the development of mobile apps that could access the web.

There is little difference between what an app icon on a smartphone does and an icon that launches a computer program does—in fact, even Microsoft's Xbox calls its programs *apps*. The differences are often structural due to decisions made during the early development stages. From a development perspective, most mobile phone apps are classified as being native, web, or hybrid apps. A *native app* is an app that resides on a device, meaning that it appears as an icon and can utilize the features of the device when the user gives it all the necessary permissions. Most native apps are device specific. For example, while there may be Android and iOS versions available, these are distinct apps designed to function solely on the platform in question. Some native apps can work even when the device is offline. Most computer programs would be considered the equivalent of native apps; however, this is not set in stone as technology advances.

In contrast, a *web app* looks like a native app, but instead of operating the same way, it is really a website coded to look like an app and run through a browser. Web apps tend not to be programmed to use the device's particular features, and they usually can be used offline only if the browser has a caching feature where it saves the pages in the app's memory. Web apps are usually not available in an app store, but directly on the web from a company or website.

A *hybrid app* combines the features and functionality of native and web apps. It relies on a browser as the web app does, wherein the developers usually build a browser directly into the app. Hybrid apps can use some of a device's built-in features, such as the camera and notifications. Because of this, hybrid apps are generally designed for a specific platform, such as Android or iOS, though they sometimes rely on cross-platform programming. Ideally, a hybrid app marries the best features and functionality of native and web apps to provide a satisfactory user experience.

Some of the most common types of apps are for messaging and gaming or entertainment. However, apps can do a lot more than messaging and gaming. The expression "There's an app for that" seems true when searching through app stores such as Google Play and the App Store. At the end of 2014, Google expanded the number of categories for apps to twenty; similarly, Apple grew to twenty-four distinct categories in the App Store. Gaming apps remained the most popular and profitable overall in 2016. With the popularity of apps and the growing usage of smartphones and tablets, developers will continue creating apps that satisfy their customers' needs.

Because of high levels of consumer demand, the app market has expanded continuously with the growing demand for mobile devices. For example, when Apple first launched its App Store in mid-2008, it had only a few apps. By early 2009, the store contained approximately 35,000 apps. By June 2016, Apple now boasts more than 2 million apps in its store (Statista 2016b). Similarly, in June 2016, Google Play offered 2.2 million apps, the Windows Store had 669,000, the Amazon App Store around 600,000, and BlackBerry World with the fewest, at 234,500 (Statista 2016a). As new technologies emerge, the current trend suggests that the number of apps will increase in order to take advantage of new device features; app producers will likely push themselves to stay on the cutting edge, ahead of their competitors.

Technology is not the only feature driving app design; there is also a human dimension, meaning that people's culture influences which types of apps are most useful and most relevant in their day-to-day lives. It is not uncommon to find mobile phones in villages or small communities that either have no electricity or share a generator to power electronic devices. Those phones may be difficult to charge, but they will still have apps, and those apps can vary from village to village or country to country.

Another factor that influences app choice has to do with how devices (in this instance mobile phones) are used culturally. In the United States, for example, the average person has one cell phone. Phones were originally used primarily by adults, but today it is becoming more and more common for middle school–aged children, generally defined as those eleven to thirteen years old, and older to possess a phone. In other parts of the world, where mobile phone usage is either popular or prevalent due to the lack of landlines, the average person often has two or more phones. From data collected at the beginning of 2014 examining the number of mobile phones per country and per capita (per 100 people), Macau, a former Portuguese colony and now a Chinese territory, ranked the highest, with 2.79 phones per person or 279 phones per 100 people. The United Arab Emirates (UAE) took second place, with 2.59 phones per person, and Hong Kong took the third position, at 2.25 phones per person. At the bottom of the list was Eritrea, with 5 mobile phones per 100 people. Second to last were the Marshall Islands, with 5.7 phones per 100 people, and Somalia ranked slightly higher, with 6.5 phones per 100 people (Indexmundi 2014). These differences are significant for understanding mobile app usage around the world because almost all smartphones contain some apps. App preferences can vary from country to country, and likely even vary on multiple phones carried by the same person.

Apps are not always different from social media or social networking sites. Rather, apps enable people to reach those sites quickly from various devices. Most social media and social networking sites have an official app available for multiple mobile phone operating systems. Four of the most popular apps related to social networking sites have been Facebook, Instagram, Twitter, and Snapchat. They are also among the most popular sites globally. However, a new international study indicates that app usage around the world has decreased, including all four of these apps, which have lost popularity in terms of downloads and time spent using them (Tuchinsky 2016). While it is unknown if the demand for apps will continue to decline, the loss of favor for some of the larger social networking sites usually means that people have shifted their preferences toward new social media sites. Specifically, young adults have been shifting away from Facebook and to other platforms in the United States, while in other countries, they are shifting away from Twitter. However, all that this means is that once these audiences have chosen a platform, they will seek the corresponding app to maintain ease of access and to receive updates. Social media, and social apps, are here to stay.

Around the World

Every region of the world has had a different experience with the internet, social media, and apps. The variations stem from different levels of technology development and telecommunications infrastructure to a wide range of government types, such as democratic, socialist, authoritarian, oligarchic, and more. In addition to economic and political considerations, human social factors have played a huge role in the adoption and use of social media, especially in terms of preferred platforms and how they connect with a person's worldview, culture, and lifestyle. Cost and availability also play a role in what gets adopted when and where, as prohibitive costs and lack of access discourage people from considering going online. People also have different activities that they prefer to do or discuss on different platforms. All these considerations lead to the conclusion that people do not use the internet or social media in exactly the same way; they are just as diverse in real life as they are in the virtual realm online around the world.

Africans are only emerging now with a strong presence online, and that presence will continue to grow as more people go online. The internet and social media arrived rather late to Africa, especially sub-Saharan Africa, than the rest of the world. Part of the delay was that internet communications technology (ICT) required specific infrastructure that was not prevalent in Africa. For example, only one undersea cable connected West Africa to the internet until 2009. Due to construction projects that lasted until 2012, additional cables were added to support the growing demand (Ward 2016). Through similar advancement projects, internet penetration advanced rapidly on the continent and most people gain access via mobile phone.

Africa's social media also grew in availability and popularity with increasing access to the internet. Across the continent, approximately 9 percent of the population uses social media (Parke 2016). South Africans, Nigerians, and Kenyans rank among the top African countries using social media and social networking platforms. Facebook is the most popular platform for communications and information sharing. It has helped achieve this ranking by investing in new ways to provide its services at low costs over low-bandwidth connections. Twitter varies in popularity depending on the country and topic; politics in particular are usually discussed over the platform, especially in, but not limited to, the English language.

Twitter has been an essential communications tool in Africa for political protests, demonstrations, and even government regime changes. Across photo-sharing platforms, many Africans prefer posting photos that they believe show the real Africa, as evidenced in hashtags such as #TheAfricaTheMediaNeverShows, or a variation thereof, which tends to trend on Instagram.

Other forms of social media in Africa have been used to revitalize the continent on a microeconomic level. Social media has been used to encourage local investments, to create and distribute microloans to small businesses and individuals, and as a burgeoning platform to develop e-commerce. In many countries, the governments now calculate iGDP (internet gross domestic product). GDP is the value of all goods and services produced within a country annually, whereas iGDP measures the internet's contribution to gains in a country's economy. In 2013, Africa's total iGDP was $18 billion and by 2025, when at least 84 percent of the continent is expected to have internet access, iGDP could reach $300 billion (Amberber 2013).

Because the continent is diverse in terms of language and culture, African choice in internet, social media, and app usage can vary. More specifically, the continent boasts more than 3,000 distinct ethnic groups that speak over 2,000 languages (Zijlma 2015). Popular world languages used online include English, Arabic, French, and Portuguese, which are completed by a host of indigenous African languages.

Asia's experience online has been vastly different than Africa's in some respects. Asia has over 50 percent of the world's population and a large percentage of the world's diversity. Countries in Asia were among the first to adopt and advance internet technologies, particularly Japan and South Korea in the early 1980s. South Korea introduced the first internet gateway with two nodes in 1982 and many Asian countries followed suit. Initially, North America and Europe had more users, but by the 2000s, Asia had surpassed these numbers and, especially in East Asia, had become a leader in internet penetration, online gaming, and mobile device usage.

One major challenge for some Asian internet users is the language barrier. As most technologies were made to be English or Roman-character compatible, many Asian languages could not be utilized online. For example, though email had been available in some Asian environments since the late 1980s or mid-1990s, the internet had not been designed to interpret and render non-Roman fonts. As an example, Google launched as a search engine in 1998. Gmail came later in 2004. In 2005, it was available in select Asian languages such as Chinese, Japanese, Thai, and Vietnamese. Several years later, it added support for Urdu and a number of languages spoken in India. Since 2014, it added Khmer, Lao, Mongolian, and Nepali. Without the ability to type and see their language on a screen, many people had high barriers to access not because they couldn't read, as literacy in many parts of Asia is very high, but because the technology itself excluded them.

To address this issue of language inequality, Asians invented a number of indigenous platforms and apps. China, for example, invented its own equivalents to Facebook, Twitter, and a number of other sites that are available almost worldwide, and then shut the international sites down to have greater control over the internal use of internet and to impose censorship. In Myanmar, several new sites have finally cracked the language font barrier and, with the country's improvement of its telecommunications infrastructure, is likely to see a huge influx of its citizens online by 2017. South Korea and Japan continue to be innovators of social media and apps, with sites like Line and KaokaoTalk starting to have a greater Asian market share.

Latin America's history with the internet and social media is both parallel to and different from that of Africa and Asia. Like most of the rest of the world, universities started the initial foray into exploring the possibilities of the internet and helped connect researchers to this new information highway. Most Latin American countries were not connected to the internet until 1997, though to put internet penetration in perspective, Uruguay, one of the early leaders in Latin American connectivity, only had 5 percent penetration in 1999.

Latin America is in an unusual position. It is considered the third-largest region for internet consumption, after the Asia-Pacific and the Middle East and Africa. It is also considered to be the region with the third-highest mobile phone penetration in the world, behind Central and Eastern Europe and Western Europe. While many countries have a very high internet penetration rate, such as Brazil and Chile, other countries in the region have lower penetration numbers (usually under 20 percent), including Nicaragua, Honduras, Guatemala, and El Salvador. There are cultural, political, and economic reasons that contribute to this disparity. However, the region shows promise for the growth of internet usage in the future, especially as many people already have access to cellular phones.

North America and Europe, unsurprisingly, continue to utilize the internet and social media frequently. In Europe, most of the northern countries of Scandinavia, as well as other developed countries in Western Europe, have fully embraced being connected online with other parts of the world. Issues that affected Europeans tend to be about privacy rights and data sharing. The European Union has passed strict laws about personal electronic data and how it is processed and retained. These laws apply to companies operating in their territory. For example, Google, a U.S.-based company, makes its money by storing as much personal data as possible on its individual users. Allegedly, this information is used to provide tailored advertisements to consumers. When content is posted online and Google archives it, it is available permanently in most of the world. In Europe, citizens have the right to ask Google to forget about an incident or even not to store their data. European law requires that Google comply with these requests, though individual countries may have more lenient or even stricter requirements. In contrast, U.S. citizens who use Google products consent to allowing Google to do whatever it wants with their information, restricted only by its broad privacy policy. In terms of privacy laws, Europe has the most advanced in contrast to other regions. One notable difference is that Europeans, by law, must opt in or consent to sharing their information. Throughout the rest of the world, consumers must actively choose to opt out to restrict the dissemination of their personal data. In addition, the options to opt out are often time-consuming or obscured on corporate webpages.

In this volume, all the world's regions are represented. Talented researchers who work on or in the region have compiled the most recent descriptions of how the internet, social media, and apps are used in their countries of focus. Through these national-level profiles, the variations of what social media and apps are being used, who uses them, and major issues surrounding e-communications are available for exploration. Together, they show how people from multiple countries and regions have tailored their usage to fit personal, cultural, and technological preferences. This is what it looks like today, online around the world.

Further Reading

Akamai. 2015. "State of the Internet." Accessed July 12, 2016. https://www.akamai.com/us/en/multimedia/documents/report/q3-2015-soti-connectivity-final.pdf

Amberber, Emmanuel. 2013. "The Internet of Africa." [Infographic] December 2. Accessed July 31, 2016. https://yourstory.com/2013/12/internet-africa/

Becker, Tyler. 2015. "The 9 Major Social Networks Broken Down by Age." April 9. Accessed July 6, 2016. https://socialmediaweek.org/blog/2015/04/9-major-social-networks-age/

Cosenza, Vincenzo. 2016. "World Map of Social Networks." Accessed July 6, 2016. http://vincos.it/world-map-of-social-networks/

Global Web Index. 2015. "Social Networking Motivations." Accessed July 31, 2016. http://www.globalwebindex.net/blog/top-10-reasons-for-using-social-media

Indexmundi. 2014. "Telephones—Mobile Cellular per Capita." Accessed July 4, 2016. http://www.indexmundi.com/g/r.aspx?v=4010

International Telecommunications Union (ITU). 2015. "Measuring the Information Society Report." Accessed July 5, 2016. http://www.itu.int/en/ITU-D/Statistics/Documents/publications/misr2015/MISR2015-w5.pdf

Kaplan, Andreas M., and Haenlein, Michael. 2010. "Users of the World, Unite? The Challenges and Opportunities of Social Media." *Business Horizons*, 53: 59–68.

Parke, Phoebe. 2016. "How Many People Use Social Media in Africa?" January 14. Accessed July 13, 2016. http://www.cnn.com/2016/01/13/africa/africa-social-media-consumption/

Sheikh, Mahnoor. 2015. "The Cost of Internet: How Does Pakistan Compare to the Rest of the World?" October 1. Accessed July 5, 2016. https://propakistani.pk/2015/10/01/the-cost-of-internet-how-does-pakistan-compare-to-the-rest-of-the-world

Statista. 2016a. "Number of Apps Available in Leading App Stores as of June 2016." Accessed July 4, 2016. http://www.statista.com/statistics/276623/number-of-apps-available-in-leading-app-stores/

Statista. 2016b. "Number of Available Apps in the Apple App Store from July 2008 to June 2016." Accessed July 4, 2016. http://www.statista.com/statistics/263795/number-of-available-apps-in-the-apple-app-store/

Truong, Alice. 2015. "The State of Internet Connectivity Around the World," December 7. Accessed July 5, 2016. https://www.weforum.org/agenda/2015/12/the-state-of-internet-connectivity-around-the-world/

Tuchinsky, Peter. 2016. "Social Media App Usage Down Across the Globe," June 2. Accessed July 31, 2016. https://www.similarweb.com/blog/social-media-usage

Ward, George F. 2016. "Social Media in Africa—A Growing Force." *Institute for Defense Analysis Africa Watch*, 11: 5–7.

Zijlma, Anouk. 2015. "Facts About Africa." August 5. Accessed July 13, 2016. http://goafrica.about.com/od/africatraveltips/a/africafacts.htm

Overview of Popular International Social Media Sites

Since the first email sent in 1971, the first bulletin board system developed in 1978, and the first website designed to connect people over the internet in 1995, social media evolved in numerous directions, and different cultures have interpreted them in a variety of ways. Globally, multitudes of social media apps and websites are available. With one to cater to every need and whim, it helps to think about the different types of platforms in terms of categories. While there are many ways this can be accomplished, no taxonomy of social media can truly account for the diversity found in the social networking landscape, nor the heterogeneous nature of the available media. Nevertheless, categories allow an examination of similar types of social media, which aid researchers and students in understanding how they function among different sectors of society. The categories established for the present volume are not intended to be exhaustive, but rather to allow the presentation of an overview of the more prominent social media found around the globe.

Social media categories began to define themselves with the introduction of CompuServe's CB Simulator, one of the first instant messaging (IM) programs. With this tool, people could communicate via text in real time. The ability to send instant messages and converse in real time is still one of the most popular ways that people in different parts of the world use the internet. Some examples of this category of social media include QQ, Skype, and WhatsApp. However, more often than not, social media is associated with the websites that allow users to build profiles and network with one another. Facebook's predecessors, SixDegrees, Friendster, and MySpace, may not have ultimately succeeded in their visions, but they paved the way for the networking websites that dominate the internet today. By far, the most

There is truly a social media app for every interest. Animal lovers can communicate with like-minded people at Catster, Dogster, and Hamster. Fans of the zombie genre can discuss books and movies at LostZombie, a site advertised as being under development in the fall of 2015. Parents with single children can set them up on dates at Date My Single Kid. Sight-impaired people can chat with other people who have a disability or their family and supporters at Blindworlds. People who like facial hair can discuss the perfect mustache at Stache Passions. Science fiction fans can find a mate at Star Trek Dating, and fans of the paranormal can plot alien sightings at UFO Social (Interlat 2015). The boundaries of social media are unlimited and there is a platform for every interest.

popular of these websites is Facebook, but others, such as Renren, VK, and Mixi, are also prominent. A media-sharing feature is common to most social networking platforms but can also comprise a category on its own. YouTube, Instagram, and YY are such instances and are representative of the diverse ways these types of websites are utilized. Modern forums can trace their roots to the early days of the internet, but have also transformed in interesting directions, such as commentary news websites like Reddit and Taringa! or bookmarking websites like Stumble-Upon and Pinterest. Blogging is an activity that naturally fits with the tools that the internet offers, and recent years have seen this category of social media reimagined as microblogging with websites such as Twitter and Weibo. In what follows, each of these categories will be examined through a sample of the more popular and influential websites preferred by social media users today.

IM and VoIP

IM is the ability to exchange text messages in real time. This idea was expanded with voice over internet protocol (VoIP), a function that allows users to speak to one another with or without video using the internet instead of phone services. This is also sometimes called *telephony* or *videotelephony*.

One of the first mobile messaging apps on the market was QQ. The first version of this program was released by the Chinese company Tencent in February 1999. Originally called OICQ, the name was quickly changed to QQ due to its similarity with ICQ, a different IM app owned by AOL. QQ is currently the most popular IM app in the world, second only to Facebook in total active social media users. In the wake of mobile cell phone proliferation, QQ was adapted for use with virtually any kind of mobile device in China, which is contributes to its popularity. In 2010, QQ officially became international after a year of beta testing when Tencent introduced the mobile messaging app to the global market in English, French, and Japanese, with later versions in Spanish, German, and Korean. Currently, QQ is accessed primarily via phone and also serves as the primary entry point for Qzone, the most popular social networking site in China. It offers games, shopping, and music in addition to its messaging function. In August 2015, QQ had 830 million active users worldwide.

Weixin, another IM app to come out of China, has also dominated the market, with just under 549 million active monthly users by August 2015. Known as WeChat in the United States, the product was first introduced to the public in January 2011 by the Chinese company Tencent. The app was designed specifically for mobile devices as an IM service. It incorporates elements found on sites like Facebook and Instagram, but also offers the unique feature of being able to hold down a button and record a message for another user. Three years after its introduction, Weixin had more than 300 million users, which is a faster adoption rate than popular social media sites in the United States, such as Facebook and Twitter.

Skype is also one of the most popular IM apps in the world and was one of the first VoIP programs with video capability. This meant that users could communicate via video over the internet for free. Skype provides other services, such as voice-only calling to telephone numbers, IM, file transfers, and conference call capabilities. Two European entrepreneurs, Janus Friis (1976–) and Niklas Zennström (1966–), introduced the app in 2003. During this time period, telephone calls could be expensive, especially to other countries, which meant that Skype being free was a huge draw for users.

The company has changed hands several times. In 2005, eBay acquired Skype for $2.6 billion with the intention of combining the software with their e-commerce franchises to create a central hub for business on the internet. At the time, Skype had 53 million registered users and the company reported over 2 million people using the product at any given time. In 2009, after years of not having been able to successfully integrate Skype into their business model, eBay decided to sell 65 percent of its share in the company for about $2 billion to a group of private investors that included the Canadian Pension Plus Investment Board. At the time, eBay reported that Skype was earning $551 million from its pay services and had a user base of about 480 million. In 2011, despite a lifetime accumulation of over $680 million in debt, several companies were interested in purchasing or partnering with Skype. In May, however, Microsoft acquired the company for $8.5 billion in cash and took on Skype's debt.

Currently, Skype has about 300 million active monthly users and is one of the top ten most popular social media platforms in the world. This is true despite competition from other VoIP products, such as the Apple program FaceTime (introduced in 2010) and Google's Hangouts (introduced in 2013). On the other hand, Facebook Messenger, the IM service with voice and video capability that Facebook introduced in 2011, has surpassed Skype in global popularity, with about 700 million active monthly users. However, part of the reason why Skype has so many active users is that it is popular all over the world. With only about 1 in 7 users from the United States, it is also widely used in India, Russia, the United Kingdom, and Brazil. In fact, many IM and VoIP apps find their popularity in countries other than the United States.

Blackberry Messenger, while it has only 100 million active monthly users, is still among the top twenty most popular social media apps in the world. Also known as BBM, Blackberry Messenger was first launched in August 2005, with voice capability introduced in 2012. According to inventor Gary Klassen (c. 1969–), BBM was the first of its kind to offer cross-carrier capability when IM was still considered a PC phenomenon. Currently, over one-third of BBM users are from India, about 1 in 5 users are located in Saudi Arabia, and more than 1 in 7 users are located in Nigeria.

In 2009, Jan Koum (1976–) and Brian Acton (1972–) released WhatsApp, an IM app that allows users to send messages internationally without incurring any costs from smartphone providers. The original idea behind the app was that smartphone users should not have to pay fees to text from phone to phone. Koum also wanted a simple product that gave users what they wanted without the advertisements and fluff that are common to many other apps. Based on these ideas, the service was initially free, but it switched to a subscription model to avoid growing too quickly. Now the first year of service is free; each year thereafter is $0.99. WhatsApp also relied on word of mouth rather than self-promotion. Koum's childhood experience in Eastern Europe with oppressive government control inspired him to create a unique registering process; instead of giving the company personal information, such as name, age, and gender, the user was only required to provide a phone number. WhatsApp's business model was successful because in 2014, when Facebook acquired the company for $19 billion, WhatsApp had over 450 million monthly users. By August 2015, this number had almost doubled to 800 million monthly users and was the third most popular social media site in the world. WhatsApp users are all over the world, with almost a fifth of them located in India. The app is also popular in Brazil, Turkey, Italy, and Spain.

The IM and VoIP app Viber is most popular in Russia, parts of Eastern Europe, and Southeast Asia. In December 2010, Israeli-American Talmon Marco (c. 1975–) launched the app to compete directly with Skype. Viber initially only offered free, web-based voice services, but a few months later, it added a texting function. Within the first 24 hours after the app was made available, it had already reached 100,000 users, and by the end of the third day, it had millions of users. Viber quickly grew to be one of the five most downloaded smartphone call and messaging apps in the world. The Japanese Internet company Rakuten announced in February 2014 that it would acquire Viber for $900 million. This would add 300 million users to Rakuten's 200 million registered user base. A year and a half later, the mobile voice and messaging app Viber ranked in the top fifteen most popular social media sites, with just under 300 million active monthly users.

Line, a South Korean IM and VoIP app, has an interesting origin. In the wake of the devastating 2011 tsunami and subsequent nuclear crisis that crippled Japan and disrupted telephone communications in the country, NHN Corp in South Korea and Naver, a Japanese search engine, decided to develop Line. The app provides an avenue for users to text and make calls with their smartphones via the internet, rather than being charged minutes or texting fees. Within a year of launch, Line users had surpassed 58 million and the app had reached top ranking in the Apple App Store in over twenty different countries, including Kuwait and Kazakhstan. While the company has plans to try and expand its market to China and the United States, over two-thirds of its users are from either Taiwan or Japan.

Social Networking

The concept of social media is most often associated with social networking websites; however, while the term *social media* refers to all the different categories, *social networking* covers a more narrow set of websites. Essentially, online social networks help connect people with similar interests or backgrounds by providing a comprehensive space where they can communicate. This most often entails users building profiles and creating lists of online connections, which represent their personal networks. These lists can be cross-referenced and used to build a user's network. Often, social networking sites will combine tools from other social media categories, such as IM or media-sharing capabilities.

While Facebook dominates the social networking category, it is nevertheless comprised of a wide variety of popular social media sites that differ from country to country. One interesting notable fact is that many of these websites have been fairly well established for about a decade or more. Exceptions to this observation include Facenama (based in Iran) and Google+. Facenama is a local alternative to Facebook in Iran because the government blocks Facebook access. Google+ is focused around communities and circles and often promotes collaboration over communication.

Some of the oldest social networking websites are business oriented, such as LinkedIn. This website was launched in May 2003 by Reid Hoffman (1967–) and his collaborators with the purpose of connecting business professionals around the world. Within the first month, the website had 4,500 members. Currently, LinkedIn has 380 million members, about a quarter of whom are active users. The website is available in twenty-four different languages, and it takes more than 8,700 full-time employees to keep it running successfully. While one in four members is from the United States, LinkedIn is also commonly used in India, Brazil, the United Kingdom, and Canada, with registered members in over 200 countries.

Other popular examples of business professional social networking websites include Xing from Germany and Viadeo from France. Xing was originally founded as openBC (Open Business Club) in 2003 in Hamburg, Germany, and went public in 2006 when it changed its name to Xing. Currently, it has over 13 million members from all over the world, including Austria, Switzerland, and India; however, over 60 percent of Xing's web traffic occurs in Germany. Viadeo, launched in 2004 by Dan Serfaty (1966–) and Thierry Lunati (c. 1965–), is currently the leading professional social network in France, with 9 million members. In total, Viadeo has over 65 million members, at least a third of whom are from China, and, while the site is most popular in France, Viadeo also sees a lot of traffic from India, Morocco, Algeria, and Russia.

Some of the more prominent social networking sites that still affect society today were established between the years of 2004 and 2006. Facebook is the primary example, which would eventually come to dominate the internet in many countries. While originally launched by Mark Zuckerberg (1984–) in 2004, the website did not expand beyond college campuses until two years later. During this time, several other social networking websites were introduced, some of which became very popular in different countries around the world. Others, which were successful for a time, were eventually outpaced by Facebook and shut down or forced to change their business model completely to stay competitive.

In January 2004, Google launched Orkut. It was named after its designer, Orkut Büyükkökten (1975–), whose intended audience was non-English-speaking youth. With at least 1,500 users within the first few weeks, Orkut became very popular in Brazil and India. However, just over a decade after it launched, on September 30, 2014, Orkut was shut down. Google made that decision in order to focus attention on other projects that had outpaced Orkut, such as YouTube, Blogger, and Google+. The latter was an endeavor by the search engine giant to launch a social networking site that combined elements of many other popular social media platforms. Google+ launched on June 28, 2011, and offered a number of functions to facilitate connecting with other people online. One of the more prominent functions was Circles, a way to organize the people on a user's friends list. Sparks was also

Joe Fernandez (1977–), the inventor of Klout, developed the idea after undergoing major jaw surgery. In 2007, when his jaw was wired shut, Fernandez was able to communicate only through Facebook and Twitter. During this time, he started thinking about how people communicate via social networking and wondered about the impact that electronic communications can have on individual reputations and on spreading messages to other people. His curiosity led to work on how to measure the impact that an individual has using these applications, especially Twitter. Once he recovered, he took several months to develop and launch Klout, a tool that applies an algorithm to a person's social media communication to determine the amount of online influence that they exert across platforms (Shontell 2014).

prominently advertised in the beginning. It was a tool to save web searches and intended to be a vehicle for starting conversations online. Hangouts was a messaging and videotelephony communication tool. Google Circles and Hangouts are still common functions on the site; however, Sparks was a rapid failure. Within a few months of launch, it was no longer emphasized and by November 2012, it had been completely removed from the website. Google's social networking endeavor was not an immediate success and due to the way that Google required Google+ membership to use other websites, such as YouTube, it was difficult to tell how many people were active users of the service. In 2015, Google finally stopped attempting to integrate Google+ into all its other products.

Other popular social networking sites launched during this two-year period focused on countries other than the United States. For example, founded in February 2004, Mixi was to become one of Japan's most successful social networking sites; during its peak success, it had around 30 million members. Facebook was introduced to Japan in 2008. Nevertheless, it had not become popular in Japan and, towards the beginning of 2011, ten times more Japanese people were members of Mixi than Facebook. This was due in part to Japanese cultural values regarding privacy. Facebook had been emphasizing real identities, where most Japanese social media websites, including Mixi, allowed users to use pseudonyms. Many Japanese internet users were uncomfortable with the idea of displaying and broadcasting their real names and the details of their lives; therefore, they shied away from Facebook. However, due to a number of factors, this trend was completely reversed by the end of 2014, when Facebook was not only the most popular social media website in Japan, with almost 34 million users, but Mixi had lost most of its membership. Mixi had never really expanded outside Japan, and today it ranks as the thirty ninth most popular website in the country, while Facebook is the tenth most popular website.

However, not all local social media is consumed by Facebook when it enters a national arena. Draugiem, out of Latvia, is a good example of a social media website that manages to stay competitive with Facebook while catering to a limited audience. Draugiem was founded in 2004 and currently has about 2.6 million registered users, with about 500,000 visitors to the website each day. The name means something like "friends" in Latvian. It is the fifth most popular website in Latvia, and three-quarters of its traffic comes from within the country. Other countries that use Draugiem tend to be those in Europe, such as the United Kingdom, Ireland, Sweden, and Germany. Draugiem has a strong presence in Hungary, where it is known by a different name, baratikor, which is a loose translation of the name into Hungarian. Draugiem also has an English version, called Frype, but this version is mostly used in Latvia, Ireland, and Germany. One interesting fact about Draugiem, which may be a factor in maintaining its popularity, is that registration to the website is by invitation only, making it feel more private and exclusive.

In 2005, two very popular social media sites were launched in China. The first, introduced by Tencent, one of the largest internet companies in the world, was Qzone. It was originally conceived as an online space for users to customize their blogs, share photos, keep diaries, and listen to music. Over the years, it transformed from a personal space to more of a social network. Currently, it is one of the five most popular social networking sites in the world, with almost 670 million active users. Then, in December, Joseph Chen (c. 1970–) founded Renren, which has become known as the "Facebook of China." It was originally called Xiaonei, or "schoolyard" in Chinese, and was intended as a platform for reconnecting with former classmates. Ranked among the 165 most popular websites in China, only 80 percent of its users are Chinese, with about one in eight originating from the United States. In 2009, the Chinese government blocked several Western websites, including Google, Facebook, Twitter, YouTube, Vimeo, and Blogspot. One consequence of this action is that the Chinese social media were allowed to flourish. In August 2015, four of the top ten most popular social media websites in the world were Chinese.

The two most popular social networking platforms in Russia both launched in 2006. VKontakte, now known simply as VK, was created by Pavel Durov (c. 1985–) and is often called "the Russian Facebook" due to the similarity between the two products. In November 2014, there were almost 55 million Russian users of VKontakte, while there were less than half that number of Russian Facebook users. Odnoklassniki, the second most popular social networking site in Russia, had just over 40 million monthly Russian users in the same time period. In August 2015, there were about 100 million monthly users of VKontakte worldwide, many of them under the age of eighteen. VKontakte tends to attract younger users, while Odnoklassniki tends to attract older users (i.e. older than forty-five). This was one of the deciding factors in Odnoklassniki's decision to rebrand in 2014 to try and attract a younger audience. The website is now known as OK.RU.

The year 2006 was also when Facebook opened up to the general public. Since that time, it has grown enormously and, as of August 2015, it had about 1.5 billion monthly users worldwide. It was also ranked the second most popular website in the world, just behind Google. Other regional websites are often touted as that region's version of Facebook; for instance, VK is the "Russian Facebook" and Renren is the "Facebook of China." In some places, Facebook and the regional social media are able to coexist or even beat Facebook, as in Latvia and Russia. More often than not, Facebook overwhelms the regional social media, as with Mixi.

In some countries, such as China or Iran, Facebook has been banned. Facenama is Iran's answer to that ban. It is just like an Iranian version of Facebook and, as of

December 2014, it was the number one social networking website in the country. There are suggestions that just as Facebook is banned from being used in Iran, Facenama has banned all non-Iranian internet protocol (IP) addresses so that the site cannot be accessed from outside Iran; however, usage statistics point to there being users in other countries, most notably Russia, the Netherlands, and Brazil. Thus, despite Facebook's incredible global penetration, its popularity is not guaranteed in all corners of the globe.

Media Sharing

When digital cameras became more prevalent than film cameras in the mid-2000s, media-sharing websites that utilized this technology started becoming more popular as well. The primary function of media-sharing websites is to provide a space to users to share their media. Unlike social networking sites that use media-sharing abilities to supplement their users' experience, media-sharing websites highlight this function as their specialty.

Flickr, launched in February 2004 by the Canadian company Ludicorp, was one of the first of these types of websites. The website allowed users to tag photos and make comments. The website was a photo management system with social network features. A year after the company launched, it was purchased by Yahoo. At the time, Flickr had 1 million registered users; by 2013, it had 87 million of them. Despite this huge increase in membership, the website was most successful in early 2009. However, by the middle of 2011, 6 billion photos had been hosted on Flickr. While this number did not even come close to the 100 billion photos then hosted on Facebook, Flickr users were concerned with quality rather than the more mundane shots of vacations and children that filled Facebook's pages. However, due to its slow transition to mobile and the introduction of other media-sharing platforms, Flickr's relevance and use has declined. In 2013, Yahoo overhauled the website and changed its pricing plan. In 2015, they revised it again and boasted 113 million members.

Where Flickr focused on pictures, YouTube focused on videos. Today, along with Facebook and Google, YouTube is one of the most popular websites in most countries and is ranked number three in the world. It was launched in February 2005, and by November of the same year, YouTube users were utilizing 8 terabytes (TB) of data every day. In the fall of 2006, Google purchased the video-sharing website for $1.65 billion. At the time, YouTube had over 700 million views a week. In 2007, the site also offered the ability for users to monetize their content, which helped pave the way for the myriad innovative ways that it is used today. By 2010, YouTube users were uploading more video in two months than all three major television networks in the United States had created in the past sixty years.

Youku Tudou, a Chinese video-sharing platform, was the result of the 2012 merger of two websites. Tudou was introduced in China in April 2005 by Gary Wang (1973–) as a video-sharing website. The name means "potato" in Chinese, and was meant to invoke the idea of the "couch potato." Tudou's model was similar to that of YouTube, in that it facilitated user-generated content; however, it also delivered its own drama series called *That Love Comes*. Youku was founded in 2007 by Victor Koo (c. 1967–), former president of the Chinese internet company Sohu. While the names may seem similar, Youku and YouTube were very different kinds of video websites. First, Youku, in Mandarin, means "good" and "cool," contrasting with YouTube's name, which promotes the idea of a channel. Second, CEO and founder Koo deliberately did not adopt YouTube's business model; instead, he chose to focus on providing alternative television to young, urban professionals by licensing shows from China, Hong Kong, and Taiwan.

In March 2012, Youku and Tudou announced a merger, with Youku acquiring Tudou for $1.1 billion in an all-stock deal. The previous rivals' new company would be known as Youku Tudou, with Victor Koo retained as the chief executive officer (CEO). Before the two companies merged, Youku was ranked as the eleventh most popular website in China and Tudou was the fourteenth most popular website. After the merger, Youku Tudou still only composed about a third of the market share of video websites in China. The resulting company eventually dominated the market; however, today it is being overtaken by more aggressive video-sharing platforms, like LeTv and iQiyi.

China's YY is another very popular video-streaming website. It was launched in 2008 by David Li (c. 1974–) and Lei Jun (1969–). Originally, it was a voice-based gaming service that allowed users to communicate with one another via either voice chat or text chat while playing online video games. Later, it added video-streaming capabilities. By 2012, YY had over 100 million channels, each which could support an excess of 100,000 users simultaneously. YY also incorporated some unique features that set it apart from similar websites in the United States. For example, users can buy virtual "roses" and give them to performers on their YY channel. The performers, in turn, can sell these roses back for real currency. Virtual classrooms are also a feature of YY, which include educational tools, such as embedded whiteboards and Microsoft PowerPoint-like slides. By 2012, YY had over 300 million registered users, and by 2015, it had 100 million active monthly users.

In October 2010, Kevin Systrom (1983–) and Mike Krieger (1986–) launched Instagram in the iPhone App Store. Within the first twenty-four hours, the app had been downloaded 25,000 times. By the third week, it had 300,000 users. Part of this instant popularity stemmed from word-of-mouth advertising by some very influential people in the tech world, including Jack Dorsey (1976–) of Twitter, who had tested the app before it became available to the public. Instagram grew into an international phenomenon. In addition to being able to upload photos and videos, users can use filters to transform the images. Two years after the app came out, Facebook bought it from Systrom and Krieger for $1 billion. The acquisition neutralized a rival for Facebook and opened up its mobile offerings. At the time, Instagram had 30 million active users; in August 2015, it had ten times that number.

Snapchat is one of the most recent media-sharing websites, and it has a twist. Photos and videos that are uploaded to the website are available only for a few seconds. The original incarnation of this ephemeral mobile messaging app was launched as Picaboo in the iOS app store in July 2011. Two months after its launch, it had fewer than 130 users. At that point, cofounders Evan Spiegel (1990–) and

Bobby Murphy (c. 1988–) decided to remove the third member of their partnership, Reggie Brown (n.d.). The original idea for Snapchat can be traced back to Brown, who was then serving as chief marketing officer. Brown brought a lawsuit against the company for one-third of its worth. In 2014, the lawsuit between Reggie Brown and Snapchat was settled out of court for an undisclosed amount. The press release regarding the settlement, however, suggests that Brown settled for less than the one-third of the company's worth that he had been seeking. At that time, Snapchat, Inc. was valued at $10 billion. Around the same time that Brown was removed from the partnership, the name was changed from Picaboo to Snapchat. Shortly thereafter, the number of Snapchat users began to rise, mostly because Snapchat marketed the short duration that the pictures lasted on the app as a social media solution to the permanence of information on other sites, and by the end of the year, it had more than 2,000 users. By the end of January 2012, it had more than 20,000 users and by the end of April 2012, it had more than 100,000 users. The majority of activity on Snapchat was taking place between the hours of 9 a.m. to 3 p.m., marking it a tool primarily used by school-aged young people.

Mark Zuckerberg of Facebook offered to buy Snapchat in 2012 for $3 billion in cash, which would have been Facebook's most expensive acquisition to date. In what some considered a bold move by Spiegel and Murphy, the offer was declined. At the time, Snapchat users were posting 50 million photos a day, compared with the 300 million photos a day shared by Facebook users. Zuckerberg responded by developing an app called Poke, which, while it had some additional features, was essentially a copy of Snapchat. Poke was designed in 12 days and launched on December 21. The day after its launch, Poke was the most-downloaded app. However, publicity from Facebook's offer garnered a great deal of interest in Snapchat. Three days after Poke's launch, on December 25, Snapchat had surpassed it in popularity and Poke rapidly fell behind, dropping out of the thirty most-downloaded apps. On May 9, 2014, Facebook finally ended its failed ephemeral messaging app, Poke, removing it from the app store. In September 2015, Snapchat introduced its first in-app purchase, Replay. Previously, all of Snapchat's features were free, including a version of Replay introduced in 2013, which allowed users to replay one message a day. The purchasable version of Replay, available for $0.99, allows users to replay up to three messages a day. In 2015, Snapchat's daily active users numbered in excess of 100 million people and the company was valued at $16 billion.

Commentary and Bookmarking Forums

Commentary forums are those that allow users to upload links or other content with the purpose of starting a conversation about that topic. These websites usually provide a voting mechanism that pushes the most popular posts to the top of a list of related posts. Bookmarking forums are less about starting a conversation and more about providing users with mini-web-browsing experiences. These websites allow users to bookmark websites that they find interesting. The links are saved on a board or user-created category that allows them to organize these bookmarks.

Their boards can be shared with and followed by other users, which provide them with a kind of personalized web-browsing experience.

The commentary and bookmarking forums are fairly diverse in terms of what the different websites offer. StumbleUpon, for example, founded by Garrett Camp (1978–) in 2001, was designed to offer a browsing experience tailored to user interests. At the time, Google was relatively new and could be difficult to navigate. The idea was to make it easier for internet users to find something cool and interesting. Users can follow other users with similar interests; however, that was about the extent of the social networking tools that StumbleUpon offered. Pinterest, a much newer platform, offers a very similar model. Users find content on the web they like and can "pin" it to boards on their profile. Meanwhile, other users can browse these boards for content that interests them. While it was launched at the end of 2009, the company had been founded several years earlier, which meant that a lot of planning and investment was already behind the website at launch. Despite this preparation, Pinterest did not take off immediately. Cofounder Ben Silbermann (1982–) claims that people didn't get it. However, when it launched in the iPhone App Store in 2011, user numbers started climbing. Today, it is ranked number sixteen in the United States, and over half of its users are from other countries, like India, the United Kingdom, Brazil, and Canada. Silbermann does not consider Pinterest a social network site, but rather a "catalog of ideas." As with StumbleUpon, users can follow other users, but outside of commenting on pins and repinning content, there are no real social networking tools.

Other websites in this category involve interacting and communicating with other users in a more bulletin board or forum fashion. Baidu Tieba is one of the first examples of this type of website. In December 2003, the Chinese company Baidu (also an internet search site), introduced its communication platform, Tieba. Baidu Tieba is a kind of bulletin board system (BBS) in which users can search keywords and discuss topics with other people who share the same interests. At its tenth anniversary, Baidu reported that Tieba had 1 billion users, 200 million active monthly users, and over 8 million bars. (A *bar* is a forum created when a term is searched so that users have a space to discuss the topic.) By 2015, Baidu Tieba had become one of the top ten most popular social networking sites in the world, with about 300 million active monthly users.

In Latin America, Taringa! is one of the most popular social media websites, trailing only Facebook. Launched in early 2004 by Argentine high school student Fernando Sanz (c. 1988–), known as Cypher, Taringa! was originally based on an already existing website, Teoti.com, which was a news-feed community where users linked content and then commented on it. Taringa! quickly became bigger than Cypher could handle, and the website was acquired in 2006 by Hernán Botbol (1981–) and his brother Matias Botbol (1978–).

Reddit, a similar website, was launched a year after Taringa! in June 2005, by Steve Huffman (1983–) and Alexis Ohanian (1983–), recent graduates of the University of Virginia. As with Taringa!, users can post content and comment on it. One of the central features of Reddit is the ability for users to vote content up or down, thereby organizing content according to user opinion. Condé Nast Publications bought Reddit in October 2006 for about $20 million. At the time, the website had 500,000 unique visitors a day. Over half of Reddit's user traffic originates in the United States, where the website is ranked tenth, but it is popular in other countries as well, including India, Canada, the United Kingdom, and Australia.

Blogging and Microblogging

Blogging is one of the earliest examples of social media; the name is derived from the term *weblog*. Blogger is one of the oldest websites to offer this service to internet users, and, as an early platform, it helped to popularize the format of online web journaling. It was launched in August 1999 by Pyra Labs, out of San Francisco, California. By 2002, Blogger had hundreds of thousands of users. The company was sold to Google in February 2003 for an undisclosed amount. Around the time of this acquisition, Blogger hosted more than a million blogs, about 200,000 of which were active. One of the benefits of using Blogger over some other blogging platforms, like WordPress, is that it gives users the option to using the Blogger domain host, Blogspot, or, for free, users can use their own domain name. WordPress charges for using a different domain name. On the other hand, WordPress, a blogging platform that was introduced in 2005, is also a content management system, which means that there are opportunities for a lot more customizability and control over how the content is delivered. Blogger is largely used by English speakers and most commonly used in the United States, but it also has a very strong presence in India, along with Indonesia, Pakistan, and Thailand. WordPress, on the other hand, has a slightly more diverse user base, with less than 20 percent of its users originating in the United States. It is also popular in India and Brazil; the second most common language used with WordPress is Spanish.

With the launch of Twitter in 2006, social media users were introduced to the idea of microblogging. That is, instead of composing an entire essay on a topic, users are limited to 140 characters in each post, on which other users can then comment. Due to the character limitation, the microblogging phenomenon is supposed to be more spontaneous than traditional blogging. Cofounder Jack Dorsey posted the first "tweet" in March 2006, writing "Just setting up my twttr." The influence of Twitter has been enormous, and as of August 2015, the website had over 315 million active monthly users. It is ranked in the top ten, both in the United States and globally. Around 500 million tweets are posted daily, and almost three-quarters of Twitter accounts are located outside the United States. In late 2015, Twitter announced that it was considering offering a feature that expands the 140-character limit; however, it had not yet finalized this mechanism.

Despite its global influence, Twitter is not the most popular microblogging platform in all countries. In China, the microblogging platform Sina Weibo is one of the top five most used websites, and while often compared to Twitter, it is really more of a Twitter-Facebook hybrid. Due to its popularity, Sina Weibo is often referred to simply as Weibo; however, as the term is Chinese for "microblog," numerous other sites also use the term in their name. Weibo was launched in 2009 by the Sina Corporation, headquartered in Shanghai and Beijing. A year later, Weibo already saw more than 5 million users. By 2013, Weibo would have over 600 million registered users, and by 2015, over 200 million active users would visit the site every month.

These five categories represent the bulk of what social media has to offer today. Their special functions and abilities often overlap, but nevertheless each kind fills a different need for social media users. What becomes most evident in examining these categories and looking at the various social media websites from each one is the fact that, while some stand out as global giants, like Facebook and YouTube, there is a wide range when it comes to what is popular in different regions of the world. The websites examined in this overview represent a small sample of some of the more prominent social media sites at the time of writing. Some social media websites and apps are able to sustain interest longer than others by staying competitive, but user preferences change over time, and new start-ups are always being introduced. This means that the landscape of social media is fluid and ever-changing.

Further Reading

Angwin, Julia. 2009. *Stealing MySpace: The Battle to Control the Most Popular Website in America*. New York: Random House.

Avalaunch Media. 2013. "The Complete History of Social Media," April 15. Accessed August 13, 2015. http://avalaunchmedia.com/history-of-social-media/Main.html

Boyd, Danah M. 2008. "Social Network Sites: Definition, History, and Scholarship," *Journal of Computer-Mediated Communication*, 13(1):210–230.

Doll, Jen. 2011. "R.I.P., Friendster, the Social Media Site of Our Relative Youth," April 26. Accessed August 12, 2015. http://www.villagevoice.com/news/rip-friendster-the-social -media-site-of-our-relative-youth-6666000

Jones, Steve. 2003. *Encyclopedia of New Media: An Essential Reference to Communication and Technology*. Thousand Oaks, CA: Sage Publications.

Kaplan, Andreas M., and Haenlein, Michael. 2010. "Users of the World, Unite? The Challenges and Opportunities of Social Media." *Business Horizons,* 53: 59–68.

Noor Al-Deen, Hana S., and Hendricks, John Allen. 2012. *Social Media: Usage and Impact.* Lanham, MD: Lexington Books.

Solomon, Brian. 2014. "From Alibaba to Weibo: Your A-Z Guide To China's Hottest Internet IPOs," March 19. Accessed August 31, 2015. http://www.forbes.com/sites/brian solomon/2014/03/19/from-alibaba-to-weibo-your-a-z-guide-to-chinas-hottest -internet-ipos/

Waters, John. 2010. *The Everything Guide to Social Media: All You Need to Know About Participating in Today's Most Popular Online Communities*. Avon, MA: F+W Media, Inc.

Chronology of Significant Events in Cyber History

1958   The U.S. Department of Defense issued Directive 5105.15 to establish the Advanced Research Project Agency (ARPA). The purpose of this agency was to stimulate and support research projects to advance the development of science and technology in the United States. ARPA was later renamed to DARPA, the Defense Advanced Research Project Agency.

1968   Bolt Beranek and Newman (BBN) was hired to construct the Advanced Research Project Agency Network (ARPANET), which connected computers at various universities around the country in order to facilitate ARPA projects. This marked the beginning of networked computers and is the predecessor of the modern-day internet.

1969   CompuServe was founded as a subsidiary of Golden United Life Insurance. At its inception, it was called Compu-Serv Network, Inc., and its original intent was as a time-sharing service. During the 1960s, many large corporations used computers in everyday business; however, as computers were still prohibitively expensive, they did not own them. Instead, these corporations would rent time on a computer owned by a company who provided a time-sharing service. CompuServe introduced two important changes to this service. The first was technology that allowed the network to transmit data over ordinary telephone lines. The second was the use of business software, which CompuServe made available for a fee, in addition to time on the computer.

1971   Ray Tomlinson (c. 1941–), an engineer employed by BBN, sent the first email from one computer to another. While the two computers sat side by side, the only connection between them was ARPANET, making this the first networked message ever sent. It has been reported that the text of this first email was "QWERTYUIOP;" however, Tomlinson has stated only that the first message was something *like* "QWERTYUIOP." In fact, he does not recall the contents of the first email.

1977   The first personal modem was offered to the public by DC Hayes Associates.

1978   Computer Bulletin Board System (CBBS), the first functional online bulletin board system, was developed by Ward Christensen (1945–) and Randy Suess (n.d.). The bulletin board system, or BBS, became one of the first real tools that personal computer (PC) users had to interact with other PC users at different locations. It offered the ability for users to create their own boards, to upload and download files, share messages on public boards, disseminate news, and even play the first online games. In the early 1990s, at the height of their popularity, more than 60,000 BBSs could be found in the United States alone.

The first multiuser dungeon (MUD) was written by Roy Trubshaw (1959–) and Richard Bartle (1960–), students at Essex University in the United Kingdom. A MUD is a program that allows multiple users from different computers to interact in a virtual environment simultaneously. MUDs were text-based and consisted of "rooms" that represented different environments. Many MUDs were used primarily for online role-playing games. MUDs and BBSs were the first examples of people socializing over computer networks and were the primary method of online communication before the arrival of the World Wide Web.

1979   On September 24, CompuServe became the first provider of dial-up service to the general public. Part of their business model included email and BBSs for consumers. At the time, it was also the only company to offer technical support to PC users. Initially, this service was named MicroNET, but was quickly renamed CompuServe Information Service (CIS).

The Source, the company that would become CompuServe's biggest rival for nearly a decade, was launched by William von Meister (1942–1995). It was the first online service in the United States to target individual consumers exclusively.

1980   CompuServe released its CB Simulator, which was one of the first real-time chat services on the internet. Initially, it offered only 40 different channels, but it became so popular that, not long after its release, it accounted for 20 percent of the traffic on CIS.

The first online newspaper article was published by the *Columbus Dispatch*.

1984   Trintex, a service that provided online connection and content to consumers was founded as the collaborative endeavor of Sears, CBS, and IBM. The "tri" in the name stood for the three companies. The idea was that Sears would provide online shopping, CBS would provide content, and IBM would provide the supporting technology. Ultimately, the endeavor failed and was replaced in 1988 with the much more successful Prodigy.

1985   The predecessor of America Online (AOL) was introduced to the public by Quantum Computer Services. It was known as "Q-link," a dial-up network service for Commodore 64 computer users.

1986   CompuServe was bought by H&R Block for $23 million.

1988   Prodigy became available to the general public, with backing from IBM and Sears. It offered two very attractive benefits over the very successful

CompuServe. Prodigy introduced a flat-rate pricing plan that allowed customers to pay a monthly fee of around $10 to $13 for five hours of access to the internet at any time. At the time, CompuServe's model was the prohibitively expensive pay-per-minute plan, which ran consumers anywhere between $5 and $12 an hour, depending on the speed of the connection. Prodigy also offered a graphical interface, which was more attractive and easier to use than the ASCII-based textual interface of CompuServe.

1989   CompuServe, at the time a subsidiary of H&R Block, bought The Source in June and closed the service down in August.

AOL was born as America Online, a revised version of an earlier network service for Macintosh and Apple II computers. Previous incarnations of this service were introduced by Quantum Computer Services, which officially changed its name to America Online, Inc. in 1991. Originally, there was some debate over the acronym *AOL*, with suggestions that *AO* was sufficient. Stephen Case (1958–), cofounder of AOL, noticed that many successful companies at the time had three-letter acronyms for their names, such as IBM and MTV; therefore, AOL was adopted. This same year, the company introduced the Instant Messenger service and coined the phrase "You've got mail," spoken in a distinctive electronic voice. AOL would become extremely successful and be the first internet company to be included in the Fortune 500.

1990   A British scientist, Tim Berners-Lee (1955–), employed at the European Organization for Nuclear Research (otherwise known as CERN) invented the World Wide Web. It was originally conceived as a vehicle to facilitate the sharing of information among universities and other research institutions. Often confused with the internet itself, the World Wide Web is software that utilizes Hypertext Markup Language (HTML), a markup language that describes the text, images, and other elements of a webpage using tags. The web browser interprets these tags and presents the webpage in a visually pleasing format. HTML introduced hypertext and hyperlinks to the internet, which allowed data to be organized in a weblike manner rather than in a linear format. This facilitates a high level of interactivity in a multimedia setting, which forms the backbone of the internet today.

1991   Berners-Lee posted the first website using HTML. While its contents have changed slightly over the years, this page is still accessible at http://info .cern.ch/hypertext/WWW/TheProject.html.

Prodigy surpassed its millionth subscriber. This same year, it became embroiled in a censorship and free speech controversy related to anti-Semitic posts and messages. This controversy prefigured the debates to come regarding freedom of expression online.

1993   CERN released into the public domain the software that runs the World Wide Web. This resulted in its growth through the developments of numerous innovators, which helped to transform it into the web with which we are familiar today.

1995   Randal Conrads (c. 1949–) created Classmates.com as a tool for reconnecting with people from high school. He later added features to promote social interaction, such as user profiles and message boards. In 2016, Classmates .com focused on collating nostalgia and serving as an online tool for organizing reunions. In 2015, Classmates.com boasted the largest directory of high school rosters, with over 57 million names, and the largest online collection of high school yearbooks, currently totaling over 300,000.

1997   Andrew Weinreich (c. 1968–) launched SixDegrees.com. This website allowed users to create personal profiles and connect with friends. A year after its launch, it added the capability of being able to browse other people's friends lists. While building profiles and listing friends were features of some websites at the time, SixDegrees.com was the first to combine the two and allow others to view them. While Classmates.com was first on the scene, it did not add these functions until much later, thus making SixDegrees.com the first recognizable social networking site.

AOL acquired CompuServe for $1.2 billion.

1999   The first version of QQ, the Chinese instant messaging (IM) service, was released by Tencent.

SixDegrees.com was bought by YouthStream Media Networks for $125 million.

Prodigy was first traded as a public corporation and partnered with SBC Communications.

2000   While SixDegrees.com had millions of registered users, its business model was not sustainable and the website was shut down. Weinreich believed that people were just not ready for it. As access to the internet was still limited for many, once people connected via the website, there was not much else to do. He has also suggested that the technology was not advanced enough yet, as digital cameras were not prevalent until years later.

2001   SBC Communications bought all Prodigy stock for a total of $384 million and then converted it back to a private company. Prodigy remained active in some form until 2011, when it discontinued support for its users.

AOL was the biggest internet provider in the United States. It was worth more than $125 billion and, at the time, earned more than $7 billion a year in revenue. AOL acquired Time Warner for $106 billion, firmly establishing the company as the largest media conglomerate in the world.

2003   Jonathan Abrams (c. 1972–) launched Friendster in March. The website was originally conceived as a competitor to Match.com, the dating website. The idea was that dating friends of friends would lead to more successful relationships than dating strangers from a dating site. Since viewing friends of friends was limited to four degrees, people began amassing as many "friends" as possible in order to expand their network.

MySpace.com was released in August to compete with sites like Friendster. The new site was able to capitalize on some alienating decisions by Friendster, such as when it deleted fake identities and considered a subscription model. While MySpace was not initially conceived as a music promoter, the website embraced the way that bands and musicians were using it. Today, MySpace has been instrumental in the careers of many musicians, such as Sean Kingston (1990–), Panic! at the Disco, and Owl City.

2004   In February, Mark Zuckerberg (1984–) introduced the first incarnation of Facebook to the Harvard University student body. Originally, it was called "thefacebook" and was intended as a vehicle for Harvard students to connect with one another. Within a month of its launch, over half the undergraduate population had registered. Shortly thereafter, the website expanded to other universities in the area; and a year later, "the" was dropped from the name.

Successful through the beginning of the 2000s, Classmates.com was bought in October by United Online, Inc. for $100 million.

2005   In February, YouTube was launched.

Tencent released Qzone, a social networking site that grew out of the IM program QQ.

2006   Jack Dorsey (1976–), cofounder of Twitter, sent the first tweet via an internal system in March: "Just setting up my twttr." By July, the full version of Twitter was introduced to the public as a type of group IM system, with a 140-character limit for its messages. While it was not the first microblogging website, it has certainly been one of the most influential.

Facebook expanded beyond university systems in September and became available to anyone with an email. Since that time, it has grown tremendously, and currently it has almost twice as many active users than any other social media website in the world.

Despite the video-sharing website YouTube's legal troubles over copyright infringement, Google purchased it in October for $1.65 billion. At the time, it was by far the company's most expensive purchase.

2009   Jan Koum (1976–) and Brian Acton (1972–) released WhatsApp, the IM app that allows users to sends messages internationally without incurring any costs from smartphone providers.

In June, China severely restricted its citizens' use of Google by blocking many of Google's communication tools, such as Gmail, Google Apps, and

Google Talk. A few weeks later, after 140 people were killed in riots that took place in Xinjiang's capital of Urumqi, China also banned Facebook and Twitter. Other foreign websites blocked in China in 2009 included YouTube, Vimeo, and Blogspot.

In December, MOL Global, one of Asia's largest internet companies, acquired Friendster for $26.4 million.

2011   In January, the Chinese company Tencent released Weixin, known as WeChat in the United States.

Friendster shut down in May and erased all user-created content. A few months later, it reemerged as an entertainment and gaming website.

2014   In March, a lengthy lawsuit between Viacom and Google finally concluded. In 2007, Viacom had filed a lawsuit asking for $1 billion for copyright infringement by YouTube over more than 79,000 videos broadcasted between 2005 and 2008. While reportedly no money was exchanged, the details of the settlement were not released to the public. However, this would not be the last lawsuit against Google for YouTube's copyright infringement.

2015   Verizon acquired AOL for $4.4 billion, making the resulting company one of the world's most prominent media technology companies.

Streaming apps grew exponentially in popularity. Aleph, an Israeli-based firm, launched Meerkat, an app that can livestream a Twitter feed, in March. That same month, Twitter launched Periscope, an app that streams in real time from a user's phone camera to the web. At the end of the year, Periscope won Apple's award for best app of 2015.

As of August, the top social media websites with the most active users were Facebook (1.490 billion), QQ (832 million), WhatsApp (800 million), Facebook Messenger (700 million), Qzone (668 million), Weixin/WeChat (549 million), and Twitter (316 million).

In September, when Instagram officially boasted 400 million monthly active users, it became more popular than Twitter.

2016   In response to the legal troubles that Apple faced when refusing to unlock an iPhone for the Federal Bureau of Investigation (FBI) after a mass shooting in San Bernardino, California, WhatsApp added end-to-end encryption to its messaging service in April for every method of communication. This meant that regardless of the device used, not even WhatsApp employees would have access to the data sent over their network, thereby making it impossible for any government or other agency to compel the company to grant access to a user's data.

In June, Microsoft purchased LinkedIn for more than $20 billion.

In July, Niantic released Pokémon Go, the first mobile app to introduce an augmented reality game for iOS, Android, and some Apple wearables.

The game was introduced in select countries, such as the United States, and was expanded to other countries, such as Brazil so that it could be used during the 2016 Summer Olympic Games.

The same month, Verizon purchased Yahoo, along with Tumblr, for approximately $5 billion.

The number of Facebook Messenger monthly active users reached more than 1 billion in July.

Tencent, owner of QQ, Qzone, WeChat, and several other online services, surpassed Alibaba, the e-commerce giant whose annual transactions exceeded that of Amazon and eBay combined, as China's most valuable tech company in August. This increase in wealth can be seen in the continued popularity of its services, such as WeChat, whose active user base grew by over 40 percent from 2015 to 2016.

On September 1, YouTube rolled out a new notification policy intended to create more transparency and goodwill among its content creators. Notifications began being sent to users when they violated the "ad-friendly" policy, which demonetized their videos for potentially offensive content, including swearing. According to YouTube, the ad-friendly policy had been in place since ads were initially run with user content. Ironically, this attempt at improving communication led to a community uproar documented via the hashtag #YouTubeIsOver. Users were angry over lost potential revenue and began rehashing old feelings of being taken for granted.

Later in September, Twitter was set to change its method of counting characters, effectively increasing its 140-character limit. Some examples included not counting quoted tweets, attached media, or user names that appear at the beginning of a reply. This decision could be an attempt to rekindle interest in the microblogging platform, as the number of active Twitter users has stagnated and even fallen over the past year.

As of September 2016, the top social media websites with the most active users were Facebook (1.712 billion), WhatsApp (1 billion), Facebook Messenger (1 billion), QQ (899 million), Weixin/WeChat (806 million), Qzone (652 million), Tumblr (555 million), Instagram (500 million), and Twitter (313 million).

A

AFGHANISTAN

Located in South and Central Asia, Afghanistan, with a population of 33 million people, is a country that has faced serious conflict, most recently during the Soviet occupation (1979–1989), a civil war in 1992 that eventually involved the North Atlantic Treaty Organization (NATO) and a U.S.-led coalition (2001–2014), and the transfer of that war effort to Afghan forces (2015–). Although conflict still plagues the nation, the Afghan government continues to make strides that have made improvements for its residents; some of its greatest successes have involved the country's telecommunications infrastructure. As of 2016, the Ministry of Communications and Information Technology claimed that 90 percent of the country had network capabilities, though this number reflected the amount of territory where people can go online, not how many users actually had access. Social media has been slowly penetrating the country as people have acquired enabled devices and used them to promote social change online; both citizens and the government use social media sites to address some of the country's most urgent needs and challenges.

Because of the ongoing conflict in the country between the Taliban, an Islamic fundamentalist group that has committed acts of terrorism, and the Afghan government, security issues have historically hindered the availability and accessibility of the internet in the country. However, Afghanistan has also seen some progress despite the fighting. In 2007, Afghan Wireless completed the construction of a "Microwave Ring" around the country, a project that ultimately provided internet access to thirty-one provinces and covered more than 250 towns, cities, and major transportation routes (Afghan Wireless 2007). The project cost millions of U.S. dollars. The ring paved the way for the introduction of mobile 3G capability. Etisalat, a telecommunications company, first announced its 3G offerings in 2012 (Capacity 2012); by the end of 2016, all mobile service providers offered 3G service (Budde 2016). Some major cities, such as the capital, Kabul, already had 4G access.

Afghans primarily connect to the internet via mobile phone. As of 2016, the country's five major providers claimed 25 million users, of which approximately 20 million could access the internet (Budde 2016). Subscription numbers have trended downward since 2015, as foreign military members working in the country have departed. The statistics from 2015 must include the foreign military presence, as statistics from late 2016 show a 30–40 percent drop in mobile subscriptions after the foreign militaries departed (Budde 2016). Nonetheless, internet penetration in the country remains low. Freedom House (2016) reported that Afghanistan's internet penetration hovered around 8 percent at the end of 2016. When Afghans connect to the internet, their most accessed sites are Google.com, Google.com.af,

YouTube, Facebook, Yahoo, BBC.com, MyWay.com (an internet portal), Wikipedia .org, Acbar.org (a job site), and Amazon.com (Alexa 2016). They access online services at some of the lowest connection and load time speeds of any other Asian nation.

Afghanistan's app usage is difficult to discern. While some top app lists include Afghan television and radio stations, a number of the apps appearing on those lists for iOS and Android include English-language chatting apps. From comments on app ratings' sites, many of these downloads were used by foreign troops deployed to Afghanistan rather than by Afghans themselves. Because of the foreign troop usage, Afghanistan's actual app usage and preferences are unclear, and no definitive lists currently exist.

In April 2016, a controversial indigenous app appeared on the market. Alemarah, an Android app created by the Taliban and briefly hosted on Google Play before it was removed, provided Pashtu-language videos and statements to the group's supporters in eastern Afghanistan. Researchers speculated that the app was attempting to provide the Taliban direct digital access to its supporters and compete with the so-called Islamic State or Daesh, a terrorist organization based in Iraq and Syria that has a large presence on social media and claims to control territory in Afghanistan that they refer to as "Khorasan province" (Guardian 2016). Because Google removed the app quickly, however, it is not known how many people downloaded it.

In 2015, Afghanistan's most used social media platforms included Facebook (98 percent), Twitter (1.3 percent), Pinterest (0.1 percent), Google+ (0.1 percent), Tumblr (0.09 percent), YouTube (0.02 percent), StumbleUpon (0.02 percent), and VKontakte (0.01 percent); other platforms, such as WhatsApp, LinkedIn, Instagram, and Paywast, had nominal use (StatsMonkey 2015). On these platforms, users are predominantly male (97.5 percent) and primarily communicate in English (62.1 percent), Dari (30.1 percent), and Pashtu (3.3 percent) (GIZ 2014). More recent statistics suggest that 37 percent of women used social media in 2016 and that this demographic is ripe for expansion, though official tallies have this number much lower. The low number of women online is a reflection of local culture, where women traditionally received little education. It may also be in part because women reportedly experience significant harassment online, especially on Facebook, the country's preferred platform. There is also an Association of Afghan Blog Writers, which connects bloggers of both genders in the country and mobilizes them to defend their rights when necessary.

Afghans are using innovative ways to promote social media in their country. Although Afghanistan is not yet known for regular, active social media use, the country held its first annual Social Media Summit in 2013. The two-day conference attracted more than 200 social media users from twenty-four of the thirty-four provinces. It focused on promoting social media as a legitimate form of communication, especially for journalism. In 2014, the summit expanded to three days and attracted more widespread attention. Its focus also shifted to how social media can be used for promoting positive social change. The 2015 conference utilized the hashtag #ASMS2015 to promote online discussion and to attract more attention to

On March 17, 2016, women from Afghanistan's Solidarity Party took to the streets of Kabul to memorialize the death of Farkhunda Malikzada. Social media captured the tragic death of Fark-hunda, leading to the arrest of her killers and the police officers who refused to interfere as she was dying. Her death and the social media surrounding it exposed some of the harsh truths about women's rights in the country and is ultimately credited with improving women's rights in the country. (AP Photo/Rahmat Gul)

the event. It also featured the Afghanistan Social Media Awards, which honored so-cial media personalities and companies in ten categories. After a public nomina-tion, voting, and then international jury deliberation, the awards recognized Safi Airways in the category of Best Social Media Advertising Campaign, Norband.com as the Best Blogger, SR Bros Entertainment as the Best YouTube Star, Malali Bashir (n.d.) as the Twitter Power User, and Wais Barakzai (n.d.) as the Citizen Journalist of the Year (Wadsam 2015). A 2016 conference did not take place, and there was no indication whether one would be held in 2017.

One area where social media has received credit for improving is women's rights. In 2015, a young woman called Farkhunda Malikzada (c. 1988–2015) was beaten to death on the streets of Kabul after being wrongfully accused of burning the Koran, the Muslim holy book, a blasphemous act. After her tragic death, videos taken of the assault helped identify her assailants so they could be brought to trial. Eleven police officers were also charged for not intervening during the riot (Kargar 2015). Despite the tragic outcome in this case, social media has played a role in combating harassment aimed at women. Harim, a local app, provides women with updates on street harassment. Social media also allows women who might other-wise be voiceless to defend their rights and have a say in the country's politics and development (Afghanistan Times 2016). The power of social media for women's rights is that it can highlight injustices and social issues that were previously not discussed in any public forum.

Another challenge facing Afghanistan is that many people are tired of the seemingly endless conflict. In a 2016 national survey, 40 percent of the people indicated that they would leave if given the opportunity, citing the fighting and the political and economic problems in the country (Guardian 2015). When they leave, most of them head to Europe, exacerbating the problem that that region is facing with refugees from Syria, other parts of the Middle East, and North Africa. To combat this problem, Afghanistan's Refugees and Repatriations Ministry initiated a social media campaign, primarily on Facebook and Twitter, encouraging its people to stay. The campaign uses high-quality pictures and promotes slogans such as "Stay with Me." The driving force behind the campaign is the reality that if everyone leaves, no one will remain behind to rebuild the country (Bezhan 2015). Thus, in appealing to Afghans' patriotism and sense of nationalism, the government wants to prevent the loss of both its highly educated and working class populations. The initial campaign has spread to a popular, citizen-driven Twitter campaign called "Afghanistan Needs You." It has also led to initiatives that promote education and jobs for unemployed young people.

Laura M. Steckman

See also: Iran; Iraq; Russia; Syria; United States

Further Reading

Afghan Wireless. 2007. "Afghan Wireless Again Creates History in Afghanistan." November 3. Accessed August 28, 2016. http://www.afghan-web.com/economy/afghan_wireless_history.html

Afghanistan Times. 2016. "Social Media Can Empower Women: Raziya Masumi." June 19. Accessed August 28, 2016. http://afghanistantimes.af/social-media-can-empower-women-raziya-masumi

Alexa. 2016. "Top Sites in Afghanistan." Accessed August 20, 2016. http://www.alexa.com/topsites/countries/AF

Bezhan, Frud. 2015. "'Stay With Me': Afghan Government Begs Citizens Not to Flee." September 22. Accessed August 20, 2016. http://www.theatlantic.com/international/archive/2015/09/afghanistan-brain-drain-migrant-crisis/406708/

Budde. 2016. "Afghanistan—Telecoms, Mobile and Broadband—Statistics and Analyses: Executive Summary." Accessed December 29, 2016. https://www.budde.com.au/Research/Afghanistan-Telecoms-Mobile-and-Broadband-Statistics-and-Analyses?r=51

Capacity Media. 2012. "Etisalat Launches 3G in Afghanistan." March 19. Accessed August 20, 2016. http://www.capacitymedia.com/Article/2997565/Etisalat-launches-3G-in-Afghanistan.html

Freedom House. 2016. "Afghanistan." October 4. Accessed December 29, 2016. https://freedomhouse.org/report/freedom-press/2016/afghanistan

GIZ, Deutsche Gesellschaft für Internationale Zusammenarbeit. 2014. "Social Media in Afghanistan: Measuring the Usage & Perceptions of the Afghan People." Accessed August 28, 2016. http://ez-afghanistan.de/fileadmin/content/news/Social_Media_251114.pdf

Guardian. 2015. "Discontent Rife in Afghanistan: 40% of People Keen to Leave, Survey Says." November 18. Accessed August 28, 2016. https://www.theguardian.com/global-dev elopment/2015/nov/18/afghanistan-survey-discontent-40-per-cent-people-keen-to -leave

Guardian. 2016. "Afghan Taliban Create Smartphone App to Spread their Message." April 3. Accessed August 28, 2016. https://www.theguardian.com/world/2016/apr/03/afghan -taliban-create-smartphone-app-spread-message

Kargar, Zarghuna. 2015. "Farkhunda: The Making of a Martyr." August 11. *BBC News.* Accessed August 28, 2016. http://www.bbc.com/news/magazine-33810338

StatsMonkey. 2015. "Mobile Facebook, Twitter, Social Media Usage Statistics in Afghanistan." Accessed August 20, 2016. https://www.statsmonkey.com/table/21279-afghan istan-mobile-social-media-usage-statistics-2015.php

Wadsam. 2015. "10 Awards Given Out in First-Ever Afghan Social Media Awards." October 21. Accessed August 28, 2016. http://wadsam.com/afghan-business-news/10 -awards-given-out-in-first-ever-afghan-social-media-awards-232/

ALGERIA

Algeria is a country located in the northern part of Africa. It has a population of nearly 39 million inhabitants and a gross domestic product (GDP) per capita of $3,316. Internet access was available to about 18 percent of the population in 2014 (World Bank Data 2014). There are no less than 6.8 million Facebook accounts and close to 400,000 Twitter handles (Baala 2014). Social media platforms are mostly in Arabic and French, although some personal blogs can be found in Tamazight (Algeria's official language, along with Arabic) or English. The majority of the country's online users access the internet from computers at home or at internet cafés. Despite the developing mobile phone market, internet is still overwhelmingly accessed via desktops.

Compared to its neighboring countries, 3G services in Algeria, first available in 2013, were introduced much later. For instance, Morocco launched its first 3G services in 2007, Tunisia in 2010, and Egypt in 2008 (Yaici 2013). However, despite such a delayed start, mobile internet usage and the access of social media through phones and mobile devices in Algeria have grown very rapidly. The Algerian Ministry of Post and Information and Communications Technologies (MPTIC) reports that in 2015, there were nearly 2 million internet users and 423,280 4G LTE users, and that 3G had surpassed 18 million subscriptions. As of summer 2016, users can expect to pay 1,000–1,600 dinars ($12.60–20) monthly for decent 3G service on their mobile phones, compared to a cost of at least 1,600 dinars monthly for internet services at home. Access to internet at home is much more limited than via mobile phones because the number of landlines is limited.

Mobile telephony subscribers had reached 45 million as of 2015, primarily through prepaid subscriber identify module (SIM) card services. After a 2000 law opened mobile telephone services to international competition, three main companies came to dominate the Algerian market: the Egyptian-owned Orascom Telecom Algerie (Djezzy), whose share of mobile telephony reached 45 percent in November 2015, and in 2016, it became the dominant provider, with 18.6 million subscribers; Qatar's Wataniya Telecom Algerie (Ooredoo), with 11.7 million; and state-owned Mobilis Algerie Telecom, the former market leader in the number of 3G subscribers, with 13 million (Mayer 2016). In order to remain at the top of mobile internet competition in Algeria, Mobilis has contracted with China's giant Huawei company for migration to 5G. The competition among these three operators is driving internet costs down for consumers and creating connectivity opportunities for Algerians.

Statistics show that Facebook is by far the most used platform for social media (97 percent), followed by Twitter (1.26 percent), Google+ (0.42 percent), Tumblr (0.12 percent), and Pinterest (0.25 percent) (StatsMonkey 2015). Accessing the internet and social media has had an important impact on Algeria's economic, political, and social spheres. Economically, easy internet access has prompted an increase in networking opportunities, the development of crowdsourcing apps, and linking young entrepreneurs from different parts of the country to their international counterparts. On a more commercial level, internet access permitted the creation of a local version of Craigslist called OuedKniss. This website is a platform for sellers to advertise their products and for buyers to contact them. While mostly used for car sales and apartment hunting, the website advertises a wide variety of services and products.

Socially and politically, internet and social media access has facilitated a strong presence of Algerian civil society organizations, with groups advocating for topics as diverse as gender equality rights, environmental protection, and anticorruption. Algerians take to Facebook and Twitter to tell the world of their disappointments, disagreements, and general reactions to both domestic and international events. During the series of popular uprisings that spread in neighboring countries Tunisia, Libya, and Egypt in 2011, Algerians used Facebook to organize protests and "rage days." While the so-called Arab Spring did not have a major influence on Algeria's regime, Algerians organized several protests and engaged in long debates asking for political reforms via social media.

Facebook and social media also have played a curious role in the latest presidential election in 2014, when slogans equating the reelection of Abdelaziz Bouteflika (1937–) with stability and security were disseminated. During the campaign, Bouteflika supporters circulated pictures of suicide bombings that occurred during Algeria's decade of turmoil in the 1990s prior to Bouteflika's presidency. This pro-Bouteflika social media campaign suggested that he was the only candidate who succeeded in putting an end to that violence, and that changing the status quo may result in triggering instability anew. In addition, several of the presidential candidates had official Facebook accounts for their campaigns, using those pages to reach out and mobilize the diaspora abroad, explain agenda points, communicate rally and event information to supporters, and call for boycotts.

Upon Bouteflika's reelection for a fourth presidential term, a civil society group opposing the election results was formed and became very active on social media. The movement, called Barakat (which translates as "Enough!"), used its Facebook page to call for sit-ins and demonstrations. Barakat's founding members are doctors, lawyers, teachers, and citizens who were not previously political activists. Although Barakat's impact has been marginal, their Facebook page collected 30,000 likes in just one month, March 2014 (Bouraoui and Benneoune 2014).

Other civil society activist groups who primarily use social media to voice their positions include the anti-fracking movement No! It used Facebook to raise awareness of the dangers of fracking by sharing videos explaining the process and its consequences. It also carried a wide campaign to change profile pictures to yellow, the official color symbolizing the antifracking campaign.

Besides Facebook, Algerians seem to use applications such as Skype, Twitter, You-Tube, Tumblr, and Viber. Frequently, Twitter is a great place to follow debates and get a sense of the topic of the day in Algerian politics. Algerian Twitter accounts often mix wit, satire, and intelligent analyses of political situations in the country. There has also been a rise in creating apps in Tamazight in order to facilitate both learning the language and accessing news content for speakers of Tamazight. One example of such apps is Azul, which was developed for Tamazight-language learners. Finally, in addition to these apps, some Algerians have personal blogs, many of which are in French and Arabic, but some are also in English. Some of these blogs are dedicated to political conversations, while others are cultural, educational, entertainment, or a mix thereof.

It should be noted that the government has a strong grip on internet access. The scandal concerning the BAC high school exit exam that arose in June 2016, when several exam questions were leaked through Facebook and Twitter, resulted in cancelling some of the exam subjects and rescheduling a retake session held in early July. For the second round of the exam, the Ministry of Education collaborated with the Ministry of Telecommunication to block access to social media during the first day of the exam to prevent cheating (Le Matin D'Algérie 2016). This disruption in internet accessibility was reported to have had a severe negative economic impact on companies and enterprises that rely on the internet for their services. While the idea that blocking access to the internet was efficient in preventing cheating remains questionable, it nonetheless showed the extent of the government's willingness and ability to control access to social media.

Algeria has a number of hacker groups and individual hackers who are notorious not only at home, but also abroad. Perhaps one of the most renowned cases is that of Algerian hacker Hamza Bendelladj (1988–), who was arrested in Thailand in 2013 and extradited to the United States to face several charges of cybercrime. Bendelladj, also known as the Algerian digital Robin Hood and "BX1," pleaded guilty to developing SpyEye, a Trojan horse program used against more than 200 bank accounts to extract millions of dollars that he sent to Palestinian nongovernmental organizations (Hatuqa 2015). Other forms of cybercrime include web defacement, which is frequent in North Africa and involves hacking into foreign websites and replacing content as a sign of protest or activism. Some of Algeria's best-known hacker groups include Algeria to the Core and the hacktivist group Anonymous Algeria. The latter was involved in hacking and defacing over forty French websites after the *Charlie Hebdo* attack in order to protest the periodical's defamatory caricatures.

Lina Benabdallah

See also: Algeria; China; Egypt; France; Libya; Morocco; Qatar; Tunisia

Further Reading

Baala, Hamdi. 2014. "L'Algérie Compte 6.8 Millions D'utilisateurs sur Facebook, seulement 37 500 sur Twitter (Étude)." June 30. Accessed July 22, 2016. http://www.huffpost maghreb.com/2014/06/30/facebook-twitter-algerie-_n_5543478.html

Bakarat Movement. 2014. "Algeria: Barakat Movement Manifesto." *Secularism Is a Women's Issue.* Accessed July 22, 2016. http://www.siawi.org/article7193.html

Bouraoui, Amira, and Karima Bennoune. 2014. "The Birth of the Barakat Movement In Algeria: Every Generation Needs Hope." Accessed July 22, 2016. https://www.open democracy.net/5050/amira-bouraoui-karima-bennoune/birth-of-barakat-movement -in-algeria-every-generation-needs-hope

Hatuqa, Dalia. 2015. "Hamza Bendelladj: Is the Algerian Hacker a Hero?" September 21. Accessed July 22, 2016. http://www.aljazeera.com/news/2015/09/algerian-hacker-hero -hoodlum-150921083914167.html

International Media Support (IMS). 2013. "Authoritarianism and Media in Algeria." Accessed July 22, 2016. https://www.mediasupport.org/wp-content/uploads/2013/07/au thoritarianism-media-algeria-ims-20131.pdf

Le Matin D'Algérie. 2016. "BAC: les Réseaux Sociaux Bloqués en Algérie depuis Samedi Soir." June 19. Accessed July 22, 2016. http://www.lematindz.net/news/21040-bac-les-reseaux -sociaux-bloques-en-algerie-depuis-samedi-soir.html

Liebelson, Dana. 2015. "Meet the Algerian Hackers Who Say They're Attacking French Websites over Charlie Hebdo." January 13. Accessed July 22, 2016. http://www.huffington post.com/2015/01/13/charlie-hebdo_n_6464318.html

Mayer, Sheldon. 2016. "Algerian Telecom Djezzy Sees Path Forward." April 5. Accessed December 29, 2016. http://africa-me.com/algerian-telecom-djezzy-sees-path-forward/

StatsMonkey. 2015. "Social Network Usage Statistics Using Mobile." Accessed July 22, 2016. https://www.statsmonkey.com/sunburst/21282-algeria-mobile-social-media-usage -statistics-2015.php

World Bank Data. 2014. "Internet Users per 100 People." Accessed July 22, 2016. http:// data.worldbank.org/indicator/IT.NET.USER.P2?locations=DZ

Yaici, Karim. 2013. "Mobile Telecoms Market in Algeria: the Untapped Potential of 3G." Accessed July 22, 2016. http://inspiremagazine.anasr.org/feature-mobile-telecoms -market-in-algeria-the-untapped-potential-of-3g/

ANGOLA

Angola, a southern African nation, gained independence in 1975, following the decolonization process initiated after the democratic revolution that took place in Portugal in April 1974. After independence, Angola was ravaged by a civil war that only ended in 2002. In one interlude of this war resolved by the Bicesse Accords in 1991, and after almost two decades of one-party rule, Angola initiated some democratic reforms with the adoption of a democratic constitution, the introduction of a multiparty system, and a legislative election that was held in 1992. The incumbent party, MPLA (Popular Movement for the Liberation of Angola), won that election with an absolute majority.

Today, Angola exemplifies a hybrid regime or a defective democracy because it blends many undemocratic with some democratic features: constraints to freedoms are significant and long-lasting, the same president has held power since 1979, and Angola has one of the world's most unequal income distributions (Diamond 2002;

Merkel 2004). Although Angola is sub-Saharan Africa's second-largest exporter of oil and the world's fifth-largest producer of diamonds, and its economy has been recently growing at an average rate of 7 percent per year, a significant part of the population still suffers from extreme difficulties. In 2012, 36 percent of the population lived below the poverty line and 26 percent lived in extreme poverty (Africa Progress Panel 2013, 20), and in 2016, the literacy rate was 71 percent (CIA 2016).

In addition to constraints to freedom and to the attempts of the political regime to control online media contents and dissemination, poverty, education, and lack of infrastructure—only 35 percent of the population have access to electricity, for example (Africa Progress Panel 2015, 16)—have been major impairments in the development of the Angolan online environment. Furthermore, what is demanded of citizens in an information society is much more than just knowing how to read and write: it involves other types of literacy related to comprehension and interpretation skills and to the ability of understanding different types of media and media genres (Salgado 2014, 111).

According to Internet World Stats (2016), internet users in Angola amounted to 29.5 percent of the population in June 2016, representing 1.8 percent of the entire African continent. However, the numbers have been growing rapidly in recent years; in 2000, there were only 30,000 users, but in 2012, 14.8 percent of the Angolan population was connected to the internet (Internet World Stats 2016). With regard to mobile phones, although the ratio is more proportional than in other African countries, more people in Angola have access to mobile phones than to bank accounts. In 2011, 47 percent of the population had a mobile phone subscription, while only 39 percent had accounts at formal financial institutions (Africa Progress Panel 2014, 115).

A few projects have aimed to address the limited access to technology and low computer literacy level in Angola. One of those projects had the goal of providing all provinces with media libraries, including in the least-populated provinces, and to distribute mobile media libraries throughout the country, while more definitive structures are not in place. These mobile structures seek to facilitate the access of the general population to new technology and allow citizens to do all kinds of research on the internet (Angop 2015). In 2015, Angola had six media libraries in Luanda, Benguela, Huambo, Lubango, Saurimo, and Soyo; six mobile media libraries scattered across the country; and a plan for at least one media center to be fully operational in each of Angola's eighteen provinces by 2017 (Deutsche Welle 2015). The Angola Media Libraries Network is integrated into the larger N'gola Digital program, which is led by the Angolan Ministry of Telecommunications and Information Technologies and focuses on providing communities with computers connected to the internet and organizing workshops on how to use computers and software at different levels of proficiency.

Another project was put in place by Movicel, an Angolan mobile phone company, that developed a partnership with Facebook to create an app called Internet.org to provide free internet access to a limited selection of websites related to news, health, employment, weather, and Facebook (O País 2015). The implementation of this kind of project and the availability of cheaper services have made internet access easier in Angola; however, the prices are still very high for most of the population. For example, one of the main Angolan providers, Tvcabo Angola, charges more than $40 for 1 Mb in 2016 and most usage is still confined to the capital city, Luanda, and to the other major cities.

The use of social media is also evolving rapidly. The trend toward the use of social networking sites was followed in Angola, but in 2009, the most popular platforms were Hi5 and Sonico, but not Facebook. These social networking websites motivated young people to use computers and the internet, and they were also an important means to reconnect the diaspora communities with their country, as well as an incentive for many citizens to give opinions about the main issues facing their country (Salgado 2011). Today, following the world trend, Facebook is the most used social network in Angola, with 3.3 million users in 2015 (Internet World Stats 2016).

Blogs have also been an important vehicle for alternative voices in the Angolan public sphere, but these social media also reflect strong inequalities in the use of and access to technology in general. Most bloggers are from the middle and upper classes and have a high academic level compared with the rest of the population; there are blogs, for example, by politicians, journalists, academic researchers, historians, and artists (Salgado 2012). Although there are many posts on political issues in the Angolan blogosphere, including criticisms of political authorities, blogs are not a privileged means for collective action and mobilization, as they are mainly used as means of individual expression. When the objective is to organize protest or support actions, other online tools are usually preferred, such as emails, short message service (SMS) messages, and social networks.

After the revolutionary events of the Arab Spring in Tunisia and Egypt in March 2011, some Angolan protesters used the internet—namely, emails and an anonymous website (http://revolucaoangolana.webs.com, which has since moved to http://novarevolucaoangolana.yolasite.com)—to announce the "New Revolution of the Angolan people" and called the Angolan people to demonstrate against the government and President José Eduardo dos Santos (1942–). The MPLA government, not willing to tolerate any comparison between Angola and the northern African countries, reacted strongly and quickly ensured that public order would be maintained at any cost, despite the fact that the Angolan constitution grants all citizens the right to demonstrate peacefully.

It is thus not surprising that the MPLA government prepared new legislation to control and restrict the online environment. In 2011, the National Assembly approved a bill criminalizing the use of the internet and mobile phones to send any type of information without the prior written consent of everybody mentioned in its contents. This new law, presented as a data protection measure, was also carefully designed to prevent protests initiated and coordinated through the internet from happening. It established imprisonment as a penalty and allowed the security forces to conduct searches and confiscate data and documents without a court order. This attempt to regulate online activity through repressive legislation is in keeping with the MPLA government's restrictive posture toward the media in general. It also shows that the number of people with internet access in the country is already enough to shake up the political situation—at least slightly. Until a few years ago, almost no one dared criticize the president in public; however, more recently, the internet and pro-democracy and freedom initiatives have encouraged more people to show their discontent.

Susana Salgado

See also: Egypt; Mozambique; Namibia; Tunisia

Further Reading

Africa Progress Panel. 2013. "Equity in Extractives: Stewarding Africa's Natural Resources for All." Accessed August 10, 2016. http://africaprogresspanel.org/wp-content/uploads/2013/08/2013_APR_Equity_in_Extractives_25062013_ENG_HR.pdf

Africa Progress Panel. 2014. "Grain, Fish, Money. Financing Africa's Green and Blue Revolutions." Accessed August 10, 2016. http://app-cdn.acwupload.co.uk/wp-content/uploads/2014/05/APP_APR2014_24june.pdf

Africa Progress Panel. 2015. "Power, People, Planet. Seizing Africa's Energy and Climate Opportunities." Accessed August 10, 2016. http://www.africaprogresspanel.org/publications/policy-papers/2015-africa-progress-report/

Angop. 2015. "Bié: Província Ganha Mediateca Móvel." *Angop: Agência Angola Press*, April 22. Accessed October 1, 2015. http://www.portalangop.co.ao/angola/pt_pt/noticias/lazer-e-cultura/2015/3/17/Bie-Provincia-ganha-mediateca-movel,099252ba-f1f8-414e-a908-7e533926d6d0.html

Central Intelligence Agency (CIA). 2016. "The World Factbook: Angola." Accessed August 10, 2016. https://www.cia.gov/library/publications/the-world-factbook/geos/ao.html

Deutsche Welle. 2015. "Acesso 'Sem Filtros' à Internet em Angola." *Deutsche Welle*, June 3. Accessed October 1, 2015. http://www.dw.com/pt/acesso-sem-filtros-à-Internet-em-angola/a-18495315?maca=bra-cb_po_globalvoices-14551-xml-mrss

Diamond, Larry. 2002. "Elections Without Democracy: Thinking About Hybrid Regimes." *Journal of Democracy*, 13(2): 21–35.

Internet World Stats. 2016. "Angola: Internet Usage and Marketing Report." Accessed August 10, 2016. http://www.internetworldstats.com/af/ao.htm

Merkel, Wolfgang. 2004. "Embedded and Defective Democracies." *Democratization*, 11(5): 33–58.

O País. 2015. "Internet Grátis Chega a Angola." June 29. Accessed October 1, 2015. http://opais.co.ao/Internet-gratis-chega-angola/

Salgado, Susana. 2011. "Angola." In *Encyclopedia of Social Networks*, edited by George Barnett. Thousand Oaks, CA: Sage Publications, 42–44.

Salgado, Susana. 2012. "The Web in African Countries. Exploring the Possible Influences of the Internet in the Democratization Processes." *Information, Communication, & Society*, 15(9): 1373–1389.

Salgado, Susana. 2014. *The Internet and Democracy Building in Lusophone African Countries.* London and New York: Routledge.

ARGENTINA

Argentina is located in the southern cone of South America along the Atlantic Ocean. In 2015, the country had a population of 43.4 million people. At the beginning of 2015, Argentina's internet penetration rate was 75 percent, making it the most connected nation in Latin America. This also represented a growth of 37 percent in active internet users from the previous year. While only 59 percent of the population used cell phones, the number of mobile connections was 62 million. This suggests that a sizable proportion of cell phone users had more than one mobile subscription. Of those mobile phone users, 71 percent of them used prepaid accounts. Despite this proliferation of cell phones, Argentine internet users tended to spend an average of forty more minutes a day on their PCs or tablets than on their mobile devices (We Are Social 2015). A total of 71 percent of these internet users were over the age of twenty-five, and the gender distribution was about even (comScore 2015).

Android devices are typically more common in Latin America than other devices; this trend holds true for Argentina. At the beginning of 2015, 75 percent of smartphones were Android devices; 13 percent of smartphones were Blackberry devices; and only 11 percent were iPhones. The lesser preference for Apple devices also holds true for tablets; only about 19 percent of tablet users owned an iPad. Tablet users in Argentina overwhelmingly tend to use their devices with Wi-Fi rather than 3G/4G (comScore 2015). Toward the end of 2015, the top free app downloaded for both Android and Apple devices was WhatsApp Messenger. For Android devices, the next most commonly downloaded free apps were Facebook Messenger, Facebook, Instagram, and Moto Traffic Race game. For Apple devices, the next most commonly downloaded free apps were Spotify, Facebook Messenger, Facebook, and Instagram (App Annie 2015).

Lionel "Leo" Messi, a star Argentine soccer player with FC Barcelona, poses with a European Golden Boot award before the FC Barecelona-RCD Mallorca match at Nou Camp Stadium in October 2010. Leo Messi has found the same type of stardom online as he has on the soccer field: his Facebook page has more "likes" than anyone else in the country of Argentina. (MaxiSports/Dreamstime.com)

The most visited website on any device was Google.com.ar, followed by Facebook, YouTube, Google.com, Mercadolibre.com .ar (an online marketplace), Live, Yahoo, Twitter, Clarín (a Spanish-language news website), and Amazon. Facebook is the most popular social media website in Argentina, followed by YouTube, Twitter, and Taringa! (Alexa 2015). The Facebook page from Argentina with the most "likes" is Leo Messi, with 80,475,301. The Facebook pages with the most fans across all age ranges in the middle of 2015 were both soccer related: BOCA, with 7,509,234 fans, and Club Atlético River Plate, with 6,049,998 fans. The next top Facebook pages were the news outlets Diario Clarín (4,521,328 fans) and Todo Noticias, or TN (4,328,168 fans). The top Facebook page, which had the fifth-highest number of fans at 4,119,293, was Quilmes Cerveza (Allin1 Social 2015).

In December 2014, the Argentine government passed Law 27.078, known as the "Argentine Digital Law." This new law governs telecommunications in the country and replaces the previous law passed in 1972. The Argentine Digital Law is an important update because it addresses technology that did not exist in 1972. The new law establishes that information and communications technologies are public services that should be accessible to all citizens. It states that this technology is necessary for social development and that access should be guaranteed for all citizens, regardless of their geographic or socioeconomic situation. To this end, the law aims to encourage competition between service providers so that consumers can choose what best fits their needs. It provides the Argentine government with the right to regulate rates for these services so that smaller companies can reach consumers even if they lack their own infrastructure. The law also established mandatory minimum transmission speeds to ensure equal access and quality everywhere in the country (Télam 2014). In order to guarantee that these objectives are realized and that all citizens receive quality access, Argentina established La Autoridad Federal de Tecnologías de la Información y las Comunicaciones (AFTIC; the Federal Authority of Information and Communication Technologies) as a democratic body with representation from different sectors of society and government. While the Argentine Digital Law initially received a mixed reception, only time will tell how successfully it can achieve its goals of national telecommunications penetration (AFTIC 2015).

Argentina is also home to Taringa!, the largest social networking website of Latin American origin. It is ranked number sixteen in Argentina and is used throughout the Spanish-speaking world. In addition to Argentina, Taringa! receives the most traffic from Mexico, Spain, Venezuela, and Colombia. The website is similar to Reddit, in that users post links or upload content and then vote to rank posts according to popularity. Today, it also offers gaming and music experiences. The website is known as a collaborative system of online interaction with the tagline "collective intelligence." It was designed in early 2004 by Fernando Sanz (c. 1988–), who was still in high school at the time. In November 2006, once the website exceeded 30,000 daily visitors and he could no longer manage it on his own, Sanz sold it for $5,000 to brothers Matías Botbol (1978–) and Hernán Botbol (1981–), who were partnered with Alberto Nakayama (c. 1981–). Over the years since the site's acquisition, the design of Taringa! has undergone several different incarnations. By 2015, the website had more than 75 million unique visitors and 27 million registered members (Duarte 2012).

In April 2015, Taringa! announced an innovative addition to the website. Taringa! partnered with Xapo, a California-based bitcoin wallet provider, to offer Taringa! users opportunities to earn bitcoins for the content they post on the website. The program, called Taringa! Creadores, combines advertisement revenue sharing (similar to YouTube's model) and incentives based on Taringa's ranking system. According to Hernán Botbol, the director of Taringa!, offering financial compensation was something that he and his partners had been considering for a while; however, due to a lack of financial infrastructure in the region, implementing such a program would have been too complicated (Casey 2015). The low overhead and ease of bitcoin transfers allow Taringa! to immediately deposit bitcoins into users' wallets instead of requiring them to wait a certain amount of time. Users can then choose to either cash the bitcoins in for regional currency, save the bitcoins as an investment, or use them for other e-commerce opportunities. Since the introduction of this program, Taringa! has seen a 40 percent increase in content creation (Young 2015).

In addition to creating content on social media sites, Argentines use them to mobilize for social justice. For example, on June 3, 2015, demonstrations were held in eighty cities and towns all over Argentina, as well as in Chile and Uruguay, to raise awareness of violence against women. The movement was advertised across social media platforms with "#NiUnaMenos." The hashtag means "Not One Less" and refers to the idea of "not one more death, not one woman less" (Télam 2015). One of the motivating factors in this movement was the death of fourteen-year-old Chiara Paéz (c. 2001–2015). While she was pregnant, Paéz was beaten to death by her sixteen-year-old boyfriend. According to La Casa del Encuentro, a civil rights organization, there were 1,808 murdered women between 2008 and 2014, which left over 2,000 children without a mother. While there are laws to protect women against this kind of violence, Ingrid Beck (n.d.), a demonstration organizer, has stated that they are very seldom used to prosecute perpetrators of violence against women (CNN Español 2015).

The #NiUnaMenos movement has garnered support from all over Argentina, and supporters have used social media to spread this issue. Then president of Argentina, Christina Fernández de Kirchner (1953–), went to Twitter to express solidarity.

Trivia Crack was the number one free app downloaded in Latin America in 2014. The game requires users to spin a wheel and then answer a question under the category where the wheel stops. The game can be played alone or with friends and is specifically designed to allow social media friends to compete against each other. The app's creator, Maximo Cavazanni (c. 1986–), has turned it into a successful Argentine game show, with negotiations regarding versions in Mexico, Brazil, and Colombia already underway (Woolley 2015). One of the app's unique features is that it has a question factory that uses the power of crowdsourcing to ensure that questions reflect its users' tastes by region. For example, the sports questions are about the sports that that country watches. In Latin America, soccer players and scores are more important than those of American football. In the same vein, the crowd performs a language check so that soccer questions concern *fútbol* or *futebol*, the regionally appropriate names for soccer.

She stated that the problem is not just a political or judicial one, but a cultural one (CNN Español 2015). The famous soccer player Lionel Messi also used social media to add his voice to the movement. On his Facebook page, he reiterated the movement motto by stating that all Argentines join in the cry "not one woman less" (BBC 2015).

Marilyn J. Andrews

See also: Brazil; Chile; Mexico

Further Reading

AFTIC. 2015. "Qué es AFTIC." Accessed September 27, 2015. http://www.aftic.gob.ar/que -es-aftic_p33

Alexa. 2015. "Top Sites in Argentina." Accessed September 27, 2015. http://www.alexa.com /topsites/countries/AR

Allin1Social. 2015. "Facebook Statistics for Argentina." Accessed September 27, 2015. http:// www.allin1social.com/facebook-statistics/countries/argentina?page=1&period=six _months

AppAnnie. 2015. "Google Play Top App Charts." September 27. Accessed September 27, 2015. https://www.appannie.com/apps/google-play/top/argentina/

BBC News. 2015. "Argentine Marches Condemns Domestic Violence." June 4. Accessed September 27, 2015. http://www.bbc.com/news/world-latin-america-33001990

Casey, Michael J. 2015. "BitBeat: Latin America Facebook Rival to Use Bitcoin to Pay for Content." *The Wall Street Journal*, April 21. Accessed September 26, 2015. http://blogs .wsj.com/moneybeat/2015/04/21/bitbeat-latin-america-facebook-rival-to-use-bitcoin -to-pay-for-content/

CNN Español. 2015. "'Ni Una Menos,' La Movilización que Pide Parar la Violencia de Género en Argentina." June 3. Accessed September 27, 2015. http://cnnespanol.cnn.com /2015/06/03/ni-una-menos-la-movilizacion-que-pide-parar-la-violencia-de-genero-en -argentina/

ComScore. 2015. "ComScore IMS Mobile in LatAm Research Study." April 20. Accessed September 27, 2015. https://www.comscore.com/Insights/Presentations-and-White papers/2015/comScore-IMS-Mobile-in-LatAm-Research-Study

Duarte, Joel. 2012. "Historia y Evolución de Taringa!" Accessed September 26, 2015. http:// www.taringa.net/posts/taringa/16286651/Historia-y-evolucion-deTaringa.html

Télam. 2014. "Promulgan Ley que Garantiza el Acceso Universal y Equitativo a Los Servicios de Comunicación." December 19. Accessed September 27, 2015. http://www.telam .com.ar/notas/201412/89431-promulgan-ley-argentina-digital.html

Télam. 2015. "Más de 200 Mil Personas Se Concentraron Frente al Congreso Para Decirle Basta a Los Femicidios." June 3. Accessed September 27, 2015. http://www.telam.com .ar/notas/201506/107330-niunamenos-mas-de-80-ciudades-argentinas-unidas -contra-los-femicidios.html

We Are Social. 2015. "Digital, Social, & Mobile Worldwide in 2015." January 21. Accessed September 27, 2015. http://wearesocial.net/blog/2015/01/digital-social-mobile-world wide-2015/

Young, Joseph. 2015. "Taringa! Social Network Sees 40% Spike in Content Creation Following Bitcoin Integration." June 7. Accessed September 27, 2015. http://cointelegraph .com/news/114492/taringa-social-network-sees-40-spike-in-content-creation -following-bitcoin-integration

AUSTRALIA

The Commonwealth of Australia consists of the Australian continent, mainland Tasmania, and over 8,000 smaller islands. Australia's population density is comparable to Canada's, at about three people per square kilometer. That said, the population is heavily concentrated in the eastern and western coastal areas of the continent, particularly in the major cities, where around 64 percent of the country's population lives. While Australia's internet penetration is somewhere between 86 percent and 99 percent, this is mostly owing to mobile technologies; 99 percent of Australians have at least one device capable of accessing the internet (Sensis 2016). The country's geography and population distribution make access to higher-speed fixed internet service somewhat difficult for those not located in major population hubs. A total of 87 percent of Australians use the internet daily (Sensis 2016), and since 2007, the status of Australia's internet infrastructure has been a controversial battleground for the country's political parties. Facebook is the most popular social media platform in Australia, at around 15 million active users. YouTube is equally popular, with around 14 million users (Sensis 2016).

Australia's internet ranks sixtieth in the world for peak connection speed and forty-eighth for average peak connection speed (Akamai 2016). In terms of relative ranking, this places Australia a little below half of its geographical peers in the Asia Pacific region, such as South Korea, Japan, Hong Kong, New Zealand, Taiwan, and Singapore. Moreover, while Akamai's report puts the average peak connection speed at around 39 mbps, this figure includes the more robust government, business, and educational access points as well as residential ones, and it is only an indicator of capacity, not of the actual speeds experienced by consumers. Australian consumers are often unable to access the maximum potential connection speed due to two main factors: insufficient residential internet infrastructure and a dearth of residential internet packages that offer the highest speeds available.

Prior to the 2007 national election, the Australian Labor Party ran on a platform that included the construction of a $43 billion fiber to the premises (FTTP) National Broadband Network (NBN) to replace the preexisting copper wire network. The project commenced in July 2009. The NBN and its high cost have become a contentious issue between Australia's political parties. Prior to the 2013 national election, the Liberal-National Coalition (Coalition) opposition government sought changes to the NBN project that included replacing FTTP with fiber to the node (FTTN), a slower but ostensibly more affordable option. Research challenges this idea, arguing that FTTP has a significantly lower operating cost that would save the government as much as $9 billion over ten years (Ferrers 2016). After winning the election in 2013, the Coalition government made further changes to the NBN project, forecasting a $29.5 billion cost and a completion date of 2019. As of August 2015, costs were expected to reach as high as $56 billion.

The project is ongoing and continues to generate controversy due to its rising costs. Leaked government documents in February 2016 revealed the project to be massively behind schedule. In May 2016, NBN Co Limited, the government-owned corporation tasked with the design, construction, and implementation of the NBN, put a temporary hold on NBN rollouts in Tasmania. NBN Co plans to provide Tasmania with satellite-delivered internet rather than the originally proposed FTTN, a significantly lower-quality service. Currently, there is a local movement in Tasmania to reverse the decision and reinstate fiber optics in Tasmania. As of January 2016, NBN Co reports some 736,052 active NBN services (NBN CO Ltd. 2016). This figure includes fiber optic, satellite, and wireless services.

Mobile technologies have proven to be overwhelmingly important to Australian internet users (Sensis Social Media Report 2016). Australia's current 4G mobile networks, owned by Telstra Corporation and Hutchison Telecommunications Ltd., offer speeds comparable to Akamai's reported peak connection speed. Access to networks via cellular devices is, therefore, competitive with currently available fixed internet services. Some users still waiting for access to the NBN have taken advantage of high-cap cellular data plans to replace their residential internet service providers (ISPs) with a mobile hotspot. Moreover, Telstra announced in February 2016 that it was working aggressively to develop an Australia-wide 5G network, which could conceivably render the NBN obsolete before it is even completed. Telstra expects to run a trial 5G network at the Commonwealth Games in 2018. Mobile computing provides a massive boost to the Australian economy and has contributed $43 billion throughout 2015 (Deloitte 2016).

For their own part, Australian consumers continue to rely heavily on mobile computing. A total of 21 percent of Australians are mobile-only internet users (but still use landline telephones), whereas 12 percent of the population have cut the cord completely and have replaced fixed internet and telephone with mobile alternatives (ACMA 2015). This may be related to the difficulty in getting access to broadband internet outside the major cities, as only 10 percent of mobile-only users were found within the major cities. In addition, 76 percent of Australians own a smartphone and 70 percent have a laptop; the popularity of these devices far outweighs desktop PCs (52 percent) and even tablet computers (55 percent) (Sensis 2016).

With regard to social media access, Australians overwhelmingly favor using their mobile devices with dedicated apps over browser-based access, with 72 percent relying on their smartphones to keep in touch with their social networks (Sensis 2016). More than 50 percent of Australians report making use of their social networking sites between one (24 percent) and five (26 percent) times per day (Sensis 2016). Facebook is the most used social networking site by Australians, with 95 percent of social networkers maintaining a Facebook profile. The next most popular site is Instagram, at 31 percent (Sensis 2016). Twitter, with only 19 percent usage by social networkers, is the most rapidly declining social media site in Australia, with 34 percent of all users who abandoned a social media platform in the previous year having abandoned Twitter (Sensis 2016).

In June 2016, Australia adopted new legislation that allowed the blocking of websites to work against online piracy. The Copyright Amendment (Online Infringement) Bill 2015 allowed rights holders to force Australian ISPs to block access to websites whose primary purpose is the facilitation of copyright infringement.

Languages are always changing, yet with the invention and popularity of social media, they are changing faster than in the past. The *Oxford English Dictionary* adds words officially to the English language by putting them to the dictionary. In 2015, the dictionary added several words that originated on social media and then caught on in people's vernacular, everyday, speech. Those words included *selfie stick* (*selfie* was added in 2013), *concern troll*, *lamestream*, *downvote*, *upvote*, *cyberwarrior*, and *in-app*, among many others (Oxford English Dictionary 2015). Social media and internet users will continue to invent new terms, and as they become popular, these words will also find their way into the dictionary as official new words.

Passage of the bill has been controversial, with opponents arguing that such legislation interferes with legitimate file-sharing as well. These concerns are not wholly unfounded, as a 2013 attempt by the Australian Securities and Investment Commission to block sites that it deemed illegitimate resulted in nearly 250,000 compliant sites being blocked as well. Despite public criticism of the bill, both of Australia's major political parties support it.

In February 2016, the American internet security company Webroot released a Threat Intelligence Brief that noted a doubling of Australian-based cyberthreats over the previous year. While Australia has been moved to "high-risk" status by the security company, it is worth noting that while 2 percent of global malicious URLs are hosted in Australia, the United States hosts around 30 percent, China 11 percent, and Hong Kong 5 percent (Webroot 2016). The report further indicates that Australia hosts around 2 percent of global phishing sites (Webroot 2016). While Australian media have made much of this "high-risk" classification, the report itself doesn't single out Australia in any particular way beyond grouping it with other top-ten high-risk countries, which include the United States, Canada, Great Britain, China, South Korea, and Japan. In February 2016, ABC News Australia suggested that the small to medium-sized businesses that comprise 95 percent of Australia's organizations are more attractive targets than the higher-risk large banks and multinational corporations, which are equipped with robust IT security.

To help combat cybercrime, the Australian government hosts the Australian Cybercrime Online Reporting Network (ACORN), at www.acorn.gov.au. ACORN is an outgrowth of the 2013 National Plan to Combat Cybercrime and is a collaborative effort between Australian law enforcement and intelligence agencies, the Attorney General, the Australian Communications and Media Authority, the Australian Competition and Consumer Commission (http://www.scamwatch.gov.au), and the Children's e-Safety Commissioner (https://www.esafety.gov.au). ACORN is primarily a reporting tool coupled with awareness-raising initiatives around computer attacks, cyberbullying, identity theft, online scams, and internet-assisted child sexual abuse.

Jeff Gagnon

See also: China: Hong Kong; Japan; Singapore; South Korea; Taiwan

Further Reading

ABC News Australia. June 2016. "Fact Check: Has Australia's Internet Speed Dropped from 30th to 60th in the World Under the Coalition?" Accessed July 4, 2016. http://www .abc.net.au/news/2016-06-21/fact-check-australias-internet-speed-rank/7509352

Akamai. March 2016. "Akamai's [State of the Internet] Q4 2015 Report." Accessed July 4, 2016. https://www.akamai.com/us/en/multimedia/documents/content/state-of-the-int ernet/q4-2015-state-of-the-internet-connectivity-report-us.pdf

Australia Bureau of Statistics. 2015. "8153.0—Internet Activity, Australia, December 2015." Accessed July 4, 2016. http://www.abs.gov.au/ausstats/abs@.nsf/0/00FD2E732C939 C06CA257E19000FB410?Opendocument

Australian Communications and Media Authority (ACMA). 2015. "Australians Get Mobile." Accessed July 4, 2016. http://www.acma.gov.au/theACMA/engage-blogs/engage-blogs /Research-snapshots/Australians-get-mobile

Australian Council of Social Service. January 2016. "Staying Connected: The Impact of Digital Exclusion on People Living on Low-Incomes and the Community Organisations That Support Them." Accessed July 4, 2012. http://www.acoss.org.au/wp-content /uploads/2016/01/Digital-Divide-Policy-Snapshot-2016-Final.pdf

Deloitte. 2016. "$43bn in Productivity and Workforce Participation Benefits." Accessed July 4, 2016. http://www2.deloitte.com/au/en/pages/media-releases/articles/australias -mobile-revolution-continues-170316.html

Ferrers, Richard. December 2016. "Value Management: Innovation 2.0 Blog: FTTP vs FTTN; When is Spending $4,000 Better Value Than Spending $2,100?" Accessed July 4, 2016. http://valman.blogspot.ca/2015/12/fttp-vs-fttn-when-is-spending-4400.html

LeMay, Renai. May 2015. "Quigley Releases Detailed Evidence Showing MTM NBN Cost Blowout." *Delimiter*. Accessed July 4, 2016. https://delimiter.com.au/2015/11/05/quig ley-releases-detailed-evidence-showing-mtm-nbn-cost-blowout/

NBN Co Ltd. 2016. "Half Year Results 2016 Presentation." Accessed on July 4, 2016. http:// www.nbnco.com.au/content/dam/nbnco2/documents/nbn-half-year-financial-results -2016-presentation.pdf

Ogilvie, Felicity, and Sam Ikin. 2016. "NBN Suspends Tasmania West Coast Rollout as Region Lobbies for Fibre Over Satellite." *ABC Premium News*, May 10. Accessed July 4, 2016. http://www.abc.net.au/news/2016-05-10/nbn-suspends-rollout-to-tasmanias-w est-coast/7400454

Oxford English Dictionary. 2015. Oxford, UK: Oxford University Press.

Sensis Social Media Report 2016. June 2016. "How Australian People and Businesses Are Using Social Media." Accessed July 4, 2016. https://www.sensis.com.au/assets/PDFdi rectory/Sensis_Social_Media_Report_2016.PDF

Webroot. 2016. "Next-Generation Threats Exposed." Accessed July 4, 2016. http://webroot -cms-cdn.s3.amazonaws.com/7814/5617/2382/Webroot-2016-Threat-Brief.pdf

B

BANGLADESH

Bangladesh is a country in South Asia that is surrounded by India. It became an independent nation in 1971, and it is one of the most densely populated countries in the world, as well as one of the poorest. Overall, internet usage is still low in the country, but it is experiencing fast growth due to the mobile phone market. Similar to a number of other countries in South and Southeast Asia, a very high percentage of Bangladeshi internet users visit Facebook. Recently, online activity has played a prominent role in Bangladeshi politics. At the same time, security, censorship, and free speech issues online have led to violence in the country in recent years.

The Bangladesh Telephone and Telephone Board granted licenses to Bangladesh's first two internet service providers (ISPs) in 1996. Liberal government policies allowed more and more ISPs to register. In 2001, the Telecommunications Act created the Bangladesh Telecommunication Regulatory Commission (BTRC) to regulate the industry. Currently, there are over 200 ISPs active in Bangladesh.

Infrastructure in Bangladesh for internet access is also affected by the submarine cable that supplies internet to the region. The fiber optic cable was constructed in 2005 by a consortium of sixteen countries, including Bangladesh and other Asian, Middle Eastern, and European nations. Over the years, it has experienced a number of outages, which have a great effect on Bangladesh since cable provides the country's only international internet line. Currently, a new cable is under construction that will offer a broadband connection and stretches from Singapore to France. It is expected to be completed in 2016.

Internet usage does not fully reflect the large number of providers available, although the internet penetration rate is beginning to rise more rapidly. As recently as 2005, Bangladesh was near the very bottom of the list for the percentage of internet users in the country as compared to the rest of the world. In 2014, this number increased to 9.6 percent, showing significant growth between 2009 and 2014 (Internet Live Stats 2016). According to the BTRC, the number of internet users increased by over 38 percent from March 2015 to March 2016, and this fast-paced growth is likely to continue as mobile internet usage expands (Bdnews.com 2016).

The BTRC also reported over 130 million mobile phone subscribers in 2016, based on the sale of prepaid subscriber identity module (SIM) cards (Bdnews.com 2016).This number has taken a downward turn recently, which may be due to a government initiative to reregister SIM cards in order to obtain more accurate counts. Smartphones currently represent a small portion of this usage, with around 20 percent of mobile phone owners utilizing them (Husain 2015).

As with other countries in the region, internet users in Bangladesh skew toward the middle and upper classes and live in urban areas, which are primarily around the capital, Dhaka. The overall low rate of internet connectivity can be attributed to a number of factors, including the large number of poor people with no expendable income, high taxes and service charges for internet usage, low literacy rates, lack of websites with Bangla-language content, and lack of infrastructure in rural areas.

One unique and innovative project to provide internet and related services to rural villages has been underway in Bangladesh since 2008 (Bouissou 2013). The program, known as Infolady, was created through a nonprofit and provides training and equipment to local women so that they can become traveling internet service providers. The women, currently about 60 in number, use bicycles to travel through villages with laptops, cell phones, and other digital equipment. Thanks to this project, remote villages that otherwise would have no internet availability are able to connect online.

The government has taken some progressive stances in the move to digital connectivity. The Prime Minister's Office developed the Access to Information (a2i) Programme in 2007 to make the move to providing public services in a digital format and to widen access. One of the current projects associated with a2i is "Digital Bangladesh by 2021." This effort represents a large technological step forward for Bangladesh in multiple ways and is predicated on the notion of widespread internet usage and availability. There has been some criticism over whether this vision is achievable and whether it will bring the kind of economic growth to Bangladesh that it promises.

Google and Facebook, which are the top two global sites, also represent the top two websites used in Bangladesh (Alexa Internet 2016). A number of other globally recognized websites are also in the top twenty for Bangladesh, including Yahoo, YouTube, Twitter, LinkedIn, Wikipedia, and Amazon. Bangladesh news sites are also represented in the top websites, holding four of the top twenty spots. One notable

Bangladesh's Info Ladies bring social media and communications technologies to Saghata, a remote farming village north of Dhaka, on September 30, 2012. The Info Ladies ride bicycles all across the countryside to ensure that all Bangladeshis in rural areas without sustained electricity or telecommunications infrastructure are still able to be connected with friends and families at home and abroad. (A.M. Ahad/AP Photo)

website, called Espncricinfo.com, ranks at number twelve. This is an ESPN subsidiary devoted wholly to news about cricket, a popular sport in many South Asian countries.

Given this interest in cricket, it may come as no surprise that cricket is well represented in the top ten Facebook pages for Bangladesh (Socialbakers 2016). The number one page is for the national Bangladesh cricket team, the Tigers, which has over 7.8 million fans. In addition, the number three and number five pages are for individual cricket players on the national team. Four of the remaining top ten spots are for Bangladesh news sites that publish either exclusively in Bangla or in Bangla and English. The ninth most liked page is Radio Munna, a digital radio station. The final two top ten most popular Facebook pages in Bangladesh are for two of the top ISPs and mobile phone operators, Grameenphone and RobiAxiata Limited. In Bangladesh, as with a number of other countries in the region, Facebook is a key source for news and information. Over 80 percent of Bangladeshi internet users are on Facebook. As reflected in the page rankings, the popular social media site is used for multiple purposes, ranging from entertainment to the dissemination of government information.

In 2013, Bangladeshi bloggers made the news when they held the 2013 Bengali blog blackout. This action was taken in response to the arrest of Asif Mohiuddin (1974–), a well-known blogger, and three other bloggers because of offensive comments on their blogs that violated the country's antiblasphemy laws. At the time, the government monitored many independent blog sites and also temporarily shut down a number of them. During the blackout, all the Bengali blogs shut down for a few days and came back online as a group with a press release demanding the release of their fellow bloggers. Ultimately, the three other bloggers were released, but Mohiuddin stayed in jail for three months.

Recently, there has been an increase in domestic terrorism in Bangladesh. One infamous example of this in 2015 was the "blogger killings," in which at least four men were killed in separate incidents because of their blogs. They included Avijit Roy (1972–2015), Ananta Bijoy Das (1982–2015), and Washiqur Rahman (1988–2015). Each of these victims was known for secular or atheist and antifundamentalist writings.

Finally, also in 2015, the government cited security concerns when they shut down the internet completely for approximately one and a half hours. When the internet was made available again, Facebook, WhatsApp, Viber, and other social media sites had been banned from access within Bangladesh. All of this began when the Bangladeshi Supreme court convicted and sentenced to death two men, Salahuddin Quader Chowdhury (1949–2015) and Ali Ahsan Mohammad Mujahid (1948–2015), for war crimes committed during the war of independence in 1971. The arrests occurred after the Bangladesh National Party lost power after ruling from 1971 to 2011; both men were active in national politics until their arrest. Based on concerns about protests and violence, the Bangladeshi government banned these social media sites for three weeks, beginning November 18, 2015 (Colhoun 2015). The Facebook ban was lifted after that time, although it remained in place for both WhatsApp and Viber. On December 13, 2015, amid contradictory reports of additional various social media bans, a state minister stated that all social media was available (Bdnews24.com 2015).

Karen Stoll Farrell

See also: India; Nepal; Pakistan; Singapore

Further Reading

Alexa Internet. 2016. "Top Sites in Bangladesh." Accessed February 22, 2016. http://www.alexa.com/topsites/countries/BD

BBC. 2015. "Bangladesh Death Sentences Lead to Facebook Ban." November 18. Accessed February 22, 2016. http://www.bbc.com/news/world-asia-34860667

Bdnews24.com. 2015. "Bangladesh Unblocks All Social Media, Communication Apps." December 14. Accessed April 18, 2016. http://bdnews24.com/bangladesh/2015/12/14/bangladesh-unblocks-all-social-media-communication-apps

Bdnews24.com. 2016. "Internet Users in Bangladesh Cross 60 Million, Says Telecoms Regulator." April 18. Accessed April 18, 2016. http://bdnews24.com/business/2016/04/18/internet-users-in-bangladesh-cross-60-million-says-telecoms-regulator

Bouissou, Julien. 2013. "'Info Ladies' Go Biking to Bring Remote Bangladeshi Villages Online," July 20. Accessed April 18, 2016. http://www.theguardian.com/global-development/2013/jul/30/bangladesh-bikes-skype-info-ladies

Colhoun, Damaris. 2015. "Social Media Censorship in Bangladesh Hints at Long-Term Problems for Publishers." December 2. Accessed February 22, 2016. http://www.cjr.org/analysis/bangladesh_social_media.php

Dhaka Tribune. 2015. "Bangladesh Outshines Neighbours in Mobile Internet Use." October 6. Accessed February 22, 2016. http://www.dhakatribune.com/business/2015/oct/06/bangladesh-outshines-neighbours-mobile-internet-use

Husain, Ishtiaq. 2015. "Smartphone Users on the Rise." Accessed January 7, 2017. http://archive.dhakatribune.com/business/2016/jan/24/smartphone-users-rise

Infolady. 2015. "Infolady." Accessed April 18, 2016. http://infolady.com.bd/; archived version available: http://web.archive.org/web/20160229213634/http://infolady.com.bd/

Internet Live Stats. 2016. "Bangladesh Internet Users." Accessed April 18, 2016. http://www.internetlivestats.com/internet-users/bangladesh/

Islam, Arafutel. 2015. "Internet Users Defy Facebook Ban in Bangladesh." November 20. Accessed February 22, 2016. http://www.dw.com/en/internet-users-defy-facebook-ban-in-bangladesh/a-18863635

Prime Minister's Office, Dhaka, Bangladesh. "Access to Information (a2i) Programme." Accessed February 22, 2016. http://www.a2i.pmo.gov.bd/

Reuters. 2015. "Bangladesh Lifts Ban on Facebook After Three Weeks." December 10. Accessed February 22, 2016. http://www.nbcnews.com/tech/tech-news/bangladesh-lifts-ban-facebook-after-three-weeks-n477981

Socialbakers. 2016. "Bangladesh Facebook Page Statistics." Accessed February 22, 2016. http://www.socialbakers.com/statistics/facebook/pages/total/bangladesh/

BOTSWANA

Botswana is a landlocked country of just over 2 million in southern Africa, sharing borders with South Africa, Zimbabwe, Namibia, and Zambia. A former British protectorate, the country gained independence in 1966, by which time it was the second-poorest country in the world (after Bangladesh). However, by the mid-1990s, Botswana had the fastest-growing economy in Africa (Mosime and Chiumbu 2010). Although the country has experienced stunted growth in the recent past following a global price decline in commodities, it is still categorized as an upper-middle-income country, with a gross domestic product (GDP) per capita of $6,360 in 2015 (World Bank 2016).There are two main languages spoken by over 90 percent of the population: Setswana and English. This relative linguistic homogeneity makes mediated communication in the country relatively less complex and more accessible to the majority of the population. Botswana has a small but fairly diversified group of mainstream media made of both state-owned and privately-owned broadcasters and print titles. However, most of these are concentrated in urban areas.

Like most African countries, Botswana belatedly joined the so-called information superhighway (Nyamnjoh 2005). For example, in 2000, there were only 15,000 people in the country with access to the internet, then a predominantly urban phenomenon. This represented a motley 0.03 percent of the population (ITU 2016). At the same time, access to mobile phones was also very limited, as the costs of both handsets and voice and data were extremely high, especially compared with fixed lines. However, by 2014, the situation in Botswana had changed drastically, thanks to the increased availability of cheap handsets and the rapid rollout of broadband in the country. In 2015, for example, there were 620,000 Batswana with active Facebook accounts, making up just over 28 percent of the population (InternetWorldStatistics 2016). Batswana with access to the internet also comprised 28 percent of the population, a figure that is set to increase in 2016.

The increase in access to the internet in Botswana also means that more Batswana are using other online social media platforms such as Twitter, WhatsApp, Instagram, and Myspace, especially young people between eighteen and twenty-four years old (Butane 2013). The country's telecommunications market is dominated by three players: the privately owned (and biggest) mobile telephone company Mascom, another privately owned mobile company called Orange Botswana (formerly Vista), and the Botswana Telecommunications Company (BTC), the state-owned, fixed-line provider. Against the background of declining subscriptions to fixed-line services, BTC recently launched a subsidiary, Be Mobile, to compete with the mobile operators and broaden access to mobile data and voice services.

A recent study (Butane 2013) of internet use by Botswana's young people showed that access to the internet was still relatively quite low and was characterized by multiple "digital divides" between and among young college students. For example, the study found that students attending universities (such as the University of Botswana) had far more regular access to the internet than those attending other, smaller, nondegree-offering institutions. Even within the smaller institutions, some students reported no access to the internet at all. It also found that in the majority of cases, young people accessed the internet predominantly for entertainment purposes. The average number of hours spent on the internet, according to the study, was five hours per week.

The history of mobile telephony in Botswana dates back to the late 1990s, when the government launched its first comprehensive telecommunications policy, which established the licensing body, the Botswana Telecommunications Authority. The first licensees, Mascom and Vista, were licensed and started offering mobile telephony services in 1998. In 1996, the Botswana government launched "Vision 2016," which, among other things, articulated an ambitious program of expanding telecommunication services to all citizens (Zaffiro 2010). Partly thanks to this program, access to mobile telephone services among Batswana has jumped from 33 percent in 2005 to more than 131 percent in 2010, an indication of multiple cell phone use and subscriptions by the people of Botswana (Lesitaokana 2014). Granted, the high level of access to mobile telephony does not necessarily reflect access to mobile internet. However, an interesting fact is that the high level of access to mobile telephony has allowed the proliferation of "tele-medicines," whereby hospitals are able to communicate with patients through mobile apps that keep track of patients' adherence to medicine-taking schedules, especially patients with chronic diseases. This is particularly important in Botswana, which has the third-highest rates of HIV/AIDS after Lesotho and Swaziland.

For the past four decades, Botswana has often been cited as a shining example of a stable democracy. However, in recent years, the government of President Seretse Ian Khama (1953–) has been accused of, among other things, stifling the free press and freedom of expression (Freedom House 2015). For example, in September 2014, the government arrested two journalists from the independent weekly paper *Sunday Standard* and charged them with sedition for publishing a story alleging that the president's earlier involvement in a car accident had been due to speeding. The paper's premises were subsequently raided by the country's intelligence services, looking for seditious material (Freedom House 2015).One of the journalists fled the country to South Africa to seek asylum, while the other was eventually released pending court trial. President Khama's critics also argue that, perhaps because of his military and security background, his government is actively involved in the widespread electronic surveillance of citizens, effectively limiting and discouraging freedom of expression on the internet (Mokone 2016).

Wallace Chuma

See also: Kenya; Namibia; South Africa; Zimbabwe

Further Reading

Butane, T. 2013. "Internet Access and Use Among Young People in Botswana." *International Journal of Information and Education Technology*, 3(1):117–119.

Freedom House. 2015. "Botswana Country Report." Accessed August 12, 2016. https://freedomhouse.org/report/freedom-press/2015/botswana

International Telecommunications Union (ITU). 2016. "Percentage of Individuals Using the Internet, 2000–2013." Accessed August 12, 2016. http://www.itu.int/net/itu_search/index.aspx?cx=001276825495132238663%3Anqzm45z846q&cof=FORID%3A9&ie=UTF-8&q=Botswana+internet+statistics+in+2000

Internet World Statistics. 2016. "Africa: Botswana." Accessed on August 6, 2016. http://www.internetworldstats.com/africa.htm#bw

Lesitaokana, W. O. 2014. "Key Issues in the Development of Mobile Telephony in Botswana (1998–2011): An Empirical Investigation." *New Media and Society,* 16(5): 840–855.

Mokone, O. 2016. "Khama Turns Botswana into a Surveillance State." August 12. Accessed August 12, 2016. http://www.sundaystandard.info/khama-turns-botswana-surveillance-state

Mosime, S. T., and Chiumbu, S. 2010. "Global Pressures, Local Disparities: Deregulation's False Promise and Botswana's Domestic Digital Divide." In *Media Policy in a Changing Southern Africa: Critical Reflections on Media Reforms in the Global Age,* edited by Dumisani Moyo and Wallace Chuma. Pretoria, South Africa: Unisa Press.

Nyamnjoh, F. 2005. *Africa's Media, Democracy, and the Politics of Belonging.* London: Zed Books.

World Bank. 2016. "GDP Per Capita: Botswana." Accessed August 12, 2016. http://data.worldbank.org/indicator/NY.GDP.PCAP.CD?locations=BW

Zaffiro, J. 2010. "Realising or Dreaming? Vision 2016, Media Reform, and Democracy in 2016." In *Media Policy in a Changing Southern Africa: Critical Reflections on Media Reforms in the Global Age*, edited by Dumisani Moyo and Wallace Chuma. Pretoria, South Africa: Unisa Press.

BRAZIL

Brazil, the largest country in South America, has a total population of 208.7 million people and an internet penetration of 58 percent (i.e. a little more than half of the population has regular access to the internet). However, once individuals have access to the internet, they remain connected. About 78 percent of active internet users are online on any device daily. Brazilians use the internet on computers and tablets for an average of five hours and fourteen minutes per day.

In fact, the country's internet penetration has steadily risen annually. As the country becomes more technology friendly, the tech culture has become deeply ingrained into Brazilian society. The country's population, especially millennials, has become reliant on the internet and social media platforms. The younger generation accounts for much of the country's active internet users and has a strong presence on social media. Brazilians have also forayed into app development, with many developers creating localized apps. Apps that offer services such as free calls and texts over the internet appeal to Brazilians. In addition, in-app content has proved to be quite profitable in Brazil.

The country currently leads Latin America in the number of social media users, with about 88 percent of young Brazilians using Facebook on their mobile phones. YouTube and WhatsApp follow behind Facebook, with 81 percent and 79 percent, respectively. Brazilians spend about an average of three hours and fifty-six minutes on the internet and three hours and eighteen minutes on social media while using mobile phones (Kemp 2016). Many of Brazil's social media users are in the younger generation, largely within the fifteen- to thirty-two-year-old age bracket. This is because Brazil's overall population is primarily made up of people under age forty. Their online presence accounts for more than half of site visits (Mason Gray 2015).

WhatsApp, an instant messaging service for smartphones, is one of the most popular social media platforms in Brazil. The free services that WhatsApp provides allow Brazilians to stay in touch with friends and family anywhere in the world. (Viorel Dudau/Dreamstime.com)

Social media has become ingrained into the Brazilian culture. In 2008, Brazil was one of the largest user bases of Orkut, a Google-developed platform. The user base was so large that Google set up an office in Belo Horizonte to navigate Brazil's legal system. In December 2015, the Brazilian internet population was in a tumult over a 48-hour ban of WhatsApp. The social media app was temporarily banned because of WhatsApp's failure to comply with a court order to provide investigators with information for a criminal court case. Once the 48-hour ban was put into effect, many Brazilian users complained about the app's suspension. Others created the hashtag #Nessas48HorasEuVou ("In these 48 hours, I will") and humorous memes, which spread across social media platforms that were still accessible. During this 48-hour period, competing messaging apps saw an influx of new accounts. Telegram, a secure messaging app that is similar to WhatsApp, attracted more than a million users during the ban. WhatsApp is very popular among the Brazilian population because of its free services to call and message family and friends within the country or abroad. The high cost of living in Brazil is often compared to the United Kingdom, except Brazilians make a third of the wages that the British earn. Having cost-effective products or free services offered by WhatsApp appeals to the Brazilian population's need for low-cost communications (BBC 2015).

Brazilian telecommunication service providers are feeling the loss of business with their customers' preferences for voice over internet protocol (VoIP) services. The country's telecom providers are urging the government to enforce a more restrictive regulation on VoIPs. Telecom companies state that apps like WhatsApp and Facebook Messenger are damaging to their businesses. This maneuver to regulate

Brazil comprises the second-largest national audience on Facebook, after the United States (Levy 2015). However, Facebook was not always Brazilians' favorite platform. In 2004, Google launched its first social media platform, a program called Orkut after its primary developer, Orkut Büyükkökten. The platform took off in Brazil in 2005, when Orkut added Portuguese-language support; it was also extremely popular in India. Brazil remained one of Orkut's strongest user bases through two major redesigns and some controversy. In 2012, after Google announced that it was no longer providing support to Orkut, Brazilian users started relying more heavily on Facebook. Orkut finally closed in 2014 (Ahmad 2014).

VoIP services is similar to the ban on VoIPs implemented by Moroccan telecommunication service providers. When the Brazilian court banned WhatsApp for 48 hours, only one of the four major telecom companies protested the decision.

If telecom providers are ever successful in banning VoIPs completely, this move could be in violation of the 2014 Marco Civil da Internet (MCI), a bill that was passed by the Brazilian Congress in 2014 with the support of President Dilma Rousseff (1947–). The MCI is the regulatory framework that provides a strong foundation for online human rights in Brazil. In essence, this law states that all Brazilian internet users have a right to net neutrality, online privacy and development, liability of internet service providers, e-government expansion, and digital inclusion. Despite having a law that covers multiple online user rights, some areas still lack suitable enforcement to ensure proper regulation. While general rules have been established, they do not guarantee online user rights, especially in light of the addendum to the so-called spy bill. The surveillance law allows telecom service providers to collect and store customers' personal information, such as emails, phone numbers, and addresses, which in turn can be turned over to the authorities without a court order. Many civil society organizations protested the adoption of this bill. Even though there is an explicit clause that any future decisions would be made in accordance with the law, civil society leaders believe that it is still a violation of online privacy and a threat to Brazilians' freedom of expression (Article 19 2015).

Experts believe that Brazil should be considered as a location for global investment. Because of the country's strong social media presence, global companies would be able to utilize the highly trafficked social media platforms. The amount of time that Brazilians spend on the internet and social media can be crucial for companies looking to break into the Brazilian market. By marketing their product and services on social media, they would be able to reach much of the Brazilian population online (Walters 2015).

Three companies have utilized social media to market their brand through unique social media campaigns that captured the Brazilian passion for soccer and used it to attract even more customers. One of them is Santadar, one of the largest banks in the world. In 2013, the bank launched a new campaign that focused on its customers. Using footage of Brazilian soccer fans at games, the bank's marketing team created unique videos highlighting the fans and equating them to the bank's customers. Knowing that more than 80 percent of Brazil's internet population uses YouTube, the videos were uploaded to the bank's official YouTube channel. Not only did the bank see positive feedback on all its social media channels, but it also increased the number of views on the bank's official YouTube channel by 6.1 million views in four months (Walter 2015).

During the 2014 World Cup, Adidas created a social media campaign that produced real-time results. Brazilians active on social media experienced a creative global campaign that featured videos, pictures, player information, and live tweets about team highlights and exciting play-by-plays. As a result, the Adidas brand was one of the most talked-about brands during the World Cup and was successful in amassing nearly 5.8 million followers on its official social media channels. The ABC Group, a company that produces plastics, launched another successful social media campaign during the 2014 World Cup. When Brazil's captain, Neymar da Silva Santos Junior (1992–), was injured during a World Cup game against Colombia, the ABC Group distributed 60,000 masks for fans to wear at the next game against Germany as a tribute to the injured player. Fans were encouraged to take selfies wearing the masks and share them on the ABC Group's social media channels. More than 12,000 pictures were shared on the company's Instagram account with the hashtag #somostodosneymar ("We are all Neymar"). These campaigns are prime examples of using Brazil's social media presence to engage audiences and market products.

Even though Brazil is the world's eighth-largest economy, the mobile device software industry did not initially provide a significant contribution to the country's growing economy. The smartphone penetration, 26 percent, was relatively small compared to the country's population in 2013. However, the Brazilian population has become increasingly tech friendly, specifically with regard to smartphones. In fact, smartphone usage has increased to over 53 percent in the country. Even though the mobile device software industry in Brazil is less developed than in other countries, like the United States where more than half the population owns a smartphone, it is quickly catching up to them. Compared to other countries in Latin America, Brazil is currently the leading country in the mobile market. In response to the country's growing mobile usage, Brazilian developers have created a number of local apps. What is unique about the apps developed in Brazil is the movement away from the "one size fits all" model. By creating localized apps, Brazilian companies engage with their consumers in more ways than traditional advertising and social media marketing do. They are targeting a specific demographic that would be interested in the companies' services or products (Ayers 2016).

With localized apps, Brazilians can download apps that service only certain cities or areas, such as São Paolo and Rio de Janeiro. There is not a single app for one service that is used by the entire country. For example, Brazil offers at least ten major taxi service apps, not unlike Uber, that service different areas of the country. EasyTaxi and 99Taxis are two popular examples. Brazilians can also download a variety of utility apps in areas ranging from food service to banking to education. If Brazilians wish to participate in the improvement of cities, they can send suggestions to the local government via an app. Tourists also have access to apps to help improve their visit to the country by providing information about navigation, city guides, hotels and motels, and restaurants (Tech in Brazil 2015a). With the arrival of the app development era, Brazil has seen the creation of more than 80,000 apps designed for the different needs or wants of consumers. This is especially helpful for small businesses looking to reach multiple audiences while tapping into a new revenue source (Ayers 2016).

Even though most Brazilians prefer the free services and products offered by apps, this does not exclude in-app purchases available on freemium apps. Freemium users are not pressured to buy the in-app content, but it is available for purchase if users choose. This method of offering services within the freemium app has proven to be highly successful. It allows consumer engagement, as well as generating more revenue. In 2015, studies found that 39.5 percent of Brazilians bought in-app content from freemium apps. This is more than double the percentage calculated for Brazilians who paid for apps. Research shows that freemium apps with in-app content are more profitable and have a higher penetration than paid apps. In-app content can come in many different forms (Tech in Brazil 2015b).

The most popular form is freemium game apps. Brazilian players download a free app but then make purchases within it to obtain necessary objects to advance a game or in exchange for app "currency," similar to games like Candy Crush in the United States. Based on 2016 projections, experts estimate about $400 million will be made through mobile gaming in Brazil. In-app content can also come in the forms of video and utility content. Many video-streaming apps have in-app content for users. Some content may be additional videos that aren't included in the freemium app and must be purchased in order to be viewed. Utility apps, such as the antivirus software AVG, are available for free; however, AVG only offers minimum protection services on the mobile phone and tablet. If users wish to have the app running at full capacity, they must make in-app purchases to get the additional services (Tech in Brazil 2015b).

Karen Ames

See also: Morocco; United Kingdom; United States

Further Reading

Article 19. 2015. "Country Report: Brazil's Marco Civil da Internet." November 5. Accessed April 6, 2016. https://www.article19.org/resources.php/resource/38175/en/country-report:-brazil's-marco-civil-da-internet

Ayers, Jamie. 2016. "Smartphone Usage and Local Apps Soar in Brazil." February 2, 2016. Accessed April 6. https://www.appmakr.com/blog/smartphone-usage-local-apps-brazil/

BBC. 2015. "Brazil Judge Lifts WhatsApp Suspension." December 17. Accessed April 7, 2016. http://www.bbc.com/news/world-latin-america-35125559

Kemp, Simon. 2016. "Digital in 2016: We Are Social's Compendium of Global Digital, Social, and Mobile Data, Trends, and Statistics." January 27. Accessed April 6, 2016. http://wearesocial.com/uk/special-reports/digital-in-2016

Mason Gray, Chloe. 2015. "5 Interesting Social Media & Technology Statistics About Brazil Globally-Minded Enterprises." May 13. Accessed April 6, 2016. https://www.sprinklr.com/the-way/social-media-statistics-brazil/

Tech in Brazil. 2015a. "101+Brazilian Apps to Have on Your Mobile Device." February 4. Accessed April 5, 2016. http://techinbrazil.com/101-brazilian-apps-to-have-on-your-mobile-device

Tech in Brazil. 2015b. "In-App Purchases in Brazil." June 19. Accessed April 5, 2016. http://techinbrazil.com/in-app-purchases-in-brazil

Walter, Ekaterina. 2015. "4 Social-Media Campaigns That Demonstrate Brazil's Passionate Digital Culture." October 22. Accessed April 6, 2016. https://www.entrepreneur.com/article/250303

C

CAMEROON

Cameroon is an African country situated north of the equator along the Atlantic coastline. It became a German protectorate in 1884. After World War I, it became a League of Nations mandate territory, with governance split between France and Great Britain until 1946, when the United Nations placed the country under French and British trusteeship. As a result of this trusteeship, Cameroon is one of the few countries in the world with both French and English as official languages. Although geographical maps and many English-speaking authors locate Cameroon in West Africa, others, notably French-speaking writers, situate the country in Central Africa. Yet, as a member of the Central African Economic and Monetary Community (CEMAC), Cameroon has officially aligned itself with the Central African region, though its main economic partner and northwest neighboring country is Nigeria, which belongs to the Economic Community of West African States (ECOWAS). With close to 24 million inhabitants, Cameroon is home to more than half of the population of the CEMAC region. The other members of CEMAC are also neighboring countries, including Gabon, the Central African Republic, Chad, the Republic of Congo, and Equatorial Guinea.

According to the 2015 Global Information Technology Report, the Networked Readiness Index, which measures the propensity for countries to leverage opportunities offered by information and communication technologies (ICTs), places Cameroon at 3 on a scale of 7 and ranks it at 126 out of the 143 countries sampled. In fact, the International Communication Union figures indicate that in 2014, 6.7 percent of users accessed the internet at home, and in 2015, the internet penetration rate in Cameroon stood at 20.68 percent (Dutta et al. 2016).

Cameroon experienced a long period of economic growth from independence in 1960 to 1985; however, like many sub-Saharan African countries, Cameroon sank into a major economic crisis from 1986 to 1994. During this period, the country was placed under Structural Adjustment Programmes (SAPs) by the Bretton Woods institutions (namely, the World Bank and the International Monetary Fund), with the aim of restoring a healthy economy through widespread liberalizations. The purpose of these liberalizations was to foster the transition from a state-led economy to a more market-coordinated one. Among the economic reforms initiated was a campaign, which began in 1995, to privatize CAMTEL, a historic telecommunications company, through international investors. However, privatization of the company has yet to occur, largely due to domestic obstacles, which have often been linked to issues of national sovereignty.

Nevertheless, while CAMTEL still enjoys a monopoly on fixed telephone lines, as well as submarine fiber optic cables and their use in deploying internet all over the country, it now shares the mobile telephony market with three other companies: Orange Cameroon, MTN Cameroon, and Nexttel Cameroon. These four companies also provide internet services, alongside Yoomee, Ringo, Saconets, cyberlink, and Matrix. Despite the number of providers, connection costs remain among the highest in Africa (Banque Mondiale 2011). In fact, prices in 2015 were so high that for unlimited internet access with relatively comfortable speed users had to pay around $55 per month, an exorbitant amount considering that the country's monthly GDP per capita was only $109.11. A survey funded by the African Economic Research Consortium, conducted in 2015 by researchers from the Group for Studies and Research in Theoretical and Applied Economics of the University of Douala, revealed that users still consider connection costs to be the key hindrance to internet use (23.79 percent of the 2,232 respondents), well ahead of lack of usefulness, which is the second obstacle (19.25 percent of respondents) (Bakehe et al. 2016).

Despite these infrastructure setbacks, Cameroon has made some strides toward becoming the telecommunications hub of Central Africa. For example, the Central African Backbone (CAB) project, which will connect Cameroon, Chad, and the Central African Republic in its first phase and will ultimately involve eleven African countries, including the Democratic Republic of Congo, the Republic of the Congo, Gabon, Equatorial Guinea, Sao Tome and Principe, Nigeria, Niger, and Sudan (World Bank 2009), has laid twelve of eighteen planned optical fiber cables along the Chad/Cameroon oil pipeline (Nana and Tankeu 2012). Cameroon is connected to the internet by two fiber optic submarine cables—namely, South Atlantic 3 (SAT-3) and West African Submarine Cable (WASC), which connect Europe to the West African coasts. Furthermore, in mid-2015, Cameroon signed an agreement with Orange Cameroon for the construction and operation of a third fiber optic submarine cable called the African Coast to Europe (ACE).

According to the International Communication Union, Cameroon witnessed a drop in the rate of its mobile-cellular phone subscriptions between 2014 and 2015, from 75.69 percent to 71.85 percent. A possible explanation for this may be found in the governmental initiative that required the compulsory registration of subscribers for security purposes that reached its peak between July 2015 and June 2016. Since most Cameroonians use mobile devices to access the internet, the drop in mobile subscriptions is likely a cause for the decrease in internet penetration in Cameroon. Indeed, 84.2 percent of respondents in the 2015 University of Douala survey indicated that they accessed the internet via mobile networks. Once connected, 82.92 percent of respondents reported logging on to social networks. To understand these figures, which seem relatively high for a developing country, consider that this survey was conducted mainly in urban areas, especially in the following towns: (1) Douala, the country's economic capital (39.9 percent of the sample); (2) Yaoundé, the country's political headquarters (34.4 percent); (3) Bafoussam, the country's third-largest city (10.59 percent); (4) Limbe, which hosts the country's sole oil refinery (7.09 percent); and (5) Buea, which hosts the very first purely Anglophone university in the country (7.85 percent). To counterbalance the possible impression of all Cameroonians being very connected to the internet, according to the 2015 Global Information Technology Report, Cameroon is still at the initial stage of content development with respect to its total population because, for instance, average YouTube views of local channels accounts for only 2.6 percent of the time users spend online.

However, Cameroonians are increasingly familiarizing themselves with social media and over-the-top (OTT) services that are web-based applications focusing on the delivery of text, audio, video, or other telecommunication services via the internet without a network operator controlling the distribution of content (Godlovitch et al. 2015). StatsMonkey, a network usage monitoring institution, shows on its website that, in 2014, the five most used search engines in Cameroon were Google (with approximately 86.43 percent market share), Yahoo (about 6.49 percent market share), Bing (about 3.35 percent), Ask Jeeves (approximately 2.35 percent) and Webcrawler (approximately 0.73 percent) (StatsMonkey 2015a). The same institution also shows that, in the same year, the social network that was mostly used by Cameroonians was Facebook with about 97.56 percent rate of usage, followed by Twitter (1.29 percent), Pinterest (0.38 percent), Vkontakte (0.22 percent), Thmbr (0.2 percent), Google+ (0.16 percent) and YouTube (0.16 percent) (StatsMonkey 2015b). This ranking corroborates the results of the University of Douala's 2015 survey, which revealed that respondents mostly cited Facebook (90.46 percent) as the social media that they used, followed by Google+ (47.96 percent), Twitter (34.99 percent), Viadeo (7.25 percent), LinkedIn (5.86 percent), Monster.lu (1.88 percent), Tudivz (1.5 percent) and Xing (1 percent). Even though statistics are still scarce, users of services such as WhatsApp, IMO, and Skype are on the rise across all social groups of the Cameroonian population. In fact, about 86 percent of respondents in the 2015 survey said that they use Yahoo Messenger, Skype, MSN Messenger, or IMO at least once a month to chat with relatives within the country or abroad.

Nevertheless, though Cameroonian users rely heavily on foreign social media (as these figures show), there are a few local initiatives, especially in the area of health. Two of the most prominent of them are cardiopad and giftedMom. The cardiopad is an innovative iPad invented by a young Cameroonian engineer, Arthur Zang (1987–), for medical use, particularly in cardiology. It is a medical tablet connected to a mobile network, capable of performing cardiac readings of a patient and transmitting the data collected all over the world via the internet to a doctor who can interpret them and advise on the appropriate treatment. It is an e-healthcare mobile device solution for developing countries suffering from an acute shortage of heart disease specialists. Another effort, with its vision of "a world free of maternal and infant deaths," giftedMom appears to be a social network–like solution for developing countries suffering from insufficient prenatal healthcare centers. Invented by young Cameroonian entrepreneur Alain Nteff (1992–), its website (www.giftedmom .org), declares: "giftedMom is a leading mobile health solutions provider in Africa. We leverage last-mile technologies to provide pregnant women and new mothers access to health information and strengthen linkages to antenatal care."

A peculiarity in the use of social media in Cameroon relates to online dating and marriage. Cameroonian girls are known for their propensity to use social media as a tool to look for husbands abroad, especially in Europe and North America. This mail-order-bride phenomenon has been the subject of many research projects, and even films, the most popular of which is *Le Blanc d'Eyenga* (Eyenga's White Man), directed by Cameroonian-born Thierry Roland Ntamack (n.d.). Their reputation in this area is so established that Cameroonian girls have been referred to as the ideal for web dating (Draelants and Tatio 2003).

All these trends are possible in Cameroon because there is extensive freedom in the use of all social media. The government has not imposed any restrictions on any of them. Consequently, these media are widely used for both noble and deviant purposes. For instance, while they enable people to share information and companies to advertise themselves and their products and services, they also serve as channels to spread rumors. With regard to the sharing and commenting of viewpoints, especially on issues relating to politics, sports, religion, or culture, Cameroonians not only use conventional social media, they also are fond of transforming online web news comment sections into de facto social media.

Despite the December 2010 enactment of law No. 2010/012 on cybersecurity and cybercrime, a major weakness of Cameroonian cyberspace appears to be the lack of a nationwide operational cybersecurity strategy. While it is on the rise, the internet penetration rate shows that almost 80 percent of Cameroon's population is excluded from the digital revolution, demonstrating that the digital divide in the country is still an important concern. According to the 2015 survey, the five key impediments to the use of internet include the high cost of internet connection, lack of usefulness, complexity of the tool, lack of guidance, and absence of security. These findings identify three key areas for ICT development policy in Cameroon—namely, classical training, vocational training in internet-related competencies, and incentives for competition between domestic internet providers (Tamokwe 2013).

Georges Bertrand Tamokwé Piaptie

See also: Angola; France; Ghana; Nigeria

Further Reading

Bakehe, Patrick Novice, Fambeu, Ariel Herbert, and Tamokwe Piaptie, Georges Bertrand. 2016. "Internet Adoption and Use in Cameroon." AERC Research Paper 609.

Banque Mondiale—Unité de la Réduction de la Pauvretéet la Gestionéconomique/Région-Afrique. 2011. "Le Réveil du Lion? Point Sur la Situation Économique du Cameroun. SpécialTélécommunications." *Cahiers économiques du Cameroun*, Numéro 1. Accessed July 25, 2016. http://siteresources.worldbank.org/AFRICAINFRENCHEXT/Resources/CahiersEconCam001FR.pdf

Draelants, Hugues, and Tatio Sah, Olive. 2003. "Femme Camerounaise Cherche Mari Blanc: Le Net Entre Eldorado et Outil de Reproduction." *Esprit Critique*, Automne 2003, 5(4). Accessed July 27, 2016. http://www.espritcritique.fr/0504/esp0504article07.pdf

Godlovitch, Ilsa, Kotterink, Bas, Marcus, Scott, Nooren, Pieter, Esmeijer, Jop, and Roosendaal, Arnold. 2015. "Over-the-top (OTTS) Players: Market Dynamics and Policy Challenges." Policy Department Economic and Scientific Policy, European Parliament. Accessed July 27, 2016. http://www.europarl.europa.eu/supporting-analyses

Nana Nzépa, Olivier, and Tankeu Keutchankeu, Robertine. 2012. "Understanding What Is Happening in ICT in Cameroon." *Evidence for ICT Policy Action, Policy Paper 2.* Research ICT Africa. Accessed July 25, 2016. https://www.researchictafrica.net/publications /Evidence_for_ICT_Policy_Action/Policy_Paper_2_-_Understanding_what_is _happening_in_ICT_in_Cameroon.pdf

Soumitra, Dutta, Geiger, Thierry, and Lanvin, Bruno. 2016. "The Global Information Technology Report 2015." *World Economic Forum.* Accessed August 6, 2016. http://www3 .weforum.org/docs/WEF_Global_IT_Report_2015.pdf

StatsMonkey. 2015a. "Cameroon Social Media Usage Statistics Using Mobile, Usage Statistics—2014." Accessed July 26, 2016. https://www.statsmonkey.com/table/21314-ca meroon-mobile-social-media-usage-statistics-2015.php

StatsMonkey. 2015b. "Cameroon Search Engine Market Share, Usage Statistics—2014." Accessed July 26, 2016. https://www.statsmonkey.com/table/18345-cameroon-search -engine-market-share-usage-statistics-2014.php

Tamokwe Piaptie, G. B. 2013. "Les Déterminants de l'Accèset des Usages d'Internet en Afrique Subsaharienne. Analyse des Données Camerounaiseset Implications Pour une Politique de Développement des TIC."*Réseaux*, 2013/4(180): 95–121.

World Bank. 2009. "Central Africa Backbone—Project Information Document (PID) Appraisal Stage. Report No. AB4611." Accessed August 9, 2016. http://www.icafrica.org /fileadmin/documents/Knowledge/World_Bank/PID-Central-Africa-BckBone-JUNE -2009.pdf

CANADA

A country that is part of the continent of North America, Canada is situated directly north of the United States and boasts a landmass of 3,560,238 square miles (Statistics Canada 2016a), which makes it the second-largest country in the world in terms of size. From statistics compiled in 2015, over 35 million people live in Canada (Statistics Canada 2016b). Countless Canadians use the internet and social media daily. Currently, Canada ranks second to the United Kingdom in internet penetration among its G8 counterparts (CIRA 2016); *G8* refers to a group of eight highly industrialized nations that also includes France, Germany, Italy, the United Kingdom, Japan, the United States, and Russia (Council on Foreign Relations 2016). Canadians are also the second-heaviest users of the internet, averaging 41.3 hours per month online, trailing only the United States. To put this in perspective, globally, the average number of hours spent online by a user was 24.6. With approximately 87 percent of Canadian households having access to the internet, Canada is most definitely one of the most "wired" countries in the world (CIRA 2016). In 2013, Canada continued to lead the world in internet use, as Canadians visited the most web pages, averaging 3,371 sites per month. In comparison, the global average for visited pages per month was 2,278 (CIRA 2016).

Unfortunately, access to the internet in Canada, and social media more specifically, remains highly dependent on where one lives, with rural areas continuing to lag behind Canadian cities. For instance, although broadband is available to all Canadians who reside in urban areas, 13 percent of those living in rural areas did not have access to the internet as of 2012 (CIRA 2016). Although this number continues to decrease yearly as more technological developments are implemented in rural areas, this disparity does hint at the "digital divide" that nevertheless also exists in Canada. Moreover, internet access varies between provinces. Whereas British Columbia and Alberta have the most households with internet access at 86 percent, followed by Ontario at 84 percent, household access in New Brunswick is the lowest at 77 percent, followed by Quebec and Prince Edward Island, both at 78 percent (CIRA 2016). In addition, the digital divide is equally manifest among those with large incomes and those with lower incomes: 95 percent of Canadians who belong to the highest income quartile are connected to the internet, as opposed to 62 percent of those belonging to the lowest income quartile (CIRA 2016).

In addition, the use of social media by Canadians remains relatively similar to most of the other industrialized Western nations. According to the Canadian Internet Registration Authority (CIRA), almost 24 million Canadians visited at least one social networking site last year. This number translates to 69 percent of the Canada's total population. Facebook remains the most popular social media site in Canada, with its Canadian users increasing by 8 percent. This means that more than 19 million Canadians log on to Facebook at least once a month. In 2013, Pinterest's Canadian users increased by 792 percent from the previous year, while Tumblr boasted a 96 percent increase in Canadian users (CIRA 2016).

In 2015, Forum Research conducted a survey among Canadians who were eighteen years and older that investigated their use of social media. It found that Facebook was the most popular social media site by far (59 percent of the respondents used it). Canadians visit Facebook approximately nine times each week. Facebook is popular with 75 percent of the youngest people surveyed, and 75 percent of those earning between $80,000 and $100,000 Canadian dollars a year use the site. In other findings, 70 percent of mothers of children under the age of eighteen are on Facebook; 65 percent of Quebec residents and 65 percent of Francophones, or French-language speakers, use it; and 68 percent of college and university graduates prefer Facebook over all other social media sites (Forum Poll 2015).

The survey found that LinkedIn was the second most popular social media site, with 30 percent of the respondents using it. Canadians visit LinkedIn approximately twice per week. LinkedIn is most popular with middle-aged Canadians (i.e. forty-five to fifty-four years of age, the so-called Gen Xers). Further, 46 percent of those earning between 100,000 and 250,000 Canadian dollars a year use LinkedIn; 36 percent of mothers of children under the age of eighteen are on LinkedIn; 39 percent of British Columbians use LinkedIn; and 48 percent of those who have received a postgraduate degree prefer LinkedIn over all other social media sites.

Forum Research findings demonstrated that Twitter was the third most popular social media site for Canadians, and that 25 percent of the respondents used that site. Canadians visit Twitter approximately five times each week, and it is popular among the youngest Canadians surveyed at 36 percent. In addition, 31 percent of those earning over 80,000 Canadian dollars use Twitter; 29 percent of all parents reporting in the survey with children use the site; 31 percent of those living in the Prairies use it; and 18 percent of college and university graduates prefer Twitter over all other social media sites.

Instagram was the fourth most popular social media site for Canadians. According to the survey findings, 16 percent of the respondents use Instagram, and Canadians visit the site approximately six times each week. Instagram is popular among the youngest Canadians, those aged eighteen to thirty-four, at 32 percent; and 18 percent of those earning less than 20,000 Canadian dollars and 18 percent of those earning between $80,000–$100,0000 Canadian dollars use Instagram. Also, 25 percent of mothers of children under eighteen are on Instagram; 22 percent of those living in the Prairies use it; and 18 percent of college and university graduates prefer Instagram over all other social media sites.

Social media itself is used by Canadians in a number of different ways. Aside from Facebook, social networking, and self-promotion, Canadians use social media for news, commerce, politics, and even games. Perhaps the most indicative use of social media for political purposes in Canada occurred in the spring of 2012. A Quebec student strike of 2012, commonly referred to as "Maple Spring" by pundits, journalists, and regular Canadians, featured the largest display of civil disobedience in Canada, as over 300,000 students protested in the streets of Montreal. The Quebec student strike began as the students' measured response to a proposed university tuition fee increase by the provincial government. It quickly escalated, with the help of social media, into a social movement that opposed neoliberalism, austerity measures, economic injustice, the criminalization of protest, and the corporatization of higher education.

Social media was one of the main tactics used by the striking students alongside developing and adhering to a united front by all three student organizations in Quebec, and the use of both diversity and the escalation of tactics. The diversity and escalation of tactics included the use of petitions, the "red square" (a red piece of cloth worn by supporters of the movement that symbolized students being in debt, or "in the red"), the casseroles (a series of nightly protests that spread to residential neighborhoods where people stood on their balconies and banged on pots and pans, a tactic borrowed from protesters in Central America who used it as a symbol of solidarity), and the MaNUfestation—otherwise referred to as "the almost nude march," and described as striking students "stripping down to their birthday suits and an astonishing array of red underwear" (Sutherland 2012). However, the role of social media during the strike should not be understated. It was used in conjunction with and in addition to the variety of other tactics utilized by the students.

As noted by Laurent Gauthier (1986–), one of the student organizers, Facebook allowed the organizers "to get instant feedback and advertise for free" (Teruelle 2016, 167). In fact, many of the student organizers were quick to credit Facebook and Twitter for helping with the movement. Another student organizer, Nadia Lafrenière (1991–), shared these sentiments, crediting both social networking sites with helping to create "spontaneous actions": "The night demos (protests/marches) were started by calls on social media, and then they happened each night . . . the slogan was 'each night until victory' . . . In the beginning, it was totally spontaneous and created through social media by the students, and then people knew it was each night, so they met at 8 pm at the same places" (Teruelle 2016, 168). In a similar fashion, student organizer Anne Marie Provost (1983) underscored Facebook's strength in supporting "spontaneous" and "rapid" organization: "The strategies used . . . I wasn't involved directly . . . but, I think that it was useful for spontaneous organization. Because Facebook events, like in the strike, it could go viral. So in two hours, you could have 6,000 attendees for a big demo. It was a good way to wrap up everybody, in one day or two days or three days go to that big demo or go to that event. It was great for rapid organization" (Teruelle 2016, 168). As evidenced by these student organizers, both Facebook and Twitter were instrumental in helping the students share information, coordinate, and organize during Maple Spring.

More recently, surfeit examples of social media being used to organize, coordinate, and spread information can be found on the Facebook page of Black Lives Matter Toronto (#BLM TO). On Sunday, July 3, 2016, despite being given the honor of being Pride Toronto's "2016 Honoured Group" and tasked with helping to lead the annual LGBTQ's Pride Parade, #BLM TO staged a sit-in midway through the event as a form of protest. Its Facebook page chronicles the group's actions that day, including demands to shut down the parade. Also, a list of the group's demands appeared alongside commentary alluding to the fact that Pride actually began as an organized response against police brutality.

In a similar fashion, grassroots organizations such as the Syrian Refugee Support Group, which is based in Calgary, Alberta, also utilize social media to their advantage. A quick scan of the group's Facebook page reveals the following

The movement Black Lives Matter Toronto participated as honored guests in the Pride Parade 2016. Part of the group's participation included a sit-in demonstration that was coordinated almost entirely over the group's Facebook page. Black Lives Matter Toronto successfully used social media as a political tool to spread its message, finding that it was the best conduit to garner support for its message. (Roberto Machado Noa/LightRocket via Getty Images)

activities that they have engaged in: the coordination of bicycle donations for Syrian refugee children; the coordination of school supplies donated for Syrian refugee children; and the coordination of a goodwill soccer match between some of the Syrians versus local Calgary police officers. Groups such as #BLM TO and the Syrian Refugee Support Group provide concrete examples of how social media is being used effectively as both a communications and a political tool by countless Canadians.

Although the stereotype of a Canadian is that of an affable individual, many Canadians are indeed political—motivated to enact meaningful social change and use social media as a communications tool to help further their respective causes. Certainly, social media is an excellent tool for social movements because of one simple yet pragmatic reason: that's where the people are.

Rhon Teruelle

See also: France; Germany; United Kingdom; United States

Further Reading

Black Lives Matter Toronto. 2016. Official Facebook Page. Accessed July 5, 2016. https://www.facebook.com/blacklivesmatterTO/?fref=ts

Canadian Internet Registration Authority (CIRA). 2016. Factbook. Accessed July 5, 2016. https://cira.ca/factbook/2014/index.html

Council on Foreign Relations. 2016. Accessed July 5, 2016. http://www.cfr.org/international-organizations-and-alliances/group-eight-g8-industrialized-nations/p10647

Forum Poll. 2015. Forum Research. Accessed July 5, 2016. http://poll.forumresearch.com/post/213/facebook-leads-in-penetration-linkedin-shows-most-growth/

Statistics Canada. 2016a. Accessed July 5, 2016. http://www65.statcan.gc.ca/acyb02/1967/acyb02_19670186004-eng.htm

Statistics Canada. 2016b. Accessed July 5, 2016. http://www.statcan.gc.ca/tables-tableaux/sum-som/l01/cst01/demo02a-eng.htm

Sutherland, A. (2012, May 4). "Nearly Nude Students Take to Montreal's Streets to Protest Tuition Hike." *National Post.* May 4. Accessed July 5, 2016. http://news.nationalpost.com/news/canada/nearly-nude-students-take-to-montreals-streets-to-protest-tuition-hike

Teruelle, R. (2016). *Social Media, Red Squares, and Other Tactics: The 2012 Québec Student Protests.* Doctoral dissertation. University of Toronto. Accessed Sept. 7, 2016. https://tspace.library.utoronto.ca/bitstream/1807/73206/3/Teruelle_Rhon_201606_PhD_thesis.pdf

CHILE

The Republic of Chile is a long, narrow country that shares the southern cone of South America with Argentina. With a population of 18.1 million, it has one of the highest internet penetration rates in Latin America and, while high-quality broadband is not evenly available in all areas, Chile has one of the fastest connection speeds in South America. Nevertheless, there is a rather stark disparity between internet access in urban versus rural areas of the country. The Chilean government has implemented programs, such as free internet access through WiFiChileGob, to help ameliorate this digital divide. YouTube and Facebook are by far the most popular social media in the country; however, WhatsApp and Line messaging services have taken hold as well.

In mid-2016, Chile had 14.1 million internet users, which totaled about 78 percent of the population. While the past five years have seen slightly more growth, in part due to government initiatives, the country's internet penetration has been steadily increasing over time, from 16.6 percent in 2000, 31.2 percent in 2005, and 45 percent in 2010 (Internet Live Stats 2016). Chile's penetration rate is one of the highest among South American countries; it is surpassed only by Ecuador at 83.8 percent and barely by Argentina at 79 percent (Internet World Stats 2016).

Average broadband speeds in Chile are some of the fastest in Latin America at 7.68 Mbps, which is twice as fast as Brazil (4.88 Mbps) and three times as fast as Colombia (2.52 Mbps) and Argentina (2.4 Mbps) (Bryan 2013). The fixed internet market, which accounts for about a quarter of internet access, is controlled largely by VTR and Movistar. Three companies dominate the mobile internet market in Chile: Claro, Movistar (which holds 40.9 percent of prepaid plans), and Entel (which holds 42.5 percent of contract plans) (Subtel 2015). In March 2015, 23.7 million cell phones were active in Chile; according to the Department of Telecommunications (Subtel), this represents a 132 percent mobile penetration rate (González 2015). At that time, 77.8 percent of internet access occurred using cell phones, primarily smartphones.

According to the Minister of Transportation and Telecommunications, the recent growth in internet penetration across the country was evidence that the government's initiatives have had a positive effect. These projects are meant to close the digital divide and make access to the internet more democratic (Subtel 2015). While Chile's internet is one of the fastest and most pervasive in South America, there remains a rather large divide between urban and rural regions. Somewhere between 10 to 15 percent of the country resides in rural areas. This low population density is largely due to the phenomenon of young people leaving these regions to seek employment in the cities. Santiago, the capital, is the primary destination for job seekers; the greater metropolitan Santiago area now contains nearly 40 percent of the country's entire population. As such, rural areas are not attractive markets for internet providers, as they are sparsely populated and are usually poorer communities. However, this disparity negatively affects the Chilean populace by further isolating politically and socially marginalized communities.

One of the initiatives of the government to improve this situation has been WiFi ChileGob, which provides free Wi-Fi hubs across the country. In 2015, it was announced that 612 of these connection hubs would be added to help increase internet access and promote digital inclusion (Subtel 2015). As of mid-2016, WiFi ChileGob has established at least four hotspots in most cities and small towns across the country (WiFi ChileGob 2016). These hotspots allow up to twenty-five people to access the internet simultaneously for thirty-minute sessions each.

YouTube is the most visited website in Chile (Alexa 2016). It plays a role in Chilean society in multiple ways. Much like the original content being produced in the United States and around the world, Chileans are creating videos. The most successful YouTube personality in Chile also happens to be the second-ranking You-Tuber (based on worldwide subscribers), just behind PewDiePie (Kang 2015). Germán Alejandro Garmendia Aranis (1990–), a comedian, musician, and video-gamer, has a YouTube site called HolaSoyGerman, with more than 29 million subscribers, and he made more than $1 million in 2015 through advertisements. His style is very frenetic, whether he is performing comedic bits or playing games. While HolaSoyGerman is more lucrative, he currently uploads more frequently to a second channel, JuegaGerman (14.6 million subscribers), on which he provides commentary for video games that he plays.

YouTube has also been a valuable tool in Chile in terms of social justice. The Mapuche are an indigenous people living in the southern regions of Chile (and Argentina). They are a marginalized people with limited political power and impoverished communities, and have experienced severe discrimination from the Chilean police force. The mainstream media in Chile often portray the Mapuche as violent and dangerous, and rarely provide the opportunity for the Mapuche to present their side. YouTube has been used for nearly a decade now as a vehicle to present video productions of Mapuche reporting their side of events that happen in the country. However, they also frequently use the site to disseminate short documentaries and other videos representing their rural communities and cultural identity. For example, the channel Adkimvn, which has a complimentary website, Facebook page, and Twitter account, hosts Mapuche-made videos ranging from trailers to news coverage to events that take place in Wallmapu (Mapuche territory). The YouTube channel Kimeltuwe, Materiales de Mapudungun (Mapuche Language Materials) focuses on posting videos for the purpose of teaching Mapuzugun, the Mapuche language. Another YouTube channel, Wetruwe Mapuche, mainly posts recordings of live Mapuche music. In many cases, these musicians are performing rap, a genre that has become popular among indigenous peoples all over the Americas to draw attention to their ongoing struggles for political recognition, rights, and equality.

Facebook is the second-most-trafficked social media website in Chile (Alexa 2016). There are about 12 million Facebook subscribers in Chile, many of whom like using social media sites on their phones, making it one of the most popular smartphone apps. A survey conducted by Motorola and Adimark GFK in Santiago reported that 46 percent of women consider Facebook their favorite app (America Digital 2004). The most popular smartphone apps were related to social networking and communications (82 percent), followed by games (70 percent) and music (66 percent).

Chileans use Facebook mainly to post pictures (62 percent), but they also enjoy using it to post links (20 percent) and videos (16 percent) (Socialbakers 2016). The vast majority of the time Chilean Facebook users interact with this content by "liking" the content (95 percent), but sometimes they also share (4 percent) or comment (1 percent). Chileans are much more likely to comment on Twitter posts (20 percent) than Facebook, although Twitter is not as popular in Chile, ranking only eighteenth in web traffic (Alexa 2016; Socialbakers 2016).

After a devastating earthquake of 8.3 magnitude in 2015 that killed at least eight people and evacuated a million more, Chileans used Facebook's Safety Check

This Mapuche woman is playing the kultrun, a traditional drum used ceremonially by machi, or spiritual leaders, which is often used online as a symbol of her people's cultural identity. Inhabiting the southern regions of Chile and Argentina, the Mapuche are a marginalized indigenous people who have endured tremendous discrimination and have historically been denied a public voice. YouTube has become a powerful tool, allowing them to challenge mainstream media representations about who they are, and to draw attention to their ongoing struggle for political recognition and social equality. (Marcelo Vildósola Garrigó/Dreamstime.com)

feature to check on friends and family to make sure that they were all right (Garske 2015). While Chile's infrastructure is designed to withstand the effects of most earthquakes, telecommunications connections are often severely affected after earthquakes, especially ones of such magnitude, making it impossible for those affected to call, text, or post a message on social media. Facebook's Safety Check allows those affected to simply click a button to let their Facebook friends know that they are safe.

WhatsApp has also become very popular in Chile; it ranks alongside Facebook as one of the most downloaded apps for smartphones (America Digital 2014). WhatsApp competes with Line to provide free texting and calling services to Chileans. In fact, half of female smartphone owners in Santiago report using either WhatsApp or Line. Interestingly, the term "WhatsApp" has been turned into a verb in Chile. One may frequently hear the phrase "Wazapéame!" (abbreviated "Wz-pme"), which means "WhatsApp me!" in Spanish (America Digital 2014).

The only nationally produced websites in Chile that rank in the top ten most popular websites are news sites: Biobiochile, Emol, and Lun (Las Últimas Noticias) (Alexa 2016). However, Taringa! (the Reddit-like forum out of Argentina) ranks fourteenth, Twitter ranks nineteenth, and Instagram ranks nineteenth.

Marilyn J. Andrews

See also: Argentina; Brazil; Colombia; Ecuador; United States

Further Reading

Alexa. 2016. "Top Sites in Chile." Accessed August 28, 2016. http://www.alexa.com/topsites/countries/CL

America Digital. 2004. "Facebook and WhatsApp Are Chile's Favorite Smartphone Apps." Accessed August 28, 2016. https://congreso.chile-digital.com/english/whatsapp-facebook-chiles-favorite-smartphone-apps/

Bryan, Mason. 2013. "Chile's Internet Speed Fastest in Latin America." *Santiago Times,* October 26. Accessed August 27, 2016. http://santiagotimes.cl/2013/10/26/chiles-internet-speed-fastest-in-latin-america/

Garske, Monica. 2015. "Facebook's 'Safety Check' Feature Used in Aftermath of 8.3-Magnitude Chile Earthquake." *NBC San Diego*, September 18. http://www.nbcsandiego.com/news/local/Chile-Earthquake-Facebook-Safety-Check-Santiago-328086861.html

González, Francisco. 2015. "Subtel: 77% de Lasconexiones a Internet Son a Través del Celular." *La Tercera,* June 13. Accessed 28, 2016. http://www.latercera.com/noticia/negocios/2015/06/655-634150-9-subtel-77-de-las-conexiones-a-internet-son-a-traves-del-celular.shtml

Internet Live Stats. 2016. "Chile Internet Users." Accessed August 27, 2016. http://www.internetlivestats.com/internet-users/chile/

Internet World States. 2016. "Internet Penetration in South America, June 30, 2016." Accessed August 28, 2016. http://www.internetworldstats.com/stats15.htm

Kang, Cecilia. 2015. "The Real Reasons Why YouTube's 5 Biggest Stars Became Millionaires." *Washington Post,* July 23. Accessed August 28, 2016. https://www.washingtonpost.com/news/the-switch/wp/2015/07/23/how-these-5-youtube-stars-became-millionaires-and-why-you-wont-be-joining-them-anytime-soon/

Socialbakers. 2016. "March 2016 Social Marketing Report Chile." Accessed August 28, 2016. https://www.socialbakers.com/resources/reports/chile/2016/march/

Subtel. 2015. "Penetración de Internet en Chile Alcanza los 64,2 Accesosporcada 100 Habitantes." Accessed August 28, 2016. http://www.subtel.gob.cl/penetracion-de-internet-en-chile-alcanza-los-642-accesos-por-cada-100-habitantes/

Worldometers. 2016. "Chile Population (Live)." Accessed August 28, 2016. http://www.worldometers.info/world-population/chile-population/

ZonaWiFiChileGob. 2016. "Bienvenido a la ZonaWiFiChileGob." Accessed August 28, 2016. http://www.wifigob.cl/

CHINA

China, one of the world's largest countries in terms of landmass, is located in East Asia and is officially known as the People's Republic of China (PRC). It has the

Many people visit an internet cafe in Chengdu, China, to surf the web and play online games. Internet cafes have become an integral part of Chinese online culture, to the extent that some customers have been known to stay for long hours, or even days, at a time. To combat online addiction and to ensure the people's safety, the government has been imposing stricter regulations on café owners. (Piero Cruciatti/Dreamstime.com)

largest population in the world with over 1.3 billion people. While the PRC was not one of the early adopters of the internet, it now boasts the world's largest online population, with over 650 million people using some 700 million connected devices. Internet usage was initially limited to a few computer research labs until the government developed its own infrastructure for official use in the early 1990s. By the latter half of the decade, the Chinese Communist Party (CCP) had decided to reap the economic benefits of cyberspace and began a massive campaign to bulk up infrastructure across the country. Since then, China's internet usage has experienced explosive growth, particularly with the advent of mobile computing and the rise of smart devices.

As the PRC's internet has expanded from a largely academic tool to an engine of economic growth, so too has the population of internet users, or "netizens" (网友). Early users were typically well-educated, urban males, even as the use of the internet expanded beyond the initial research labs. This was mainly due to the expenses associated with acquiring a computer and an internet connection; it was not until the rise of internet cafés and drastic increases in disposable income that the demographics of China's online community began to reflect that of the country as a whole. Lowered barriers to entry have allowed for the exponential growth of internet usage among lower-income groups, particularly among young students.

While wealthier knowledge brokers still tend to be the heaviest internet users, young gamers between the ages of 19 and 25 make up the bulk of social media users and thus tend to have the most influential voices on online forums (CNNIC 2015).

Even now, however, Chinese netizens are more likely to be young males living in medium-sized to large cities. Despite having a staggering number of internet users, less than half of China's population has access to the internet. The lack of internet penetration in rural areas in particular is primarily due to the CCP's control over the nation's internet infrastructure. Six state-owned internet operators strictly control access to internet bandwidth. Three telecom corporations (China Telecom, China Unicom, and China Mobile), along with three government networks (China Science and Technology Network, China Education and Research Network, and China International Economy and Trade Net), jointly maintain a monopoly on the country's backbone of fiber optic networks. State control has allowed the PRC to invest billions in rapidly developing this network while still maintaining control over the spread of information.

Much of China's online activity actually takes place within a closed, hierarchical intranet. The largest portion of this network is dedicated to domestic traffic between users of the same internet operator. For information to be shared between the operators (and thus across the entire country), internet connections have to be routed through one of three network access points (NAPs). Finally, any information that enters or leaves the country has to connect through one of the three gateways operated by Telecom and Unicom in Beijing, Shanghai, and Guangzhou. These consecutive levels of centralized network access, along with a centrally channeled connection to the global internet, create multiple points for the government's censors to limit (and even block) certain activities.

This censorship of the Chinese internet is a complex issue. The CCP maintains a strict control on anything that can potentially mobilize social forces and undermine the authority of the state. Along with the state's family planning policies, the censorship of the government is one of the greatest examples of the Chinese government's direct involvement in the lives of its citizens. While other countries have had to address burgeoning social movements and the various issues arising from a hyperconnected society, the CCP has both limited the impact that the internet has had on Chinese netizens and used it to strengthen its own control. Early analyses of Chinese netizens indicated that they were more politicized and open to information that had not been filtered through government sources. As a result, many external observers have expressed the hope that the internet would be a tool to help democratize China by providing unfettered access to information and allowing previously unheard voices to become very vocal (Lei 2011). However, the CCP has proven very adept at addressing these issues and manipulating online discourses in its favor.

Part of the reason that the CCP has been so adept at controlling the internet is that its censorship is both reactive and proactive. At its most basic level, the Chinese internet is shielded from foreign influence by the Golden Shield Project (金盾工程)—often referred to as the "Great Firewall of China"—a series of hardware and software elements built into the nation's internet infrastructure that filter out key words and websites that have been identified as potentially subversive. While internet usage is largely monitored and controlled by the Ministry of Industry and Information Technology, the Ministry of Public Security and the Publicity Department of the CCP provide guidance on what information users should and should not be allowed to access. Domestically, this system is backed up by legislation that has established licensing requirements for websites to operate legally and holds the owners responsible for maintaining proper content (Feng and Guo 2013). As with more traditional media outlets, fines and prison sentences against violators have compelled many internet users to police themselves.

As impressive as China's system of blocking unfavorable content is, its system of editing existing content and promoting certain ideals is even more extensive. In addition to blocking specific websites, censorship software actively scans articles, blog posts, and even comment threads against an ever-changing list of sensitive terms to identify and edit out potentially harmful statements. While the exact parameters of what this software is looking for is not public knowledge, researchers have found that it is nuanced enough to allow limited criticism of government policy, while deleting any potential references to calls for collective action (King et al. 2014). The government has become even more proactive in recent years by hiring internet commentators to publish websites, articles, and even comments on online forums that promote a positive view of the CCP. Known as the "50 Cent Party" (五毛党) for the supposed fee that they are paid per internet post, these commentators are a new way for the government to guide public opinion. The exact number of paid commentators is unknown; however, since large websites are required to train their own internal teams to direct online discussions in order to operate legally, they are estimated to be in the tens of thousands. The government has also started to sponsor the creation of a parallel "Water Army" (网络水军), where profiles designed to resemble normal users are actually programmed to post preapproved statements on various social media (Yu, Asur, and Huberman 2015). While most savvy Chinese netizens are scornful of these paid commentaries and are quick to call out posts that they think are sponsored by the CCP, the sheer number of such users serves to muddy any potential discourse critical of the government.

It is difficult to determine the extent of the impact of censorship on the nature of discourse within China's online forums. Early analysis of Chinese netizens indicated that they were more critical of government and willing to voice dissent; however, the "democratization" of the internet has allowed a number of previously unheard voices to become very vocal, many of which approve of the CCP and its policies. At the same time, the government's most ardent critics have had to become more nuanced in vocalizing their concerns in order to evade censors. Initially, this was largely accomplished through various linguistic tricks. Netizens often used homophones or pidgin English to replace potentially offensive words with more benign terms that censorship software would often let pass. Nevertheless, increasingly sophisticated censorship techniques, alongside a constantly updated list of key terms, have limited the effectiveness of this tactic, with even seemingly innocuous terms like "Grass Mud Horse," "River Crab," and "Jasmine" blocked temporarily due to their connection to various protests (Byer 2015).

Eventually, this method fell out of favor, and most netizens who wish to escape the Great Firewall use virtual private networks (VPNs) to access international sites through remote logins or the Tor anonymity network to disguise their identity. Even though both of these methods are frequently subject to blocking by the Great Firewall, they have led to an essentially hollowing-out of the Chinese internet. Those who are content to operate within the confines of China's censors remain the largest users of local blogging services and social media sites, while dissidents have become a more globalized community that prefer overseas diaspora forums, where they can access unfiltered news sources and freely share their ideas with a receptive audience. While this has had the intended effect of removing critics, it has also served to create an overblown sense of government support, even for unpopular public policies.

This separation of netizens into a more global-minded community, interested in the international internet, and those willing to operate within the Chinese intranet has had a profound impact not only on the nation's political discourse, but also on the cyber economy. As mentioned earlier, the fact that the demographics of Chinese netizens skew younger has had an impact on how the internet is used in China. The vast majority of users now access the internet through mobile devices like smartphones and tablets, and they are also active users of social media. Limited access to international sites like Facebook and Google have led domestic companies to create a native set of websites meant to mimic the functions of their foreign counterparts. Some of these indigenous sites have formidable audiences; for instance, Tencent QQ, an instant messaging service coupled with online shopping and microblogging, claims 830 million active users—nearly triple the number of Twitter users in China. Even niche sites like Jiayuan, a dating service, and Douban, for hobbyists, claim over 100 million users (Millward 2013).

The creation of a robust, native intranet has played well into the CCP's attempts to shift the nation's economic reliance upon export-driven growth to domestic consumption. Even though the PRC was able to weather the 2009 financial crisis, the Great Recession proved that Chinese manufacturing was extremely vulnerable to shocks to the international economy. Bolstering domestic spending is seen as a way to continue to drive China's economic growth, and the cyber economy is seen as the best way to convince Chinese shoppers to spend their savings. Online retailers have proven critical in shaping the attitudes of Chinese consumers; brands are aware that internet-savvy shoppers research for the best deal and no longer adhere to a single brand. Furthermore, e-commerce holidays like Singles' Day, on November 11, have proven enormously successful, bringing in profits that dwarf those of similar events in the United States.

Government departments like the previously mentioned Ministry of Information Technology and Industry are not only bolstering these online marketplaces through new business-friendly regulations, they are also pushing these companies

The first Chinese social media world record was set by Lu Han (1990–), a Chinese pop star. Back in September 2012, Lu Han posted about his love for Manchester United. His post received more than 13 million responses through August 2014, the highest response rate to date on China's Weibo platform. The *Guinness Book of World Records* confirmed the achievement, making Lu Han the first-ever recipient of a verified Chinese social media world record. Lu Han is a former member of the Asian boyband EXO and is now working on his acting career (Lynch 2015).

to revamp China's nascent tech sector. Much as manufacturing was able to lift millions of Chinese citizens out of poverty, recent government white papers on cyber policy have suggested that building up the nation's tech industry could create an even larger middle class. In addition to writing new regulations regarding online banking and real estate transactions, there has even been a push to expand internet coverage by further extending existing infrastructure into rural areas and simplifying the language used in online documents.

This dichotomy of attempting to limit the social impact of the internet while still reaping its economic benefits carries over to the CCP's distinction between promoting a secure domestic intranet and exploiting flaws in the global internet to launch cyberattacks against foreign countries. China, like many other nations, has been engaging in cyber espionage for years; some of the culprits are believed to be private businesses attempting to get ahead of their foreign competition, while others are explicitly backed by the government. In 2011, the People's Liberation Army (PLA) established a cyberwarfare group whose nominal purpose is to defend Chinese servers from foreign attacks. However, the U.S. government has officially declared China-based cyberattacks to be advanced, persistent, and ongoing; and given the fact that recent attacks have targeted the U.S. military's advanced systems, it is largely assumed that they emanate from the PLA (Brownlee 2015).

Cyberspace in China is a rapidly changing place, full of contradictions. Even as the government promotes the expansion of the internet to an ever-wider audience, young male students remain the most vocal users. Although the CCP purposefully limits expression on online forums, it also believes that the freedom of information provided by the internet has the potential to drastically revamp its economy by bolstering consumerism and a nascent tech sector. Despite the fact that China fiercely guards its domestic intranet from foreign meddling, the Chinese military has proven itself willing and able to conduct serious cyberattacks against other countries. While the PRC may be able to keep these competing interests in balance for the time being, the sheer growth of its online community means that it will eventually have to address these very pressing concerns.

Jonathan Dixon

See also: Japan; South Korea; Vietnam

Further Reading

Brownlee, Lia. 2015. "China-Based Cyber Attacks on US Military are 'Advanced, Persistent, and Ongoing': Report." Accessed December 13, 2015. http://www.forbes.com/sites/lisabrownlee/2015/09/17/chinese-cyber-attacks-on-us-military-interests-confirmed-as-advanced-persistent-and-ongoing/

Byer, Kelly. 2015. "Jumping the Great Firewall: Social Media Among China's Youth." Accessed December 13, 2015. http://internationalstorytelling.org/china/scaling-the-great-internet-wall/

China Internet Network Information. 2015. "Statistical Report on Internet Development in China." Accessed December 13, 2015. http://www1.cnnic.cn/IDR/ReportDownloads/201507/P020150720486421654597.pdf

Feng, Guangchao Charles, and ZhongshiGuo, Steve. 2013. "Tracing the Route of China's Internet Censorship: An Empirical Study." *Telematics and Informatics*, 30: 335–345.

King, Gary, Pan, Jennifer, and Roberts Margaret E. 2014. "Reverse-Engineering Censorship in China: Randomized Experimentation and Participant Observation." *Science*, 345: 1–10.

Lei, Ya-wen. 2011. "The Political Consequences of the Internet: Political Beliefs and Practices of Chinese Netizens." *Political Communication*, 28: 291–322.

Millward, Steven. 2015. "Check out the Numbers on China's Top 10 Social Media Sites." Accessed December 13, 2015. https://www.techinasia.com/2013-china-top-10-social-sites-infographic/

World Bank. 2015. "China." Accessed July 30, 2016. http://data.worldbank.org/country/china

CHINA: HONG KONG

Officially called the Hong Kong Special Administrative Region of the People's Republic of China, Hong Kong was a British colony until 1997 and is now an autonomous territory located south of mainland China. With 7.2 million residents, Hong Kong is the world's fourth most densely populated sovereign territory. Rooted in a society that had a free-market system due to its being a British colony, Hong Kong takes pride in its long-standing colonial history of press freedom and significant degree of civil liberties. Hong Kong's transition to Chinese government control in 1997 required the region to transform its political and social institutions rapidly. Its mass media, previously a symbol of modern capitalism, has gradually needed to adapt itself, both culturally and ideologically, to accommodate China's more restrictive social and political policies (Lee 2000). These changes in media organizations' political positions, however dramatic, did not occur overnight after being officially returned to mainland China. Constrained by the promise of one country, two systems, the Chinese government did not continuously interfere with Hong Kong, and it chose not to impose the same degree of media censorship on the territory that it instilled across the rest of the country (Fung 2007).

Internet user growth across the Asia-Pacific region directly correlates with the growing popularity of smartphones and their rapid assimilation regionwide. However, the growth in Hong Kong has been saturated over the years. As of 2016, Hong Kong has 4.8 million smartphone users, a figure equal to 80.7 percent of the population. While the penetration rate is slightly behind South Korea, Taiwan, and Japan, it is about equal to the top three in the region (Emarketer 2016). As a result of high-speed internet and advanced telecommunications infrastructure implemented in the city, 96 percent of Hong Kong smartphone users browse the internet daily via mobile—the highest rate in Asia. As for the region's use of various languages, traditional Chinese (the complicated form of Chinese used for Hong Kong's Chinese dialects, such as Cantonese) remains the most popular, followed by English, traditional Chinese (the complicated form of Chinese used to represent Taiwan's Chinese dialects, such as Mandarin), and subsequently, simplified Chinese used in mainland China (Go-globe 2016).

As a Special Administrative Region, Hong Kong is exempt from the internet restrictions imposed on the rest of mainland China, particularly the Great Firewall

In August 2016, hundreds gathered in the Tin Shui Wai district of Hong Kong to play *Pokémon Go*. This augmented reality game for mobile devices, released just a month earlier by Niantic, caught on rapidly and became popular all over the world. (Yiu Tung Lee/Dreamstime.com)

and site blocking. Because of the looser internet policies, social media and app penetration is more diverse and competitive than that of North America. Facebook, along with other social network platforms, are blocked in China, but they are accessible in Hong Kong. Social media giants in China, such as Ren-Ren (literally meaning "people" in Mandarin Chinese, this site is akin to Facebook) and SinaWeibo (akin to Twitter), are also extremely popular in Hong Kong due to its increasing integration with China. With such a wide variety of social network platforms to keep up with, it is normal to see people enjoying their daily activities, such as cycling or walking, while texting in Hong Kong. And it is common to hear a public service announcement for people to observe their surroundings rather than their mobile phones while performing tasks such as riding an escalator.

Despite a wide range of social media platforms available to people in Hong Kong, Facebook is overwhelmingly the most popular social network. A TNS survey determined that Hong Kong has at least 4.4 million people on Facebook, equal to more than half of the area's population using the platform (Go-globe 2016). Almost half of these users, or approximately 44 percent, utilize Facebook as their first point of contact for keeping up with important news. Of the platform's 4.4 million active users, around 70 percent (3.1 million people) log on to Facebook daily. Most users stay online for at least 30 minutes per unique visit. WhatsApp earns the second spot; however, the tenure of this position remains to be seen. Since Facebook acquired WhatsApp, the company has initiated changes that appear to be loosening WhatsApp's historical policy of excluding advertisements; these changes, along with the site's usage in local protests against the government that have placed the site under scrutiny, make it unclear if WhatsApp can retain its position as the second most popular platform. WhatsApp is currently the preferred site for public relations and media personalities working in Hong Kong's Chinese-language media arena (Emarketer 2016).

SinaWeibo is the third most popular social platform. The site's popularity stems from the fact that China blocks Western social media platforms on the rest of the mainland through its Great Firewall. SinaWeibo facilitates social media communications with the largest audience of Chinese people (Stecklow 2016). The Chinese government's preference for only allowing indigenous platforms has allowed them to dominate the world's most populous nation. Along these same lines, SinaWeibo, the Chinese equivalent of Twitter, functions all across China, whereas Twitter is accessible only in less restricted areas such as Hong Kong (Go-globe 2016). However, SinaWeibo's high penetration rate in Hong Kong is not solely a product of Chinese internet policy; it is also correlated to the fact that it is highly used by well-known and followed personalities, ranging from celebrities to government officials and business elites. In addition, sites like Tencent's WeChat (called "Weixin," or "micro-message" in Mandarin Chinese), is at its simplest form a multifunctional platform that includes social media and voice communications, text services, and online gaming, similar to other platforms. It also has some unique functionality, such as allowing users to join large groups of up to 500 and allowing them to access "city services," which are available to users located in specific urban areas (Emarketer 2016). The platform has become quite significant, as it has successfully integrated with a banking and payment system, referred to as the "wallet," that connects users' credit cards to their accounts and allows them to pay for services via their WeChat accounts. In some areas, users can make and pay for doctor's appointments and ride-sharing services.

With Hong Kong having such significant internet and social media penetration rates, the social media landscape is always in flux. Hong Kong, in particular, is in a unique space at the crossroads of Western and Chinese social media and communications innovations, where citizens can enjoy the best of both worlds. In some ways, Hong Kong has access to more online communications media than other parts of the world, and the trends from the region, while similar to other Asian nations at the moment, could markedly change in the future. Twitter, for instance, opened a Hong Kong office in early 2015; Line, the largest social media player in Japan, South Korea, and Taiwan, has slowly gained popularity; and LinkedIn has slowly come to dominate the business and career online space.

Hong-Chi Shiau

See also: China; China: Macau; Japan; South Korea; Taiwan; United Kingdom

Further Reading

Emarketer. 2016. "Asia-Pacific Boasts More than 1 Billion Smartphone Users: China, the World's No. 1 Smartphone Market, Is Beginning to Mature." Accessed August 1, 2016. http://www.emarketer.com/Article/Asia-Pacific-Boasts-More-Than-1-Billion-Smartphone-Users/1012984#sthash.WtABffED.dpuf

Fung, Anthony. 2007. "Political Economy of Hong Kong Media: Producing a Hegemonic Voice." *Asian Journal of Communication*, 17(2), 159–171.

Go-globe. 2016. "Social Media Usage in Hong Kong: Statistics and Trends: 2016." Accessed August 1, 2016. http://www.go-globe.hk/blog/social-media-hong-kong/

Lee, Chin-Chuan. 2000. "The Paradox of Political Economy: Media Structure, Press Freedom, and Regime Change in Hong Kong." In *Power, money, and media: communication patterns and bureaucratic control in cultural China,* edited by Chin-chuan Lee. Evanston, IL: Northwestern University Press, 238–336.

Socialbakers. 2016. "June 2016 Social Marketing Report Hong Kong Social Media Use in Hong Kong: Statistics and Trends." Accessed July 17, 2016. https://www.socialbakers.com/resources/reports/hong-kong/2016/june/

Stecklow, Sam. 2016. "How China's Biggest Social Network Works with the Government." *New York Magazine,* March. Accessed July 17, 2016. http://nymag.com/selectall/2016/03/how-weibo-works-with-the-chinese-government-to-censor.html

CHINA: MACAU

Macau, also spelled Macao, is a Special Autonomous Region of the People's Republic of China, located across from Hong Kong on the Pearl River Delta. The significance of Macau can easily be underrated, which is why this section singles out Macau to shed light on its unique history and social political environment. Here are three facts about Macau that might not be common knowledge (BBC 2013):

- With a size thirty-nine times smaller than Hong Kong, Macau is the most densely populated region in the world.
- Macau was the last remaining European colony in Asia until late 1999, and it had long been administered by the Portuguese Empire.
- Macau has maintained one of the most open economies worldwide since being reintegrated into China in 1999.

Macau's economy thrives on tourism, specifically gambling, which was legalized in the nineteenth century and has continued to be an integral source of government revenue. The tax revenue from Macau's robust gambling industry has enabled the government to institute a number of social welfare programs, including the initiative to provide all Macau citizens with fifteen years of free education (Barboza 2007). However, Macau's Chinese casino capitalism comes at a price: while the gambling industry "stimulates economic growth, provides employment, and strengthens the post-colonial state," it also widens the income gap between the rich and the poor and enables addictive gambling (Lo 2009, 19).

Against the backdrop of the aforementioned dialectical social political tendencies, Macau reportedly has the highest "media density" in the world—nine Chinese-language dailies, three Portuguese-language dailies, two English-language dailies, and about half a dozen Chinese-language weeklies and one Portuguese-language weekly. Macau is probably the most saturated smartphone market across the world. The majority of the country's internet users go online via mobile phone (Socialbakers 2016). Also heavily influenced by its casino capitalism, its social media system has been the most diverse—with Weibo and Zenzen for its connection with Chinese visitors, and Facebook, Twitter, and YouTube as well, with additional growing user bases on Viber, WhatsApp, Line, WeChat, and BeeTalk outside the Chinese circle (Socialbakers 2016). According to a Creativity, Solution, Growth (CSG) research report, one conclusion suggested that despite the popularity of smartphones, individuals' intention to purchase a new one in the near future is low, possibly due to the lack of new innovations in the market (CSG 2013). By 2013, most Macau residents possessed at least one smartphone, while tablet ownership reached half of all households. Consumers spent an average of 28 percent of their time online—around forty-seven hours per week (CSG 2013). Social media browsing via smartphone is becoming as frequent as browsing the internet overall. The use of tablets for online activities is still not evident in 2016.

Hong-Chi Shiau

See also: China; China: Hong Kong; Taiwan

Further Reading

Barboza, David. 2007. "Asian Rival Moves Past Las Vegas." *New York Times,* January 24. Accessed August 2, 2016. http://www.nytimes.com/2007/01/24/business/worldbusiness/24macao.html?_r=0

BBC News. 2013. "Macau Profile—Media." *BBC News,* January 21. Accessed August 8, 2016. http://www.bbc.com/news/world-asia-pacific-16599928

Creativity, Solution, Growth (CSG) Research Report. 2013. "Macau Smartphone Market Research." Accessed August 1, 2016. http://www.csg-worldwide.com/csg/wp-content/uploads/2014/01/Macau-Mobile-and-Social-Media-Report-Dec-2013.pdf?phpMyAdmin=d1bb9c7d2d321d9948e75af396f89348

Lo, Sonny. 2009. "Casino Capitalism and Its Legitimacy Impact on the Politico-Administrative State in Macau." *Journal of Current Chinese Affairs*, 38, 1:19–44.

Socialbakers. 2016. "Macao Facebook Page Statistics." Accessed August 2, 2016. https://www.socialbakers.com/statistics/facebook/pages/total/macao/

COLOMBIA

Colombia is a country of roughly 48 million people situated in the northwest portion of South America. In the mid-1990s, the internet user base in the country grew slowly, but it has subsequently grown to half the population. Many users still access the internet from community centers, but mobile access is growing and government support of internet use is both generous and pervasive. Personal and journalistic freedoms are strongly protected, and yet sanctions for defamation and libel are severe.

Internet connectivity first came to Colombia in June 1994, when a handful of corporations and universities created the INTERRED–CETCOL (Red Nacional de Ciencia, Educación y Tecnología) that connected the United States to the University of Los Andes also called Uniandes (Lu 2010). In 1995, Colombia had 0.2 internet users for every 100 citizens. That number slowly rose to 1.1 in 1998 and then doubled by the year 2000 to 2.2. Between the years 2001 and 2005, the percentage of internet users rose from 2.9 percent to 11.0 percent. A steady increase in the number of internet users continued, as it rose to 15.3 percent in 2006, 21.8 percent in 2007, and ultimately 36.5 percent in 2010 (World Bank 2016).

Vive Digital (Live Digital), a national campaign to expand the use and access to the internet throughout Colombia, was launched in 2010. It had a drastic and beneficial effect on the scale of internet usage. Broadband connections went from 3.1 million installations in 2010 to 9.9 million installations by the middle of 2014. Small business and medium sized business use of the internet rose from 7 percent in 2010 to more than 60 percent in 2014, and Colombian municipal internet usage grew from 17 percent to 96 percent in that same period (Vega 2015). The total number of internet users went from 40.4 percent in 2011 to 52.6 percent by 2015 (World Bank 2016).

Infrastructure in rural areas was key to the Vive Digital campaign. Nearly 900 internet access centers (community centers with access to the internet) and 7,621 kiosks (computers installed in family homes, community centers, drugstores, schools, or shops) were installed from 2010 to 2015 (Vega 2015). Community centers, known as Los Centros Tecnológicos Comunitarios, are buildings situated in rural areas to bring the internet to citizens who live outside cities or urban areas.

Accessibility for the disabled helps to bring the internet to those with visual or hearing impairments. Over 77,000 free licenses for screen readers and text-magnifying software were provided to visually impaired users, and the hearing impaired were provided with centers to help them communicate with text chat or video (Vega 2015).

Despite these improvements in infrastructure, a marked slowdown in the expansion of internet to the remaining population persisted in 2014, partly because only 11 percent of households had personal computers (Freedom House 2014). In 2015, though, internet penetration reached 53 percent in spite of the slowdown. Barriers to greater access stem from socioeconomic factors such as lack of funds or access to developed urban areas. Infrastructure still has not reached all rural areas, and technology prices are still beyond the reach of much of the population (Freedom House 2015).

Facebook may be helping break through some of these barriers. According to Mark Zuckerberg (1984–), chief executive officer (CEO) of Facebook, only 50 percent of Colombians have access to the internet. Therefore, Facebook introduced its Internet.org app in Colombia, making it the first Latin American country to receive it to give more Colombians the opportunity to gain access to the internet (Murphy 2015). The average internet speed in 2015 was 3.7 Mbps, and a majority of internet users accessed the internet from cybercafés or education centers (Freedom House 2015). However, increased home and mobile use has reduced people's need to use these public facilities (ICT Ministry 2013).

Mobile internet may be the key to allowing the remaining 47 percent of citizens who currently do not have access to the internet to go online, as mobile phones are increasingly used to access the internet, but the disparity between those mobile users who have subscription plans and those who have pay-as-you-go plans may indicate that socioeconomic factors still play a role in limiting internet access where land-based digital subscriber line (DSL) service cannot reach. In 2015, approximately 5.6 million mobile internet users (12 percent of the entire population of Colombia) had subscription plans, and 21.4 million mobile internet users had pay-as-you-go plans (Freedom House 2015). A 2014 survey of the people who did not have internet in their homes revealed that 44 percent of them list high prices as the reason why they do not have access. The average price per month for internet access is $19.80, yet the minimum legal monthly wage is $260.00, which means that internet access is too costly for people in the lowest socioeconomic class (Freedom House 2015).

Cost is not the only barrier, though. More than half of Colombia's landmass is comprised of the southern regions of Amazonas, Vaupés, Vichada, Guainía, and Guaviare, which are rural areas. Even though a small percentage of the Colombian population lives in that southern region, the indigenous languages spoken there are not accounted for in internet apps or websites (Freedom House 2015).

Of all internet websites, Facebook and Twitter are widely used social media. Colombia ranks fourteenth in the world in number of Facebook users (15 million) and the city of Bogotá ranks as the ninth city in the world in terms of internet use, with 6.5 million users (MINTIC 2016). The city of Medellín has 2.5 million users, Cali has 1.6 million users, and Barranquilla has 1.2 million users. Between May 2014 and May 2015, the total number of Facebook users nationwide grew by 2 million. By the year 2015, the total number of Facebook and Twitter users reached 24 million (Abad 2015). Approximately 6 million internet users use Twitter, which has become the premier method of communication for political campaigns, sports, and government officeholders. This usage places Colombia above France and Germany in terms of Twitter popularity (MINTIC 2016).

The gender divide among Facebook users is even at 50 percent, and most users (30 percent) are between the ages eighteen and twenty-four. Nearly 100 percent of Facebook and Twitter users speak Spanish and only 8.75 percent speak English. Those who use Facebook and Twitter via a mobile device primarily do so with an Android-powered phone (16 million), while much smaller numbers use iOS-based phones (1.5 million) and BlackBerry devices (1.5 million) (Abad 2015).

Governmental regulation of content is limited. The Colombian Ministry of Information and Communications Technologies has decreed that the only content to be blocked is child pornography, and internet service providers (ISPs) are required to prevent child pornography from being made available online. While the government itself does not often order the removal of content from view, court orders sometimes result in content removal when it is deemed to violate fundamental rights. For example, content from the Fuerzas Armadas Revolucionarias de Colombia (FARC) guerrilla group has been removed or restricted at times. Conversely, the seminal court case of *Guillermo Martínez (n.d.) v. Google* in 2013 involved the newspaper *El Tiempo*, after it reported that Martínez was part of a mafia group. Charges were dropped against him, but when he searched for his name on Google, the original mafia allegation still appeared. He sued to have it removed, but Colombia's Constitutional Court ruled that Google was not responsible for the content that it indexed in its search engine (Freedom House 2014). In spite of these relative freedoms, 47 percent of respondents to a national survey of journalists reported that they avoid publishing certain information online due to fear of aggressive action against them, and 35 percent of respondents feared losing their jobs. Furthermore, 25 percent of respondents feared pressure from government entities, and 57 percent believed that local governments pressure the media through government advertising (Freedom House 2015).

In 2015, President Juan Manuel Santos (1951–) put forward a National Development Plan (Plan Nacional de Desarrollo, or PND) which left out language found in Article 56 of its predecessor stating that telecommunications authorities should regulate neutrality on the internet. *Internet neutrality* is the principle that the government or ISPs should allow access to all content without favoring or blocking based on sources, particular products, or websites visited (Peñarredonda 2015). After fervor and uproar, however, a debate ensued and Article 56 ended up being preserved.

Articles 20 and 73 of Colombia's National Constitution guarantee freedom of expression, prohibit prior restraint, and protect the liberty and independence of journalistic activity. However, Colombia still enforces its criminal penalties for defamation—even speech online. The penal code allows punishment of any third party who publishes, reproduces, or repeats an insult or libel. A private citizen was convicted of libel for an anonymous insult to a public utility company manager in 2014. After posting an insulting comment about the manager to *El Pais*'s online comment section in 2008, the government tracked him down using his internet protocol (IP) address. Colombia's Constitutional Court rejected his appeal in February 2015. Yet, overall, the freedoms enjoyed in Colombia are fairly strong (Freedom House 2015).

Jeffrey M. Skrysak

See also: Brazil; Ecuador; Peru; Venezuela

Further Reading

Abad, Daniel. 2015. "Estadísticas de Facebook y Twitter en Colombia (2015)." Accessed March 13, 2016. https://www.latamclick.com/estadisticas-de-facebook-y-twitter-en-colombia-2015/

Freedom House. 2014. "Freedom on the Net 2014—Colombia." Accessed March 11, 2016. https://freedomhouse.org/report/freedom-net/2014/colombia

Freedom House. 2015. "Freedom on the Net 2015—Colombia." Accessed March 11, 2016. https://freedomhouse.org/report/freedom-net/2015/colombia

ICT Ministry. 2013. "ICT Trimestral Bulletin." December. Accessed March 12, 2016. http://colombiatic.mintic.gov.co/602/w3-article-4992.html

Lu, Tania. 2010. "Historia de Internet en el Mundo y Su Llegada a Colombia." January 12. Accessed March 13, 2016. http://tanialu.co/2010/01/12/historia-de-internet-en-el-mundo-y-su-llegada-a-colombia/

MINTIC. 2016. "Colombia es Uno de los Países con Más Usuarios en Redes Sociales en la Región" Accessed March 13, 2016.http://www.mintic.gov.co/portal/604/w3-article-2713.html

Murphy, David. 2015. "Facebook Rolls out Free Internet.org App in Colombia." January 15. Accessed March 14, 2016. http://www.pcmag.com/article2/0,2817,2475278,00.asp

Peñarredonda, Jose Luis. 2015. "Neutralidad en la Red: ¿En Riesgopor el Plan de Desarollo?" Accessed March 13, 2016. http://www.enter.co/cultura-digital/colombia-digital/neutralidad-en-la-red-colombia-plan-de-desarrollo/

Vega, Diego Molano. 2012. "Inforgrafia, Historia de Internet en Colombia." May 17, Accessed March 13, 2016. http://www.slideshare.net/DiegoMolanoVega/infografa-historia-de-internet-en-colombia

Vega, Diego Molano. 2015. "Colombia's Internet Advantage." May 7. Accessed on March 13, 2016. http://www.americasquarterly.org/content/colombias-internet-advantage

Venegas, María del Rosario Atuesta. 2003. "Los CentrosTecnológicos Comunitarios, una Opciónpara el Acceso a la Tecnología en la ZonasRurales." Accessed on March 13, 2016. http://publicaciones.eafit.edu.co/index.php/revista-universidad-eafit/article/download /917/822

World Bank. 2016. "Internet Users (per 100 People)." Accessed March 14, 2016. http:// data.worldbank.org/indicator/IT.NET.USER.P2/countries/1W?page=3&display =default

CUBA

Cuba is an island in the Caribbean Sea, just south of Florida. The main language spoken there is Spanish. The population is 11.39 million; 64 percent of Cubans are between fifteen and fifty-nine years of age, with the median age being forty-one. In 1959, rebels led by Fidel Castro (1926–2016) overthrew the Cuban government and the Communist Party has governed the island ever since. In 1960, the United States placed an embargo on Cuba, which meant that no trade could be exchanged between the countries. Until December 2014, relations between the two countries had been hostile, and little political interaction had taken place over the past fifty years. As of late 2015, the trade embargo remained in place.

On December 17, 2014, after a phone conversation between President Barack Obama (1961–) of the United States and President Raúl Castro (1931–) of Cuba, the two countries announced that they would reestablish diplomatic relations. This was the first exchange between the presidents of these two countries since Vice President Richard Nixon met with Fidel Castro a few months after the Cuban Revolution (1953–1959). The decision to restore relations between countries included plans to reopen the U.S. embassy in the capital city of Havana, to reduce trade and travel restrictions, and release certain political prisoners (DeYoung 2014). One of the prisoners released was Alan Gross (1949–), a government contractor who had been arrested and imprisoned by Cuban officials five years earlier for attempting to destabilize the Cuban government. He had been working with the U.S. Agency for International Development (USAID) and had been in Cuba to secretly distribute internet equipment to Jewish community groups (DeYoung 2014; Ruane 2014). After their phone conversation, President Obama commented publicly that Cuba had also promised to expand internet access throughout the country, although President Castro did not include this statement in his public address to the Cuban people. In April 2015, the two presidents met for the first time in person when they held a one-on-one meeting during the Summit of the Americas in Panama (Lee 2015). While the two countries still disagreed on many issues, these meetings represented the first substantive attempt in the past fifty years to normalize relations between Cuba and the United States.

Cuba has restrictive internet practices that correspond mainly with access to the internet, such as prohibiting home internet connections, requiring official approvals for office-based connections, prohibiting satellite connections, and controlling and monitoring all access points. As of 2014, only about a quarter of the

A woman stops to check her cell phone in Cuba's historic neighborhood of La Habana Vieja. In a country where internet is strictly controlled, access to popular sites such as Twitter is restricted. Nevertheless, Cubans have found ways around these measures by using a system called "speak-to-tweet," which allows people to use their cell phones to record a message that is converted to text and posted to restricted sites. (Lembi Buchanan/Dreamstime.com)

population could get online, and many of these users had to deal with outdated technology and slow connections. Once online, however, the censorship was actually less comprehensive than it is in other places in the world, like North Korea or China. Websites like Facebook and Twitter are not officially banned; however, other websites are blocked, such as YouTube or those written by Cuban journalists that have been deemed antigovernment.

Several factors contribute to the low internet penetration in the country. Until 2008, for example, it was illegal for private citizens to own computers. In 2011, there were fewer than 783,000 computers in the whole country (Franceschi-Bicchierai 2014). In 2014, almost all internet traffic occurred on desktops or laptops (We Are Social 2014). Home connections are very rare, and only approved jobs, such as academics, doctors, or government officials, have access to the

internet at work. There are cybercafés where Cubans can pay to connect online; however, it can be prohibitively expensive, at $6–$10 an hour, while an average monthly salary in Cuba is only $20. At cybercafés, computers are made available to tourists, who have access to the entire internet, and there are computers for Cuban citizens, who have access only to state-approved websites. In either case, the connection speed can be very slow, which prevents most Cubans from doing much other than checking their email (Franceschi-Bicchierai 2014). Due to embargo restrictions and lack of internet penetration, there are not many statistics available on how Cubans use the internet and social media.

Once online, Cubans utilize some of the most globally popular social media sites. Among the most popular sites in Cuba are Facebook, Twitter, Google+, Pinterest, and Vkontakte. The first two are by far the most popular; of these top social media sites, Facebook receives 87.44 percent of the traffic and Twitter receives 10.44 percent. The other websites all receive less than 1 percent of the traffic apiece (StatsMonkey 2015).

In addition to these internationally popular websites, the Cuban government is involved in providing local social media to those Cubans who use the internet. For example, EcuRed is the Cuban version of Wikipedia. However, EcuRed had only 78,000 articles as of 2014, and unlike Wikipedia, its editing is not open. There are only a few government-approved editors who write articles for the website (Franceschi-Bicchierai 2014). Interestingly, EcuRed has a rather large following in Mexico, from which about a third of the website's traffic occurs (Alexa 2015). EcuRed was popular enough to inspire a mobile app called EcuMobil.

The Cuban government also introduced Red Social in December 2011, which was supposed to be the Cuban Facebook. The site mirrored that of Facebook and even had Facebook in the domain name: facebook.ismm.edu.cu. However, this website was short-lived; it no longer exists. In September 2013, the Cuban government announced another attempt at launching a social network. This was called La Tendedera, which would be available from only Joven Clubs (youth centers). As of December 2014, the website was still in existence, but it had not reached anything close to the popularity of Facebook (Franceschi-Bicchierai 2014; Freedom House 2014).

In 2010, the U.S. government introduced to the Cuban people a social media site akin to Twitter. The intention was to help promote democracy in the region; however, the Associated Press uncovered evidence suggesting that the U.S. government actually planned to use it as a tool to incite unrest and potentially "renegotiate the balance of power between the state and society" (Arce, Butler, and Gillum 2014). It was commissioned by the U.S. Agency for International Development (USAID), but when it was introduced, there was no mention of this origin. It was called ZunZuneo, which is Cuban slang for a hummingbird's tweet. Within six months of its launch, it had almost 25,000 subscribers.

As ZunZuneo grew in popularity, USAID strategized about how to maintain the website without disclosing its involvement. By March 2011, ZunZuneo had 40,000 members, at which time USAID decided to cap the number of subscribers at a relatively low number to avoid attracting too much attention. Nevertheless, in the middle of 2012, the website vanished. It had been rerouted to a children's website, and later Cuban intelligence informed users that the website had been blocked. Some time after this occurrence, USAID announced that the program ran out of money and ZunZuneo was abandoned. Users of ZunZuneo were left feeling as though a vacuum existed where the social network had once been (Arce, Butler, and Gillum 2014).

In the wake of reopened relations between Cuba and the United States, technology companies are looking into expanding their market to the island. This includes the actual Twitter, which has stated that it is in talks with the Cuban government to introduce the messaging app to the country. Due to low internet penetration, the company plans to allow Cubans to send tweets via text messages from their cell phones (Scola 2015). However, Cubans have already found a way to do this. It is a system known as "speak-to-tweet." A Cuban can call a phone number in the United States and record a message. That message is converted to text and shared over Twitter or Facebook. However, this method can be expensive, as it can cost up to $1 per message (Franceschi-Bicchierai 2014). Other methods of working around Cuba's restrictions include setting up illegal dial-up connections and sharing access.

Despite Cuba's tightly controlled internet, many activist bloggers and digital journalists have become internationally recognized for their writing on daily life in Cuba and their opinion pieces on the Cuban government. This movement is expanding, despite Cuba's attempts at discouraging it. One of the best-known Cuban bloggers is Yoani Sánchez (1975–). She has been blogging since 2007, when she started her blog, Generation Y (http://www.14ymedio.com/blogs/generacion_y/). She would pose as a tourist to access the unrestricted internet at cybercafés in order to publish her pieces. In 2008, she was named one of the 100 most influential people in the world by *Time* magazine (Hijuelos 2008). She is married to Reinaldo Escobar (1947–), the editor-in-chief of 14ymedio, an online newspaper that they began together in May 2014.

Sánchez is also an active Twitter user. Over the years, the Cuban government has harassed, repressed, and detained Cubans who are viewed as dissident or who are openly critical of the government. On at least two occasions, Sánchez has been arrested for her activities. In October 2012, she and her husband were arrested as they traveled to cover the story of a Spanish man on trial for a fatal car crash that killed another activist. They were released after being held for thirty hours (Orsi 2012). In December 2014, Sánchez and her husband were arrested again, along with several other people, in order to prevent them from taking part in an event at Havana's Revolution Square, where artists were scheduled to perform. The Cuban government had called the event a "political provocation." The arrests were part of a pattern of the Cuban government taking preemptive action to deter activities that it deemed politically dissident. The detainees were released the next day (Trotta 2014).

Generally, in situations like this, the Cuban government does not hold people for very long; however, there are cases of activist artists and writers receiving lengthy prison sentences. This happened with Angel Santiesteban-Prats (1966–), who was tried and convicted in December 2012 on suspicious charges of home violation and injuries (Reporters Without Borders 2014). Santiesteban-Prats is a dissident blogger who has been openly critical of the Cuban government. He started his blog, The Children That No One Wanted (Los hijos que nadie quiso) (https://blogloshijosquenadiequiso.wordpress.com/), in 2008. After serving two-and-a-half years of a five-year sentence, he was released July 17, 2015 (14ymedio 2015). In addition to Generation Y, 14ymedio, and Los hijos que nadie quiso, other notable independent digital journalism outlets have appeared, including On Cuba, Voces Cubanas, and the *Havana Times* (Fortes 2015).

Marilyn J. Andrews

See also: China; Mexico; North Korea; United States; Venezuela

Further Reading

14ymedio. 2015. "Blogger and Activist Angel Santiestaban Released from Prison." July 17. Accessed September 29, 2015. http://www.14ymedio.com/englishedition/Blogger-Activist-Santiesteban-Released-Prison_0_1819018095.html

Alexa. 2015. "Ecured.cu." Accessed September 29, 2015. http://www.alexa.com/siteinfo/ecured.cu

Arce, Alberto, Butler, Desmond, and Gillum, Jack. 2014. "US Secretly Created 'Cuban Twitter' to Stir Unrest." April 4. Accessed September 29, 2015. http://bigstory.ap.org/article/us-secretly-created-cuban-twitter-stir-unrest

DeYoung, Karen. 2014. "Obama Moves to Normalize Relations with Cuba as American is Released by Havana." December 17. Accessed September 29, 2015. https://www.washingtonpost.com/world/national-security/report-cuba-frees-american-alan-gross-after-5-years-detention-on-spy-charges/2014/12/17/a2840518-85f5-11e4-a702-fa31ff4ae98e_story.html

Fortes, Heidi. 2014. "Forbidden Freedom: Yoani Sánchez' Struggle for an Open Cuba." July 3. Accessed September 29, 2015. http://www.cjfe.org/forbidden_freedom_yoani_s_nchez_s_struggle_for_an_open_cuba

Franceschi-Bicchierai, Lorenzo. 2014. "The Internet in Cuba: 5 Things You Need to Know." April 3. Accessed September 29, 2015. http://mashable.com/2014/04/03/internet-freedom-cuba/#Y6WAwUWVEkk5

Freedom House. 2014. "Cuba." Accessed September 29, 2015. https://freedomhouse.org/report/freedom-net/2014/cuba

Freedom House. 2015. "Cuba." Accessed September 29, 2015. https://freedomhouse.org/report/freedom-world/2015/cuba

Hijuelos, Oscar. 2008. "Yoani Sánchez." May 12. Accessed September 29, 2015. http://content.time.com/time/specials/2007/article/0,28804,1733748_1733756_1735878,00.html

Lee, Carol E. 2015. "Barack Obama and Cuban President Raúl Castro Hold Historic Meeting." April 11. Accessed September 29, 2015. http://www.wsj.com/articles/barack-obama-and-cuban-president-raul-castro-hold-historic-meeting-1428783765

Orsi, Peter. 2012. "Yoani Sanchez, Cuban Blogger, Released from Jail." October 6. Accessed September 29, 2015. http://www.huffingtonpost.com/2012/10/06/yoani-sanchez-cuba-blogger_n_1945283.html

Reporters Without Borders. 2014. "Dissident Blogger Completes Year in Detention." February 28. Accessed September 29, 2015. http://en.rsf.org/cuba-dissident-blogger-completes-year-28-02-2014,45939.html

Ruane, Michael E., Shapiro, T. Rees, and Shapira, Ian. 2014. "Amid Jubilation by Relatives and Friends, a 'Hannukkah Miracle' for Alan Gross." December 17. Accessed September 29, 2015. https://www.washingtonpost.com/local/amid-jubilation-marylander-alan-gross-is-released-from-cuban-incarceration/2014/12/17/e257c56e-8607-11e4-9534-f79a23c40e6c_story.html

Scola, Nancy. 2015. "Twitter Wants a Way into Cuba." June 17. Accessed September 29, 2015. http://www.politico.com/story/2015/06/twitter-cuba-social-media-119086

StatsMonkey. 2015. "Mobile Facebook, Twitter, Social Media Usage Statistics in Cuba." Accessed September 29, 2015. https://www.statsmonkey.com/table/21330-cuba-mobile-social-media-usage-statistics-2015.php

Trotta, Daniel. 2014. "U.S. Condemns Cuba's 'Practice of Repressing' After Activists Detained." December 31. Accessed September 29, 2015. http://www.reuters.com/article/2014/12/31/us-cuba-usa-dissidents-idUSKBN0K81DN20141231

We Are Social. 2014. "Digital in the Americas." Accessed September 26, 2015. http://wearesocial.net/blog/2014/06/social-digital-mobile-americas/

DENMARK

The internet, information and communications technologies (ICTs), and social media have become ordinary yet ubiquitous parts of everyday life in Denmark. As most Danes enjoy comparatively low-priced high-speed connections, they have become network ready and mediatized heavy users of social media. As a Scandinavian country located in northern Europe, Denmark is a small constitutional monarchy with a parliamentary democracy, as well as a member of the European Union (EU).

Even though Denmark is considered a small country, with a population of 5.7 million people, territory of 43,000 square kilometers, and only one border with Germany, the country's size has never been a limiting factor with regard to development. Danes often take pride in being citizens in an open, equal, and homogeneous welfare state built on a highly industrialized and modern society with organized labor, liberal markets, high standards of living, a public sector that facilitates a comprehensive health system, a high educational level, a public service and government-funded media system, and nationwide infrastructure. A total of 95.2 percent of Denmark's population have Danish citizenship, speak Danish, and have the same cultural understanding; even though Denmark is a secular society, 83 percent of the population are members of the Danish National Church, which is Lutheran (Nielsen 2016). Although Danes do not always share the same world views, their background, their way of living and traditions, and their values are often very similar.

Although the penetration of internet connectivity is high, the adaptation of ICT is ubiquitous, and social media is an integral part of everyday life, it seems that Denmark is on the verge of an even greater expansion of internet and mobile connectivity, which lead to a more diversified use of social media, new forms of communications, and access to more streaming services within the foreseeable future. However, the promise of a greater and more transformative digital potential is not without challenges for legacy media, political life, and, last but not least, equality and homogeneity within the Danish population.

Danish internet penetration has been stable and high for many years, and, according to Akamai's State of the Internet report in 2016, it is ranked among the best in Europe, as well as globally (Belson 2016). This is not surprising in any way. Connectivity and internet penetration has been on the political agenda for years and is regarded as a strategic part of the development of the Danish infrastructure. In February 2015, most of the parties in the Danish parliament, Folketinget, signed an agreement on the future digitization of Denmark, in which the government is committed to secure high-speed broadband of at least 100 Mbit/s for all households and business, as well as to secure a complete penetration of mobile broadband in all of Denmark by 2020 (Erhvervs- ogVækstministeriet 2015). The agreement's goal is to prepare Danish businesses for structural changes in production in the near future, thereby securing economic growth and supporting and promoting changes in the educational system in order to equip students with digital skills and facilitate adequate learning methods.

Current Danish internet penetration was, in 2015, well above the average penetration within the European Union of 83 percent, and ranged somewhere between 92 percent and 97.4 percent of Danish households for fixed broadband solutions (Elkjær and Tassy 2016, 7; the European Commission 2015, 2; Schrøder and Ørsten 2016, 54; Kulturstyrelsen 2015, 5). It is worth noting that penetration has decreased since 2013, whereas the penetration of smartphones and mobile broadband has grown significantly in recent years. In 2011, smartphone penetration was at 33 percent, but by 2015, its penetration had increased to 77 percent (Lauterbach 2015, 13).

According to the World Economic Forum's yearly report on Global Information Technology, the Danes are the most network-ready people when it comes to individual skills (2016, 88), which they use online to communicate via email, use phone services such as voice over internet protocol (VOIP), search, shop and trade, do their banking, download software, and consume streaming media from services like Netflix, HBO, iTunes, and Spotify (Lauterbach 2015, 17ff). And then, of course, Danes have embraced social media in large numbers and to a degree where it is often described as an ordinary and even integral part of everyday life in Denmark (Rossi, Schwartz, and Mahnke 2016). The overall penetration of social media in Denmark ranges from 65 percent to 76.7 percent of the population for people aged twelve and older with at least one social media profile (Lauterbach 2015; Kulturstyrelsen 2015, 5).

Facebook is, by far, the largest social media platform in Denmark, both in terms of penetration and active usage, and apart from Facebook and YouTube, which have a high penetration but comparatively low usage, most social media sites in Denmark appear small and limited to specific segments of the population. Some social media platforms, like Pinterest, Vimeo, and Tumblr, have an estimated penetration in Denmark of 1 percent or less. The fact that a social media site has a low penetration rate or is used fragmentarily in Denmark does not indicate that it is not important, however. The use of Twitter in Denmark is a good example: It has a very low penetration and low usage, but it is still an important social media site in the extensive exchange of news and views between Danish politicians, journalists, and legacy media, and, as such, it is often referred to as a "media bubble."

For many Danes, Facebook is the social media of choice, and in terms of penetration, the platform has almost 3.6 million active users (Runge 2016) of a total Danish population of 5.7 million people (Danmarks Statistik 2016a). In 2006, Facebook was introduced in Denmark, but the platform's breakthrough happened after it was translated into Danish in 2007, and since then, Facebook's growth in

Facebook: Active Danish users 2010, 2013, and 2016

Facebook: Active Danish users 2010, 2013, and 2016 (Damgaard Nielsen, 2010; 2013; Runge, 2016).

Denmark has been remarkable. In 2008, Lisbeth Klastrup (1970–), a researcher of information technology (IT) at the University of Copenhagen, noted that 460,000 Danes used Facebook (Damgaard Nielsen 2012). By 2010, a comparative study showed that the number of Danes active on Facebook was above 2.2 million (Damgaard Nielsen 2010), and when the study was replicated in 2013, the number had risen to 2.9 million (Damgaard Nielsen 2013).

More women than men use Facebook—51.4 percent are women and 48.6 percent are men (Runge 2016)—and this statistic is similar to the composition of the Danish population, where 50.3 percent are women and 49.7 percent men (Danmarks Statistik 2016b). The Facebook users' locations are also representative of the Danish population, with more than 40 percent being in five Danish cities: the capital, Copenhagen, and the major cities Aarhus, Odense, Aalborg, and Esbjerg (Facebook 2016). There is, however, a considerable, but not surprising, gap when it comes to age: the Danish population is getting older, as current life expectancy is high—82.5 years for women and 78.6 for men (Danmarks Statistik 2016c), and the older generation has more members than the younger generation. The reverse is true on Facebook, however, where young people constitute the majority.

The embrace of social media is not without its challenges, both in a practical and a mundane sense, but also on a structural level. In a practical sense, users still have to adapt to the ever-changing affordances and limitations on social media. On a more structural level, problems and challenges are increasing with the Danish media system. For legacy media, social media may be a valuable channel to distribute content, but it can also be expensive, as large quantities of advertising directed toward social media platforms and search engines can carry high costs. Thus, social media poses a substantial threat to legacy media's earnings. Another, and perhaps more important, challenge for legacy media is that social media is rendering legacy media redundant.

On Facebook and Twitter, politicians have demonstrated over and over that social media is an efficient way to bypass journalists and editors of legacy media, and some of the more prominent politicians on social media are even able to boast larger audiences than most national Danish newspapers and television. In addition, YouTube, along with streaming media services, like Netflix and HBO, seem to have made a substantial impact on Danish television (e.g. overall viewing has dropped by twenty-nine minutes a day from 2010 to 2015). That may not sound like much, but it becomes clear how massive the transition really is when one regards the change among young people, where the young viewers between twelve to eighteen years old have reduced viewing by a massive seventy-one minutes a day to only sixty-five minutes (Christensen 2016).

Although the flux stirred by social media can raise challenges, bypassing the news media or losing viewers are matters that naturally go to the very foundation of legacy media, and, ultimately, they may erode the legitimacy of legacy media and diminish its position within the Danish media system.

Troels Runge

See also: Finland; Germany; Iceland; Poland

Further Reading

Belson, D. 2016. "Akamai's State of the Internet." Q1 2016 Report, Volume 9 / Number 1. Accessed August 5, 2016. https://content.akamai.com/PG6575-q1-2016-soti-connecti vity-report.html

Christensen, D. 2016. *DR Medieforskning: Medieudviklingen 2015*. Copenhagen: Danish Broadcasting.

Damgaard Nielsen, L. 2010. "Facebook Statistik: Sådaner Den Danske Befolkning Fordelt Efter Alder OgKøn." Accessed August 5, 2016. http://www.nettendenser.dk/2010/03/18 /facebook-statistik-sadan-er-den-danske-befolkning-fordelt-efter-alder-og-k%C3%B8n/

Damgaard Nielsen, L. 2012. "Sådan Blev Facebook en Succes." Danish Broadcasting Corporation. Accessed August 5, 2016. http://www.dr.dk/nyheder/indland/saadan-blev -facebook-en-succes

Damgaard Nielsen, L. 2013. "Facebookstatistik 2013 for Danmark: Sådan er Befolkningen Fordelt." Accessed August 5, 2016. http://www.nettendenser.dk/2013/01/25/facebook -statistik-2013-for-danmark-sadan-er-befolkningen-fordelt/

Danmarks Statistik. 2016a. "Population in Denmark." Accessed July 29, 2016. https://www .dst.dk/en/Statistik/emner/befolkning-og-befolkningsfremskrivning/folketal

Danmarks Statistik. 2016b. "FOLK1A: Folketal den 1. i Kvartalet Efter Område, Køn, Alder Og Civilstand." Accessed July 29, 2016. https://www.statistikbanken.dk/statbank5a /SelectVarVal/Define.asp?MainTable=FOLK1A&PLanguage=0&PXSId=0&wsid=cftree

Danmarks Statistik. 2016c. "FaldiKvindersLevetid for Første Gang i Over 20 år." Nyt Fra Danmarks Statistik. February 16. Accessed July 29, 2016. http://www.danmarksstatistik .dk/da/Statistik/NytHtml?cid=20889

Elkjær, Kamilaa, and Tassy, Agnes. 2016. "It-anvendelseiBefolkningen—EU-sammenligninger 2015." Danmarks Statistik. Accessed July 29, 2016. http://www.dst.dk/Site/Dst/Udgiv elser/GetPubFile.aspx?id=20743&sid=itanvbefeu2015

Erhvervs- ogVækstministeriet. 2015. "Aftale: Vækstplan for Digitalisering." Aftale: Erhvervs-
    ogVækstministeriet. Accessed August 5, 2016. https://www.evm.dk/aftaler-og-udspil
    /15-02-26-aftale-om-vaekstplan-for-digitalisering
European Commission. 2015. "Denmark, October 2015." Excerpted from Special Euroba-
    rometer 438: E-Communications and the Digital Single Market. October 2015." Euro-
    pean Commission: Brussels. Accessed August 5, 2016. http://ec.europa.eu/information
    _society/newsroom/image/document/2016-22/eb_ecommunications_and_the_digital
    _single_market_dk_en_15839.pdf
Facebook. 2016. "Audience Insights." Accessed July 19, 2016. https://www.facebook.com
    /ads/audience-insights
Hjarvard, S. 2013. *The Mediatization of Culture and Society.* New York: Routledge.
Kulturstyrelsen: MediernesudviklingiDanmark. 2015. "SOCIALE MEDIER. BRUG, INTER-
    ESSEOMRÅDER OG DEBATLYST." Kulturstyrelsen. Accessed July 30, 2016. http://
    slks.dk/fileadmin/user_upload/dokumenter/medier/Mediernes_udvikling/2015
    /Specialrapporter/Sociale_medier/PDF-filer_dokumenter/SPECIALRAPPORT_2015
    _SOCIALE_MEDIER_ENDELIG.pdf
Lauterbach, T. 2015. "It-anvendelseiBefolkningen—2015." Danmarks Statistik. Accessed
    July 29, 2016. http://www.dst.dk/da/Statistik/Publikationer/VisPub?cid=20737
Nielsen, B. 2016. "Danmark." Den Store Danske, Copenhagen: Gyldendal. Accessed
    August 4, 2016. http://denstoredanske.dk/Danmarks_geografi_og_historie/Danmarks
    _geografi/Danmark_generelt/Danmark
Rossi, L., Schwartz, S., and Mahnke, M. 2016. "Social Media Use & Political Engagement in
    Denmark, Report 2016." Decidis Research Group at the IT University Copenhagen. Ac-
    cessed July 29, 2016. https://blogit.itu.dk/decidis/2016/03/10/slides-decidis-survey/
Runge, T. 2016. "Danskerneog Facebook 2016, Copenhagen: Digitalkommunikation.net."
    Accessed August 8, 2016. http://digitalkommunikation.net/danskerne-og-facebook
    -2016/
Schrøder, K. C., and Ørsten, M. 2016. "Denmark." *Digital News Report 2016,* Reuters Insti-
    tute for the Study of Journalism. Accessed July 29, 2016. http://www.digitalnewsreport
    .org/survey/2016/denmark-2016/
Slots-og Kulturstyrelsen. 2016a. "Mediernesudviklingi Danmark 2016: Internetbrugogen-
    heder." Slots-og Kulturstyrelsen. Accessed July 29, 2016. http://slks.dk/mediernes-ud
    vikling-2016/internetbrug-og-enheder/
Slots-og Kulturstyrelsen. 2016b. "MediernesudviklingiDanmark 2016: TV." Slots-og Kul-
    turstyrelsen. Accessed July 29, 2016. http://slks.dk/mediernes-udvikling-2016/tv/

ECUADOR

Ecuador is a country of roughly 15 million people situated in the northwestern portion of South America. It currently has a large internet user base, but that was not always the case. Internet penetration lagged behind the rest of the world but has accelerated in recent years to the point where today 75 percent or more of the population regularly uses the internet and social media. Many users access the internet from a desktop PC, but over half of them connect using a laptop or mobile device. Few domestic social media services are offered or widely used. Most Ecuadorans prefer to use Facebook, WhatsApp, Twitter, Skype, and YouTube. In spite of the popularity of social media and the large percentage of internet users, crackdowns on freedom of speech have had a chilling effect on social media usage.

The growth of the popularity of social media has lagged behind most other countries, and it is paired with the growth of infrastructure and access to technology. From 1996 to 1998, only 0.1 of 100 people used the internet in Ecuador. Those numbers rose to 0.8 people per 100 in 1999 and 1.5 percent in 2000. Between 2001 and 2006, the percentage of internet users rose from 2.7 percent to 6.0 percent (World Bank 2016).

In 2010, Ecuador accounted for 1.5 percent of total internet users in South America. In that year, the internet was used by 16 percent of the population after experiencing an extreme growth rate of 1,211 percent from 2000 to 2010. The most popular search engine in 2010 was Google, with 96.89 percent of users preferring it over other search engines (Global Search Engine Marketing 2011). In 2011, internet users grew to 27.2 percent of the population, and by the year 2014, the percentage of citizens who used the internet skyrocketed to 74.4 percent (Internet World Stats 2014).

Mobile users are also at an all-time high for Latin American countries in 2015, with 14 percent of citizens using 3G+4G devices, 26 percent using 2G, 23 percent

---

In Los Rios, Ecuador, an officer with the Transit Authority had an unusual experience in early 2016. After traffic had stopped, the traffic officer discovered a sloth clinging to a metal pole by the side of the road, appearing as if it wanted to cross. The Transit Authority took pictures of the scared sloth before and during the rescue. When those photos hit Facebook, they went viral; one photo alone generated over 800,000 "likes." After the rescue, a veterinarian checked the sloth and determined it fit to return to its habitat.

using voice only, and 36 percent not using a mobile device at all (GSMA Intelligence 2016).

Key improvements to infrastructure have helped the growth of mobile phone penetration in spite of state-owned intervention in the marketplace. State-owned companies are sometimes given preference over privately owned entities. In 2012, the state-owned CNT Mobile (Corporación Nacional de Telecomunicaciones) was awarded a 4G spectrum, while its competitors Movistar and Claro were refused one. Yet Telefónica deployed approximately 100 solar-powered base stations to improve network coverage for rural areas of Ecuador (GSMA Mobile 2014).

Even though mobile infrastructure continues to grow, landline broadband provided by internet service providers (ISPs) continues to have a strong presence in the country, with CNT Mobile accounting for 51.1 percent of the market, Grupo TV Cable accounting for 29.2 percent, and Claro serving the remaining 9.4 percent (ITU 2013). In 2015, a $300 million investment project neared completion: The Pacific Caribbean Cable System (PCCS) will provide 100-Gbps service to internet users in the country (El Telegrafo 2015).

Facebook is a very commonly used social media site. According to an April 2015 poll by the Ecuadorian Instituto Nacional de Estadisticas y Censos (INEC), 98 percent of Ecuadorian citizens over the age of twelve have a Facebook account. Approximately 96.8 percent of Guayaquil residents have a Facebook account, and 97.9 percent of Quito residents have a Facebook account (Sandoval 2015). Guayaquil, a coastal city in the province of Guayas, is the largest city in the country, eclipsing the population of the capital, Quito, by a few hundred thousand residents. Guayaquil also has eclipsed Quito in terms of the percentage of social network usage. Twitter is quite popular; 29.3 percent of Guayaquil residents and 21.4 percent of Quito's residents use it. WhatsApp is even more popular; 48.6 percent of Guayaquil residents and 37.4 percent of Quito's residents use it. Further, 15.1 percent of Guayaquil residents and 17.5 percent of Quito residents use Skype.

All social media users reported their reason for using each software or website. According to those results, Facebook is used primarily for entertainment (53.4 percent) and reading the news (18.1 percent), followed closely by keeping in touch with family and friends (14.7 percent). Conversely, Twitter is used primarily for reading the news (53.1 percent) and entertainment (12.3 percent). Roughly 30 percent of social media users used a desktop computer to access websites or software; meanwhile, 21.7 percent used a laptop, 6 percent used a tablet, and 36.1 percent used a smartphone (Sandoval 2015).

In spite of the overwhelming use of portable devices and laptops to access social media, an overwhelming 78.4 percent of users claim that they access social media only while they are at home. Approximately 13.5 percent access social media while they are at work (Sandoval 2015).

Home usage may explain why Gmail, Google's email service, is the most popular email provider in the country. Gmail's current market share is 66 percent, which far overshadows the 15.81 percent market share of Microsoft Outlook, 5.67 percent market share of Zimbra, and 4.21 percent market share of Hotmail (Datanyze 2016).

Yet in spite of these statistics, censorship and fear of government reprisal have had a chilling effect on social media usage. In July 2011, the editor of the newspaper *El Universo*, Emilio Palacio (1954–), was found guilty of libel for publishing an article online and in its print edition implying that President Rafael Correa (1963–) was a dictator and a liar. All four men accused, including Palacio and three top executives at the newspaper, received three-year jail sentences, and the newspaper was fined $40 million. In order to bring the libel suit, Correa used an obscure part of Ecuador's penal code, Article 230, which prohibits disrespect of the president (Lauderbaugh 2012, 165). In November 2012, an Ecuadorian blogger named Paul Moreno (n.d.–) identified a security hole in the nation's national identity database and posted information about it on the internet. The victim of Moreno's hacking test was Correa himself. The blogger was arrested and held for forty-five days, which caused a popular outcry. After the outcry, President Correa ordered him released from prison (Conan 2012)

Furthermore, President Correa passed a $42 tax on all international online purchases in October 2014. The goal was to prevent Ecuadorian citizens from bypassing their own local stores and sellers in favor of dealing with international retailers (Goldman 2015).This was a continuation of the encroachment on freedom of speech, keeping internet users from feeling that they could comment openly about the president or the government via a bill that regulates media in Ecuador. Article 474 of the penal code (Código Orgánico Integral Penal) was eventually rescinded by the government, but only after large public outcry (Freedom House 2014).

Article 16.2 of the Ecuadoran constitution guarantees "universal access to information technologies and communication" (Constitute Project 2016) yet laws have been passed that require all content to be properly verified or else the content creators would be fined. This provision, Article 11 under the Organic Law on Communications, was used to fine a humorist, and the newspaper *El Universo* for publishing a cartoon satirizing a police raid on the home of Francisco Villavicencio (n.d.–). Journalist Juan Carlos Calderon (c. 1972–) received death threats for launching a digital magazine in 2013 that exposed fraud in state institutions (Solano 2013). In January 2014, a government representative allegedly targeted and threatened a writer and former TV news anchor Carlos Vera (1955–) on Twitter. Vera subsequently abandoned his blog Polificcion and online political debate (Freedom House 2014). A law passed in June 2013 created two media regulatory bodies, the Information Authority and the Communications Regulation Council, which can levy fines on media outlets or writers who are found guilty of slander or discrimination. This also extends to speech on social media.

Jeffrey M. Skrysak

See also: Colombia; Ecuador; Peru

Further Reading

Conan, Matt. 2012. "Blogger Jailed After Password-Hacking Ecuador's President." December 3. Accessed March 11, 2016. http://www.wired.com/2012/12/security-post-lands-ecuadorian-blogger-in-jail/

Constitute Project. 2016. "Ecuador Constitution of 2008." Accessed March 11, 2016. https://www.constituteproject.org/constitution/Ecuador_2008.pdf

Datanyze. 2016. "Gmail Market Share in Ecuador." March 13. Accessed March 13, 2016. https://www.datanyze.com/market-share/email-providers/Ecuador/gmail-market-share

El Telegrafo. 2015. "Ecuador Tendrá Internet con 100 Megas de Velocidad." May 4. Accessed March 13, 2016. http://www.eltelegrafo.com.ec/noticias/economia/1/ecuador-tendra-internet-con-100-megas-de-velocidad

Freedom House. 2014. "Freedom on the Net 2014—Ecuador." Accessed March 11, 2016. https://freedomhouse.org/sites/default/files/resources/Ecuador.pdf

Global Search Engine Marketing. 2011. "Internet and Search Engine Usage by Country." Accessed March 11, 2016. http://ptgmedia.pearsoncmg.com/images/9780789747884/supplements/9780789747884_appC.pdf

Goldman, Alex. 2015. "Ecuador's President Will Respond to You on Twitter." May 21. Accessed March 13, 2016. http://digg.com/2015/ecuadors-president-does-all-his-own-tweets

GSMA Intelligence. 2016. "Connected Society—Content in Latin America: Shift to Local, Shift to Mobile." February. Accessed March 13, 2016. http://www.gsma.com/latinamerica/resources

GSMA Mobile. 2014. "The Mobile Economy—Latin America 2014." Accessed March 12, 2016. http://www.gsmamobileeconomylatinamerica.com/GSMA_Mobile_Economy_LatinAmerica_2014.pdf

Internet World Stats. 2014. "Ecuador Internet Usage Population and Telecom Reports." May 29. Accessed March 10, 2016. http://www.internetworldstats.com/sa/ec.htm

ITU Telecommunication Development Bureau. 2013. "International Internet Connectivity in Latin America and the Caribbean." March. Accessed March 12, 2016. https://www.itu.int/en/ITU-D/Regulatory-Market/Documents/Internationalpercent20Internetpercent20Connectivitypercent20inpercent20Latinpercent20Americapercent20andpercent20thepercent20Caribbean.pdf

Lauderbaugh, George. 2012. *The History of Ecuador*. Santa Barbara, CA: Greenwood.

Sandoval, Carla. 2015. "Facebook es la Red Social másUtilizada en el Ecuador y No TieneCompetencia." April 15. Accessed March 11, 2016. http://www.elcomercio.com/tendencias/facebook-redessociales-ecuador-inec-usuarios.html

Solano, Gonzalo. 2013. "Journalists: Death Threats, Intimidation Mark Rising Hostility against Press in Ecuador." December 7. Accessed March 12, 2016. http://www.foxnews.com/world/2013/12/07/journalists-death-threats-intimidation-mark-rising-hostility-against-press-in.html

World Bank. 2016. "Internet Users (per 100 People)." Accessed March 14, 2016. http://data.worldbank.org/indicator/IT.NET.USER.P2/countries/1W?page=3&display=default

EGYPT

The most populous Arab country, Egypt has historically played a significant role in the Middle East, setting trends across the region in media and culture. While the golden age of Egyptian cinema has long passed, the country remains a major influence, even as the media of communication and consumption have changed. Despite the relatively late adoption of internet and social media, persistently low

An Egyptian man uses a laptop in Tahrir Square during an anti-government demonstration in 2011. Social media played a pivotal role in the uprising that led to the Egyptian Revolution in the same year, culminating in the end of President Hosni Mubarak's thirty-year rule. (Joel Carillet/iStockphoto.com)

penetration rates, and slow internet speeds, Egypt's social media landscape is particularly dynamic. This was most evident in the prominent role that social media played in the country's 2011 revolution; however, it also manifests itself in the frequency and ways in which Egyptians use social media platforms to collect, communicate, and express ideas.

While the numbers remain low, internet usage has spiked dramatically in recent years, as the government has worked to implement policy and achieve universal access. In 2002, the Ministry of Communications and Information Technology launched the Free Internet Initiative, providing internet at the same price as local phone calls. At the same time, the ministry and other private and public stakeholders implemented plans to make PCs and internet connection available to all households and communities. With the majority of the population living in rural areas, the aims of these projects include raising computer literacy levels and reducing the digital divide between these areas and urban centers, particularly the capital, Cairo (Ministry of Communications and Information Technology 2015).

The effect of these and other initiatives reflect clearly in the numbers. In 2000, Egypt had less than half a million internet users. By 2004, that number had jumped to 1.5 million, and in 2008, there were almost 14 million internet users (Vargas 2012), and by 2016, that number had spiked to over 45 million, representing a penetration rate of approximately 50 percent (Abdulla 2016). Still, the numbers remain low relative to the overall population. And despite government efforts, the numbers indicate the infrastructure is still lacking. According to the Internet Society, an organization that tracks and measures internet access and speeds, Egypt ranked 127th out of the 144 countries measured for download speeds (Internet Society 2016).

With a population of over 90 million (as of 2016) and a median age of around twenty-five, Egypt's young people are becoming increasingly engaged on digital media platforms. A high youth unemployment rate, which has risen steadily since well before the 2011 uprising (World Bank 2016), has inevitably affected the frequency and ways in which users engage on these platforms.

The 2011 uprising that led to the end of former president Hosni Mubarak's (1928–) thirty-year rule was dubbed by many as a Facebook, Twitter, or social media revolution due to the pivotal role that these platforms played in organizing and coordinating the protests that ultimately led to the change in regime. In the years leading up to the so-called Arab Spring, which saw mass protests across the Middle East and North Africa that in most cases resulted in regime change, civil war, or mass suppression, many scholars had begun to identify the potential impact of the internet and social media as an instrument for collective action. This action originally took the form of blogging, eventually shifting to more interactive websites like Facebook and Twitter.

It is widely believed that what tipped the balance and helped spur the events of 2011 in Egypt was the death of a young man named Khaled Said (1982–2010), who had been placed in police custody the previous year. Years of police brutality, growing social inequality, and inaccessibility of many basic goods and services had led to wide-scale disillusionment with the government. In June 2010, after Said's death, a young Google executive named Weal Ghonim (1980–) started a Facebook page called Kulena Khaled Said ("We are all Khaled Said"). The page, with 32 million "likes," 1.9 million active users, and widespread engagement by Egyptians, ended up being one of the main organizing platforms for the January 25 revolution (Abdulla 2016). It was largely through exposure to Facebook pages like this that protests in the form of silent stands were organized; in many silent stands, protestors wearing black clothes chained themselves together and stood silently, sometimes reading a Koran or Bible, standing at least five meters apart so as not to break Egyptian laws outlawing mass assemblies. As a result, as early as January 26, 2011, the Egyptian government began shutting down websites, including Facebook and Twitter, to minimize the spread of the unrest.

For the last several decades, Egypt's social and political system has provided less and less space for youths to have a voice. As the young population has grown, they have become increasingly disenfranchised. Educated but unable to find employment, and without access to decision makers to have their grievances heard, social media increasingly became a place for young people to engage. However, as quickly as this new space for dissent emerged, it has in recent years become riskier to use.

On June 30, 2013, exactly one year after the inauguration of Mohamed Morsi (1951–), Egypt's first democratically elected president and a member of the Muslim Brotherhood, mass protests erupted as millions of Egyptians gathered in the streets, demanding his immediate removal for what many viewed as violations of democracy taken during his time in office. These protests, widely recognized as the largest in the country's history despite varying estimates, were also organized via social media for the most part. On the evening of July 3, General Abdel Fatah el-Sisi (1954–), announced that Morsi had been deposed.

In the period following the removal of Morsi, several measures were taken that were widely viewed as limiting freedom of expression. While social media and other digital platforms remained among the most open environments to voice dissent because they are harder to regulate and individuals are harder to track, these new measures have affected how Egyptians engage online. In November 2013, under the governance of interim president Adly Mansour (1945–), a controversial protest law was instituted that significantly limited freedom of assembly and provided security forces with broader reach to disperse protests forcibly. While the law more directly targeted offline engagement, it inevitably influenced the type of online organization that had been so pivotal in the three years prior.

While this and other actions indirectly affected social media activity in the country, it was announced in 2014 that broad surveillance of the most widely used social media platforms would be introduced. With the technical support of a private company, SEE Egypt, operating under the umbrella of the larger U.S.-based cyber-security firm Blue Coat, the Ministry of Interior announced that it would begin broad surveillance of social media platforms, including Facebook, Twitter, Skype, WhatsApp, and Viber (Frankel 2014).The official purpose of this surveillance was to combat terrorism, portrayed as a necessary measure in light of numerous threats and attacks attributed to both members and supporters of the banned Muslim Brotherhood (who officially gained status as a terrorist organization in late 2013) and local affiliates of the so-called Islamic State organization. However, it was widely reported that these measures were also taken to protect the moral fabric of Egyptian society and monitor the online activity of certain groups. This type of surveillance ultimately affects Egyptians active on social media in terms of how they engage. As a result, self-censorship has been observed and discussed widely, as the spaces for political discourse have shrunk.

According to the Middle East Media Survey, an annual publication of Northwestern University in Qatar, in 2015, Egyptians were most active on Facebook, WhatsApp, and YouTube, representing 87 percent, 58 percent, and 49 percent of internet users, respectively. The same survey also found that of internet users in Egypt, 50 percent use online platforms to increase contact with friends and family, whereas 26 percent use them to connect with individuals with the same religious beliefs, and 29 percent use them to connect with individuals with the same political beliefs.

Despite slow internet speeds and still low penetration rates, those Egyptians using social media are extremely active. According to the social media analytics company Socialbakers, as of 2012, Egypt was among the most "Facebook-addicted" countries in terms of posting (second only to Brazil).

Egyptians have long been known for their colorful sense of humor. Limits on many forms of freedom of expression, combined with social and political grievances, have fostered a tradition of humor as an outlet for much of society. With the emergence of new media, this has extended beyond jokes told between friends and family in coffee shops and homes to online memes. These can vary from mocking major events to making light of some of the inefficiencies and oddities in day-to-day life in the country, often opening discussions about the boundaries of tasteful comedy. For instance, when a man hijacked a plane in May 2016, forcing it to land in Cyprus (allegedly so he could reunite with his estranged wife), many Egyptians immediately jumped on the opportunity to make light of the situation, posting images and jokes on Facebook and Twitter. Some used Photoshop to manipulate the Egyptair logo, changing the tagline to "Egyptair: We Fly You to Your Love," while others quipped that the hijacker was unable to email his wife due to slow internet connectivity in the country. There are countless examples of similar viral memes, from an Egyptian woman caught on camera saying "Shut up your mouse, Obama" in 2014, to donkeys breaking into Cairo International Airport, to Photoshopped images making light of the devastating flood in Alexandria in late 2015.

The introduction of the internet, and more specifically social media, into Egyptian society has had a massive impact. While it initially created new space for discussion and opposition, this space has been made more restricted since. Still many analysts and media scholars see this shift as irreversible, particularly in a country like Egypt, with consistently growing penetration rates and a young population.

Sarah El-Shaarawi

See also: Brazil; Libya; Syria; Tunisia; Yemen

Further Reading

Atef, Mageg, and Frenkel, Sheera. 2014. "Egypt Begins Surveillance of Facebook, Twitter, and Skype on Unprecedented Scale." *Buzzfeed*, September 17. Accessed July 30, 2016. https://www.buzzfeed.com/sheerafrenkel/egypt-begins-surveillance-of-facebook -twitter-and-skype-on-u?utm_term=.rd7VxKrA6#.adzV63pLm

Arab Republic of Egypt Ministry of Communications and Information Technology. 2015. "WSIS+10 Overall Review of the Implementation of the WSIS Outcomes Profiles of Progress." World Summit on the Information Society. Accessed July 30, 2016. http:// www.mcit.gov.eg/Upcont/Documents/Profiles-of-Progress.pdf

Arab Social Media Influencers Summit. 2015. "Arab Social Media Report." Accessed July 30, 2016. http://dmc.ae/img/pdf/white-papers/ArabSocialMediaReport-2015.pdf

Faris, David. 2013. *Dissent and Revolution in a Digital Age: Social Media Blogging and Activism in Egypt*. London: I. B. Tauris.

Hughes, Thomas, and Mubarak, Emad. 2014. "Censorship in Egypt: Online and Offline." *Madamasr*, November 30, Accessed July 30, 2016. http://www.madamasr.com/opinion /politics/censorship-egypt-online-and-offline

International Labour Organization. "Unemployment Youth Total." World Bank. Accessed July 30, 2016. http://data.worldbank.org/indicator/SL.UEM.1524.ZS

Internet Society. "Global Internet Maps." Accessed July 30, 2016. http://www.internetsociety .org/map/global-internet-report/?gclid=Cj0KEQjwt-G8BRDktsvwpPTn1PkBEiQA -MRsBYsc_ntfS1okNJGbS4GkuFVVtoUz1_SG2qlXJwEfuucaArfI8P8HAQ#download -speed-fixed

Khamis, Sahar, and Vaughn, Katherine. 2011. "Cyberactivism in the Egyptian Revolution: How Civic Engagement and Citizen Journalism Tilted the Balance." *Arab Media & Society*, 14. Accessed July 30, 2016. http://www.arabmediasociety.com/?article=769

Northwestern University in Qatar. 2015. "Media Use in the Middle East, 2015." Accessed July 30, 2016. http://www.mideastmedia.org/survey/2015/

Socialbakers. 2012. "10 Most Facebook-Addicted Countries." October 16. Accessed July 30, 2016. https://www.socialbakers.com/blog/961-10-most-facebook-addicted-countries-on-facebook

Vargas, Jose Antonio. 2012. "Spring Awakening: How an Egyptian Revolution Began on Facebook." *The New York Times*, February 17. Accessed July 30, 2016. http://www.nytimes.com/2012/02/19/books/review/how-an-egyptian-revolution-began-on-facebook.html?_r=0

World Bank. 2016. "Egypt: Unemployment, Total (% of Labor Force)." Accessed January 8, 2017. http://data.worldbank.org/indicator/SL.UEM.TOTL.ZS?locations=EG

EL SALVADOR

Located along the Pacific coast of Central America, the Republic of El Salvador is nestled between Guatemala and Honduras. The tiny country of over 6 million inhabitants endured twelve years of a devastating civil war, from roughly December 10, 1981, to December 15, 1992. At the war's conclusion, the country embarked upon a path of reconciliation and economic reconstruction, including the country's informational infrastructure, which has led to one of Central America's smallest and most densely populated countries ranking higher in networked readiness than many other Central American countries (World Economic Forum 2016). Today, despite the legacy of years of war and a history of devastating national disasters in the form of hurricanes and earthquakes, there are approximately 2.352 million internet users in El Salvador (Internet Live Stats 2016).

Salvadorans abroad, especially in the United States, play a large role in the economic and technological development of the country. In response to the civil war, many Salvadorans fled to other countries, such as Guatemala, Mexico, Honduras, and the United States. Today, over 2 million people of Salvadoran origin live in the United States. In addition to contributing to the global marketplace as users of the internet and consumers of mobile communications technology, these expatriates engage in the practice of sending "remittances," the transfer of money by a person in one country to family and friends in another, back to family members in El Salvador. Remittances sent from the United States to El Salvador have increased every year for decades. At the end of the civil war, Salvadorans received less than $1 billion in remittances; however, this total nearly tripled in the succeeding fifteen years (Gammage 2007). In 2013, the country received $3.965 billion (90 percent of which originated from the United States), which accounted for 16.5 percent of El Salvador's annual gross domestic product (GDP) (Pew Research Center 2013). By 2015, this total had increased to $4.27 billion (Trading Economics 2016). Some have suggested that remittances have become one of the most effective tools for reducing poverty and inequality in El Salvador.

Remittances go far beyond the personal support of friends and relatives. Some of these funds go toward purchasing computers for classrooms in their former communities. In other cases, money has been donated to build or rebuild soccer stadiums. In Intipuca, a former cotton farming community with fewer than 10,000 inhabitants whose industry collapsed during the economic crash of the 1960s, new business has been introduced as a result of money that flows into the town through foreign remittances. For example, there are five money transfer agencies, a travel agency booking flights to the United States, and an internet café (Elton 2002). While the U.S. dollar had been used alongside the Salvadoran Colon for years in communities like Intipuca, on January 1, 2001, the U.S. dollar became the official currency of El Salvador.

In accord with the policy of reconciliation and reconstruction established after the conclusion of the civil war, Armando Calderón Sol (1948–), who became El Salvador's first democratically elected president on June 1, 1994, began advocating for privatized telecommunications companies in an effort to stimulate the economy and spur the development of the country's information infrastructure, paving the way for internet access. Unfortunately, development efforts suffered severe setbacks as a result of the devastation that followed Hurricane Mitch in 1998 and a severe earthquake in 2001. Reports of people waiting ten years to have a telephone installed may provide answers as to why fixed-line telephony penetration had only reached 15 percent by 2013 (ITU 2013). On the other hand, mobile telephony penetration, which had reached 19.3 percent in 2004, had grown enormously to a rate of 136 percent by 2013 (Internet World Stats 2014; ITU 2013). Despite access to mobile telephones, access to the internet remained limited. By the year 2000, there were 68,428 internet users in El Salvador. During the next five years, the number of internet users more than tripled, to 249,783. In 2013, only 6 percent of the population had access to mobile broadband and national internet penetration had only reached 23 percent (ITU 2013).

Many individuals working to enhance and sustain El Salvador's information infrastructure believe that technologies such as the internet, personal computers, wireless or mobile communication, along with increased access to education, are the answers to El Salvador overcoming its "lost decade" and become an active participant in the global network. For example, Salvadoran poet and novelist Manlio Argueta (1935–) returned to El Salvador after twelve years of exile in Costa Rica. Argueta believed that books, reading, and libraries were essential elements in the reconstruction of El Salvador. He also recognized the value of the internet and libraries as major components of El Salvador's efforts to move forward (McPhail 2000).

As of January 1, 2016, internet penetration in El Salvador was approximately 38 percent of the population (Internet Live Stats 2016). These figures do not take into account the millions of Salvadorans abroad that use the internet to maintain regular contact with friends, relatives, and business associates. Similar to other countries in Central America, the majority of internet users in El Salvador access the internet with their cell phones. Initially, the majority of internet users lived in one of El Salvador's fourteen largest cities, like San Salvador, with a population of 1.7 million people. However, with the advent of foreign investments from countries like France and the United States, the introduction of mobile technology offering free Wi-Fi, and social media like Facebook and Twitter in predominantly rural areas, these demographics changed drastically. In fact, although El Salvador is the smallest and perhaps one of the most impoverished countries in Central America, by the end of 2008, it had the largest number of cell phones per person in the region, with 6.6 million for a population of 5.8 million people, the estimated number of residents at that time (Gutierrez 2009).

Several Salvadoran daily news outlets dominate the rankings of most visited websites in El Salvador: namely, El Blog, El Salvador, La Prensa Gráfica, and La Página (Alexa 2016). The top two social networking websites are Facebook, dominating social media usage with 92.6 percent penetration; and Twitter, with 6.89 percent penetration (StatsMonkey 2015). Not only does Facebook allow users to connect with family and friends, upload photos, and share links, the platform has recently added a new feature that allows users to send and receive money online via a secure connection. Other social media sites used in El Salvador are YouTube and Google+. Top Facebook pages reflect overall internet usage, with some of the most visited sites belonging to popular Salvadoran news outlets. Salvadorans' reasons for using the internet are communicating, watching and uploading videos, sending remittances, taking classes, searching for information, and promoting local businesses.

In addition to the typical uses of El Salvador's most popular social media website, many organizations and programs have found Facebook to be the most successful method of reaching target audiences. For example, Carana Corporation's nongovernmental organization (NGO) project, the Improving Access to Employment Program, aimed at young Salvadorans seeking employment found success only after creating a Facebook page. In the two-year lifespan of the project, their Facebook page grew from 800 to 65,000 fans, and by 2014, forty companies were sharing multiple job opportunities every week (Taurasi 2014).

Other uses of social media have been less positive for Salvadorans. For some time, gang violence in El Salvador has been a major national issue; in the first three months of 2016, there was one murder per hour in the country on average (Harris et al. 2016). Access to social media allowed gang members to glorify gang life through images of violence and other and symbols like music, modes of dress, and behavior that romanticizes the life of the outlaw in society. In June 2016, El Salvador's director of prisons, Rodil Hernández (n.d.–), announced that inmates at prisons filled with gang members had been accessing Wi-Fi signals to use social media sites such as WhatsApp to control gang activity outside the prison, including extorting victims and issuing death warrants. Hernández did not explain how inmates had been able to access the internet but explained that in the past, family members had aided inmates by moving close to the prisons to provide them with internet signals (Alonso 2016; López 2016).

John G. Hall and Marilyn J. Andrews

See also: France; Honduras; Mexico; United States

Further Reading

Alexa. 2016. "Top Sites in El Salvador." Accessed August 21, 2016. http://www.alexa.com/topsites/countries/SV

Alonso, Luis Fernando. 2016. "El Salvador Inmates Using WhatsApp for Extortion: Official." *InSight Crime*. Accessed August 21, 2016. http://www.insightcrime.org/news-briefs/el-salvador-inmates-using-whatsapp-extortion-official

Elton, Catherine. 2002. "Remembering Their Roots; Emigrants to U.S. Maintain Strong Ties to Hometown." *Washington Times*, March 26. Accessed August 21, 2016. https://www.questia.com/article/1G1-84182418/remembering-their-roots-emigrants-to-u-s-maintain

Gammage, Sarah. 2007. "El Salvador: Despite End to Civil War, Emigration Continues." *MPI: Migration Policy Institute*. Accessed August 21, 2016. http://www.migrationpolicy.org/article/el-salvador-despite-end-civil-war-emigration-continues

Gutierrez, Raúl. 2009. "El Salvador: Central America's Leader in Cell Phone Use." January 14. Inter Press Services News Agency. Accessed August 21, 2016. http://www.ipsnews.net/2009/01/el-salvador-central-americarsquos-leader-in-cell-phone-use/

Harris, Dan, Desiderio, Adam, Milliman, Jenna, and Effron, Lauren. 2016. "In El Salvador, the Murder Capital of the World, Gang Violence Becomes a Way of Life." *ABC News*. Accessed August 21, 2016. http://abcnews.go.com/International/el-salvador-murder-capital-world-gang-violence-life/story?id=39177963

International Telecommunications Union (ITU). 2013. "El Salvador Profile." Accessed August 21, 2016. https://www.itu.int/net4/itu-d/icteye/CountryProfileReport.aspx?countryID=210

Internet Live Stats. 2016. "El Salvador Internet Users." Accessed August 21, 2016. http://www.internetlivestats.com/internet-users/el-salvador/

Internet World Stats. 2014. "El Salvador." Accessed August 21, 2016. http://www.internetworldstats.com/am/sv.htm

López, Jaime. 2016. "Reos Siguen Comunicándo se por Internet." *Noticias de El Salvador*, June 24. Accessed August 21, 2016. http://www.elsalvador.com/articulo/sucesos/reos-siguen-comunicandose-por-internet-116913

McPhail, Martha E. 2000. "After the War in El Salvador." *American Libraries,* 31:1. Accessed August 21, 2016. https://www.questia.com/read/1G1-59013163/after-the-war-in-el-salvador

Pew Research Center. 2013. "Remittances Received by El Salvador." Accessed August 21, 2016. http://www.pewhispanic.org/2013/11/15/remittances-to-latin-america-recover-but-not-to-mexico/ph-remittances-11-2013-a-08/

StatsMonkey. 2015. "El Salvador Social Media Usage Statistics Using Mobile." Accessed August 21, 2016. https://www.statsmonkey.com/sunburst/21339-el-salvador-mobile-social-media-usage-statistics-2015.php

Taurasi, Lynda-Marie. 2014. "Social Media for Development: Making the Most of Facebook." *The Guardian*, June 2. Accessed August 21, 2016. https://www.theguardian.com/global-development-professionals-network/2014/jun/02/social-media-facebook-case-studies

Trading Economics. 2016. "El Salvador Remittances." Accessed August 21, 2016. http://www.tradingeconomics.com/el-salvador/remittances

World Economic Forum. 2016. "Network Readiness Index." Accessed August 21, 2016. http://reports.weforum.org/global-information-technology-report-2015/network-readiness-index/

ESTONIA

Estonia, located on the northeastern edge of the European Union, is a tiny Baltic state that has become known as one of the most wired and digitally advanced societies in the world. Currently, "Estonia has a technological ecosystem in which almost every regular daily activity of public life provides the impetus for its transformation into an e-service" (Kotka, Alvarez del Castillo, and Korjus 2015, 2), earning the country the label of "e-Estonia." The country's success story of transitioning from a postsocialist country to a modern democracy has been built upon the "internetization" of the society, the result of a partnership among a forward-thinking government, a proactive information and communication technology (ICT) sector, and a tech-savvy population of 1.3 million people.

When the country regained independence in 1991 after nearly fifty years of Soviet rule, Estonia's new leadership initiated an ambitious agenda for administrative reform, aiming to build a tech-savvy society that would be competitive in the world. By the end of 2015, 939 institutions use the x-road system (i.e. a technological and organizational environment enabling secure internet-based data exchange between information systems), daily in Estonia and 1,723 different e-services, including digital identification, digital signature, e-voting, electronic tax filing, online medical prescriptions, internet voting, online banking, and mobile parking were made available to Estonian citizens (Introduction of X-Road 2016). For instance, filing an online tax declaration takes on average five minutes, participating in elections via internet can be done in ninety seconds, and it takes about fifteen minutes to create a new company.

The Estonian ICT success story is largely made possible due to the nature of its infrastructure, which was designed as a decentralized system. In 2000, the Estonian Parliament passed the Digital Signature Act, which made the digital signature equivalent to a handwritten one. The law also enabled the development of electronic identity (ID) cards that would allow Estonian citizens and residents to sign documents digitally and to use private and governmental e-services that require secure authentication. The first electronic IDs were issued in January 2002 and have become compulsory for all citizens are as they are equally valid for both digital and physical identification. Since 2011, the citizens of Estonia can also officially identify themselves with a mobile ID and use various public and private sector e-services via cell phones. For example, mobile-ID owners can provide digital signatures, apply for a driver's license, access and process real estate data and documents, apply for a personal loan, purchase and manage insurance, file court cases, pay for bills, etc. Currently, more than 1 million active ID cards have been issued and more than 75,000 people use mobile IDs (ID.ee 2016).

By the year 2000, internet access had been established as a basic human right in Estonia; nevertheless, most Estonians at the time were still unable to afford computers or internet connections in their homes. In fact, at the beginning of the new millennium, less than one third of Estonians had used the internet (Kalkun and Kalvet 2002). Furthermore, in addition to the digital divide in terms of access, the majority of the population lacked the motivation and skills to use the ICT. In order to address these issues, various public-private partnership projects emerged that aimed to promote the spread of internet use among the Estonian population. Two of the most well known projects of the time were the Tiger Leap project, which helped to bring computers and provide internet access to all schools in Estonia, and the Look@World Project, which helped to create hundreds of internet access points all around Estonia, thereby enabling 10 percent of the adult population in Estonia to learn about the use of ICTs.

By September 2015, 88.4 percent of the Estonian population used the internet (Facts about e-Estonia 2016). Internet penetration was highest among the youngest age groups—reaching 100 percent in the sixteen to twenty-four-year-old age group, followed by 99 percent among twenty-five to thirty-four-year-olds (Ait 2016). Eurostate Information Society indicators reveal that 88 percent of the households and 97 percent of the enterprises in Estonia have internet access (eGovernment in Estonia 2016). Furthermore, free and speedy wireless internet access is available almost everywhere around the country—not only in public transport, parks, pubs and restaurants, airports, and bus stations, but internet can also be accessed from a forest or while lying on the beach.

In comparison to the other European Union countries, where only 41 percent of the individuals who use the internet use it for interacting with public authorities and only 39 percent of individuals use the internet for obtaining information from the public, the respective numbers for Estonia are 81 and 79 percent (eGovernment in Estonia 2016). Estonia's success in modernizing its public sector and providing transparent governance has turned the country into a champion of e-governance, which has helped to provide a model for developing similar e-solutions for many countries in the world.

Estonian success in this realm is built on the understanding that various e-government systems used in Estonia, such as e-tax, online health-registry, or e-law, help to create an atmosphere of openness and trust between the state and the people that it serves. Furthermore, all levels of administration in Estonia, from the national leadership to local councils, use various e-solutions. For instance, since 2000, the Estonian government has been using the e-Cabinet, which allows the ministers to prepare for cabinet meetings, conduct them, and review minutes entirely without paper. As the e-Cabinet system uses web-based software and audiovisual equipment, ministers can participate in the meetings remotely.

Estonia was also the first country in the world to introduce nationwide internet voting. E-voting was first practiced on the local government elections in 2005, and two years later, the country also implemented the first online parliamentary elections in the world. In ten years' time, Estonia has held eight elections during which citizens have been able to cast their legally binding vote either via internet or, since 2011, via mobile phone. The number of internet voters has grown considerably—while only every fiftieth vote was cast over the internet during the first internet enabled elections in 2005, every third vote was cast online during the European Parliament elections in 2014 (Vassil 2016, 3).

Estonia also has a striving start-up culture that has sometimes been referred to as "Europe's Silicon Valley" or "the European Union's Delaware." In fact, it has been suggested that Estonia produces more start-ups per head of population than any other country in the European Union. For instance, Estonian programmers have been behind creating such digital brands as Hotmail, Skype, Transferwise, and CrabCad.

One of the first social networking sites in the world was also created in Estonia; on May 1, 2002, a national-language based dating and communication website called Rate was launched. The site provided its users with various opportunities for communication, self-expression, and identity-creation (e.g. uploading photos, rating the photos and sending comments to other users, socializing in the chatroom, keeping a blog, reading news stories, etc.). The site was immensely popular among the young, but also among the adult population as of the year 2007, there were more than 300,000 registered users of the site, which meant that every third Estonian was a user of Rate (Siibak & Ugur 2010). Currently, the most popular social media site in Estonia by far is Facebook, with 58 percent of the Estonian-speaking and 39 percent of the Russian-speaking population using the site (Seppel 2015). Other popular social media sites among Estonians are YouTube, Instagram, and LinkedIn, whereas among the Russian-speaking population, Odnaklassniki and VKontakte are most often used. Research suggests that the majority (72 percent) of Estonian internet users engage in social media, with 51 percent of the users claiming to create content on the sites, and 15 percent acting as lurkers (Murumaa-Mengel, Pruulmann-Vengerfeldt, & Laas-Mikko 2014). The most active social media users are young people, fifteen- to twenty-nine year olds, where the usage is close to 100 percent, but the number decreases for older age groups.

In 2014, Estonia launched a new e-residency program with the aim to offer a transnational digital identity to every citizen of the world. E-residents receive an e-ID that, in addition to other opportunities, enables them to establish an Estonian company and manage it from anywhere in the world. However, in contrast to the digital ID of Estonian residents, the digital-ID of e-residents does not confer Estonian citizenship, tax residency, free access to Estonia or the European Union, nor is it valid as a travel document. By August 2016, more than 12,400 e-residency applications from 133 countries had been issued (e-residents dashboard, 2016). The aim of the program is to have more than 10 million e-residents all over the world by 2025.

The multitude of e-services, including the e-Cabinet system, and the comfortable and effective online lifestyle that these offer to the public is why Estonia has invested heavily in its cybersecurity infrastructure. In fact, a secure 2,048-bit encryption powers Estonia's electronic-ID, digital signature, and X-road-enabled systems to provide built-in safety and security to every single Estonian e-government and IT infrastructure system. Nevertheless, in April 2007, Estonia made international headlines by becoming the first nation in history to fall under a large-scale cyberattack; some scholars have referred to it as the first cyber war in history. The distributed denial of service (DDoS) attacks, which were reportedly perpetrated by politically motivated Russian hacktivists, lasted for several days, during which websites of various Estonian ministries, banks, government, and news agencies were hacked and defaced. The Estonian government's decision to relocate the Soviet World War II memorial known as the Bronze Soldier from its original place to the Tallinn Military Cemetery has been identified as the possible trigger both for rioting on the streets of its capital, Tallinn, by the Russian-speaking individuals, as well as cyberattacks. Estonia's experience in handling the cyberattacks of 2007 has positioned the country as a leader in cybersecurity. Its model of building national

Tallinn, Estonia, hosts the NATO Cooperative Cyber Defence Centre of Excellence, a center that conducts training and research on cybersecurity for its twenty-eight members. Estonia is not only working at the forefront of cybersecurity; the country itself is sometimes nicknamed e-Stonia because of its progress on e-Government and e-Residency. Many other countries are keeping an eye on the country's efforts to learn how they might be incorporated into their own policies. (AP Photo/NIPA, Timur Nisametdinov)

defensive cybersecurity capability systems has been studied by many countries of the world. Since 2008, Tallinn has been home to the Cyber Defense Centre of North Atlantic Treaty Organization (NATO).

Andra Siibak

See also: Finland; Russia

Further Reading

Ait, Jaanika. 2015. "Internet Use among People Aged 65–74." *Quarterly Bulletin of Statistics Estonia*, 3: 94–99. Accessed June 26, 2016. http://www.stat.ee/90737

Anthes, Gary. 2015. "Estonia- a Model for e-Government." September 14. Accessed July 5, 2016. http://incorporate.ee/news/estonia-a-model-for-e-government

e-Estonia. 2016. Accessed July 5, 2016. https://e-estonia.com/

e-residents dashboard. 2016. Accessed August 13, 2016. https://app.cyfe.com/dashboards /195223/5587fe4e52036102283711615553

European Commission. 2016. "E-Government in Estonia." Accessed 26 June, 2016. https:// joinup.ec.europa.eu/sites/default/files/ckeditor_files/files/eGovernment%20in%20Es tonia%20-%20February%202016%20-%2018_00_v4_00.pdf

Herzog, Stephen. 2011. "Revisiting the Estonian Cyber Attacks: Digital Threats and Multi-national Responses." *Journal of Strategic Security*, 4 (2):49–60. Accessed July 4, 2016. http://scholarcommons.usf.edu/cgi/viewcontent.cgi?article=1105&context=jss

ID.ee. "ID.ee." Accessed July 5, 2016. http://id.ee/?lang=en&id=36881

Information System Authority. 2016. "Introduction of X-Road." Republic of Estonia. Accessed August 13, 2016. https://www.ria.ee/en/introduction-of-xroad.html

Kalkun, Mari, and Kalvet, Tarmo. 2002. "Digital Divide in Estonia and How to Bridge It." Accessed July 4, 2016. http://www.praxis.ee/wp-content/uploads/2014/03/2002-Digital -divide-in-Estonia.pdf

Kotka, Taavi, Alvarez del Castillo, Carlos Ivan, and Korjus, Kaspar. 2015. "Estonian e-Residency: Redefining the Nation-State in the Digital Era." University of Oxford Cyber Studies Programme, Working Paper Series, 3, 1–16. Accessed June 26, 2016. http://www.politics.ox.ac.uk/materials/publications/14883/workingpaperno3kotkavargasko rjus.pdf

Murumaa-Mengel, Maria, Pruulmann-Vengerfeldt, Pille, and Laas-Mikko, Katrin. 2014. "The Right to Privacy as a Human Right and Everyday Technologies. Methodology and Results of the Study." Accessed August 13, 2016. http://www.humanrightsestonia.ee/wp /wp-content/uploads/2014/12/ENG-Study-IV-part-Methodology-and-results-of-the -study.pdf

Seppel, Külliki. 2015. "Meediajainfoväli [Media and Information field]." Accessed August 13, 2016. http://www.kul.ee/sites/kulminn/files/7peatykk.pdf

Siibak, Andra, and Ugur, Kadri. 2010. "Is Social Networking the New Online Playground for Young Children? A Study of Rate Profiles in Estonia." In *High-Tech Tots: Childhood in a Digital World*, edited by Ilene M. Berson, and Michael J. Berson. Information Age Publishing, 125–152.

Vassil, Kristjan. 2016. "Introduction." In *E-voting in Estonia: Technological Diffusion and Other Developments over Ten Years (2005–2015)*, edited by Mihel Solvak and Kristjan Vassil. Johan Skytte Institute of Political Sciences: University of Tartu. Accessed June 26, 2016. http://skytte.ut.ee/sites/default/files/skytte/e_voting_in_estonia_vassil_solvak_a5 _web.pdf

ETHIOPIA

The Federal Democratic Republic of Ethiopia is situated in the horn of Africa. It is located east of Sudan and South Sudan, west of Djibouti and Somalia, south of Eritrea, and north of Kenya. Ethiopia continues to have one of the lowest internet penetration rates in the world. In mid-2016, Ethiopia's internet penetration was approximately 3 percent of a population estimated to be 94 million. More recently, the majority of residents living in major Ethiopian cities have begun accessing the internet through mobile phones. Internet users in Ethiopia mainly use Facebook, and YouTube, with growing user bases on Viber, Facebook Messenger, Imo, WhatsApp, and Tango. To date, there are no viable alternative domestic platforms that are positioned to prominently address local needs.

For many centuries, Ethiopia was under imperial rule; later it was governed by a communist dictatorship until the current ruling party, Ethiopian People's Revolutionary Democratic Front (EPRDF). The EPRDF is a coalition of ethnically based political parties that took power in the early 1990s. Since taking power, the government has instituted multiyear economic development agendas that have included an emphasis on telecommunication services. Most notably, the government instituted a five-year "growth and transformation plan" to upgrade existing networks to accommodate emerging information communication technologies (ICTs), and to improve and maintain the quality of existing ICT infrastructure. As a result, the cost of owning mobile phones continues to decrease. Ethio Telecom, the sole government-owned telecommunications operator in the country, offers varying levels of connectivity alternatives, including 4G, 3G, EVDO, and CDMA 1X, in monthly packages from 1 GB to 100 GB, and at prices ranging from $8 to $440, respectively. The average retail price of a smartphone in Ethiopia is $262, which is about 47.6 percent of per capita income (KPCB Internet Trends 2016). Although the cost of mobile ownership and services are significantly lower than before, they remain out of reach for many Ethiopians whose livelihood depends on subsistence farming.

To improve infrastructure, the Ethiopian government has been working with the Chinese telecommunication technology companies ZTE Corporation, Huawei Technologies, the Chinese International Telecommunication Construction Corporation, and more recently with the Swedish telecom group Ericsson, to transform existing networks and increase network capacity to accelerate the rate in which people living in rural parts of Ethiopia acquire access to ICT services. The resulting expansion is estimated to have reached a total of 85 percent land coverage, and to have increased the service capacity to 59 million mobile subscribers (Ethio Telecom 2013). The increase in subscribers is expected to accelerate the development of mobile devices and services that support local languages.

Although there are more than 80 languages spoken in Ethiopia, Amharic is the lingua franca (working language) of the federal government. Amharic uses non-Latin scripts that are sometimes referred to as *abugida*, which contain more than 250 individual characters. It has been a challenge for Ethiopians who cannot speak English to use the internet, or even operate computers and mobile phones. However, companies such as Microsoft and Google are actively working to change this by localizing their web-based and desktop software. For example, as part of its local language program, Microsoft Corporation has released an Amharic version of its Windows operating system and Microsoft Office suite, its flagship productivity software. The company teamed up with Ethiopian language experts and technology professionals, as well as the Ethiopian Information Communications Technology Development Agency (EICTDA), to complete the project (SAHLE 2010).

Nevertheless, using their own language on the internet continues to be the biggest obstacle facing internet users in Ethiopia. For example, Facebook, the largest social network in the world, does not support any of Ethiopia's languages. Ethiopian investors and entrepreneurs are beginning to address this issue. In 2011, TANA Communications began assembling one of the first mobile phones to use Amharic as its default language. Since then, additional mobile device manufacturers have begun assembling mobile phones that support the local languages in Ethiopia.

Government restriction over internet startups is slowly and selectively loosening. As a result, internet companies with limited reach have emerged. One such company is ICE Addis, founded by Markos Lemma (1984–) and Oliver Petzoldt (c. 1976–), which has helped build services such as LOCALLY, a location-based social networking platform that aims to connect businesses and services with their local communities in Ethiopia and the African continent. Karta is another web service that focuses on creating digital mapping solutions for Ethiopia. Karta seeks to solve the problem that many Ethiopians face navigating around their cities due to a lack of street name signage and general unfamiliarity with the concept of street names among city dwellers. Instead of using street names to locate a specific place, Karta uses local administrative structures such as subcities, districts, and house numbers. Although the two services have social components, neither of them directly compete with social networks such as Twitter and Facebook.

Facebook continues to be the most popular social media platform in Ethiopia. In 2014, there were approximately 1,960,000 Facebook users in Ethiopia (Riaga 2014). Ethiopian social media users tend to use Facebook somewhat differently than in neighboring countries and the world. Facebook in Ethiopia is used as a source of entertainment and news. In fact, seven of the other top ten pages are all dedicated to either news or entertainment: DireTube, Yehabesha, Yegna Tube, Yeneta Tube, BeteTube, EthioTube, and Zehabesha (Socialbakers 2016). Still, many Ethiopians use Facebook for social engagement and community building as well. Social engagement and discussions tend to be different depending on the age group involved. While younger Ethiopians tend to be more interested in current affairs and less interested in addressing or discussing sensitive old grievances, older Ethiopian Facebook users are more likely to bring up historical injustices (Gagliardone et al. 2016).

The Ethiopian government maintains strict control over internet and mobile technologies. As a result, it has the capability to monitor, restrict, and disrupt information that flows through the internet, including day-to-day communication of Ethiopians taking place via ICTs. Due to long-standing government monopoly of telecommunication services, international watchdogs claim that the Ethiopian government actively limits access to information while keeping tabs on activities that it considers a threat, often without any "oversight from independent legislative or judicial mechanisms that would typically ensure that surveillance capabilities are not misused" (Human Rights Watch 2015). In fact, the Ethiopian government's online practices were recently labeled as some of the most restrictive and repressive in the world. The Committee to Protect Journalists, Freedom House, and Human Rights Watch claimed that the Ethiopian government abuses its power to silence online speech. These organizations claim that Ethio Telecom, in tandem with the Information Network Security Agency (INSA), routinely blocks critical news and opposition websites. The blockage is especially intensified during the lead-up to elections. The government's approach to internet filtering generally entails hindering access to a list of specific internet protocol (IP) addresses or domain names at the level of international gateways. Deep-packet inspection (DPI), a censorship strategy used by China, Iran, and Kazakhstan, was also employed in the beginning of 2012 (Runa 2012).

The Ethiopian government maintains that the restrictions and limitations imposed on selected sources of information on the internet are justified and consistent with Ethiopia's constitution. However, in addition to restricting internet activities in the country, INSA targets individuals and groups living outside Ethiopia. For example, in 2013, 2014, and 2015, Ethiopian Satellite Television Service (ESAT) employees residing in the United States received a spyware file concealed as a Microsoft Word document file through Skype. When installed, the spyware would grant access to files, emails, passwords, Skype calls, and keystrokes made on the infected computer via a remote control system. Citizen Lab, a research laboratory that focuses on ICTs, human rights, and global security, linked the attacker to Ethiopia's INSA. Testing conducted by researchers at Citizen Lab linked the spyware to an Italian private security group known as Hacking Team, implying that the INSA attacker obtained access to the Hacking Group spyware and utilized it (Citizen Lab 2015). The Ethiopian government claimed that its targets in the United States and elsewhere in the world were threats to national security. In some instances, according to the government, the individuals belonged to terrorist organizations seeking to overthrow the Ethiopian government by force.

Ethiopia does not have prominent citizen hacker groups. However, political activists use social media (and, more specifically, Facebook) to conduct their activism. Often, they share photos and videos of police brutality with commentary and propaganda that aims to grab attention of Ethiopians living abroad. On May 31, 2016, Oromo activist groups leaked the Ethiopia Higher Education Entrance Exam on social media before it was administered to force the Ethiopian government to comply with their demands. The activists reportedly asked the government to postpone the national exam because for the previous five months, schools in the Oromia region were closed due to protests. These protests had resulted in widespread conflict between government forces and students, which in turn resulted in violence and the arrests of a large number of high school students in the Oromia region. After posting images of the examination, the activists said that they were forced to do so because the Ethiopian government declined to comply with their demands, which they insisted were reasonable. As a result of the leak, the Ethiopian ministry of education had to suspend the exam indefinitely and vowed to bring those involved in the leak to justice (Fantahun 2016).

Solen Feyissa

See also: China; Iran; Italy; Kazakhstan; United States

Further Reading

Citizen Lab. 2015. "Hacking Team Reloaded? US-Based Ethiopian Journalists Again Targeted with Spyware." Accessed June 5, 2016. https://citizenlab.org/2015/03/hacking -team-reloaded-us-based-ethiopian-journalists-targeted-spyware/

Ethio Telecom. 2013. "Press Release: Ethio Telecom Concluded Vendor Allocation for Launching Telecom Expansion Projects." Accessed June 5, 2016. http://www.ethiote lecom.et/?q=node/158

Fantahun, Arefayné. 2016. "Matric Exam Postponed After Paper 'Leak.'" *Ethiopia Newspaper.* Accessed July 31. http://www.ethiopiaobserver.com/2016/05/matric-exam-postpo ned-after-paper-leak/

Garliardone, Iginio et al. 2015. "MECHACHAL: Online Debates and Elections in Ethiopia. From Hate Speech to Engagement in Social Media." October 2. Accessed July 31. http:// papers.ssrn.com/sol3/papers.cfm?abstract_id=2782070

Human Rights Watch. 2015. "Ethiopia: Hacking Team Lax on Evidence of Abuse." Accessed June 4, 2016. https://www.hrw.org/news/2015/08/13/ethiopia-hacking-team-lax-evidence-abuse

Runa. 2012. "Deep Packet Inspection: An Update on the Censorship in Ethiopia." Accessed June 4, 2016. https://blog.torproject.org/category/tags/deep-packet-inspection

Socialbakers. 2016. "Ethiopia Facebook Page Statistics." Accessed June 5, 2016. http://www.socialbakers.com/statistics/facebook/pages/total/ethiopia/

F

FINLAND

Finland is a Nordic country that borders Sweden on the west, Norway on the north, and Russia on the east. The country is officially called the Republic of Finland. It became a part of the European Union in 1995, and the Eurozone in 1999. According to a 2015 report from Internet World Stats (2016), 93.5 percent of the population of Finland uses the internet. The people of Finland believe that having access to the internet is crucial to becoming an information society. They believe that every citizen should have equal rights to access information via the internet. Consequently, on July 1, 2010, Finland became the first country in the world to make broadband a legal right for every citizen. That means it is mandatory for telecommunications companies to give every person in the country access to a 1 Mbps broadband connection (BBC News 2010).

Although Finland lags behind the European average for mobile device usage, approximately 70 percent of the Finnish population owns a mobile device, including smartphones and tablets (Deloitte and Touche 2014). The rate of adoption of tablets is about 33 percent (Deloitte and Touche 2014). Consumers between the ages of thirty-five and forty-four are the biggest users of tablets (Deloitte and Touche 2014). Smartphones are very popular among eighteen to forty-four-year-olds, and 60 percent of consumers over the age of sixty-six still use basic phones (Deloitte and Touche 2014). Finnish consumers frequently use smartphones for instant messaging (IM), short message service (SMS), and voice calls. Most of the IM services are cheaper and have higher functionality than SMS. However, people in Finland continue to use SMS and voice calls.

The use of IM services is more popular among young consumers. They use IM services to send and receive videos, photos, and text messages. Apps are commonly used by consumers between the ages of eighteen and fifty-four, typically for online shopping and playing games. Despite their popularity and high usage, there is an overall decline in the number of apps that people download and purchase. In a typical month, 40 percent of Finns do not download any new app, and purchasing an app is very uncommon (Deloitte and Touche 2014). Likewise, people commonly use their smartphones to play games, but only 6 percent of smartphone owners spend money purchasing games (Deloitte and Touche 2014).

Finnish people between the ages of sixteen and sixty-four are fast adopters of mobile devices that allow them to access the internet anywhere, anytime (Statistics Finland 2014). Consequently, they heavily rely on the internet for their everyday activities. According to Statistics Finland (2014), 86 percent of the population between the ages of sixteen and eighty-nine are internet users. Finns rely on the internet significantly for banking, social networking, online shopping, IM, job searching, and submitting job applications. The internet use for these activities is steadily increasing; however, online purchasing is more common in large towns than in semiurban and rural municipalities.

There is a steady growth in the use of social networking services in Finland. Over 51 percent of the population between the ages of sixteen and eighty-nine follow social network and cloud services for storage and data processing (Statista 2016c). More and more people are avoiding external storage devices and use cloud servers to save their files. Facebook, YouTube, WhatsApp, and Google+ are the most popular social media in Finland, followed by blogs and online forums (Statista 2016c). Since the adoption of smartphones, Facebook usage has increased significantly. Twitter, Pinterest, Foursquare, and LinkedIn are not as popular as other forms of social media in Finland. While businesses use Facebook for advertising, entertainment, and casual discussion, people in Finland also use it as a platform to voice their opinions to bring political and social changes to the country.

Manufacturing companies and other business enterprises leverage the power of online social media to communicate with their customers, employees, and partners. The use of social networking sites is very common in industries. Companies also rely on video sharing, online discussion forums, and wikis to reach out to their customers. While many Finnish businesses are still using social media for traditional marketing purposes such as raising brand awareness, increasing traffic to the website, improving sales, generating leads, and recruiting, some companies have started using social media for customer service. The most common use of wikis is by employees to communicate internally and also to collaborate with partners. Companies that steadily use social media have strict guidelines in place, including strategies for internal and external use and policies to monitor information flow. Despite the steady use of social media, Finnish companies believe that they have not figured out an efficient way to measure the return of investment for using social media.

Schools and universities in Finland routinely use online learning platforms. Educators believe that such platforms allow students to study collaboratively outside the classroom. Their availability has influenced the instructional goals and strategies of teachers. Educators have integrated these tools into their curricula in order to encourage students to discuss and work collaboratively on their assignments. Students can upload their individual homework and download learning materials from their instructors any time. The most common learning platform used by Finnish schools and higher-education institutions is Moodle. Universities use a social networking platform called Yammer to facilitate interactions between students and faculty. Yammer, now owned by Microsoft, is a free social networking service used for private communications within organizations.

The use of social media in the country's 2011 parliamentary election was a novel development in the history of Finnish politics. This was the first time that social media sites such as Facebook, YouTube, and Blogger were used in the election campaigns. However, studies have found that social media had little effect on people's voting decisions (Khaldarova et al. 2012; Strandberg 2012). This is because candidates used social media as a broadcasting platform rather than a tool to engage with their constituents. After the election, members of parliament closed their personal accounts on Facebook, Twitter, Blogger, and other social media sites that they used for campaigning. This caused disappointment in people who searched for them after the election. Evidently, social media served as an interactive platform that was used by the candidates for one-way mass communication. Thus, the innovation of using social media in the parliamentary election was ineffective, with minimal influence on people's voting decisions.

The government of Finland is very active in controlling and regulating information flow on the internet. Censorship laws are implemented for the welfare of the citizens, especially children and youth. The Finnish National Bureau of Investigation (NBI) has partnered with Finnish internet service providers to maintain a secret list of sensitive websites that should be blocked. The blocking list initially focused on foreign child pornography sites but has been extended to a number other sites since then. The law also allows the government to censor certain other foreign sites, as well as social media advertisements on online gambling and alcohol. The law does not prohibit Finnish alcohol companies from maintaining their websites. However, they must delete or block any customer reviews of alcohol products.

In 2008, Matti Nikki (n.d.–), a software developer and internet activist, protested the government's action to censor websites. He created a website called lapsiporno.info, which listed many of the sites from the Finnish internet censorship list. The government did not object to the site until Nikki turned his website into a clickable list of censored sites (Puolamaki 2008). In 2008, NBI blocked lapsiporno.info because of its clickable links to child pornography websites (Puolamaki 2008).

Since most Finnish people use the internet for their everyday activities, they show an ongoing concern about cybersecurity breaches. A survey conducted by Statista (2016b) showed that although 91 percent of the respondents say that their social media and email accounts have never been hacked, 50 percent of the respondents were concerned about possible breaches (Statista 2016a). This affirms Finns' alertness about and awareness of such possible threats.

To address those concerns, the government of Finland published a government resolution on January 24, 2013, called "Cyber Security Strategy." It defines key goals and guidelines that can be used nationally to control deliberate or inadvertent threats in the cyber domain, respond to cyber disruptions, and recover from them (Security Committee, 2015). This program focused on developing the Cyber Security Center, the Central Government 24/7 Information Security Operations, security networks for encrypted data transfer and administration, police capabilities for responding to cybercrime, legislation associated with the cyber domain and cybersecurity, and research education programs to improve awareness among its citizens (Security Committee, 2015). Later that year, the Ministry of Foreign Affairs announced that foreign intelligence agents have infiltrated the country's government communication systems. This cyber espionage activity is currently under investigation.

Anamika Megwalu

See also: Denmark; Estonia; Iceland; Russia

Further Reading

BBC News. 2010. "Finland Makes Broadband a 'Legal Right'." Accessed August 4, 2016. http://www.bbc.com/news/10461048

Deloitte and Touche. 2014. "Mobile Consumer 2014: The Finnish Perspective." Accessed August 4, 2016. http://www2.deloitte.com/content/dam/Deloitte/fi/Documents/technology-media-telecommunications/Global%20Mobile%20Consumer%20Survey%202014_medium.pdf

Internet World Stats. 2016. "Finland: Internet Usage Stats and Telecom Reports." Accessed August 4, 2016. http://www.internetworldstats.com/eu/fi.htm

Khaldaova, Irina, Laaksonen, Salla-Maaria, and Matikainen, Janne. 2012. "The Use of Social Media in the Finnish Parliament Elections 2011." *Media and Communications Studies Research Reports* 3: 1–32. Accessed August 4, 2016. http://www.helsinki.fi/crc/Julkaisut/SoMe_Elections.pdf

Puolamaki, Kai. 2008. "Finnish Internet Censorship." Accessed August 4, 2016. https://effi.org/blog/kai-2008-02-18.html

Security Committee. 2015. "Cyber Security Strategy." Accessed August 4, 2016. http://www.turvallisuuskomitea.fi/index.php/en/component/k2/38-cyber-security-strategy

Statista. 2016a. "How Concerned Are You About Experiencing or Being a Victim of Your Social Media or Email Account Being Hacked?" Accessed August 4, 2016. http://www.statista.com/statistics/498080/levels-of-concern-over-the-hacking-of-social-media-or-email-accounts-in-finland/

Statista. 2016b. "How Often Has Your Social Media or Email Account Been Hacked?" Accessed August 4, 2016. http://www.statista.com/statistics/498095/finland-frequency-of-experiences-of-social-media-or-email-accounts-being-hacked/

Statista. 2016c. "Share of Social Media Users in Finland in 2015, by Platform." Accessed August 4, 2016. http://www.statista.com/statistics/540705/social-media-usage-in-finland-by-platform/

Statistics Finland. 2014. "One Half of Finnish Residents Participate in Social Network Services." Accessed August 4, 2016. http://www.stat.fi/til/sutivi/2014/sutivi_2014_2014-11-06_tie_001_en.html

Strandberg, Kim. 2012. "A Social Media Revolution or Just a Case of History Repeating Itself? The Use of Social Media in the 2011 Finnish Parliamentary Elections." *New Media and Society,* 15(8): 1329–1347.

FRANCE

Located on the European continent, north of the Iberian Peninsula and across the channel from the United Kingdom, France is a nation of approximately 64.5 million people. Though internet penetration in the country has reached around 86.4 percent, the French have been slow to adopt the internet and social media due to arguments that it could damage their culture and tarnish their strong privacy regulations. The slow acceptance was also partially based on the popularity of Minitel, a precursor to the internet that dominated France's telecommunications infrastructure for almost thirty years. In spite of the challenges to internet and social media usage, the French utilize many international, European, and French-based social media and social networking sites. Usage is growing, especially among the lower and the elder age brackets, though members of those age groups frequent different sites.

The internet arrived in France relatively early in the 1990s, but it faced many challenges that slowed its adoption. Initially, internet service providers (ISPs) could not find subscribers willing to sign up for their service. French critics of the internet levied a number of complaints against the internet, particularly the use of email, including accusations that the technology was available only in English and it threatened to destroy the French language. Other criticisms focused on how the internet invited chaos into French society, threatening its political, educational, and social traditions. By 1997, America Online's French office, one of the earlier providers offering internet access in France, could barely find subscribers. At this time, only 10 percent of French households had computers—an extremely low number compared to Canada and the United States, where more than 35 percent of all households already possessed one (Schiller 1996). In fact, it appeared initially that the French might reject the internet as a valid means of communication.

French resistance to the internet was not, however, a bias against the adoption of new and emerging technologies. In fact, France had already adopted the Minitel online technology decades earlier. Minitel, a precursor to the World Wide Web, looked like an early computer model but was attached to a telephone landline. Installed in the late 1970s as a limited provincial experiment, over 1 million Minitel terminals existed in people's homes by 1985, and the number increased to more than 9 million terminals by 1999 (Lichfield 2012). Through Minitel, the French could search for information, make travel reservations, read news, and chat, among other services. Like its successor, people developed a special shorthand to communicate in French over the system that is comparable to modern-day "text speak." French companies had planned to export the Minitel technology abroad, but they were unsuccessful. Though the United Kingdom and the United States developed similar systems, none of them gained and maintained the public's attention, including Minitel. Eventually, the internet offered more services and more content in the French language, which enhanced its appeal. In 2012, Minitel met a quiet end, as the internet far surpassed its capabilities, and with greater agility and speed.

Once the internet gained acceptance, social media and social networking followed quickly. However, not everyone is excited about social media penetration. Specifically, the French government discourages social media usage, at least to some extent. France's laws pertaining to social media use and how the data can be used are sometimes quite restrictive. In 2012, the government passed the Employment Law Review, which not only permitted but encouraged prospective employers to use social media to assess applicants.

At the same time, the French believe that internet access is a basic human right, but that it also must preserve high standards related to user privacy. In fact, the French government fined Google in 2011 for illegally collecting user data without permission (Pfanner 2011). A more recent, controversial privacy law passed in 2016 allows grown-up children to sue their parents for posting pictures of their younger selves online (Chazan 2016). The French take privacy seriously, and violations of privacy are just as serious online as offline. Privacy concerns fuel some of the main objections to social media use.

Despite the concerns surrounding social media use in France, it is nonetheless on the rise for all age groups. In January 2016, Génération Numérique conducted a survey of eleven- to eighteen-year-olds and their social networking preferences. Among this age group, 37 percent claimed Facebook as their most used platform, followed by Snapchat at 27 percent and Instagram at 17 percent. Males used Facebook more frequently than females; for eleven- to fourteen-year-olds, the rates were 62.6 percent and 47.2 percent, respectively. The trend held for fifteen- to eighteen-year-olds, with 93.1 percent of males surveyed using the network, as opposed to 82.6 percent of females (eMarketer 2016b). Females overwhelmingly preferred Snapchat and Instagram over Facebook.

The same trend continues for older users, who are flocking to social networks. For people over eighteen years of age, Facebook was the most popular site. Members of both genders preferred the platforms in the same order and to almost the same degree; following Facebook, they listed Google+, the French-language Copains d'avant, Twitter, and LinkedIn as their favorite sites (eMarketer 2016a). These surveys mostly corroborate other studies on French social media, which show that France has 31 million users on Facebook, 24 million on YouTube, 12.8 million on Twitter, 10.7 million on Snapchat, 10 million on Instagram, 10 million on Google+, 8.8 million on LinkedIn, 5.6 million on WhatsApp, 3.5 million on Viadeo, and 2.7 million on Pinterest (Alexitauzin 2016). These numbers reflect total registered accounts rather than unique individuals because some French citizens have accounts on multiple platforms.

France also has a popular indigenous platform called Skyrock. As of August 2016, Skyrock was ranked 146 in France, frequented by the most users in France (56.5 percent), India (7 percent), Belgium (4.9 percent), Algeria (4.9 percent), and Morocco (2.2 percent) (Alexa 2016). Skyrock's website contains a live counter to show the types of content available. As of August 2016, the site contained more than 657 million articles, 26 million blogs, 585,780 secret blogs (one accessible only to designated VIPs, or friends), 18 million profiles, 1.1 billion hearts (equivalent to Facebook's "like" feature), and 4.5 billion comments. In addition to these statistics, Skyrock features image and video sharing and hosts an active chat function (Skyrock 2016). The site also has a connection to Skyrock FM, a popular online radio station, and offers multiple apps to users. In addition to Skyrock and the major international platforms, French users can be found in smaller numbers on Windows Live Profile, MySpace, Amsterdam-d Before Buddies, Badoo, and Trombi.

Since 2015, French social media has attracted local and international attention as the country suffered several major tragedies. In January 2015, armed men wreaked havoc in the country, starting with shootings at the office of *Charlie Hebdo* magazine (BBC 2015). The shootings occurred because the gunmen believed that the magazine's satirical images promoted blasphemy against Islam. In the wake of that tragedy, social media users across the globe responded, showing support through the hashtags #JeSuisCharlie and #IAmCharlie (Martinson 2015). In November 2015, simultaneous attacks on various locations across Paris, including bars, restaurants, a stadium, and a concert hall, pushed Facebook to launch a feature called Safety Check, which let people check during a public crisis event whether their loved ones survived. The hashtag #PorteOuverte (Open Door) allowed people to find refuge throughout the aftermath of the Paris attacks. In July 2016, another radical Islamist attack occurred in Nice, when a man inspired by online militant jihadist propaganda killed more than eighty people. Similar to the earlier attacks, people around the world offered support through #JeSuisNice and #PorteOuverte or #PorteOuverteNice.

After each of these tragic events, social media functioned to inform French police, locals, and people all over the world about what was happening (Sharkov 2016). People not only shared information over the networks, but also used it as a mechanism to inform and report on useful details pertaining to the tragedy and its wake.

Laura M. Steckman

See also: Algeria; Canada; Germany; India; United Kingdom; United States

Further Reading

Alexa. 2016. "Site Overview: Skyrock.com." Accessed August 18, 2016. http://www.alexa .com/siteinfo/skyrock.com

Alexitauzin. 2016. "Combien d'Utilisateurs des Réseaux Sociaux en France de Facebook, Twitter, Instagram, LinkedIn, Snapchat, YouTube, Google+, Pinterest, WhatsApp, Viadeo." [Infographie] June 22. Accessed July 15, 2016. http://www.alexitauzin.com /2013/04/combien-dutilisateurs-de-facebook.html

BBC. 2015. "Charlie Hebdo Attack: Three Days of Terror." January 14. Accessed August 18, 2016. http://www.bbc.com/news/world-europe-30708237

Chazan, David. 2016. "French Parents 'Could be Jailed' for Posting Children's Photos Online." March 1. Accessed August 18, 2016. http://www.telegraph.co.uk/news/world news/europe/france/12179584/French-parents-could-be-jailed-for-posting-childrens -photos-online.html

eMarketer. 2016a. "Social Networking on the Rise Among France's Older Web Users." March 2. Accessed July 15, 2016. http://www.emarketer.com/Article/Social-Networking -on-Rise-Among-Frances-Older-Web-Users/1013650

eMarketer. 2016b. "Young Social Network Users in France Love Facebook, Snapchat." March 9. Accessed July 15, 2016. http://www.emarketer.com/Article/Young-Social -Network-Users-France-Love-Facebook-Snapchat/1013682

Lichfield, John. 2012. "How France Fell out of Love with Minitel." June 8. Accessed August 18, 2016. http://www.independent.co.uk/news/world/europe/how-france-fell -out-of-love-with-minitel-7831816.html

Martinson, Jane. 2015. "Charlie Hebdo: A Week of Horror When Social Media Came into Its Own." January 11. Accessed August 18, 2016. https://www.theguardian.com/media /2015/jan/11/charlie-hebdo-social-media-news-readers

Pfanner, Eric. 2011. "Google Faces French Fine for Breach of Privacy." March 21. Accessed August 18, 2016. http://www.nytimes.com/2011/03/22/technology/22privacy.html?_r=0

Schiller, Suzanne. 1996. "The Internet in France." Accessed August 18, 2016. http://besser .tsoa.nyu.edu/impact/f96/Projects/sschiller/howard.htm

G

GEORGIA

The Republic of Georgia is a small country located at the intersection of Eastern Europe and Asia. Georgia became part of the Soviet Union in 1921 following an invasion by Bolshevik Russia and remained part of it until 1991. Upon its independence, the country fell into political and economic crisis, culminating in the 2003 Rose Revolution. Despite a less than inauspicious beginning and the brief Russo-Georgian War in 2008, Georgia has achieved significant economic and political progress, joining the Council of Europe and the Organization for Security and Cooperation in Europe (OSCE) and becoming a member of the European Union's Free Trade Area in 2014. The country's progress can also be seen in the sphere of internet development. Internet use in Georgia has grown rapidly in recent years, with a sharp increase in usage among both citizens and government agencies. The internet is primarily utilized for social networking, although many also access the internet for news and, increasingly, for social activism. Despite some democratic backsliding in recent years, Georgia remains the only country in the Caucasus without significant censorship and internet restriction.

As of 2015, approximately 49 percent of Georgia's population of 4.8 million had access to the internet, compared to only 10 percent in 2008 (ITU 2016). This rapid growth is due in large part to a proliferation of social network use among citizens and increasing internet use among politicians and government agencies. Nearly half of Georgia's population are internet users, with over one-third (35 percent) accessing the internet every day (CRRC 2015). The most active internet users among Georgia's population are located in the capital of Tbilisi.

Despite rapid growth in internet usage, internet is still a luxury item for many individuals with low income or located outside urban areas. As of 2014, when average monthly wages were less than $325 (Trading Economies 2017), the average cost per month for an internet connection was $20 (Freedom House 2015). Interestingly, despite the high rate of mobile phone penetration, which had reached 125 percent in 2014, the majority of internet users gain access from a home computer or laptop (82 percent) rather than through mobile internet (12 percent) (Caucasus Research Resource Center 2013). The low rate of mobile internet use may be due to its high cost, which is often double that of broadband internet. This reliance on broadband connections limits internet availability in rural areas, where broadband infrastructure is underdeveloped.

According to a 2015 survey for Transparency International, the majority of internet users in Georgia (75 percent) report that they access the internet primarily for social networking (CRRC 2015b). Over 40 percent use the internet to search for information or news, with over one-fifth (21 percent) of all users stating that the internet is their primary source for information. Other popular internet activities include downloading and listening to media (23 percent), sending emails (23 percent), and playing online games (10 percent).

Social media has become the primary reason for accessing the internet in Georgia. Nearly nine out of ten regular internet users (individuals who use the internet at least once a month) frequent one of the four most popular social media websites—Facebook, Twitter, Odnoklassniki, or VKontakte—at least once a week (CRRC 2015a). Facebook is the most popular social networking website; 79 percent of regular internet users visit Facebook at least once a week. Indeed, it has become the most visited website in the country, overtaking popular search engines such as Google (Alexa Internet 2016).

Social media users can be divided into three categories: those that use only Russian networks, those that use only U.S. networks, and those that use a mix of both. Individuals who only use Russian networks make up the smallest group (9 percent), followed by U.S. networks (36 percent). A plurality (43 percent) of internet users use both U.S. and Russian social media networks on a regular basis (CRRC 2015b). U.S. network users are typically more highly educated, urban, and have a higher income than their counterparts using mixed and Russian networks. Given the overwhelming popularity of social media, it should come as no surprise that social networks have become a platform for social and political activism. Activists and civil organizations utilize these networks to communicate with supporters and post calls for action.

The internet has also become a popular tool for government agencies and politicians, who connect with citizens through social media and build new online services to disseminate information to constituents. As of 2016, more than seventy electronic government services from fifty-two state institutions have been integrated into a central online portal (https://www.my.gov.ge). Through this portal, citizens can request information from public institutions, obtain services, and make payments.

There is limited evidence of internet censorship in Georgia, due in part to provisions put in place after the Rose Revolution that prevent the government from censoring internet access. In addition, the majority of internet service providers (ISPs) are privately owned businesses, which limits the government's ability to control the internet. While there was an increase in censorship in the last few years of President Mikheil Saakashvili's (c. 1967–) rule (which ran from 2004 through mid-November 2013), recent years have seen a decrease in internet restrictions. Nevertheless, there have been some isolated incidents of censorship. For example, on March 14, 2016, access to YouTube was restricted around the country in the wake of anonymous threats to leak sex tapes of journalists and opposition politicians (Civil Georgia 2016). Despite these and other rare incidents, internet use in Georgia remains largely free.

Hannah S. Chapman

See also: Kazakhstan; Kyrgyzstan; Russia; Ukraine; Uzbekistan

Further Reading

Alexa Internet, Inc. 2016. "Top Sites in Georgia." Accessed June 1, 2016. http://www.alexa
.com/topsites/countries/GE

Caucasus Research Resource Center (CRRCa). "Caucasus Barometer 2013 Georgia." Accessed June 1, 2016. http://www.crrccenters.org/caucasusbarometer/

Caucasus Research Resource Centers (CRRCb). "Caucasus Barometer 2015 Regional Dataset (Armenia and Georgia)." Accessed June 1, 2016. http://www.crrccenters.org/caucasus barometer/

Civil Georgia. 2016. "Politicians, Journalists Threatened with Sex Tape Leak." March 14, 2016. Accessed June 1, 2016. http://civil.ge/eng/article.php?id=29040

Freedom House. 2015. "Freedom on the Net: Georgia." Freedom House. Accessed May 31, 2016. https://freedomhouse.org/report/freedom-net/2015/georgia

International Telecommunications Union (ITU). (2016). "ICT Statistics 2016—Internet." Accessed June 2, 2016. http://www.itu.int/ITU-D/ICTEYE/Indicators/Indicators.aspx

Trading Economics. 2017. "Georgia Average Monthly Wages." Trading Economics. Accessed January 19, 2017. http://www.tradingeconomics.com/georgia/wages

GERMANY

Germany, a country located in central Europe, is a nation teeming with 80.69 million people. Germans value security and protection, especially with regard to personal information. The country's internet and social media usage is highly reflective of these values. Despite the global trends of digital media consumption, Germans look to traditional regional media as credible sources of information. Germans use a mix of global and locally-based social media platforms and apps, such as WhatsApp, Threema, Mytaxi, and Xing. To keep up with its audience, Germany's *Bild-Zeitung* tabloid even created a user interface with Facebook Messenger. The failure of some U.S. apps, such as Uber, also reflects Germany's preference to use German services (or at least services that reflect German values), as discussed below.

With 71.73 million active internet users, the country has an internet penetration of 89 percent. Despite the high internet penetration, the country's social media penetration is relatively low (36 percent). German users spend an average of three hours and twenty minutes a day on the internet using a computer or tablet, with only an hour of that time used for social media. About 77 percent of Germany's internet users access the internet on a daily basis (Kemp 2016). Conservative values may have shaped Germany's internet and social media usage, but Germans are slowly changing their online activities. About 86.7 percent of Germany's online users utilized search engines frequently or occasionally, followed closely by checking personal emails, at 86.4 percent. Considering how protective Germans are of their personal information, online commerce has surprisingly become the third most popular online activity in Germany. About 73 percent of internet users shop online for practical needs (eMarketer 2015).

An average of 81 percent of Germans between the ages of eighteen to thirty-four have a strong social media presence (Cohen 2016). While Facebook dominates other global social media providers, Germans prefer to use WhatsApp. About 39 percent of German internet users favor WhatsApp, followed closely by

Facebook at 38 percent (Kemp 2016). Even though Germany has a stronger presence on WhatsApp for instant messaging (IM), the situation is changing. Germans are not swapping WhatsApp for Facebook, but rather for another global app called Threema. Created in Switzerland in 2012, Threema offers Germany the security that many messaging apps cannot guarantee. The app appeals to Germans' need for protection and security by ensuring end-to-end encryption and security features for its users. Account users do not use their phone numbers, but rather a unique ID, affording people added protection and anonymity. Compared to WhatsApp's 800 million monthly users worldwide, Threema has only 3.5 million monthly users. Despite this huge disparity at the global level, Threema is topping Germany's paid download charts, and the user base increased by 900 percent between mid-2014 and mid-2015 (Acharya 2015).

As social media becomes the go-to for many internet users around the world for social connections, news, videos, and other activities, Germany is still rebelling against the global trend with regard to digital media consumption (Walpole 2015). Germans view traditional media as credible sources of information with journalistic standards, whereas digital media can share conflicting information and sources. About 40 percent of German internet users ranked daily newspapers as the most reliable media available, while other internet users cited television and radio as the most dependable source of information. Media sources on the internet (specifically Google and the German-based news magazine *Der Spiegel*) were least trusted, with about 14 percent of Germans giving them credibility (eMarketer 2014).

As of January 2016, only 10 percent of Germans trusted U.S. internet services. This distrust is directed toward not only the United States, but also other foreign entities (only 22 percent of Germans trusted other European internet services). German society values consumer protection, safety, and privacy for its citizens, meaning that this distrust is largely related to the recording of personal, communication, and miscellaneous data with or without the user's consent (eMarketer 2016). In 2014, Google was warned, and consequently fined, for its methods of collecting personal information. Google routinely collects users' personal data to build interest groups that marketers use for target-specific ads. From their perspective, Google was creating nearly comprehensive personal records of their consumers for its benefit, but Germany did not share that view. Germany's data protection commission ordered the global company to change its methods. While data collection is not illegal, Google was not getting the consumers' "explicit and informed

The 2014 World Cup Final, in which Germany defeated Argentina 1-0, set a world social media record. While the players were on the field, spectators tweeted over 35.6 million times using the #BRAvGER hashtag. After the German player Mario Gotze (1992–) scored the only goal of the game, Twitter users sent more than 580,000 tweets per minute, or around 9,667 tweets per second—the highest volume of posts per minute ever seen over the platform (Chase 2014). The *Guinness Book of World Records* now credits the game as the most tweeted event in history.

consent," nor was Google giving them the opportunity to object to the data collection (Barr 2014). While the majority of Germans accept such procedures for security measures, a growing number believe that the federal government is not doing enough to protect its citizens' personal online data (an increase from 28 percent in 2013 to 32 percent in 2015) (eMarketer 2016).

There are two examples of German-created social media platforms and apps that are the by-products of Germany's caution about foreign options. The first is the German social networking platform Xing. In 2006, Xing was initially established as the Open Business Club (openBC) in Hamburg, Germany. Since its inception, Xing has accumulated nearly 10 million users in Germany and 15 million worldwide. Companies and freelance workers utilize the social media network to find talent, search for jobs, and make connections. Xing also allows account users to start or join discussion forums about shared interests or the latest business trends. Many regard Xing as a German counterpart to LinkedIn. Xing's popularity in German-speaking countries may also be attributed to the social network's commitment to privacy. Because Xing is a German service, it reflects German values. Xing allows account users to choose their privacy settings. Members can decide on how much personal information is available for public consumption and what is kept private (Vitaud 2015). Xing also boasts that all personal data is not kept abroad, but rather stored in servers located only in Germany. Security for all members is highly valued by the German social media network (Xing 2016).

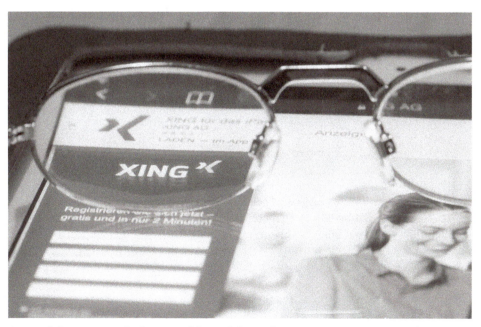

A pair of glasses sit perched on top of the mobile app for Xing, a German social media network akin to LinkedIn. Xing is extremely popular in Germany where it was designed to meet German regulations and privacy laws, which are stricter than in the United States. (Tomnex/Dreamstime.com)

The second example is Germany's Mytaxi app. When Uber expanded to Germany, they were entering a tightly-regulated transportation market. Uber ultimately failed because of its unwillingness to adapt to German regulations for consumer protection and safety. Instead, Uber's lobbyists insisted that Uber was a tech company, not a taxi firm, contending that it should not be subjected to Germany's rules for transportation. However, this strategy did not work. As some experts have noted, an American approach to a foreign market will not always succeed, which can be seen with this example in Germany. As of 2016, Uber is operating in only two major cities, Berlin and Munich (Vitaud 2015).

The Mytaxi app arose to take the place of Uber. Developed by a Hamburg tech startup, Intelligent Apps, in 2010, it shares a similar concept to Uber, but instead of having a single company hiring drivers who then offer their services, existing taxis from multiple taxi companies can use the app. This option has enabled nearly 45,000 affiliated taxis in more than forty German cities to find customers. The app has over 10 million downloads by consumers (Mytaxi 2016).

These two examples showcase Germany's reluctance to utilize global social media platforms and apps. Their preference for German services is prevalent throughout the country and can be seen by many locally-created internet services. However, this does not mean that all Germans use only German products. *Bild-Zeitung,* a widely popular German tabloid, has utilized the internet to reach a wider audience, created an innovative interface for users, and developed additional services and functionalities in the form of downloadable apps. After noticing much of its online audience switching to Facebook for their daily news, *Bild* experimented with a new user interface through the Facebook Messenger app in early 2016. Initially, active Facebook users messaged the Bild Ticker to receive news and information in German about the German soccer league's transfer window, the portion of the year when teams can transfer players. Later, the Bild Ticker expanded to provide other news tidbits. This experiment enabled *Bild* to stay connected with its audience and provide content faster through another internet platform (Hazard Owen 2016).

Karen Ames

See also: France; Poland; United Kingdom; United States

Further Reading

Acharya, Sarmistha. 2015. "Germany's Popular Threema Secure Mobile Messenger Eyes U.S. Market." June 22. Accessed March 16, 2016. http://www.ibtimes.co.uk/germanys-pop ular-threema-secure-mobile-messenger-eyes-us-market-1507311

Barr, Alistair. 2014. "Google Runs Afoul of Germany Privacy Laws." October 1. Accessed May, 2016. http://blogs.wsj.com/digits/2014/10/01/google-runs-afoul-of-german-priv acy-laws/

Cohen, David. 2016. "Internet Users in Emerging Markets Flock to Social Networks." February 22. Accessed March 11, 2016. http://www.adweek.com/socialtimes/pew -research-center-smartphone-ownership-internet-usage-emerging-economies/634596

eMarketer. 2014. "Germany's Digital Teens Trust Traditional Media More." December 14. Accessed May 15, 2016. http://www.emarketer.com/Article/Germanys-Digital-Teens -Trust-Traditional-Media-More/1011732 eMarketer. 2015. "Digital Commerce Is the Norm as Germany's Internet Culture Matures." April 27. Accessed May 13, 2016. http://www.emarketer.com/Article/Digital-Commerce-Norm-Germanys-Internet-Culture-Matures/1012399

eMarketer. 2016. "Internet Users in Germany Hope for Balance of Safety and Privacy." March 4. Accessed March 14, 2016. http://www.emarketer.com/Article/Internet-Users-Germany-Hope-Balance-of-Safety-Privacy/1013664

Hazard Owen, Laura. 2016. "Axel Springer's Bild Is Testing News Delivery via Facebook Messenger." January 14. Accessed March 15, 2016. http://www.niemanlab.org/2016/01/axel-springers-bild-is-testing-news-delivery-via-facebook-messenger/

Kemp, Simon. 2016. "Digital in 2016: We Are Social's Compendium of Global Digital, Social, and Mobile Data, Trends, and Statistics." January 27. Accessed March 10, 2016. http://wearesocial.com/uk/special-reports/digital-in-2016

Mytaxi. 2016. "About Mytaxi." Accessed March 15, 2016. https://us.mytaxi.com/jobs/about-mytaxi.html

Vitaud, Laetitia. 2015. "Freelancers in Germany: Why Global Gig Economy Platforms Find It So Hard to Succeed." November 19. Accessed March 15, 2016. https://medium.com/switch-collective/freelancers-in-germany-why-global-gig-economy-platforms-find-it-so-hard-to-succeed-134ccdff651#.na1m3165v

The Local. 2015. "Why Educated Germans Avoid Social Media." December 22. Accessed March 14, 2016. http://www.thelocal.de/20151222/why-do-educated-germans-avoid-twitter

Walpole, Jennifer. 2015. "Traditional vs. Digital Media Consumption Rates Shifting." October 4. Accessed May 12, 2016. https://theamericangenius.com/business-news/traditional-vs-digital-media-consumption-rates-shifting/

Xing. 2016. "Privacy Policy and Settings." Accessed March 15, 2016. https://www.xing.com/help/help-and-faq-2/general-stuff-55/privacy-policy-and-settings-147

GHANA

Ghana is a West African nation along the coast of the Atlantic Ocean and the Gulf of Guinea. It lies between the Ivory Coast and Togo, with Burkina Faso to the north. The official language spoken in the country is English; however, there are several other national languages. The population distribution between urban areas and rural areas is about even, with slightly more people living in the cities. Ghana has 27.4 million people, many of whom are very young. The median age of a resident in 2015 was only twenty-one years old; 39 percent of the population was less than fourteen years old, and all but 6 percent of Ghanaians were under the age of sixty. This affects the ways in which Ghanaians use the internet, as older people tend to prefer the traditional media that have been around longer, like radio and television, whereas the internet is more likely to be used by those in the eighteen- to twenty-four-year-old category. In fact, internet users in the latter group were just as likely to use the internet as they were to use the radio or television (Balancing Act 2014).

The internet and social media landscape in Africa is changing rapidly. A decade ago, there was virtually no connectivity in the whole region, but today, there are millions of internet users. Ghana, for example, had internet penetration of only 1 percent in 2006. By 2013, this had risen to 21 percent of the population; however, internet connection accessibility varied in different parts of the country. The

Central, Upper Eastern, Upper West, and Western regions had some of the lowest rates of penetration. Connection speeds also tended to be very slow, which made using data-heavy websites like YouTube difficult. This has not affected that site's appeal, however—YouTube is the fifth most popular website in the country (Alexa 2015). Some notable ways in which Ghanaians use the internet in a unique fashion include using national websites instead of social media sites like Facebook or Twitter for news gathering, and placing a strong emphasis on music.

Ghanaians are accessing the internet mainly with mobile devices. In 2006, only 26 percent of the population owned mobile phones and only 5 percent owned a computer. By 2013, 88 percent of Ghanaians owned mobile phones, but only 6 percent owned a smartphone. In 2014, about three-quarters of Ghanaians intended on upgrading to a smartphone in the near future. When they do, the features that will be most important to them are the ability to play music, access the internet, play video, and have a touch screen (Balancing Act 2014). The rise of computer ownership over the past several years has progressed much more slowly. As of 2014, about a fifth of the nation owned a laptop, making Ghana one of the highest-ranking African countries for computer penetration. Tablets, however, were still not very common, with national ownership at only about 6 percent (Balancing Act 2014).

The most common free apps downloaded from the Google Play Store were WhatsApp Messenger, IMO free video calls and chat, Opera Mini web browser, Facebook Messenger, Facebook, Holy Bible (King James version), Viber, 360 Security–Antivirus Boost, Photo Grid–Collage Maker, and Instagram (AppAnnie 2015). This usage suggests that Ghanaians are mostly using their Android devices for communication via messaging apps, and for taking and sharing photos. The fact that the Bible was one of the most common apps reflects Ghanaian cultural values, which are overwhelmingly Christian. This was reflected in religious ideas about the internet noted in 2014; some believe that social media was a waste of time and that its use would undermine morality. A more common belief regarding the internet was that it would damage Ghanaian culture; however, even those with fears about the internet saw it as a way for Ghana to become globally competitive (Balancing Act 2014). The priority of Ghanaian culture is evident in how many websites of Ghanaian origin rank in the top twenty, such as GhanaWeb, My Joy Online, Pulse, Peace FM, and Tonaton (Alexa 2015).

With 85 percent of the Ghanaian people owning radios and listening to it every day, it makes sense that many people will receive their news via that medium. Radio, television, and friends and family were the most dominant ways in which Ghanaians obtained their news. However, members of the younger generation (between eighteen and twenty-four years old), who were generally more likely to use the internet than older generations, were just as likely to get their news from the internet as any other source. Men were also slightly more likely to seek out news on the internet than women (Balancing Act 2014). Of note, several of the more popular websites originated in a radio station, such as Joy Online, Peace FM, and Citi FM. While the internet is catching up as a source of news, it is important to note that Ghanaians do not necessarily use social media sites like Facebook to collect this information. Instead, they prefer news websites, such as Ghana Web, My Joy Online, and Pulse (Alexa 2015; Buzz Ghana n.d.).

The most popular websites visited in any category are Google.com.gh, Google, Facebook, Yahoo, YouTube, Ghana Web, Ask, My Joy Online, Pulse, Twitter, and Peach FM. The number of Ghanaian Facebook users has quadrupled over the past few years. In August 2010, there were only 621,000 Facebook accounts; in March 2014, there were 2.4 million accounts. In 2015, six of the top ten most popular Facebook pages were for companies in telecommunications or related industries: Airtel Ghana, MTN Ghana, Tigo Ghana, Vodafone Ghana, Samsung Mobile Ghana, and Samsung Ghana. The other four top pages were Goal.com Ghana (news), e.TV Ghana, UTV Ghana, and AllSports.com.gh (Allin1 Social 2015). Other social media platforms include Twitter, Pinterest, Tumblr, Google+, Stumble Upon, Reddit, and LinkedIn (StatsMonkey 2015). While they are not as competitive as these Western social media, some African social media sites have a small user base in Ghana, including Eskimi, Mxit, 2Go, and biNu (Southwood 2014).

For several years now, Ghana has had the goal of becoming an "information hub" of West Africa (UNIDIR 2013). As part of this goal, education reforms began in 2007 with the intent of improving the country's education system at all levels. One of the stipulations of these reforms was that all primary and secondary students receive information and communications technology (ICT) training. This allowed the younger generation to develop computer literacy skills, which included internet use, and learn how to apply these skills in everyday life.

Part of this goal has been realized, as most institutions have made some computers available to students. Despite this achievement, very few of these computers were connected to the internet, and students far outnumbered the computers in each school (Mereku et al. n.d.). Nevertheless, this initiative continues and schools are improving their ICT resources. In 2015, the University of Ghana began its first distance-learning program with the aid of a Chinese-sponsored project that was introduced in 2012. The vice chancellor of the university, Ernest Aryeetey (c. 1957–), stated that the distance learning program was just the first step toward elevating the university to a research-intensive institution that can aid in Ghana's development (University of Ghana 2015).

In another step to become an information hub, Ghana passed the Electronic Communications Act and the Electronic Transactions Act in 2008. These laws helped establish Ghana's cybersecurity program. At various stages over the past several years, the Ghanaian government has revised its strategies for combating cybercrime (UNIDIR 2013). However, as of 2011, Ghana had developed a bad reputation for being the origination of cybercrime, which posed a threat to national security (Warner 2011). The National Information Technology Agency set forth a new strategy in 2012 that included external collaboration, as well as plans to establish a Computer Emergency Readiness Team for each major hub within the country (UNIDIR 2013).

Marilyn J. Andrews

See also: Cameroon; Nigeria; Senegal

Further Reading

Alexa. 2015. "Top Sites in Ghana." Accessed September 28, 2015. http://www.alexa.com /topsites/countries/GH

Allin1 Social. 2015. "Facebook Statistics for Ghana." Accessed September 28, 2015. http:// www.allin1social.com/facebook-statistics/countries/ghana

AppAnnie. 2015. "Google Play Top App Charts." Accessed September 28, 2015. https:// www.appannie.com/apps/google-play/top/ghana

Balancing Act. 2014. "A Detailed Snapshot of Africa's Emerging Internet and Social Media Space—The Users and What They Are Doing." September 19. Accessed September 28, 2015. http://www.balancingact-africa.com/news/en/issue-no-724/top-story/a-detailed -snapshot/en

Buzz Ghana. n.d. "10 Most Visited Websites in Ghana." Accessed September 28, 2015. http:// buzzghana.com/visited-popular-websites-ghana

Hill, Laurence. 2015. "Ghanaian Culture." Accessed September 28, 2015. http://www .drumafrica.co.uk/articles/ghanaian-culture

Mereku, D. K., Yidana, I., Hordzi, W., Tete-Mensah, I., Tete-Mensah, W., and Williams, J. B. n.d. "Ghana Report." Accessed September 28, 2015. http://www.ernwaca.org/panaf /pdf/phase-1/Ghana-PanAf_Report.pdf

Southwood, Russell. 2014. "CIMA—The Changing Media Landscape in Sub-Saharan Africa." October 29. Accessed September 30, 2015. http://www.slideshare.net/newsbunny/ media-internet-and-social-media-landscape-in-subsaharan-africa

StatsMonkey. 2015. "Social Network Usage Statistics Using Desktop in Ghana." Accessed September 30, 2015. https://www.statsmonkey.com/sunburst/21587-ghana-desktop -social-network-usage-statistics-2015.php

United Nations Institute for Disarmament Research (UNIDIR). 2013. "The Cyber Index: International Security Trends and Realities." Accessed September 28, 2015. http://www .unidir.org/files/publications/pdfs/cyber-index-2013-en-463.pdf

University of Ghana. 2015. "University of Ghana Inaugurates Distance Education ICT Project Under the Chinese Phase 2 ICT Project." February 4. Accessed September 28, 2015. https://www.ug.edu.gh/news/university-ghana-inaugurates-distance-education -ict-project-under-chinese-phase-2-ict-project

Warner, Jason. 2011. "Understanding Cyber-Crime in Ghana: A View From Below." *International Journal of Cyber Criminoloigy*, 5(1): 736–749.

GREECE

Greece is a country in southern Europe, bordered by Albania, Macedonia, Bulgaria, Turkey, and the Mediterranean Sea. In a country of nearly 11 million people, social media—the milestone of the Web 2.0 era—has totally transformed both interpersonal and corporate communication by establishing the notion of a *produser,* a new media user who acts simultaneously as a content producer. Within the business context, social media is becoming "an ideal, cost effective, interactive and, above all, targeted communication tool for every organization thanks to its high consumer engagement offered through the eWOM (electronic word of mouth) and web-interactivity" (Belenioti et al. 2015, 2). Despite the popularity of social media in Greece, little is known about users' behavior on these sites.

Where does Greece stand in terms of internet and social media usage? When it comes to internet usage, Digital Scoreboard by European Commission, Directorate General for Communications Networks, Content & Technology (DG CONNECT) (2016), concludes that with an overall connectivity score of 0.43, the country ranks twenty-sixth among European Union (EU) countries. Despite Greece's progress in terms of connectivity (i.e. access of households to fixed broadband), it still performs worse than most other EU countries. Specifically, while 99 percent of Greek households have access to fixed broadband, 34 percent still do not yet subscribe to it. In addition, high-speed Internet (at least a 30-Mbps download speed) is available to only 36 percent of homes—far below the EU average of 71 percent. Thus, Greece, ranking twenty-sixth out of the twenty-eight EU member states, belongs to the "falling behind" cluster of countries according to the Digital Scoreboard classification (European Commission/DG CONNECT 2016).

All in all, the European Commission (EC) identifies two major challenges that Greece needs to overcome. First is the penetration of the level of mobile broadband access required for Greece to reap the full potential of the digital economy; only 66 percent of households subscribe to fixed broadband and there are only 44 mobile broadband subscriptions per 100 citizens. Second is the improvement of information technology (IT) infrastructure and the establishment of fast networks (i.e. Next-Generation Access [NGA] networks) providing at least 30 Mbps. To date, NGA coverage in Greece is only half of the average EU level.

User behavior depends on a series of demographic factors, such as socioeconomic status, or behavioral factors, such as the conduit used to access the internet (Hargittai 2008) or information literacy (Chiang et al. 2010). Each user has different motivations for using the internet. For instance, Papacharissi and Rubin (2000, 189) have identified five different motives: "utility, pasttime, information seeking, convenience, and entertainment."

Examining Greece's information literacy, the European Commission also concluded that it still lagged in terms of both digital skills and trust in the digital economy, such as the integration of digital technologies by businesses and online shopping transactions. Consequently, Greece's score is lower than the EU average. In terms of the percentage of individuals with basic or better digital skills, regardless of employment, Greece ranks twenty-first out of twenty-eight, revealing a slow pace of internet acceptance. Similarly, Greeks fall behind in terms of online public services, even though a small but encouraging proportion (37 percent) has utilized online public administration services. Conversely, Greeks show a better performance, ranking eleventh out of twenty-eight in relation to digital specialized skills linked to software for content manipulation.

Likewise, Whiting (2013) has identified ten reasons for using social media: social interaction, information seeking, passing the time, entertainment, relaxation, communicatory utility, convenience utility, expression of opinion, information sharing, and surveillance/knowledge about others. Consistent with the findings of Papacharissi and Rubin (2000) and Whiting (2013), Belenioti et al. (2015) classified Greek social media users into three segments based on reason of use: "Information

Seekers," "Operational and Psychological Boost Benefits Seekers," and "Communication Seekers."

An online questionnaire (conducted by the author) examining patterns of Greek users on internet and social media found that, of the initial cohort of 280 respondents, 47 percent were men and 53 percent were women, with ages ranging from eighteen to over sixty-five years old. In terms of internet usage, the findings showed that the majority of Greeks (57.4 percent) have integrated internet use into their lives on a daily basis. Heavy internet users belong to the groups aged twenty-five to thirty-five years old (27.8 percent) and eighteen to twenty-four years old (23.3 percent). Light internet users comprise a minority (3 percent) and belong to the different age groups as follows: from eighteen to twenty-five years old (1.5 percent), from twenty-five to thirty-five years old (1.1 percent), and from thirty-six to fifty years old (0.4 percent). Regardless of their demographics, on average, Greek users have a middle-to-high educational level. In the same vein, "heavy users" are shown to have a high educational level, with 31.9 percent of the users having a college degree and 18.4 percent holding a master's degree. Finally, 2.7 percent of light users have only a secondary level of education.

Greeks (54.8 percent) surf the web more than thirteen hours per week. Moreover, they use the internet for exchanging messages (mean, or $m=4.14$), for information seeking for job or educational issues ($m=4.10$), for general information seeking ($m=3.92$), for watching sports online ($m=3.92$), for listening to e-radio ($m=3.70$), for entertainment ($m=3.43$), for seeking tourist information ($m=3.34$), for downloading music files ($m=3.15$), and for chatting ($m=3.12$). Surprisingly, Greeks rarely use the net for e-banking activities ($m=2.08$).

When it comes to social media usage, despite its small size, Greece ranks forty-third globally in internet penetration at 35.45 percent, with 3.9 million users (Socialbakers 2012). Roughly half of Greeks are social media users. Recent studies also conclude that a massive increase of social media usage followed the sharp, increasing penetration of the internet. Greeks spend more than 80 minutes per day networking, as half the population is registered on at least one social media platform (Kassimi 2015). According to ELTRUN (2012), Facebook seems to be the most favorite social media platform, used by 77.29 percent of the country's population. Three years later, ELTRUN showed again that Facebook remains the premier social media site, followed by YouTube and Twitter.

These results suggest that social media experience depends both on the users' age and the type of social media. For instance, experienced social networking users are women between eighteen and twenty-four years old, while experienced blog users are women aged twenty-five to thirty-five years old. Most experienced video-gamers are men aged eighteen to twenty-four years old. Representative examples of the most popular virtual games are World of Warcraft, Dotta, and Call of Duty. Furthermore, ELTRUN (2015) concluded that social media is an important part of business strategy, as Greeks tend to express their consumer attitudes and interact with brands via these platforms.

Moreover, research conducted by the Greek agency Focus Bari (January–March 2015) reported that usual patterns of "networking" by Greek users consisted of playing games, while the average time spent on social media websites was calculated at 82.2 minutes per day. On average, Greeks who are experienced users of information and communication technology (ICT) and social media fall between eighteen and thirty-five (62.6 percent) years old and have high educational background and upper-middle-class status (Belenioti 2015). They primarily prefer social networking sites, content-sharing communities, blogs, and finally games. They usually log in via PCs. The majority of Greeks has three accounts and they update content on a daily basis. Within content communities, most users have only one account, and they prefer to consume content rather than create it (Belenioti 2015).

In terms of the frequencies of visits, social networking sites have the highest percentage use on a daily basis (38.9 percent), then content communities (32 percent), blogs (26 percent), and finally, virtual games/worlds (3.1 percent). Interestingly, the usual frequency of the use of blogs and virtual games/worlds is one or two times per week (32.6 percent).

Social networking sites, especially Facebook, dominate social media in Greece, notably for young people and regular web users. In contrast, other social media applications, such as content communities, blogs, and virtual games/worlds, have fewer active users or less user penetration. Finally, Greeks exhibit passive behavior in terms of content-sharing sites, choosing to consume content and sometimes comment one or two times a week, rarely creating content to share. These results are likely to be related to the popularity of the social networking sites in question and their multidimensional benefits, such as ease of use, interactivity, and multitasking applications (e.g. chatting, sharing files, playing games, and job and information seeking).

Zoe-Charis Belenioti

See also: Hungary; Italy; Romania; Turkey

Further Reading

Belenioti, Zoe-Charis. 2015. *A Snapshot of Greek Social Media Users,* 1–15. http://meanin gacrossmedia.mcc.ku.dk/proceedings/.

Belenioti, Zoe-Charis, Andronikidis, Andreas I., and Vassiliadis, Chris. 2015. "Classifying and Profiling Social Media Users: An Integrated Approach." *Proceedings of the Second European Conference on Media and Mass Communication,* 183–204. http://iafor.org /archives/proceedings/EuroMedia/EuroMedia2015_proceedings.pdf.

Brandtzæg, Petter Bae, Heim, Jan, and Karahasanović, Amela. 2011. "Understanding the New Digital Divide—A Typology of Internet Users in Europe." *International Journal of Human-Computer Studies,* 69(3): 123–138.

Bruns, Axel. 2007. "Produsage." *Proceedings of the 6th ACM SIGCHI Conference on Creativity and Cognition,* 99–106.

Chiang, I., Huang, Chun-Yao, and Huang, Chien-Wen. 2009. "Characterizing Web Users' Degree of Web 2.0-ness." *Journal of the American Society for Information Science and Technology,* 60(7): 1349–1357.

Constantinides, Effthymios, Alarcón del Amo, María del Carmen, and Lorenzo Romero, Carlota. 2010. "Profiles of social networking sites users in the Netherlands." *18th Annual*

High Technology Small Firms (HTSF) Conference, May 25–28, 2010, Enschede, the Netherlands.

DeKay, S. 2009. "Are Business-Oriented Social Networking Web Sites Useful Resources for Locating Passive Jobseekers? Results of a Recent Study." *Business Communication Quarterly,* 72(1): 101–105. http://doi.org/10.1177/1080569908330378

ELTRUN. 2015. "Annual Resereach for Social Networking." Accessed May 5, 2016. http://www.eltrun.gr/wp-content/uploads/2016/01/%CE%94%CE%B5%CE%BB%CF%84%CE%AF%CE%BF-%CE%A4%CF%85%CF%80%CE%BF%CF%85DigitalMKT2016.pdf

European Commission (EC). 2016. "Digital Scoreboard." Accessed May 5, 2016. https://ec.europa.eu/digital-single-market/en/scoreboard/greece#3-use-of-internet

Focus Bari Survey, "Social Media in Greece (January–March 2015)." Accessed May 8, 2016. http://www.socialmediaworld.gr/presentations_2015/tsami.pdf

Hargittai, Eszter. 2007. "Whose Space? Differences Among Users and Non-Users of Social Network Sites." *Journal of Computer-Mediated Communication,* 13(1): 276–297.

Kassimi, Alexandra. 2015. "Half of Greeks Engage in Social Media." September 9. Accessed May 5, 2016. http://www.ekathimerini.com/201228/article/ekathimerini/community/half-of-greeks-engage-in-social-media

Katz, Elihu, Blumler, Jay G., and Gurevitch, Michael. 1973. "Uses and Gratifications Research." *Public Opinion Quarterly,* 37(4): 509–523.

Papacharissi, Zizi, and Rubin, Alan M. 2000. "Predictors of Internet Use." *Journal of Broadcasting & Electronic Media,* 44(2): 175–196.

Socialbakers. 2012. "Facebook Statistics." Accessed June 2012. http://www.socialbakers.com/facebook-statistics/

Urista, Mark A., Dong, Qingwen, and Day, Kenneth D. 2009. "Explaining Why Young Adults Use MySpace and Facebook Through Uses and Gratifications Theory." *Human Communication,* 12(2): 215–229.

Whiting. A 2013. "Why People Use Social Media: A Uses and Gratifications Approach." *Qualitiative Market Reasearch: An International Journal,* 16(4): 362–369. http://doi.org/10.1108/QMR-06-2013-0041

HONDURAS

Honduras is one of the largest countries in Central America. It has coastlines on both the Pacific Ocean and the Caribbean Sea, and is bordered on the west by Guatemala and El Salvador and on the east by Nicaragua. While Honduras has the lowest internet penetration rate in Central America (Internet World Stats 2016), the number of people with access through mobile connections has significantly increased since the 2011 infrastructure investment of Claro, a Honduran mobile service provider. However, fixed-line connections remain prohibitively expensive. The constitution of Honduras provides for freedom of speech, but the violence that has flooded the country since the 2009 presidential coup has forced many online users to censor their content in order to protect themselves.

Of the country's nearly 8.2 million people, only 21.5 percent had access to the internet in 2016 (Internet Live Stats 2016). This represents only a 5 percent increase from the previous year and actually marks a trend of very slow penetration growth, or even sometimes shrinkage—for instance, from 2012 to 2013, the number of Hondurans using the internet actually decreased 0.3 percent. Throughout most of the 2000s, penetration was slow, largely owning to the fact that very few Hondurans owned computers. According to the National Institute of Statistics (Instituto Nacional de Estadística, or INE), only 19.4 percent of households had a computer (Rivera 2016). It has been within only the past five years, with the proliferation of smartphones, that internet access has not required a computer.

While growth in penetration has been slow, 2011 saw an enormous jump of 45 percent in internet users (Internet Live Stats 2016). This is most likely due to the purchase of the mobile phone provider Digicel that year by America Movil, Claro's parent company, and subsequent investment of $150 million into mobile telecommunications infrastructure in Honduras (La Tribuna 2011). In 1996, there were only about 2,276 cell phones in the entire country; however, by 2016, the cell phone penetration rate had reached 90 percent or more (Rivera 2016).

Claro and Tigo Communications are the two competing providers that control the prepaid mobile service market. Tigo controls 60 percent, while Claro controls 39 percent; Hondutel, the state-run provider, controls only about 1 percent of the market (Rivera 2016). Recently, a representative of Tigo reported a 50 percent smartphone penetration rate, and the vast majority of smartphone users use Android (Rivera 2016). Mobile internet connections are significantly less expensive than fixed-line connections, which can run anywhere from $14 to $22 a month for speeds ranging from 256 kbps to 5Mbps (Rodriquez 2014). In contrast, mobile connections offer not only lower prices, but more varied plans, ranging from about

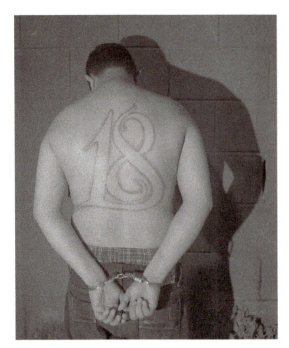

A leader of the notorious Central American gang Barrio 18, Cesar Vladimir Montoya Climaco, was arrested on July 28, 2015. While freedom of speech is constitutionally guaranteed in Honduras, horrific gang violence presents a constant threat to those who would speak against them, leading online journalists and bloggers to self-censor for their own safety. (AP Photo/Salvador Melendez)

$0.35 to $0.44 a day to about $4.37 for fifteen days' worth of service (Rodriquez 2014). Considering the average monthly income in 2015 was only $189 (World Bank 2016), it is difficult for many Hondurans to afford data plans for their smartphones even at these low prices, let alone fixed-line internet connections.

The websites with the most traffic in Honduras are similar to those of other Latin American countries; for example, YouTube sees more web traffic than Facebook in Colombia, Chile, Ecuador, Peru, and Venezuela. In Honduras, with the exception of La Prensa, a news website that ranks number nine (Alexa 2016), the most trafficked sites are almost entirely of foreign origin. With YouTube in first place, the rest of the top ten most trafficked websites, in order of popularity in 2016, were Google.hn, Google, Facebook, Yahoo, Amazon, Live, Wikipedia, and Blogspot. Other social media sites, such as Taringa! (a Reddit-like forum out of Argentina), WhatsApp, Instagram, and Twitter are less popular, ranking at seventeen, nineteen, twenty, and twenty-four, respectively. On mobile devices, Facebook is the most used website for social networking; in 2015, it comprised 83.8 percent of social media usage. Twitter ranked a distant second, with 11.32 percent usage; Pinterest ranked third, with 3.61 percent usage; and other social media sites, such as Google+, Reddit, StumbleUpon, and Tumbler, comprised the remaining 1.27 percent (StatsMonkey 2015).

As stated previously, while the Honduran constitution guarantees freedom of speech, due to the horrific gang violence and organized crime that characterizes the country today, many journalists and bloggers self-censor in order to keep themselves and their families safe. In 2015, the murder rate in Honduras was 169 in every 100,000 people, one of the highest in the world (Crilly 2015), and about double what it had been just three years earlier (Pressly 2012).

The 2009 coup d'état of Honduran president Manuel Zelaya (1952–) is often cited as a major cause in the violent reality of the country today. After Zelaya was ousted, there were several political killings, which often involved private citizens

Latin Americans often turn to social media for social activism. In Panama, the advertising agency P4 Oligvy and Mather used the TV program *Telemetro Reporta* to spread "El Hueco Twitter," or the "Tweeting Pothole." In a collaborative effort, the group placed motion sensors into Panama City's worst potholes. When the motion sensors activated, they sent signals that initiated tweets directed at the country's Ministry of Public Works. These tweets used gentle and humorous language to remind the municipal government to fix the dangerous road conditions. The account generated public interest and led to improving public works (Matchar 2015).

who spoke out against the new government. In addition, Honduras had to deal with pressures from the Mexican drug cartels and the pervasive spread of gangs throughout Central America. Today, gangs such as MS-13 and Barrio 18 present a constant threat on the streets of Honduras.

Since 2009, the Honduran government has not successfully addressed the human rights violations and attacks that occurred in the wake of the coup. Between January 2010 and November 2011, unknown attackers killed at least twelve journalists, and many more received death threats or threats directly at their families (Human Rights Watch 2012). Very few of these murderers, if any, were held accountable for their crimes. Most of the journalists killed had spoken out against the 2009 coup or had frequently reported on corruption in the country.

Marilyn J. Andrews and John G. Hall

See also: Argentina; El Salvador; Mexico; United States

Further Reading

Alexa. 2016. "Honduras." Accessed August 28, 2016. http://www.alexa.com/topsites/countries/HN

Crilly, Rob. 2015. "The Majority of Homicides Are Young People. It's So Sad." *The Telegraph,* August 3. Accessed August 28, 2016. http://www.telegraph.co.uk/sponsored/lifestyle/honduras-gangs/11701324/honduras-murder-rate.html

Humans Rights Watch. 2012. "World Report 2012: Honduras, Events of 2011." Accessed August 28, 2016. https://www.hrw.org/world-report/2012/country-chapters/honduras

Internet Live Stats. 2016. "Honduras Internet Users." Accessed August 28, 2016. http://www.internetlivestats.com/internet-users/honduras/

Internet World Stats. 2016. "Internet Usage and Population in Central America." Accessed August 28, 2016. http://www.internetworldstats.com/stats12.htm

La Tribuna. 2011. "Claro Invests $150 Million in Honduras." December 5. Accessed August 28, 2016. http://www.centralamericadata.com/en/article/home/Claro_Invests_150_Million_in_Honduras

Pressly, Linda. 2012. "Honduras Murders: Where Life Is Cheap and Funerals Are Free." *BBC,* May 3. Accessed August 28, 2016. http://www.bbc.com/news/magazine-17870673

Rivera, Juan Carlos. 2016. "El 30% de Hondureños se Conecta a Internet." *La Prensa,* February 8. Accessed August 28, 2016. http://www.laprensa.hn/honduras/927465-410/el-30-de-hondure%C3%B1os-se-conecta-a-internet

Rodriguez, Luis. 2014. "¿Cómo elegir el servicio de Internet más barato en Honduras?" *El Heraldo,* July 4. Accessed August 28, 2016. http://www.elheraldo.hn/economia/609408 -216/como-elegir-el-servicio-de-internet-mas-barato-en-honduras

StatsMonkey. 2015. "Honduras Social Media Usage Statistics Using Mobile." Accessed August 28, 2016. https://www.statsmonkey.com/sunburst/21368-honduras-mobile-social -media-usage-statistics-2015.php

World Bank. 2016. "Honduras." Accessed August 28, 2016. http://data.worldbank.org/ country/honduras

HUNGARY

Hungary is a country located in Central and Eastern Europe, with approximately 9.8 million inhabitants, and a gross domestic product (GDP) of $26,256 per capita (OECD 2016). The country is a member of the European Union, the North Atlantic Treaty Organization (NATO), and the Organisation for Economic Co-operation and Development (OECD).

According to Eurostat (2015) data, 76 percent of households in Hungary have access to the internet. The Hungarian Central Statistical Office (Központi Statiszti-kai Hivatal, or KSH) reported in 2013 that the overwhelming majority of internet connections were broadband. A Gemius (2014) report showed that the population of Hungarian internet users had reached 5 million. This translates to over 60.78 percent internet penetration, in line with the European average and higher than the Central and Eastern European average. Further, 86 percent of users access the internet on a daily basis.

Many internet users (22 percent) are based in the country's capital, Budapest, which is the most developed region in Hungary. Of the rest, 52.8 percent live other urban areas, while 25.2 percent live in villages. This pattern illustrates how the digital divide is structured in Hungary: most of the internet users are concentrated in western Hungary, in the capital, and in the larger cities, while the smaller settlements, particularly in the eastern part of the country, have less connection to the network. According to the Gemius data set, 51.2 percent of users are women and 48.8 percent are men. Over half the online audience (52 percent) is between eighteen and thirty-nine years old.

Several internet service providers (ISPs) operate in the country; among the biggest players are Magyar Telekom (a subsidiary of Deutsche Telekom), UPC, Invitel, and DIGI. The estimated size of the Hungarian ISP market is about $580 million. The majority of Hungarians access the internet through mobile devices: by the end of 2015, 67 percent of internet subscriptions were mobile internet subscriptions (KSH 2016).

Mobile phone penetration is relatively high in the country; since 2007, the number of subscriber identity module (SIM) cards has outnumbered the population (Pinter 2008). The Hungarian National Media and Infocommunications Authority (Nemzeti Média-és Hírközlési Hatóság, NMHH) reported in March 2013 that the number of active SIM cards in the country was 11.5 million, which translates to a 116.1 percent penetration. According to Think with Google data, smartphone penetration in the country in 2013 was 34.4 percent. This number grew to 62 percent in 2015 (eNET 2015a). Three providers share the market: Magyar Telekom (45.87 percent), Telenor (31.48 percent), and Vodafone (22.65 percent).

The most popular web browser is Google Chrome (46.9 percent), followed by Firefox (31.39 percent) and Opera (7.4 percent). The global players Google and Facebook dominate all statistics regarding online content and media. Around 90 percent of the online audience is on Facebook, and almost all Hungarians use some services provided by Google. Google dominates the search market with no competitors, and is leading the online mapping, video streaming, and email markets as well. The top three most visited pages in Hungary are Facebook.com, Google (both the English and Hungarian versions) and YouTube.com; domestic domains trail these in popularity. Twitter, Instagram, LinkedIn, Tumblr, Foursquare, and Snapchat are also used, but those platforms are not nearly as popular as the leading ones (Alexa 2016; Similarweb 2016).

As of April 2016, the most visited Hungarian sites were Blog.hu (blogging platform), Origo.hu (news and services), Jófogás.hu (e-commerce), Index.hu (news), 24.hu (news), NLCafe.hu (entertainment), Blikk.hu (tabloid news), Hvg.hu (news), Startlap.hu (link portal), and Árukeresö.hu (e-commerce) (Gemius 2014). There were 2,656,411 total cross-platform visitors for Blog.hu, the most popular, and 1,509,902 visitors for Árukeresö.hu, which was in tenth place.

The only social media in Hungary that is not overtaken by international platforms is blogging. The country's largest blogging platform, Blog.hu, is owned by the media company CEMP, which also publishes the Index.hu news site. CEMP established itself as the market leader by developing a particular content management strategy closely linked to Index.hu. In 2007, a so-called blog cage was introduced on the Index.hu main page, which featured the most popular posts written by users and placed them between editorial content. The move resulted in the increasing visibility of user-generated content and eventually led to the emergence of a particular blogging ecosystem in which a news site (Index.hu) dominated the blogosphere by selecting, featuring, and editing posts.

Another uniquely Hungarian chapter of social media history was presented by iWiW, one of the first social networking sites of its kind in the world. Launched in 2002, iWiW (then simply WiW) quickly grew into the most popular social networking site in Hungary, essentially accommodating the whole internet user population of more than 4 million people by 2008. The site was purchased by Magyar Telekom in 2006 for more than $3.5 million, but due to several financial and technical difficulties, the company shut down the site for good in 2014.

Now Hungarians overwhelmingly use Facebook for social networking and increasingly for media consumption. Trends show that Facebook is actually taking over as a main source for news distribution, significantly affecting user attitudes toward the nature of news. As a 2015 study by NRC Market Research Ltd., commissioned by the Association of Hungarian Content Providers (Magyarországi Tartalomszolgáltatók Egyesülete, or MTE), has shown, the average Facebook user in Hungary spends eighty-six minutes on Facebook per day, and 46 percent of them tend to read news items only in their Facebook feeds, with no clickthrough to the source of the content. According to the eNET-Telekom "Report on the Internet Economy" in 2015, 83 percent of internet users believe that being online can be interpreted as acceptable family programming, or activities in which an entire family can participate, that may include social media usage, online shopping, listening to music, or watching movies (eNET, 2015b).

In the 2015 Freedom of the Net report by Freedom House, Hungarian freedom of the press status for broadcast and print media was listed as only "partly free"; however, Hungarian internet scored a total of 24 points, with an overall assessment of "free." There is no evidence for governmental blocking of social media/information and communication technology (ICT) apps, of access to internet, of filtering, of monitoring content, nor of censorship. In its 2016 Government Requests Report, Facebook reported that in the second half of 2015, Hungarian governmental bodies requested the data of 224 users, 178 times altogether. Facebook deemed valid 42.13 percent of the requests and provided some information to the government.

Social media, online news industry, and internet policy is highly intertwined with politics in Hungary (Reuters 2016; Toth 2012). After a series of government-related scandals (such as alleged pressure put on news sites by a government official and opaque public money being channeled to favored media outlets), the online media market has been undergoing a constant transformation, with several new players appearing in the market and experimenting with new business models and content strategies (Tofalvy 2015).

A recent international internet and social media policy development that is related to a Hungarian case is the decision of the Strasbourg court on February 2, 2016, regarding intermediary liability. In *MTE and Index.hu Zrt v. Hungary,* the European Court of Human Rights ruled that an online news portal cannot be held responsible for offensive or vulgar comments posted by its readers, as that would violate Article 10 of the European Convention on Human Rights. This important decision marked a total reversal of the same court's 2015 decision in *Delfi v. Estonia,* in which a website was held responsible for reader comments (MTE 2016).

Another recent internet policy issue that made headlines was the series of mass demonstrations in Budapest and across the country during the fall of 2014, which followed the governmental announcement of an "Internet tax" based on personal data traffic. Soon after the protests, the government withdrew the proposal and later started a new program aimed at surveying user attitudes on the internet, which has often been criticized as unprofessional, biased, and part of government propaganda.

Tamas Tofalvy

See also: Germany; Poland; Romania

Further Reading

Alexa. 2016. "Top Sites in Hungary." Accessed July 4, 2016. http://www.alexa.com/topsites/countries/HU

eNET. 2015a. "Breakthrough in Mobile Net Usage: Half of Hungarian Internet Users Keep the Web in Their Pockets." Accessed July 4, 2016. http://www.enet.hu/news/break through-in-mobile-net-usage-half-of-hungarian-internet-users-keep-the-web-in-their -pockets/?lang=en

eNET. 2015b. "Internet in Hungary: (Dis)connecting Families?" Accessed July 4, 2016. http://www.enet.hu/news/internet-in-hungary-disconnecting-families/?lang=en

Eurostat. 2015. "Level of Internet Access—Households." Accessed July 4, 2016. http://ec .europa.eu/eurostat/web/information-society/statistics-illustrated

Facebook. 2016. "Government Requests Report." Accessed July 4, 2016. https://govtrequests .facebook.com/

Freedom House. 2015. "Freedom of the Net—Hungary." Accessed July 4, 2016. https:// freedomhouse.org/report/freedom-net/2015/hungary

Gemius. 2014. "Facts about Hungarian Internet Users." Accessed July 4, 2016. https://www .gemius.com/agencies-news/facts-about-hungarian-internet-users.html

Központi Statisztikai Hivatal (KSH). 2016. "Hungarian Central Statistical Office: Statistical Reflections—Telecommunications, Internet, Q4, 2015." Accessed July 4, 2016. https:// www.ksh.hu/docs/eng/xftp/gyor/tav/etav1512.pdf

Magyarországi Tartalomszolgáltatók Egyesülete (MTE). 2016. "Online News Sites Are Not Liable for Vulgar Comments: The ruling in *MTE and Index.hu Zrt v. Hungary* leads the way to a new era after *Delfi v. Estonia.*" Accessed July 4, 2016. http://mte.hu/mte_index _v_hungary/

Organisation for Economic Co-operation and Development (OECD). 2016. "Country Statistical Profile: Hungary." Accessed July 4, 2016. http://www.oecd-ilibrary.org/economics /country-statistical-profile-hungary_20752288-table-hun

Pinter, Robert. 2008. "Development of the Hungarian Information Society in the Last Decade. Hungarian Country Report, 1998–2008." BUTE-UNESCO Information Society Research Institute (BME-ITTK)—GKIeNET Ltd. Accessed July 4, 2016. http://www.ittk .hu/netis/doc/textbook/Hungarian_country_report_final.pdf

Reuters Institute. 2016. "Digital News Report—Hungary." Accessed July 4, 2016. http://www .digitalnewsreport.org/survey/2016/hungary-2016/

Similarweb. 2016. "Top 50 Sites in Hungary for All Categories." Accessed July 4, 2016. https://www.similarweb.com/country/hungary

Tofalvy, Tamas. 2015. "Newcomers in the Hungarian Online Media Market—Business Models, Content Strategies and Media Pluralism." European University Institute (EUI), Centre for Media Pluralism and Media Freedom (CMPF). Accessed July 4, 2016. http:// cmpf.eui.eu/Documents/SS15/TamasTofalvyNewcomersinHungarianonlinemedia market.pdf

Toth, Borbala. 2012. "Mapping Digital Media: Hungary." Open Society Foundation. Accessed July 4, 2016. https://www.opensocietyfoundations.org/reports/mapping-digital-media -hungary

ICELAND

Iceland is a Scandinavian country located in the Atlantic Ocean. In 2015, its population numbered around 329,000 inhabitants. An estimated 98 percent of Iceland's population has internet access, making it the most connected European nation (Robert 2015). The internet plays a large role in Icelandic society; it has functioned as a vital conduit in improving communications between the government and its citizens, and has become a platform for social activism. In fact, the government utilizes social media innovatively and places few restrictions on freedom of expression. The Icelandic people pride themselves on having one of the freest, most unrestricted media in the world.

As of 2015, Icelanders preferred international social media sites, although they frequented pages featuring local content. They predominantly used the Facebook platform. Icelanders were also active on Reddit, Pinterest, Twitter, Tumblr, StumbleUpon, Google+, LinkedIn, Instagram, Snapchat, and Y Combinator. A very small number appeared on vKontakte, the Russian version of Facebook. Younger users, particularly thirteen- to fifteen-year-olds, spent an average of four hours a day on social media; they preferred to communicate via Facebook, Instagram, and Snapchat (Robert 2015). Across age groups, Facebook was the top platform. On Facebook, the most popular page during 2015 was Iceland Foods. With 376,498 followers, the page had more followers than the country's total population. After Iceland Foods, the top pages were Icelandair, Nói Síríus, Maurizio Moletta, and WOW Air (Allin1 Social 2015). The preference for international social media platforms and local content held true when it came to Iceland's most accessed websites, which as of late 2015 were Facebook, Google, Google.is, YouTube, Mbi.is, Visir.is, Dv.is, Ja.is, Pressan.is, and Amazon (Alexa 2015).

Iceland's app preferences differed slightly from those of other European nations. The country's iOS users had a fascination with apps aimed at American celebrities, while Windows phone users took a more practical approach with local news and document reading apps. The country's most popular free iOS apps, listed in descending order, were Pop the Lock, Kylie Jenner Official App, Kim Kardashian West Official App, Snapchat, and the Kendall Jenner Official App. The top paid apps were Minecraft: Pocket Edition, Photomyne, Fantasy Premier League 2015, Monopoly, and the Big Day Event calendar. For Windows phones, the top free apps were OneDrive, Facebook Messenger, Facebook, Adobe Reader, and Skype. The top paid apps were Ruv.is fréttir, Mbl.is fréttir, Stupid Test, Fridagr, and Fruit Ninja (AppAnnie 2015). From the preferred apps, Icelanders liked to download games in addition to more practical programs.

Depicted here is an Icelander named Guðmundur, who is part of the world's first human search engine. "Ask Guðmundur" is Iceland's most successful social media endeavor to date. Staffed by volunteers from Iceland's seven regions, the service provides answers about the county and its culture via Facebook, Twitter, and YouTube. (Inspired by Iceland via AP Images)

The major trends in Icelandic social media revolved around how Iceland's government has used social media in many innovative campaigns to entice tourists and to promote civic participation. One of the most interesting initiatives is Ask Guðmundur (#AskGudmundur), the world's first human search engine designed to answer questions about the country and its culture. Iceland chose representatives from its seven regions—all named Guðmundur or Guðmunda, the female variation on the name—to answer all the questions posed via Facebook and Twitter (Iceland Naturally 2015). For some questions, the answers appeared in specially-produced YouTube videos. The initial campaign was a successful experiment, leading to the release of Guðmundur 2.0 in September 2015. The implementation of a human-run search engine has been one of Iceland's most successful social media efforts to date. The human search engine concept has provided tourists with a wealth of firsthand, credible information and placed a personal spin on how Iceland markets to its visitors.

A less successful, government-sponsored social media effort was to attempt to harness the people's voices on Facebook and Twitter to crowdsource and redraft the national constitution. In 2008, Iceland's economy crashed due to overextended banks. People protested outside Parliament, demanding that the government ensure that the economy recover. In an effort to solicit the people's opinions on how the government should implement reform measures, the government established a Facebook page where the people posted their ideas and comments. Those posts helped inform the government what the public wanted. Based on these posts, face-to-face discussions with Icelanders, and a voting process, Iceland drafted a crowd-sourced constitution. However, the new constitution stalled in 2013 with the country's change in government (Landemore 2014). The effort, while unusual and unsuccessful, showed how social media increased political participation and transparency. The people provided inputs through a variety of channels, showing that social media did contribute to promoting democracy in Iceland.

In September 2015, social media became one of the channels through which the people could influence the government over the refugee crisis. As Europe experienced a wave of refugees from Syria, the Middle East, and Eastern Europe, Iceland had announced that it would admit up to fifty refugees between 2015 and 2016, with fifty being the legal cap as to the number of refugees permitted to enter the country. In response, citizens created the Facebook page "Syria Is Calling" to pressure the government into raising the number of permitted refugees. By September 2015, more than 16,000 Icelanders had responded to the page, many of whom offered to open their homes to refugees. People unable to provide shelter offered Syrian refugees food, plane tickets, and general support (Mendoza 2015). The Icelandic effort was popular abroad, as well as at home. Shortly after its creation, the Facebook page "Open Homes, Open Hearts" opened in the United States, with similar goals as the Icelandic page. An already-existing Facebook page, "Refugees Welcome," received renewed attention as well; the page connects German hosts with refugees needing shelter as they start through the political asylum process. The "Syria Is Calling" site planned to continue spreading messages about the Syrian refugee plight until the government increased the annual national allotment for refugees.

Laura M. Steckman

See also: Finland; Germany; Russia; Syria; United States

Further Reading

Alexa. 2015. "Top Sites in Iceland." Accessed September 19, 2015. http://www.alexa.com /topsites/countries/IS

Allin1 Social. 2015. "Facebook Statistics for Iceland." Accessed September 19, 2015. http:// www.allin1social.com/facebook-statistics/countries/iceland?page=1&period=six _months

AppAnnie. 2015. "Windows Phone Top App Charts." Accessed September 19, 2015. https:// www.appannie.com/apps/windows-phone/top/iceland/

Iceland Naturally. 2015. "Iceland Launches 'Ask Guðmundur,' the World's First Human Search Engine." April 28. Accessed September 19, 2015. http://icelandnaturally.com /article/iceland-launches-ask-gudmundur-worlds-first-human-search-engine

Landemore, Hélène. 2014. "We, All of the People: Five Lessons from Iceland's Failed Experiment Creating a Crowdsourced Constitution." July 31. Accessed September 19, 2015. http://www.slate.com/articles/technology/future_tense/2014/07/five_lessons _from_iceland_s_failed_crowdsourced_constitution_experiment.html

Mendoza, Jessica. 2015. "Icelanders Use Facebook to Prompt Government to Welcome Refugees." September 3. Accessed September 19, 2015. http://news.yahoo.com/icelanders -facebook-prompt-government-welcome-refugees-204253122.html

Robert, Zoë. 2015. "Teenagers Spend Four Hours Per Day on Social Media." February 6. Accessed September 18, 2015. http://icelandreview.com/news/2015/02/06/teenagers -spend-four-hours-day-social-media

INDIA

Surrounded by its South Asian neighbors, Pakistan, Nepal, and Bangladesh, India is a country of 1.2 billion people, second in population only to China. As of December 2015, India has the second-highest number of internet users in the world; only China has more. Mobile internet access is growing rapidly in the country, though still unavailable in many rural areas. With increasing technical support for the Hindi language and a sizable English-speaking population, language barriers are still a challenge in reaching more users. Rapid growth and a sheer massive number of users have led many of the largest international technology companies and smaller venture capital groups to make strides in providing services in India.

The Internet became available in India in 1995; however, slow speeds kept the growth of internet use at a minimum for the next decade. By 2005, the government moved toward broadband access, but it was still defined by relatively slow speeds. In 2010, 3G and 4G services finally became available in India, bringing a simultaneous rise in internet users.

In 2014, the federal government launched "Digital India." One piece of this project is BharatNet, which aims to bring broadband access to more areas and to connect many rural areas to the internet. Currently, the majority of users in India do not have regular broadband access. Laying the optical fiber network, however, is still running on a delayed schedule. In early 2016, the Telecom Regulatory Authority of India (TRAI) recommended creating a private-public partnership for this project.

While the overall internet penetration rate in India remains low at about 30 percent, there has been massive recent growth. As of December 2015, there were estimated to be over 400 million users, which is a 49 percent increase over the previous year, and the second-largest internet user base in the world (Verma 2015). While this rapid growth has been affected by the availability of high-speed internet services, it is primarily explained by the rapid increase in mobile use, which is in turn related to falling smartphone prices and low data rates. Due to competition between mobile operators, data rates for mobile phones are now available in India for as little as $0.17 for 30 MB (McLain 2015).

While smartphones in India still typically cost $100 to $150 at a minimum, in February 2016, the world's cheapest smartphone was launched in India by a company called Ringing Bells. The Freedom 251 smartphone costs about $4 to purchase and made international headlines upon its release (Chappell 2016). The phones have not yet arrived into the hands of customers, and some have concluded that the entire project is a scam. The company is currently under investigation: however, if the phones become available, they have the potential to make internet access available to a large portion of the Indian population that is not yet online.

Cybercafés in many towns in India have long provided computers with internet access for a fee. Some have been in business for twenty years, providing access to a larger portion of the population that could not afford personal computers and internet access. These are on the decline as smartphones and faster internet speeds become increasingly affordable and available.

As with many other countries in South Asia, internet access is still biased toward urban areas and higher incomes. Rural areas are seeing recent large increases in internet use, primarily through mobile phones. By October 2015, rural internet users reached 108 million, a 77 percent increase over the previous year (Verma 2015).

The many languages and multiple scripts used in India are also a factor in internet use. An estimated 10 percent of the population speaks English as a second language. Hindi is the most widely spoken language in India; however, many states also have their own state-level official languages. The 2001 census in India found that there were 29 languages across the country that had at least 1 million native speakers. Most of these languages do not have support in the online environment, as the majority of them have their own script. Google has made some recent strides to accommodate this. Hindi was added to Google Translate in 2007, with five more Indian languages (Bengali, Gujarati, Kannada, Tamil, and Telugu) added in 2011. Between 2013 and 2016, Marathi, Punjabi, Malayalam, and Sindhi were also added. At the end of 2014, Google launched the Indian Language Internet Alliance to increase the amount of content available on the internet in Indian languages.

The Indian version of Google (google.co.in) is the number one website accessed in India (Alexa Internet 2016). While the main search page is in English, searches can also be conducted in Hindi, Bengali, Telugu, Marathi, Tamil, Gujarati, Kannada, Malayalam, and Punjabi. Users are able to search using their own language, as well as the script itself, rather than a transliteration. Google.com, the company's main page and the version used in the United States, ranks second.

This doubling up of the Indian and U.S. versions of websites also occurs with Amazon, with the two coming in at number five and number seventeen in popularity, respectively. Although e-commerce is still a small market, reflecting only about 15 percent of internet users in India, it is also growing rapidly, and other websites in the top twenty reflect this growth. One estimate of e-commerce spending saw a 27 percent jump, from $81 billion to $103 billion at the end of 2015 (Mukherjee and Malviya 2015). Beyond Amazon, two other major e-commerce sites are in the top twenty websites for India: Flipkart at number seven and Snapdeal at number thirteen.

The remaining websites that make up the top ten in India are all well known globally: Facebook, YouTube, Yahoo, Wikipedia, LinkedIn, and Twitter (Alexa Internet 2016). However, there are some variations in the kinds of websites found that rank eleventh through twentieth. In eleventh place is Indiatimes.com, which is a news portal covering world news, entertainment, and videos among other subjects. Stackoverflow, coming in at number fifteen, is a website devoted to questions and answers about computer programming, and reflects the strong information technology (IT) industry in India. Three other sites are national banking sites. Physical banks in India are primarily located in urban areas, and only about 60 percent of the population has a bank account. The rise of three banking websites to the top twenty reflects the growth of the internet in daily life activities. Also reflecting this task-oriented approach to the internet is the Indian Railway website, which can be used for viewing schedules and online ticket booking, comes in at number twenty.

Police departments around the United States have started developing unofficial social media officers and units. At the San Francisco Police Department, Officer Eduard Ochoa (n.d.–) has the unofficial nickname of "the Instagram Officer." Part of his job is to examine social media—and not just Instagram—to look for evidence of criminal activity (Wong 2015). Similarly, the Indian police force, which already has dedicated units to monitor social media, set up an official Social Media Lab in 2013. Specially trained personnel monitor public sentiment to prevent public protests and riots. While critics argue that the lab promotes censorship and cracks down on internet activism, the police claim to focus on maintaining order and public safety.

Facebook, ranking as the third most used website in India, represents the second-largest user base for the social media site. Unlike some of India's neighboring countries, whose rankings reflect both entertainment and news-based preferences, India's Facebook users clearly focus on entertainment in their engagement with this major social media site. While the most liked Facebook page in India is that of Narendra Modi (1950–), the prime minister of India, the remaining top nine are entertainment- or sports-related (Socialbakers 2016). Five of them are Bollywood actors and actresses, one is a Bollywood musician and music producer, and one is a Bollywood playback singer. Rounding out the top ten are two Indian cricket players and the Indian Cricket Team's page.

Recently, the TRAI was prompted to review net neutrality issues in relation to Facebook's Free Basics program. The Free Basics program offers free mobile access to Facebook, Wikipedia, and a number of other health and news sites that have agreements with Facebook to provide content. The format of the sites is for the purpose of easy access and lower bandwidth.

While India does not have any laws regarding net neutrality, the TRAI announced a ruling in February 2016 that effectively barred the Free Basics program from running in India, by stating that there could not be discriminatory fees based on different kinds of content for data services. They ruled that Facebook's service does not provide free and open internet access, which effectively shut down the program in India immediately.

With the massive growth of internet users in India, other internet and technology giants are looking toward India as well. In January 2016, Google made free wireless internet available at Mumbai's train station. This is the first in a series of planned openings at train stations across the country, and the company has implemented the program in concert with the Indian government. Users at any platform of the train station can freely access one hour of high-speed wireless internet, with ongoing access at a slower speed. Netflix joined this growing crowd in February 2016, when they began to offer streaming services there for the first time.

Censorship of specific websites, of certain types of website content, and even of the internet itself has been on the rise in India. Four different state-level governments have shut down the internet within their respective states in 2014 and 2015. The federal government has blocked various Uniform Resource Locators (URLs) and, more recently, certain kinds of content, such as pornography or those potentially related to terrorist groups, on an increasing basis over the past eight years. Currently, India has blocked the most Facebook pages of any country. According to Facebook, they blocked over 15,000 pieces of content on their site in the first six months of 2015 (Facebook 2015).

Karen Stoll Farrell

See also: Bangladesh; China; Nepal; Pakistan

Further Reading

Alawadhi, Neha. 2015. "India's Internet User Base 354 Million, Registers 17% Growth in First 6 Months of 2015: IAMAI Report." September 3. Accessed February 22, 2016. http://articles.economictimes.indiatimes.com/2015-09-03/news/66178659_1_user-base-iamai-internet-and-mobile-association

Alexa Internet, Inc. 2016. "Top Sites in India." Accessed February 22, 2016. http://www.alexa.com/topsites/countries/IN

Burke, Jason. 2015. "India Supreme Court Strikes Down Internet Censorship Law." March 24. Accessed March 17, 2016. http://www.theguardian.com/world/2015/mar/24/india-supreme-court-strikes-down-internet-censorship-law

Chappell, Bill. 2016. "'World's Cheapest Smartphone' Goes on Sale for 4$ in India." February 18. Accessed March 17, 2016. http://www.npr.org/sections/thetwo-way/2016/02/18/467220085/-world-s-cheapest-smartphone-goes-on-sale-for-4-in-india

Crabtree, James. 2016. "India Internet: Laying the Foundations." January 14. Accessed March 17, 2016. http://on.ft.com/1TYe7jH

Facebook. 2015. "Government Requests Report: India January 2015–June 2015." Accessed March 17, 2016. https://govtrequests.facebook.com/country/India/2015-H1/

Goel, Vindu, and Isaac, Mike. 2016. "Facebook Loses a Battle in India over Its Free Basics Program." February 8. Accessed February 22, 2016. http://www.nytimes.com/2016/02/09/business/facebook-loses-a-battle-in-india-over-its-free-basics-program.html?ref=technology

McLain, Sean. 2015. "India to Become World's Second-Largest Internet User Base." November 17. Accessed April 20, 2016. http://www.wsj.com/articles/india-to-become-worlds-second-largest-internet-user-base-1447776612

Mukherjee, Writankar, and Malviya, Sagar. 2015. "India's Ecommerce Market to Breach 100 Billion Mark by FY20: Goldman Sachs." October 26. Accessed April 20, 2016. http://economictimes.indiatimes.com/industry/services/retail/indias-ecommerce-market-to-breach-100-billion-mark-by-fy20-goldman-sachs/articleshow/49532128.cms

Reisinger, Don. 2015. "India's Internet Population Could Soon Rival China's." November 18. Accessed February 22, 2016. http://fortune.com/2015/11/18/india-internet-population/

Socialbakers. 2016. "India Facebook Page Statistics." Accessed February 22, 2016. http://www.socialbakers.com/statistics/facebook/pages/total/india/

The Hindu. 2014. "Google Initiative to Promote Indian Languages on Web." November 3. Accessed February 26, 2016. http://www.thehindu.com/sci-tech/technology/internet/google-launches-alliance-to-promote-indian-languages-on-web/article6561353.ece

Verma, Shrutika. 2015. "India on Course to Overtake US Next Month in Internet User Base." December 2. Accessed February 22, 2016. http://www.livemint.com/Politics/9Vipq3XmcfQuhJRMBleuwL/Indias-Internet-users-set-to-increase-49-to-402-million-by.html

Zachariah, Reeba, and Vipashana, V. K. 2014. "Are Cybercafes Logging Off?" September 1. Accessed March 17, 2016. http://timesofindia.indiatimes.com/tech/tech-news/Are -cybercafes-logging-off/articleshow/41393447.cms

INDONESIA

Indonesia is an archipelagic nation that consists of more than 17,000 islands. It sits between the Indian Ocean, the South China Sea, and the Pacific Ocean. Australia lies to its south, and the Philippines to the north-northeast. Indonesians love social media and have become one of the highest social media content producers in the world. In fact, scholarship on Indonesian culture often suggests that the country has an online social-networking addiction (Lim 2013, 636). Approximately 73 million people, or 29 percent of the country's population, access the internet. Of those, an estimated 62 million utilize social media; the majority of social media users fall into the thirteen- to thirty-four-year-old age range (Jakarta Post 2015). Indonesians are active on Facebook, Twitter, YouTube, Pinterest, Tumblr, Reddit, StumbleUpon, Instagram, Wordpress, WhatsApp, Line, Blogspot, and are often very active on public and private forums.

Across the country, Facebook is an extremely important platform. In fact, Indonesian users fall into Facebook's top five rankings by nation. Approximately 57.6 percent of Indonesian users are male and 42.4 percent are female; the site is most popular with eighteen- to twenty-four-year-old males (Allin1 Social 2015). In general, Indonesian users post on Facebook about their activities. Parents post about their kids' accomplishments, while adult men post about their activities and popular culture (Lim 2013, 639). Facebook became popular because of its ease of accessibility, its chat and messenger functions, and its photo- and video-uploading features. Most users access the platform via mobile phone, and in fact, Indonesians use more mobile phones to access the site than any other country in the world. Prior to Facebook's rise in popularity, many Indonesians used MySpace. It was not until the late 2000s that Facebook's greater functionality won over a large portion of Indonesians and caused them to shift away from MySpace.

Twitter is the second most popular mainstream platform in the country. Indonesians tweet to inform followers about their activities, common among local celebrities; to support sports teams; to galvanize people around political issues, such as to garner campaign support during elections as occurred during the presidential elections in 2014; to share news; and to promote and organize social and civic activism. For example, Indonesia's top five hashtags of 2014 were #Halamadrid, in support of the Real Madrid soccer team; #GGMU, in support of Manchester United; #PrayForGaza, to show solidarity and support for the people in Gaza and Palestine; #AkhirnyaMilihJokowi (or "Finally Pick Jokowi"), during the presidential elections; and #YNWA (or "You Never Walk Alone"), which is the slogan for Liverpool FC (Lukman 2014). These popular hashtags demonstrate some of the ways that Indonesians use Twitter.

Twitter names a "Golden Tweet" each year, for the message that received the most retweets. In 2014, the Golden Tweet was Ellen DeGeneres's star-studded selfie photo at the Oscars. The runner-up tweet came from Indonesia, as part of the over 95 million tweets produced during the presidential election. Denny Januar Ali (1963–), a political consultant, tweeted that Indonesia's netizens could ensure a win for the presidential and vice presidential candidates, Joko Widodo (1961–) and Jusuf Kalla (1942–), solely by retweeting his message (Lukman 2014). The tweet received more than a million retweets and, in the end, the candidates did win. Popular support on Twitter was not the reason why they succeeded; however, the outpouring of support on social media was indicative of how much mass support they had and provided a solid indication that they were contenders for their respective political offices.

Joko Widowo, popularly known as Jokowi, was one of Indonesia's presidential candidates in 2014. During his election campaign, he was so popular that the campaign hashtag #AkhirnyaMilihJokowi (#FinallyPick-Jokowi) was one of the year's top five most active Twitter hashtags. Indonesia boasts one of the world's largest Twitter-using populations. (Garudeya/Dreamstime .com)

Apps are an important part of Indonesian social media usage, and apps that enable chat functions or provide real-time information are preferred. On Google Play, the top free apps are BBM (BlackBerry), Facebook, Facebook Messenger, the UC Browser, and the WhatsApp Messenger. The top paid apps are Poweramp, Flightradar24, Afterlight, League of Stickman-Samurai, and Link2 SD Plus. Data on iOS

If Twitter were a nation, Jakarta, Indonesia, would be its capital. The city first received the accolade in 2012, and as of August 2015, continues to hold the title. Jakarta maintains this status because its citizens produce the highest volume of Twitter posts of anywhere else in the world. The city's love for Twitter has grown so significant that in March 2015, Twitter opened its first office in Southeast Asia right in Jakarta, the heart—or virtual capital—of the Twitter nation (Russell 2014).

top apps are similar. The top free apps include Go-Jek, BBM (BlackBerry), Instagram, Line, and WhatsApp Messenger. Popular paid apps are Pastel Keyboard Themes, Afterlight, Flightradar24, Free Music Pro (an MP3 player), and Minecraft Pocket Edition (AppAnnie 2015). The importance of apps and social media is visible when viewing Indonesia's most accessed webpages. As of September 2015, those pages were Google; Google.co.id; Facebook; YouTube; Blogspot; Yahoo; Detik, a national news site with a forum; Kaskus, arguably the world's largest social media forum; Wordpress, and Ask.com (Alexa 2015).

Because social media has become a key communications channel in Indonesia, local entrepreneurs and computer programmers have attempted to "Indonesianize" the environment. They have a record of developing local social media platforms designed to integrate local culture with all the features that Indonesians consider vital to maintaining their social networks. In 2015, Indonesians developed and launched more than twenty-five distinct local apps and platforms. One of the more promising platforms is Kwikku, intended to compete directly with Facebook and Twitter. Kwikku's primary features include the ability to use stickers and smiley faces, an e-learning feature, an email system, a blogging capability, file sharing, a business and marketing function, and an internal payment system. The platform also supports local languages (Tuliside 2015).

Other Indonesian social media platforms developed or modified in 2015 included Adandu, Akkucintasekolah, Salingsapa, Ayobai, Kenalanyuk, Indoface, Jomblo, Blucool, Smallboy, Paseban, Digli, Indofesbuk, Temanku, FB.co.id, Fupei Friends, Goesmart Sosial, Kiber, Kombes, Mobinessia, MyPulau, Ruangmuslim, Getfolks, Zeetal, and Sebangsa (Rame 2015). Indonesia's demand for social media consumption indicates that there will always be developers seeking to invent Indonesia-specific programming. Unless that demand subsides, there is a high probability that local apps and platforms will be released to meet social media users' needs and preferences.

Forums and discussion boards are another form of social communication where Indonesians are active. One of the largest forums in the world operates in Indonesia, catering primarily to Indonesians and Malaysians, whether regionally based or part of the diaspora. Kaskus was founded in Seattle, Washington, in November 1999 by a group of Indonesian college students. It grew in popularity and quickly rose to become Indonesia's largest forum. In 2008, the site owners chose to move its hosting location from the United States to Indonesia. By 2012, the site switched domains to an Indonesian server and modified its URL (Pardana n.d.). At the time of the switch, the site claimed to have over 6 million unique members. In the past few years, the site has met the changing demands of its users by developing its own mobile site and app to facilitate ease of use. It has also expanded the site to include online commerce and other features in addition to the discussion forum. The story and success of Kaskus have captured the Indonesian people's interest to the point that film producers planned to start filming a movie about the site and its developers in March 2015 (Republika 2014). Other forums have achieved popularity in the country, though none has been as phenomenal as Kaskus.

Indonesians use social media to promote social and civic activism. Older users tend to utilize blogs to discuss and convey information, while younger users tend to prefer shorter messages passed through Facebook and Twitter. The power of so-cial media is that messages can be dispersed throughout multiple networks and communities that would normally have loose ties, in contrast to the strength of physical social networks, the primary channels for Indonesian social communica-tions (Lim 2013, 643).

Ultimately, social media messaging can reach more people more quickly, which can lead to mobilization. Through that information, communities and individuals can act to support preferred causes. For example, Prita Mulyasari (1977–), a young mother of two, wrote an email explaining about the poor care she received during a 2008 hospital stay. The hospital obtained a copy of the email and had her arrested for cyberdefamation. The courts convicted her in 2009; they ordered prison time and a fine of over $20,000. In response, Facebook users started a page called Coins for Prita and asked people to donate their change to help her pay what they viewed as an unfair fine imposed on a victim of mistreatment. Seeing the public response in support of Mulyasari, the hospital decided to drop its civil suit against her. In 2012, the courts overturned her prison sentence on appeal.

In this instance, social media activism and people's willingness to right a per-ceived injustice succeeded. However, most social media activism does not achieve this level of activism and commitment. As one scholar wrote, most social media campaigns receive lots of clicks, or passive support, but often fail to reach any sort of critical mass that leads to actual change (Lim 2013, 646). However complex the transformation from social media to real-world activism is, there are people who work online and offline to improve Indonesian society.

Laura M. Steckman

See also: Australia; Malaysia; Spain; United Kingdom; United States

Further Reading

Alexa. 2015. "Top Sites in Indonesia." Accessed September 5, 2015. http://www.alexa.com/topsites/countries/ID

Allin1 Social. 2015. "Facebook Statistics for Indonesia." Accessed September 5, 2015. http://www.allin1social.com/facebook/country_stats/indonesia

AppAnnie. 2015. "iOS Top App Charts." Accessed September 5, 2015. https://www.appannie.com/apps/ios/top/indonesia/?device=iphone

Jakarta Post. 2015. "Internet Users in Indonesia Reach 73 Million." March 10. Accessed September 5, 2015. http://www.thejakartapost.com/news/2015/03/10/internet-users-indonesia-reach-73-million.html

Lim, Merlyna. 2013. "Many Clicks but Little Sticks: Social Media Activism in Indonesia." *Journal of Contemporary Asia,* 43: 636–657.

Lukman, Enriko. 2014. "Here Are Some of Indonesia's Most Memorable Tweets in 2014." December 11. Accessed September 5, 2015. https://www.techinasia.com/twitter-year-in-review-2014-indonesia/

Pardana, Igusti. n.d. "Sejarah Kaskus." Accessed August 31, 2015. http://www.academia.edu/9018872/Sejarah_kaskus

Rame, Ayo Kita Bikin. 2015. "25 Situs Media Sosial Asli Buatan Indonesia Terbaru 2015." February 25. Accessed June 20, 2015. http://ayokitabikinrame.blogspot.com/2015/02/25-situs-media-sosial-asli-buatan_13.html

Republika Online. 2014. "Sejarah Berdirinya Kaskus Diangat ke Layar Lebar." November 1. Accessed September 5, 2015. http://www.republika.co.id/berita/senggang/film/14/11/01/necuio-sejarah-berdirinya-kaskus-diangkat-ke-layar-lebar

Tuliside. 2015. "Kwikku.com—Sosial Media Asli Buatan Indonesia." January 21. Accessed June 20, 2015. http://www.tuliside.com/2015/01/kwikkucom-sosial-media-asli-buatan.html

IRAN

Known as Persia until 1935, the Islamic Republic of Iran is a Middle Eastern country located north of the Persian Gulf and the Arabian Peninsula and south of the Caspian Sea. It is bordered to the west by Azerbaijan, Armenia, Turkey, and Iraq, and to the east by Turkmenistan, Afghanistan, and Pakistan. In 2016, the country's population had reached 82.8 million (Internet World Stats 2016b). While historically Iran has ranked as having among the lowest internet penetration rates in the Middle East, recent years have seen a surge in the number of Iranians becoming connected. To some degree, this may be attributed to the development of new projects, supported by funding allowed by the 2016 lifting of sanctions against the country, aimed at improving telecommunications infrastructure. In part due to the rampant censorship practiced in the country, several indigenous social media sites have found favor among Iranian internet users. Nevertheless, studies suggest that the young population routinely find ways around national restrictions to access international social media sites.

By mid-2016, Iran's 56.7 million internet users accounted for its 68.5 percent penetration rate (Internet World Stats 2016a). This represents substantial growth over the previous fifteen years, from 250,000 users in 2000 (3.8 percent), to 7.5 million users in 2005 (10.8 percent), to 33.2 million users in 2010 (43.2 percent), to 46.8 million users in 2015 (Internet World Stats 2016b). The wide-ranging internet penetration rates in the Middle East—from 97.4 percent in Qatar to 24.7 percent in Yemen—average to 57.4 percent across the region, placing Iran slightly above average (Internet World Stats 2016a).

Over two-thirds of Iranians have access to a computer, and 39 percent of them surf the web daily (BBG 2012). However, Iran's younger generation, the majority of whom are under the age of thirty, seems to be far more tech- and social media–savvy than the general population. According to a research study conducted in 2014 by Iran's Center for Research and Strategic Studies, 67.4 percent of the country's younger generation uses the internet (primarily for chatting), and 70 percent of respondents aged fifteen to twenty-nine years old used software to bypass the government's filters to access foreign social media sites like Facebook and YouTube (TeleGeography 2014).

As internet penetration in Iran has increased, censorship practices have become more comprehensive, particularly since the 2013 presidential elections. This censorship comes in many different forms: the blocking of websites, the throttling of encrypted protocols, bandwidth limitations, and content filtering. Of Alexa's top 500 most common websites worldwide, nearly half of them are blocked from being accessed inside the country, including social media sites like Facebook and Twitter (Lee 2013). Most of these websites contained adult content; however, a high number of these blocked websites fell under the categories of art, society, and news.

Iran has also used throttling practices to discourage the use of certain websites and types of encrypted protocols. These websites are allowed to run at only 20 percent of the network's full capacity (Lee 2013). Some of these websites were not only throttled, but access was cut off after sixty seconds. This practice is claimed to be only temporary; it was used most prevalently during the 2013 elections as a tactic to minimize political dissent. For years, bandwidth in the country was actively limited to just 128 kbps, which is only about twice as fast as a dial-up connection and about one-fiftieth the speed of a typical connection in the United States (Lee 2013). However, projects based on new investments in the country to improve information technology infrastructure may eliminate such slow connection speeds.

More recently, a censorship tactic called "smart filtering" has been implemented. This procedure entails the blocking of selective content on websites rather than blocking entire websites. It is in part a reaction to government officials' concerns about being able to censor access to the internet effectively, especially in light of the rapid rise in internet users in Iran. While this technique has been a decade in the making, its efficacy is currently minimal. Smart filtering capabilities are limited to content produced inside Iran, but government concerns are largely focused on foreign websites and apps, whose servers are encrypted and located outside the country. This makes it nearly impossible for Iran to execute its smart filtering on these websites.

Some of the censorship in Iran has come in the form of arrests and prosecutions of Iranians who have used social media in ways deemed threatening by the government. For example, Atena Farghadani (1987–), a political activist and cartoonist, was arrested in August 2014 for allegedly insulting government officials and spreading propaganda with a cartoon that she published depicting parliament members as animals as a response to government measures regarding reproductive rights. Farghadani was released a few months later, only to be arrested again shortly thereafter and sentenced to twelve years in prison. She was eventually released in May 2016 and intends on remaining in Iran (Cavna 2016). Other examples include the prosecution of Ali Ghazali (n.d.–), editor of two online newspapers, *Baztab Emrooz* and *Ayandeh Online,* for allegedly inciting public anxiety with posts made to Facebook (Freedom House 2015).

In tandem with Iran's attempts at controlling internet content within the country, several Iranian companies, such as SabaIdea, have been developing indigenous social media to compete with foreign rivals. One of the most popular of these is Facenama, launched in 2011. Despite the similarity in name to Facebook, Facenama's format is more a discussion group than a social media site. In 2015, Facenama had about 2 million users and was ranked the ninth most popular website in Iran (Jafari 2015). However, its popularity has been significantly waning; as of August 2016, the website was only the forty-third most popular in Iran (Alexa 2016b).

Another popular indigenous social media site in Iran is Cloob. In 2015, it had about 2.5 million users and was ranked the twenty-ninth most popular website in Iran (Jafari 2015). Cloob offered discussion groups, photo sharing, blogs, and chat rooms. Interestingly, despite operating within the confines of Iranian laws, content on Cloob has been subjected to filtering at least twice since its launch in 2004. As with Facenama, Cloob's popularity has waned of late; as of August 2016, it was only the sixty-fifth most popular website in the country (Alexa 2016b). In both cases, the users of these social media were overwhelmingly college-educated male users accessing the websites from home (Alexa 2016a, 2016b).

Despite competition with the Iranian search engine Yooz, Google remains the most popular website in Iran, followed by Varzesh3.com (a sports news website), Yahoo, Digikala (an Iranian online shopping website), and Aparat (an Iranian photo-sharing website) (Alexa 2016c). While three of the top five most popular websites in Iran are locally based, foreign social media also remains popular in Iran. In August 2016, Instagram was the ninth most popular website in Iran (Alexa 2016c). While Instagram is not blocked by the Iranian government, it is still subject to smart filtering, and some Instagram accounts (such as celebrities) are inaccessible without using a virtual private network (VPN). Before being banned in 2009, Twitter ranked the fourth most popular website in the country (Jafari 2015). Today, despite its censorship, Twitter is occasionally used by Iranian government officials. Perhaps most surprising is the number of Iranian Facebook users. While no official statistics can be gathered, estimates

In Isfahan, Iran, a Shia mullah, or Islamic cleric, uses his cell phone while around town. Restrictions on internet usage is common in Iran; however, Android users take advantage of free apps such as VPN-Hotspot Shield to circumvent government controls by switching IP addresses to those located outside the country. (Jens Tobiska/Dreamstime.com)

suggest that there are at least 4.5 million Iranians using this banned social media site (Jafari 2015).

According to the Telecommunications Company of Iran (TCI), Telegraph is the most popular free app downloaded in the Google Play Store for Android devices, with 37.5 percent of Iranians eighteen years and older stating that they actively use the app (Financial Times 2016). The second most popular free Android app is VPN-Hotspot Shield, which allows users to change their internet protocol (IP) address to one that is located outside the country. This is how many Iranians get around the filters set up by government in order to access foreign social media sites, such as Facebook and Twitter. In third and fourth place were WhatsApp and IMO, respectively; their free calling and video-chatting tools likely contributed to their popularity. The fifth most popular free Android app was AndroDumper, which allows users to hack Wi-Fi passwords so that they may illegally use others' internet connections (Financial Times 2016).

While most Iranians have access to mobile phones, network service provider competition in the country opened only in 2014, before which 3G was the primary option for many cell phone users. While 4G networks are now offered by the three primary mobile providers, MCI, Rightel, and Irancell, they are limited to the major cities of Iran. Maximum internet speeds vary widely from company to company: Rightel's maximum speed is 5 Mbps, MCI's maximum speed is 8 Mbps, and Irancell's maximum internet speed is 25 Mbps. While mobile technology is on the rise and is the primary method of connecting to the internet in Iran, in 2015, mobile internet penetration was only 38.7 percent. Other common methods of connection were asymmetric digital subscriber line (ADSL), with 22 percent penetration; dial-up, with 9.23 percent penetration; and fiber optics, with 8.13 percent penetration (Azali 2015).

Facebook was initially blocked during the 2009 presidential elections after international outcry over a video posted there of Neda Agha-Soltan (1983–2009) being shot to death during a protest (AP 2016). While Facebook remains blocked, certain Facebook-owned apps, such as Instagram and Telegram, are readily available in Iran. In fact, Telegram, currently ranked the twelfth most popular website in Iran (Alexa 2016c), was widely used during the spring 2016 parliamentary elections by candidates who found the messaging app to be a quick and convenient way to spread their campaign platforms. With over 6,000 candidates running, the Telegram messaging app allowed lower-profile candidates access to cheap advertising and helped Iranian voters keep track of election trends (BBC Trending 2016).

Marilyn J. Andrews

See also: Afghanistan; Iraq; Kazakhstan; Syria; Turkey

Further Reading

Alexa. 2016a. "Site Overview: Cloob.com." Accessed August 27, 2016. http://www.alexa
    .com/siteinfo/cloob.com

Alexa. 2016b. "Site Overview: Facenama.com." Accessed August 27, 2016. http://www.alexa.com/siteinfo/facenama.com

Alexa. 2016c. "Top Sites in Iran." Accessed August 25, 2016. http://www.alexa.com/topsites/countries/IR

Associated Press (AP). 2016. "In Iran Elections, Getting Votes Means Going to Social Media." *Daily Mail*. Accessed August 27, 2016. http://www.dailymail.co.uk/wires/ap/article-3462193/In-Iran-election-getting-votes-means-going-social-media.html

Azali, Mohammad Reza. 2015. "Internet Penetration in Iran." [Infographic] *Techrasa*. Accessed August 27, 2016. http://techrasa.com/2015/11/20/infographic-internet-penetration-iran/

BBC Trending. 2016. "Punchy Politics on Social Media as Iranians Go to Polls." Accessed August 27, 2016. http://www.bbc.com/news/blogs-trending-35662000

Broadcasting Board of Governors. 2012. "BBG Data Show Internet, Satellite Usage in Iran at All-Time Highs." Accessed August 23, 2016. https://www.bbg.gov/2012/06/12/bbg-data-show-iran-internet-satellite-usage-at-all-time-high/

Cavna, Michael. 2016. "Cartoonist Atena Farhadani, Sentenced for Satirizing Government as Animals, Is Freed in Iran." *Washington Post*. Accessed August 25, 2016. https://www.washingtonpost.com/news/comic-riffs/wp/2016/05/04/cartoonist-atena-farghadani-sentenced-for-satirizing-government-as-animals-is-freed-in-iran/

Financial Tribune. 2016. "Top 10 Android Apps in Iran." Accessed August 27, 2016. http://financialtribune.com/articles/sci-tech/39700/top-10-android-apps-iran

Freedom House. 2015. "Freedom on the Net: Iran." Accessed August 23, 2016. https://freedomhouse.org/report/freedom-net/2015/iran

International Campaign for Human Rights in Iran (ICHRI). 2016. "Iran to Spend $36 Million on Internet 'Smart Filtering,' To No Avail." Accessed August 24, 2016. https://www.iranhumanrights.org/2016/02/iran-will-spend-36m-on-smart-filtering/

Internet World Stats. 2016a. "Internet Usage in the Middle East." Accessed January 21, 2017. http://www.internetworldstats.com/stats5.htm

Internet World Stats. 2016b. "Iran." Accessed January 21, 2017. http://www.internetworldstats.com/me/ir.htm

Jafari, Hamed. 2015. "Even Our President Is More Social Than You!" *Techrasa*. Accessed August 27, 2016. http://techrasa.com/2015/08/30/iran-even-president-digs-social-media/

Lee, Timothy B. 2013. "Here's How Iran Censors the Internet." *Washington Post*. Accessed August 24, 2016. https://www.washingtonpost.com/news/the-switch/wp/2013/08/15/heres-how-iran-censors-the-internet/

TeleGeography. 2014. "70% of Iranian Internet Users Bypass Government Web Restrictions." Accessed August 24, 2016. https://www.telegeography.com/products/commsupdate/articles/2014/09/11/70-of-iranian-internet-users-bypass-government-web-restrictions/?__hstc=43953530.ddcd5ebd8416e2dcd5bea7b432f0da41.1472083410850.1472083410850.1472083410850.1&__hssc=43953530.2.1472083410852&__hsfp=2722755842

TeleGeography. 2016. "Iran to Boost Internet Speeds, Cooperate with Kazakhstan." Accessed August 24, 2016. https://www.telegeography.com/products/commsupdate/articles/2016/05/17/iran-to-boost-internet-speeds-cooperate-with-kazakhstan/?__hstc=43953530.ddcd5ebd8416e2dcd5bea7b432f0da41.1472083410850.1472083410850.1472083410850.1&__hssc=43953530.3.1472083410852&__hsfp=2722755842

IRAQ

Iraq is located in the Middle East, bordered by Turkey, Iran, Kuwait, Saudi Arabia, and Syria. It currently has a population of around 37.7 million. Arabs comprise 75 percent of the population, followed by Kurds, at 16 percent, and other minority ethnic groups. The majority religion is Shia Islam. After several decades of sectarian unrest and violence imposed by terrorist groups, which followed decades of heavy authoritarian government control, Iraq continues to slowly expand its telecommunications infrastructure to include a growing mobile market. As access rises, more people are adopting social media and using it to communicate with friends and access news. Along with the benefits of social media, it also has a down side, as opponents of the state use it to promote radical messages and violence against non-Sunni Muslims.

Iraq's use of the internet started very slowly, and only in recent years has it expanded and become more widespread. Prior to 2003, under the rule of Saddam Hussein (1937–2006), the government viewed the internet as a threat. While it did allow extremely limited access, usage was also highly controlled and restricted. In 2000, approximately 12,500 Iraqis could access the internet. By 2011, the number was estimated to be around 325,000 people, achieving around a 1.1 percent internet penetration rate (Rashid 2012). Barriers to access included government restrictions, high bandwidth cost, inadequate telecommunications infrastructure, and lack of providers (at one point in the mid-2000s, the country was primarily connected via one satellite terminal). As the mobile phone market expanded and costs lowered around 2008, many Iraqis used mobile phone connections to connect to the internet, often to communicate with friends and family abroad or to access news.

From the early days of the internet in Iraq, one form of internet media that could sometimes break through the government's control was the blog. Salam Pax (a pseudonym) was the author of Iraq's first major blog, entitled "Where Is Raed?" Salam Pax started writing in 2002, posting about his experiences before and during the beginning of the Iraq war (2003), and later under the U.S.-led forces (2003–2011). The blog focused on what it was like to be Iraqi during that period and quickly gained followers in the United States and Europe. Salam Pax was also a controversial blogger because many Iraqis speculated that he was not Iraqi, or even located in Iraq—that he was just someone using their country to garner attention for himself. Later, it came out that the blogger was indeed Iraqi and worked as a translator (Rashid 2012). Salam Adbulmunem (1973–), became one of the country's most notable bloggers, and many of his posts have been turned into or included in academic books.

Today, Iraq has one of the highest, most robust local broadcast media markets in the Middle East. With hundreds of new publications, television channels, and radio stations both online and offline in the country, most Kurds and Iraqis have an extensive assortment of news choices (BBC 2016). For the purpose of internet accessibility, home cable packages have become ubiquitous. The collapse of the authoritarian government eliminated the severe regime control over Iraq's media and social media, escorting in an era of risky social media pluralism, with individuals all over the nation starting dozen of television and radio stations, as well as hundreds of newspapers. Nevertheless, constant sectarian struggle and government regulation of media and social media policies hamper the people's access to dependable, unbiased news and social media content. When they are online, Iraq's top ten most accessed websites in 2016 are YouTube, Google.iq, Facebook, Google .com, Yahoo, Xendan.org (a news website), Dwarozh.net, Nrttv.com, Blogspot, and Twitter (Alexa 2016). These sites show an interest in social networking, news consumption, blogging, and using the internet for finding information.

There is some debate surrounding whether the internet and social media have been fundamental tools in Iraq's internal struggles. Specifically, the mass demonstrations that traversed the Middle East in early 2011 emphasized the distinctive power of digital social media tools, modern information communication technologies (ICTs), and networks. The influences of these technologies spread worldwide, affecting developed and developing countries. The uprisings in the Middle East and North Africa known as the "Arab Spring" initially signaled a change of authoritarian regimes and the promise of democracy, especially in Libya, Egypt, Tunisia, Syria, and Yemen.

Iraq also experienced a series of protests against the government, though they predated the Arab Spring by at least a year. The people, including the country's Sunni, Shia, and Kurdish populations, demanded basic social services, utilities, and access to education. In 2011, they used Facebook to coordinate protests on February 25, also known as the "Day of Anger." Despite the fact that the Arab Spring pointed to novel, innovative uses of mass online communication, it ultimately did not succeed in its political goals. In Iraq, social media sites were not a major influencer or catalyst in the protests, despite their use as an organizing tool. Overall, the Arab Spring did not have a major effect on Iraq, which was also complicated by the country's attempted transition to democracy and the battle against terrorist groups such as Al Qaeda.

Despite some correlation between social media and the Arab Spring in Iraq, social media nonetheless plays an important role in communication and self-expression, especially among Iraq's younger population. Iraqis have adopted a variety of international social media platforms. As of 2015, Iraqis preferred Facebook (97.15 percent), Twitter (1.67 percent), Google+ (0.78 percent), Pinterest (0.16 percent), YouTube (0.15 percent), Tumblr (0.07 percent), and StumbleUpon (0.01 percent), and there are a number of other sites with a small Iraqi user base as well (StatsMonkey 2015). The top Facebook accounts in Iraq are Kadim al Sihir (1957–), a musician whose net worth is around $10 million, with over 12 million followers; and the Barbie Beauty Center, with over 8 million followers (Socialbakers 2016). No detailed information currently exists for Iraqi top app preferences.

The conflict in Iraq has caused internet and social media connectivity to be unreliable. At times, government blocks or heavy censorship have also affected access. In spite of the challenges to staying connected, Iraqis have become adept at identifying new communications channels. For example, after a terrorist organization known as the Islamic State or Daesh usurped territory in Iraq and Syria in June 2014, the Iraqi government experimented with internet blocking. Their policies attempted to counter the group's avid use of social media for recruitment, financing, and for spreading its radical ideology, rather than blocking patriotic Iraqi citizens wholesale. However, the blocks affected all users, especially those on Facebook and Twitter.

Immediately, the Iraqis responded by downloading Firechat, a peer-to-peer mesh networking app that can communicate through Wi-Fi, Bluetooth, and radio receptors embedded in most smartphones and does not require an internet connection. On June 13, 2014, Iraqi internet protocol (IP) addresses had downloaded the app 6,600 times. By late June, after the terrorist group seized Iraqi territory, Iraqi IPs downloaded the app more than 40,000 times in Baghdad alone (Hern 2014). Since that time, Firechat has allowed Iraqis to communicate with one another even during periods when internet access is limited or nonexistent. Around the same period, and for similar reasons, Iraqis adopted the Whisper app, a social media site that allows users to post anonymous messages (Sanchez 2014). Some users used it to discuss the violence, while others used it to ask questions and share secrets without revealing the source of their information.

On the other side of the conflict, Daesh actively uses social media to promote its so-called Islamic State. The group is widely known for its messaging efforts on Twitter, as well as an active propaganda campaign that spans Facebook, YouTube, JustPaste.it, blogs, group websites, and a host of other platforms. Daesh is also active on the so-called deep and dark web—namely, the places that the average user cannot access because they are not indexed by search engines or because they need users to supply appropriate credentials before they allow access. The group is also active on sites and apps that promote security and enhanced encryption, such as WhatsApp and more recently, Telegram, a site that requires registration and membership to view most of its channels (Harris 2015).

Daesh's presence on numerous websites, social media platforms, and apps has made it difficult to counter. Companies such as Facebook and Twitter actively suspend accounts of the group and its supporters as soon as they are identified. For example, from August 2015 to February 2016, Twitter announced it had suspended more than 125,000 Daesh-supporting accounts (Koh 2016). However, suspensions were not very effective, as users could create new accounts to continue spreading propaganda. More recently, the group's supporters continue to switch away from Twitter to sites such as Telegram, making it more and more difficult for the Iraqi government and the international coalition trying to stop their online activities.

Laura M. Steckman and Susan Makosch

See also: Egypt; Iran; Syria; Tunisia; Turkey

Further Reading

Alexa. 2016. "Top Sites in Iraq." Accessed August 24, 2016. http://www.alexa.com/topsites/countries/IQ

BBC. 2016. "Iraq Profile—Media." May 2. Accessed January 20, 2017. http://www.bbc.com/news/world-middle-east-14546541

Harris, Shane. 2015. "This Is ISIS's New Favorite App for Secret Messages." November 16. Accessed August 24, 2016. http://www.thedailybeast.com/articles/2015/11/16/this-is -isis-new-favorite-app-for-secret-messages.html

Hern, Alex. 2014. "Firechat Updates as 40,000 Iraqis Download 'Mesh' Chat App in Censored Baghdad." June 24. Accessed August 24, 2016. https://www.theguardian.com /technology/2014/jun/24/firechat-updates-as-40000-iraqis-download-mesh-chat-app -to-get-online-in-censored-baghdad

Koh, Yoree. "Twitter Suspended 125,000 ISIS-Related Accounts in Six Months." February 5. Accessed August 24, 2016. http://blogs.wsj.com/digits/2016/02/05/twitter-suspended -125000-isis-related-accounts-in-six-months/

Rashid, Saif. 2012. "An Overview on Internet in Iraq." Accessed August 24, 2016. http:// www.academia.edu/1952319/Internet_in_Iraq

Sanchez, Nick. 2014. "Iraqis Turn to Whisper App for Anonymous Social Networking." June 18. Accessed August 24, 2016. http://www.newsmax.com/TheWire/iraqis-turn-to -whisper-app/2014/06/18/id/577811/

Socialbakers. 2016. "Iraq Facebook Page Statistics." Accessed August 24, 2016. https://www .socialbakers.com/statistics/facebook/pages/total/iraq/

StatsMonkey. 2015. "Mobile Facebook, Twitter, Social Media Usage Statistics in Iraq." Accessed August 24, 2016. https://www.statsmonkey.com/packedcircle/21375-iraq-mobile -social-media-usage-statistics-2015.php

IRELAND

The Republic of Ireland, a country that takes up all but the northern portion of an island off the coast of Britain and Wales, has a population of 4.7 million people. Due to the government's efforts to increase telecommunications infrastructure and attract international businesses, Ireland is heavily connected to the internet and boasts among the fastest speeds in Europe. The Irish use a variety of social media and networking sites, spending significant amounts of time online. In more recent years, Ireland's privacy laws, considered looser than other European countries, have allowed international firms to transfer data around the world. The data storage and privacy laws have implications for not only Irish users, but also for those of other nationalities. As such, there are signs that Ireland will need to revisit its policies and practices quickly to bring them more in line with other European nations.

In January 2016, most Irish social media users had multiple accounts across different platforms and networks. Predominantly, they used Facebook (63 percent), Skype (47 percent), Facebook Messenger (52 percent), Viber (42 percent), Twitter (31 percent), LinkedIn (28 percent), Google+ (25 percent), Snapchat (25 percent), Instagram (22 percent), Pinterest (15 percent), Tinder (5 percent), Vine (3 percent), and Tumblr (2 percent) (Dyer 2016). Irish users also utilized YouTube; in 2013, more than 1.3 million people went on the site. These preferences are also reflected in Ireland's most viewed websites, which mirror other European societies with a high social media penetration; these include Google.ie, Google.com, YouTube, Facebook, Twitter, Wikipedia, LinkedIn, Yahoo, Amazon.co.uk, and Live.com (Alexa 2016). Clearly, Ireland has a thriving cyberculture, and both social media and the internet have become ingrained in the Irish lifestyle.

Irish social media users spend a great deal of time online. In a survey of Irish broadband users, 36 percent said they spent more than twenty hours online per week; 26 percent spent eleven to twelve hours, 24 percent spent five to ten hours, and 14 percent spent fewer than four hours online. Those surveyed said that they used the internet for multiple reasons, such as emailing, web surfing, social networking, online banking, online shopping, watching video, listening to music, making voice over internet protocol (VOIP) calls, downloading media, and gambling online (Krishna 2015). These results reflect the attitudes and behaviors of adult users online.

In Ireland, it is common for children as young as age nine to be active online. They prefer tablets over smartphones, laptops, and other internet-enabled devices. On their tablets, they not only do schoolwork, but also play games and use apps. In a 2016 study, data revealed that children are not generally aware of internet etiquette and safety. According to these findings, 54 percent claimed that they had spoken to at least one stranger online. The most difficult issue that they contended with was cyberbullying. Between 24 to 34 percent of elementary school students admitted that they had experienced it or personally knew someone who had (Zeeko 2016). Most parents had been unaware that so many children faced cyberbullying prior to the release of the study. Efforts are underway to teach parents, children, and educators how to use the internet more safely and how to respond to cyberbullying incidents.

Over the last decade, the Irish government has pushed to make the country a data hub and technology center of Europe. These attempts have been successful, as international tech firms and companies such as Google, Dell, Airbnb, Facebook, Adobe, LinkedIn, Microsoft, and Apple, among others, have established headquarters in Ireland to oversee their business in Europe, the Middle East, and Africa. One reason that the Irish government has been successful at attracting foreign businesses is that it offers a larger tax break than most other European nations. In Ireland, international companies pay a 12.5 percent tax, versus 23 percent in the United Kingdom or 33.3 percent in France (Mirani 2013). From a financial standpoint, these firms save corporate money; on the other hand, they create jobs for Ireland's young, skilled tech workforce.

Another reason that international companies, especially tech firms, prefer Ireland is that the country's regulations are looser than in the rest of Europe. When tech firms have undergone legal challenges in the past, the Irish system has generally found in their favor, even at times when it seemed clear that other European countries would not have done so. One major complaint, filed by Maximilian Schrems (c. 1987–), an Austrian law student, alleged that Facebook and other companies were using Ireland to transfer his personal information illegally from Europe to the United States. By 2013, two of Schrems's complaints had made it to the Irish data protection commissioner (DPC), who decided in favor of Facebook. Schrems initially dropped his complaint with the DPC, but in 2015, the European Court of Justice (ECJ) ordered the DPC to address his case—and that of other consumers of the content (Hunt 2016).

Some analysts speculate that the issue, and the demand for authorities to look more closely at corporate data storage and transmission practices in Ireland, came about because of growing concerns that organizations such as the National Security Agency (NSA) in the United States were using the data to spy on people; indeed, Schrems definitively stated that these revelations were what prompted him to act (Gibbs 2015). Schrems is moving forward with the case. He has also started a class-action style lawsuit against Facebook in Austria and has submitted complaints with two other European courts. The final ruling in Ireland, especially if it overturns Ireland's previous legal support to companies such as Google and Facebook, could have ramifications for all tech firms nationwide.

Laura M. Steckman

See also: France; Germany; United Kingdom; United States

Further Reading

Alexa. 2016. "Top Sites in Ireland." Accessed August 28, 2016. http://www.alexa.com/topsites/countries/IE

Dyer, Tris. 2016. "Social Media Usage in Ireland 2016." January 28. Accessed August 28, 2016. http://www.leadingsocial.net/blog/social-media-ireland

Gibbs, Samuel. 2015. "Max Schrems Facebook Privacy Complaint to Be Investigated in Ireland." October 20. Accessed August 28, 2016. https://www.theguardian.com/technology/2015/oct/20/max-schrems-facebook-privacy-ireland-investigation

Hunt, Gordon. 2016. "New Data Transfer Woes Looming for Tech Companies in Ireland?" May 27. Accessed August 28, 2016. https://www.siliconrepublic.com/enterprise/irish-data-protection-commissioner

Krishna. 2015. "How Irish People Spend Their Time Online in 2015." Accessed August 28, 2016. https://www.krishna.me/2015/internet-and-mobile-use-in-ireland/

Mirani, Leo. 2013. "The Reason American Tech Firms Like Ireland Isn't Just the Low Taxes." September 13. Accessed August 28, 2016. http://qz.com/124133/the-reason-american-tech-firms-like-ireland-isnt-just-the-low-taxes/

Zeeko. 2016. "Every Primary School in Ireland Will Get a Digital Copy of the Zeeko Internet Safety Guide!" January 13. Accessed August 28, 2016. http://zeeko.ie/uncategorized/every-primary-school-in-ireland-will-get-a-digital-copy-of-the-zeeko-internet-safety-guide/

ISRAEL

Israel (officially the State of Israel), established in 1948, is located in the Middle East. In comparison to neighboring countries, Israel enjoys significantly higher internet penetration rates. In the years 2014–2015, local and global measurements estimated the internet user population in Israel to be as high as 71–74 percent of the overall Israeli population, compared to an average of 38.3 percent internet users in the rest of the Middle East. While traditional types of mass media in Israel, such as newspapers, television, and radio, established themselves in close association with nationalistic efforts and governmental offices, the internet marks the opening of the Israeli communications market to global communication technology and media content. Private companies introduced internet connection to Israeli society in the early 1990s, and although it is considered a relatively small market in the global internet economy, Israel maintains a prominent role in multiple internet arenas, including hardware technology improvements, mobile application developments, and cybersecurity advancements.

Similarly to its advent in the United States, the internet was first used in the Israeli media market by academic institutions. The Hebrew University and the Weizmann Institution were the first to facilitate network connection for scholars arriving from the United States in the early 1980s, temporarily connecting them to global networks, but at a high cost. In one of the most comprehensive account of internet evolution in Israel, John (2008) follows the diffusion of computer networks since the 1980s. The arrival of the internet at that time, he argues, is linked with the greater Jewish immigration narrative, in which Jewish immigrants returning to their homeland (i.e. Jewish academics immigrating to Israel) brought the need for internet connection with them. These technological pioneers are referred to as "the cosmopolitans" in the processes of internet diffusion in Israel. By 1985, Israel was the third country, after the United States and the United Kingdom, to attain a state code suffix (namely, co.il); however, computer networks were open only to academics.

In 1992, with the creation of the first Israeli website (of the Hebrew University), the internet became partially available to nonacademic users if they held a license authorized by the Israeli Ministry of Communication. In the mid-1990s, newly established local internet providers, such as Bezeq International, Cellcom, and Smile, extended internet access to the general public. Israelis quickly adopted the internet. Around 2000, there was a spike in usage (from 4.4 percent internet users in 1997 to 20.9 percent in 2000) following the implementation of asymmetric digital subscriber line (ADSL) technology. Internet adoption in Israel has continued to rise to this day.

In terms of infrastructure, the connection of Israel to global internet networks was made possible in 1990 by the creation of an underwater cable (EMOS-1) that connected Israel to Greece, Turkey, and Italy. Technological advancements in later years led to the use of faster and broader internet connectivity via fiber optic cables. Today, most Israeli internet traffic passes through cables owned by both domestic and international companies. To gain internet connectivity, Israeli consumers have to subscribe to two separate companies—the services of both an internet infrastructure supplier and an internet service provider (ISP). Two companies provide access to the infrastructure: Bezeq and Hot, while multiple companies act as ISPs, such as Netvision, Orange, Smile, and Internet Zahav. As of 2014, Israel was graded as the fifth-fastest internet connection in the world in terms of broadband transmission, with a connection as fast as 47.7 Mbps.

In addition to being a highly connected country, Israel has produced many influential digital software products and applications. For example, one of the first instant messaging (IM) platforms, ICQ, was developed in 1996 by Mirabilis, an Israeli company, and bought in 1998 by America Online for $407 million, the highest price paid at the time for an Israeli company. This was one of many examples of small Israeli companies developing impressive high-tech abilities that were then purchased by larger (usually U.S.) companies, such as Waze (bought by Google) and LinX (bought by Apple). The Israeli National Cyber Bureau (INCB) estimates that around 10 percent of all cyberspace transactions around the world involve Israeli companies. Experts referring to the state as a "Start-Up Nation" and the "Silicon Wadi" have accurately characterized Israel's significant contribution to global digital media.

When it comes to cyberattacks and data abuse online, Israel is considered to be a leading country in terms of cybersecurity, awarded by European initiatives with a top cyberdefense grade (Benoliel 2014). In fact, Check Point, one of the first and leading international data security companies, was established in Israel in 1993. Alongside about 250 commercial cybersecurity companies operating in Israel nowadays, the Israeli prime minister office manage the INCB since 2011. This governmental branch is in charge of advising and advancing national defense initiatives in the cyber field. With the goal of defending the Israeli civic cybersphere as well, government officials in Israel declared in late 2014 the establishment of a national operative cyberdefense authority, designed to work side by side with the INCB. This move came about as a result of the high rate of international cyberattacks on Israeli users and institutions, estimated by state officials at 1000 attacks per minute.

With regard to mobile phone adoption, Israel holds one of the highest adoption rates in the world, second only to Italy. Until the mid-1980s, cellular communication in Israel was used for military and transportation needs (such as aircraft and sea craft communications). In 1986, Pelephone established the first public mobile phone service, followed by Cellcom, established in 1994, and Partner in 1998. As early as 2002, the number of mobile phone subscribers in Israel surpassed that of landline service subscribers. Only two years later, in 2004, Israel had 95.45 mobile phone subscribers for every 100 inhabitants—almost twice as many mobile phone subscribers as in the United States (Lemish and Akiba 2005). In 2009, 91.8 percent of Israeli households possessed mobile phones, and in 2013, more than 9 million cellular phones were in use (1 million more than the Israeli population), with more than half of the Israeli population owning a smartphone.

Mobile communication, including internet connectivity, in Israel is provided today by five main companies: Pelephone, Cellcom, Partner, HOT, and Golan Telecom. Mobile devices are being used for web surfing, checking and sending emails, downloading and using applications, logging on to social networking sites, and

---

In May 2011, Israeli couple Lior Adler (c. 1974–) and Vardit Adler (c. 1976–) named their daughter Like after being inspired by the "like" feature on Facebook. The Facebook site allows users to click a button to support, or like, another user's content. After determining that no one else in Israel had the name Like, the Adlers decided to use it. They felt that the name was modern and original. The couple also commented that they viewed the name as being similar to the word *ahava*, which is Hebrew for "love," and that the name would convey how much they loved their daughter (BBC 2011).

utilizing location-based services. One of the most popular mobile applications used in Israel is WhatsApp, which 92 percent of smartphone users downloaded to their devices in 2013 (Canetti 2015). Most Israelis use their cell phone to stay tuned to the news and communicate with family and friends constantly. Today, Israeli users enjoy multiple social media platforms, most prominently Facebook, Twitter, and WhatsApp. In fact, in 2011, Israeli internet users were declared record holders for use, with as much as 11.1 hours per month spent on social networking sites.

Aside from heavy consumption of international websites and social media, Israeli users also enjoy a variety of local content online. For example, the news website Ynet.co.il is a popular Israeli site, with 88.3 million views in the last quarter of 2015; and the news and entertainment site walla.co.il earned 67.4 million views. Israeli mass media outlets, such as radio, newspapers, and television, also contribute content to online venues (such as Nana10 of Channel 10), which is consumed alongside original web content such as Quickie (2003), Pini (2010), and YouTube national stars such as Srutonim. When it comes to e-commerce, 91 percent of Israelis shopped online in 2015, and many use the internet for peer-to-peer shopping via websites like yad2.co.il, for price comparisons on websites like zap.co.il, and for mobile shopping in applications like Shufersal.

The relatively small population of Israel (around 8 million) is divided into several sectors, with each sector exhibiting unique internet adoption patterns and uses. For example, the Russian community in Israel is internet savvy, with the website ok.ru rated as one of the top ten most visited sites in Israel. In contrast, the Arab sector in Israel was much slower to connect to internet services due to lack of infrastructure in Arab cities and villages, as well as lower financial and educational resources. Today, a digital divide remains in place, with 63.6 percent of the Israeli-Arab population obtaining access to the internet, compared with 73.9 percent of Israeli Jews (Ragnedda and Muschert 2013). At the same time, access to the internet has allowed more freedom of expression among Israeli Arabs, a larger consumption of Arab media content online, and a growing presence and involvement in the mediated global Arab sphere.

A different example for sectorial internet adoption in Israel is that of the religious ultra-Orthodox, a community that uses enclaved media produced by and for them. Initially, many ultra-Orthodox leaders vocally rejected the internet and even saw it as dangerous due to its lack of communal supervision. As a result, this sector has developed specific means of internet diffusion, which allow people to use the internet in a religious, supposedly safe way. For instance, many ultra-Orthodox Jewish people use "kosher" phones, provided by companies such as Nativ, and "kosher" internet filters, such as Internet Rimon. These specialized devices block indecent and religiously problematic content and allow the individuals of this sector to use new media technologies for work and religious purposes (Soffer 2014).

One prominent problem related to mobile devices, the internet, and social media in Israel is that of language. This seems to be a cross-sectorial issue in Israel, pertaining to all subcommunities. Although English is widely used in Israel, the two dominant languages are Hebrew and Arabic, both written from right to left with non-Latin letters. The Unicode system standard of encoding characters enables users in Israel to surf the web, use mobile devices, and use social media in their own language instead of English.

Ruth Tsuria and Aya Yadlin-Segal

See also: Greece; Italy; Turkey; United Kingdom; United States

Further Reading

Benoliel, Daniel. 2014. "Towards a Cybersecurity Policy Model: Israel National Cyber Bureau Case Study." *NCJL & Tech,* 16:435.

Canetti, Nurit. 2015. "Israel's cellphone addiction." *Al-Monitor,* March 19. Accessed March 17, 2016. http://www.al-monitor.com/pulse/originals/2015/03/israel-mobile-phone-addiction-youngsters-whatsapp-technology.html

Dahan, Michael. 1999. "National Security and Democracy on the Internet in Israel." *Javnost-The Public,* 6(4): 67–77.

John, Nicholas. 2008. "The Arrival of the Internet in Israel: The Local Diffusion of a Global Technology." PhD dissertation, Hebrew University. Accessed March 25, 2016. http://sociothink.com/njohnphd.pdf

Lemish, Dafna, and Cohen, Akiba A. 2005. "On the Gendered Nature of Mobile Phone Culture in Israel." *Sex Roles,* 52(7–8): 511–521.

Mesch, Gustavo S. 2001. "Social Relationships and Internet Use among Adolescents in Israel." *Social Science Quarterly,* 82(2): 329–339.

Ragnedda, Massimo, and Muschert, Glenn W. 2013. *The Digital Divide: The Internet and Social Inequality in International Perspective.* New York: Routledge.

Soffer, Oren. 2014. *Mass Communication in Israel: Nationalism, Globalization, and Segmentation.* Oxford, U.K.: Berghahn Books.

ITALY

Italy is a southern European country, and a founding member of the European Union (EU). With 60,656,000 inhabitants (National Institute of Statistics 2016a), it is the fourth most populous EU member state, after Germany, France, and the United Kingdom. Currently, in terms of connectivity, Italy ranks second to last among EU countries (European Commission, 2016). On the other hand, Italy has a high penetration of mobile phone subscriptions (134 percent), and a large number of social media users compared to the whole internet user population.

Compared to other EU countries, Italian internet penetration appears to be below average: according to the National Institute of Statistics (Istituto Nazionale di Statistica–ISTAT 2016b), only 60.2 percent of Italians (six years old and older) were internet users in 2015. According to European Union (EU) statistics, when only considering the population between six and seventy-four years of age, Italian internet users reached 68 percent, while the EU average was 81 percent (Eurostat 2015). In 2015, people using the internet "every day" were 40.3 percent, while people using it "one or more times a week" were 16.8 percent (National Institute of Statistics 2016b). Digital inequalities in Italy, at an individual level, appear to be related to age, education, socioeconomic status, geographical factors (i.e. citizens living in northern Italy are more connected than those living in southern Italy); among people over fifty-five years of age, there is also a gender gap, which has been overcome with the younger generations.

The Agenda Digitale Italiana, promoted by the Agenzia per l'Italia Digitale (AGID), a governmental agency devoted to the diffusion of digital technologies in the country, has set several goals with regard to internet diffusion and to digital skills in order to fill the gap between Italy and the other EU countries. For instance, according to AGID (2016), in 2014, 23 percent of Italian citizens accessed e-government services, while the EU average is 46 percent; and 35 percent used e-commerce, while the EU average is 63 percent.

In 2015, among Italian internet users, 75.9 percent used emails, 56.1 percent accessed at least one social network site, 52.8 percent played a game or downloaded multimedia files, and 52.5 percent read news online (National Institute of Statistics 2016b). According to Alexa.com, as of January 2016, the most popular websites in Italy were Google.it, Facebook, YouTube, Google.com, Amazon.it, and Wikipedia.org. The most read online newspaper was the digital version of *La Repubblica* (Repubblica.it). Google is the prevailing search engine, while Bing covers only a very small share of the market.

With regard to connectivity, the Digital Economy & Society Index (DESI) scores concerning Italy are particularly low, putting the country second to last among EU countries. More specifically, while broadband coverage appears higher than EU countries' averages, broadband penetration, as well as connection speed, are currently not satisfactory (European Commission 2016). Mobile phones are widespread in Italy; specifically, mobile subscriptions account for 134 percent of the population, while mobile phone penetration is 84 percent (meaning that several Italians own more than one mobile phone subscription); 62 percent of the adult population owned a smartphone in 2015, and 48 percent were mobile internet users (We Are Social 2016). In the same survey, 84 percent of mobile connections were prepaid, 16 percent were postpaid, and 75 percent were broadband (3G and 4G); in most European countries, prepaid connections prevail (We Are Social 2016). The average daily use of the internet via mobile phone is two hours ten minutes (We Are Social 2016). While 40 percent of the Italian population use social media through mobile devices, this number is constantly growing.

As of January 2016, the most popular mobile apps among smartphone users thirteen years old and over are WhatsApp, Google, Facebook, Skype, Microsoft Outlook, Instagram, Amazon, and Yahoo (Cosenza 2016). As already mentioned, according to ISTAT, 56.1 percent of Italian internet users accessed at least one social network site in 2015. More specifically, the leading social media platform was Facebook, which reached 28 million active users per month in June 2016 (Cosenza 2016). Among them, 52 percent were male and 48 percent were female; and 10 percent were aged thirteen to nineteen years old, 26 percent were aged twenty to twenty-nine years old, 23 percent were aged thirty to thirty-nine years old, 21 percent were aged forty to forty-nine years old, 12 percent were aged fifty to

The @ sign dates back to sixteenth-century shorthand. Today, it is known as the "at sign," or the "commercial at" in some business circles. However, those names have come about only recently, with the first electronic usage in 1971. Until the symbol had an accepted name, people around the world had to describe the symbol in order to name it. In different parts of the world, people have referred to the "at sign" as the "crazy A" (Bosnian), "little dog" (Russian and Armenian), "monkey's tail" (Dutch), "spider monkey" (German), "small snail" (Italian), "pig's tale" or "elephant's trunk" (Danish), "worm" (Hungarian), and "strudel" (Hebrew) (Specktor 2014).

fifty-nine years old, and 8 percent were sixty years of age or older (We Are Social 2016). Facebook started spreading in Italy as a "niche" platform, and it witnessed its first relevant growth (moving from few hundred thousand users to more than a million) during the summer of 2008. Despite the prevalence of users in their twenties and thirties, the number of overall users (and the high penetration of the platform) has turned Facebook into a "mainstream" environment, which represents, for a significant number of users, one of the most relevant social media experiences.

The second most popular social media platform in Italy is WhatsApp, followed by Facebook Messenger, Google+, Twitter, Instagram, and Skype (We Are Social 2016). The average daily use of social media is one hour and fifty-seven minutes, accounting for about half of overall internet usage (four hours and five minutes) (We Are Social 2016). In 2015, concerning the youngest sectors of the population (users aged twelve to seventeen years old) 75 percent had a Facebook account, 36 percent an Instagram account, and 29 percent a Twitter account. In general terms, young people appear to appreciate Instagram more than the general population. Snapchat is also spreading rapidly among younger users, but official data are not available (or relevant) yet. Social media platforms are currently used in Italy for a variety of purposes.

Besides private citizens, who share content with friends and acquaintances, politicians, brands, celebrities, and media outlets are increasingly using social media to reach their audiences and encourage people's engagement. In this regard, Facebook is the most used platform due to its popularity, but several forms of professional use also occur on Twitter (primarily with regard to political communication and media outlets) and on Instagram (with a growing share of small- or large-scale influencers, such as those in the field of fashion, design, and food). After decades of broadcast television playing a central role in the public debate, especially with the influence of Silvio Berlusconi (1936–), television network tycoon and former prime minister, Italian politicians in recent years have started to use the internet intensively for political communication, with a special focus on social media. For instance, the recent Movimento 5 Stelle (Five Stars Movement), founded in 2009 by former comedian Beppe Grillo (1948–), used social media and Web 2.0 platforms for organizing its activities and spreading its messages. On the other hand, current prime minister Matteo Renzi (1975–), along with several ministers and local level authorities, are diffusively using social media, including short live videos, to communicate with citizens.

Francesca Comunello and Simone Mulargia

See also: France; Germany; Greece; Spain; United Kingdom

Further Reading

Agenzia per l'Italia Digitale (AGID). 2016. "Coalizione per le Competenze Digitali." Accessed July 29, 2016. http://competenzedigitali.agid.gov.it/content/cittadini

Cosenza, Vincenzo. 2016. "Vincos.it." Accessed July 29, 2016. http://vincos.it

European Commission. 2016. "Digital Economy & Society Index (DESI)." Accessed July 29, 2016. https://ec.europa.eu/digital-single-market/en/desi

Eurostat. 2016. "Internet Use by Individuals." Accessed July 29, 2016. http://ec.europa.eu/eurostat/web/products-datasets/-/tin00028

Ipsos Public Affairs. 2015. "Safer Internet Day Study 2015: I Nativi Digitali Conoscono Veramente il Loro Ambiente?" Accessed July 29, 2016 http://images.savethechildren.it/IT/f/img_pubblicazioni/img263_b.pdf?_ga=1.237351299.243225816.1469802512

National Institute of Statistics (Istituto Nazionale di Statistica–ISTAT). 2016a. "Indicatori demografici." Accessed July 29, 2016a. http://www.istat.it/it/files/2016/02/Indicatori-demografici_2015.pdf?title=Indicatori+demografici++-+19/feb/2016+-+Testo+integrale+e+nota+metodologica.pdf

National Institute of Statistics (Istituto Nazionale di Statistica–ISTAT). 2016b. "I.Stat: Your Direct Access to Italian Statistics." Accessed July 29, 2016. http://dati.istat.it

We Are Social. 2016. "Digital in 2016." Accessed July 29, 2016. http://www.slideshare.net/wearesocialsg/digital-in-2016

JAMAICA

Jamaica is located in the Greater Antilles in the Atlantic Ocean, south of Cuba. It is the third-largest Caribbean island with a population of 2.8 million. Until recently, it had a low internet penetration rate of 3.1 percent due to a monopoly held by the British Cable & Wireless Communication Group from 1988 to 2000. But recorded statistics in 2016 showed the penetration rate to be 43.4 percent, with 1.2 million users (Internet Live Stats 2016). The most popular domestic platform is Facebook, followed by Google+, Instagram, Twitter, YouTube, and LinkedIn with users accessing the internet several times a day (D&N Research Network 2015).

The liberalization of the industry in 2000 brought about fierce competition, bringing major new telecommunication carriers. One of the two leading players was Digicel, a new telecom company owned by Irish billionaire Denis O'Brien (1958–), which launched in Jamaica in March 2001. It introduced handsets and services at almost half the price of Cable & Wireless. Digicel exceeded its first-year expectations in mere months, surpassing the customer base rate of its rival. Flow, the other major player, owned by Columbus Communications International, began operating in 2006. It has invested millions of dollars in its broadband network.

The arrival of these carriers had an enormous impact on the social, economic, entertainment, and educational system of the island. It led to a rapid growth in internet connectivity, which benefited from the installation of a local internet exchange point (IXP) in Kingston, the capital. The increased demand for new licenses led to phenomenal growth in the number of cellular mobile subscribers and the amount of internet penetration.

In 2010, Dekal Wireless, a Jamaican company, was founded, and together with technology from Altai Technologies, Hong Kong, they developed Jamaica's first SuperWiFi broadband network, which was realized in 2011. It provided internet access to more than half of the population. In a strategic move, Cable & Wireless acquired Dekal Wireless at the end of 2014. Furthermore, in 2015, it announced a merger with Flow, acquiring 100 percent of Columbus International at the cost of $1.85 billion. It rebranded itself as LIME, an acronym standing for "Landline, Internet, Mobile, Entertainment." By 2015, LIME and Flow completed their merger, with the goal of maintaining a competitive industry by boosting the country's internet penetration rate (Golding 2011).

Customers reaped the benefits, enjoying reduced rates for entry-level broadband packages, faster internet speeds, and a range of new services. A reduction in rates in October 2004 saw prices drop from $66 per month to $45, with a further reduction in 2005 to $29.95. This led to a 488 percent growth in broadband subscribers in a short span of time. By 2016, the entry-level broadband package included access to the internet, home phone, and cable TV at $32.95 per month.

The education system benefited from the liberalization as well, as internet services were introduced into schools. The Ministry of Education and the Universal Service Fund created the Tablets in School (TIS) pilot program in 2006, which was initiated at the cost of $1.4 billion. It was implemented in 38 educational institutes across the island and provided tablets to 24,000 students. It proved highly successful, and in 2016, the initiative was extended to all teacher training colleges (Angus 2016).

The merger of Flow and LIME in 2015 brought about a slew of internet-related problems, such as dropped calls, service disruption, overall poor service quality, and billing problems. In addition to infrastructural problems, internet access was affected by the stealing of copper cable, because it could be sold off to make ammunition for the illegal arms market (Jamaica Observer 2015). Internet crime was also on the rise with the sending of sexually-explicit messages and photographs using mobile phones.

WhatsApp is one of the most popular mobile apps with Jamaicans. In 2014, when it introduced its voice over internet protocol (VoIP) services, mobile users were accessing the network to enjoy free calls. The ISPs collectively blocked access to WhatsApp, Viber, and Numbuzz. Following a backlash from mobile users, Cable & Wireless resolved the issue by signing a deal with Brian Acton (1972–), cofounder of WhatsApp, and became the first Caribbean provider to exclusively offer data bundles via its telecommunication companies of Flow and LIME (C&W Communication 2016).

Cybercrime became even more prevalent in 2013. There were widespread reports of persons taking photos of schoolgirls and imposing their heads on other bodies as a means of blackmail. The level of cybercrime grew rapidly and included scamming, electronic fraud, online harassment, cyberstalking, and the posting of obscene material. The laws in force under the 2010 Cybercrimes Act became outdated, and the Jamaican Constabulary Forces were unable to prosecute people held in connection with cybercrimes.

The government established the Communication, Forensic, and Cybercrimes Unit (CFCU) to monitor online activity to maintain national security while it was updating its data protection legislation (Gunn 2016). Cybercrime escalated to the point that governmental ministries, agencies, tertiary institutions, and private institutions were hacked. A focused group consisting of cybersecurity and forensics experts, known as the Cyber Incidents Response Team (CIRT), was constituted in 2014 to monitor Jamaica's internet activity for warnings or threats (Gunn 2016). Further preventive measures were taken in 2015 with the launch of the National Cyber Security Strategy to strengthen and protect the country's information and communication technology (ICT).

Nadia Ali

See also: China: Hong Kong; Ireland; United Kingdom

Further Reading

Angus, Garfield L. 2016. "Jamaica Information Service. Tablet in Schools Project for all Teachers Colleges." Accessed April 15, 2016. http://jis.gov.jm/tablet-in-schools-project -for-all-teachers-colleges/

C&W Communication. 2016. "Cable & Wireless & WhatsApp Sign Partnership." Accessed April 13, 2016. http://www.cwc.com/news-and-media/press-releases/cable-wir eless-whatsapp-sign-partnership.html

D&N Research Network. 2015. "Jamaica Chamber: Jamaica's Social Media Usage." Accessed April 11, 2016. http://jamaicachamber.org.jm/wp-content/uploads/2015/09/Social-M edia-Usage-Jamaica-Insider-Sept2015.pdf

Golding, Paul. 2011. "Telecommunications in Jamaica: Monopoly to Liberalized Competition to Monopoly (2000–2011)." September 20. Accessed April 14, 2016. http://www .globdev.org/files/Shanghai%20Proceedings/18%20PAPER%20Telecommunica tions%20in%20Jamaica%20Sept%202.pdf

Gunn, Tomeica. 2016. "Cyber Incident Response Team Fully Equipped and Operational." Accessed April 13, 2016. http://www.jamaicaobserver.com/news/Cyber-Incident -Response-team-fully-equipped-and-operational_50027

Internet Live Stats. 2016. "Jamaica Internet Users." Accessed May 4, 2016. http://www .internetlivestats.com/internet-users/jamaica/

Jamaica Observer. 2015. "Scrap Metal for Guns." June 7. Accessed January 20, 2017. http:// www.jamaicaobserver.com/news/Scrap-metal-for-guns_19086694

Schwab, Klaus. 2014. "World Economic Forum. The Global Competitiveness Report 2014– 2015." Accessed April 9, 2016. http://www3.weforum.org/docs/WEF_GlobalCompeti tivenessReport_2014-15.pdf

JAPAN

Japan is an island country in East Asia with a population of over 126 million people. Over 114 million people, or 90 percent of the population, are connected to the internet through fiber optics, asymmetric digital subscriber line (ADSL), community antenna television (CATV, or cable television), or cable internet, or through a mobile network such as WiMAX, making Japanese internet users more than 7 percent of the internet users in Asia (Miniwatts 2015). Although they trail behind South Korea and Hong Kong in speed, the Japanese enjoy fast connection speeds when accessing sites such as Google, YouTube, 2ch, Yahoo, Facebook, and Amazon. The internet is not tightly regulated by the government, leaving it vulnerable to security threats such as hacking and drug trafficking. The heavy use of the internet and mobile devices is blamed for the decline of Japanese who can properly write Kanji, one of the three alphabets of the Japanese language. This issue is expected to escalate as Japan prepares for the 2020 Tokyo Olympics with faster networks and more free access points.

In 1996, major companies Nippon Telegraph (also referred to as NTT) and Fujitsu began offering internet service provider (ISP) services. Mobile internet service and high-speed broadband were later introduced in the late 1990s and early 2000s. Fiber to the x (FTTH), fiber optic internet, was introduced in 1999

In Japan, passengers on the metro system use mobile devices while in transit. Japan has incorporated wireless communications into its culture; in the late 1990s the country became famous for the invention of emoji, which means "picture" + "character" in Japanese. Emojis have become a worldwide craze, and are in high demand across most social media platforms and chat apps. (Phuongphoto/Dreamstime.com)

and saw growth in the early 2000s. In 2008, Japan was the world leader in the number of fiber-connected homes at 13.2 million, followed by the United States at just above 6 million (Jackson 2009). As of 2011, the broadband market share was dominated by NTT East and NTT West, followed by cable television networks and Softbank, while the FTTH market was mostly controlled by NTT East and West (Sugaya 2012). Mobile internet is also ubiquitous in the island nation.

Japan was the first country where mobile data revenues exceeded voice revenue. This trend is expected to continue, as the Japanese market is filled with the latest smartphones, tablets, and other devices, though it should be noted that mobile internet was popular in Japan before the introduction of smartphones (Freedom House 2013). The Japanese enjoy surfing the internet at high speeds. In the third quarter of 2015, Japan's average connection speed was 15 Mbps, with a peak average connection speed of 78.4 Mps (Akamai 2015). In early 2013, So-Net, a Japanese service provider owned by Sony, launched Nuro, an FTTH service providing 2 Gbit/s internet to certain areas in Japan, which at the time was the world's fastest service.

Freedom of speech and the press is respected on the internet, and Japan's internet status has been reported as "free" by the Freedom on the Net Report of 2013 (Freedom House 2013). The Japanese internet has been described as self-regulatory,

Japanese parents are proud of their babies and love to share their photos over social media. In early 2016, one mother noticed that her baby's arms resembled a popular type of bread sold at local 7-Elevens. She took a picture of the baby's arm next to the loaf. Other parents copied her to show how their children's arms measured up. The "loaf and arm" meme follows several other social media trends to compare babies to food. In late 2015, Japanese celebrity Masahiro Ehara (c. 1983–) took pictures showing how his kids' faces could look like rice balls. Shortly thereafter, other parents copied him and posted cute pictures online. The trend even spread to pets, where the phenomenon was called "mochi mochi" after the popular rice cake snack.

a stark contrast to other East Asian countries such as China, North Korea, and even South Korea. While political censorship is not directly controlled, an old law that banned campaigning before an election caused many bloggers to delete content posted before the campaign period for fear of violation. However, this law has been repealed in the new day of internet and social media under Prime Minister Shinzo Abe's rule (2012–) (Huffington 2013).

Illegal material such as obscene content and child pornography are filtered by ISPs on a voluntary basis, and parental filter options may be set up by users. In 2007, parental filters were mandated by the Ministry of Internal Affairs and Communications; however, groups such as the Japan Internet Providers Association and the Movement of Internet Active Users lobbied against them (Aizu 2009). Online gaming is also a point of contention due to the potential for child abuse and possible exposure to online gambling. As a result, game developers Gree and DeNA Mobage placed caps on online virtual purchases for minors (Freedom House 2013). Regarding piracy, illegal transfer of music or films is punishable by prison terms and large fines.

Cybercrime is increasing in severity, and organized crime groups such as the Yakuza are expected to become involved in the cybercrime economy. 2ch, which is the short name for 2channel, a bulletin board site similar to 4chan in the United States and one of the most visited sites in Japan, is often used as a port for illicit sales and criminal behavior due to a policy of user anonymity. Police cracked down on a drug trafficking ring after a group of ten sold $1.4 million worth of drugs through the 2channel site. Criminals have also turned to online shopping sites as a way to traffic drugs. Smugglers hacked an online shopping site and used customer order information to send illegal substances. The buyers were then contacted by the smugglers, who said that the wrong package had been shipped and to forward the parcel to another address, fooling naïve shoppers into unknowingly help smuggle drugs (Ryall 2014).

The Japanese government faces serious security threats from hacking both nationally and abroad. Chinese gangs have been accused of hacking into Japanese bank accounts, and in 2012, nineteen Japanese websites were attacked during a time of increased tensions between Japan and China resulting from a land dispute. Foreign hacker groups such as the LuckyCat hacker group and Lizard Squad are suspected of carrying out attacks on Japanese websites. In addition, it is suspected that the hacker group Anonymous was responsible for shutting down Prime Minister Abe's personal website in retaliation for resuming whale hunting activities in the Antarctic (Soble 2015). The Japan Pension Service has also admitted that hacking resulted in 1.25 million cases of personal data being leaked (Otake 2015). In the international forum, Abe has referred to the hacking and theft of intellectual property by China while working to create a cybersecurity pact with U.S. president Barack Obama (Bennett 2015).

Social media is prevalent in Japan. Social networking and chatting apps are the most commonly installed, followed by gaming and video streaming apps (eMarketer 2015). There are about 25 million Facebook members and Twitter users grew from 18 million in 2013 to 22.5 million in 2014, with numbers continuing to rise (Bennett 2015). Japan has repeatedly broken the tweet-per-second record, and many attribute the popularity of the app to the ease of expressing oneself within Twitter's 140-character limit through the logographic Japanese language—in which one character can represent a complicated word.

While kanji may help the Japanese be more expressive on their microblogs, many traditionalists are pointing the finger at social media and other technologies for the degeneration of the language. It is believed that as these technologies make it easy for users to generate kanji characters, many are losing the ability to write them by hand. A survey by the Agency for Cultural Affairs of the Japanese Ministry of Education found that two thirds of the respondents believe that they are losing the ability to hand-write kanji due to automatic generation (Ghosh 2012).

Social media has had other negative impacts in Japan. During periods of increased political tension with other nations such as China and South Korea, social media has been used as a tool to spread hate speech. While free speech is protected in Japan, there are no antidiscriminatory laws to protect minority groups such as the 500,000 ethnic Koreans in Japan. Online harassment is often done anonymously with harassers taking advantage of forums such as 2ch to make personal attacks. Only recently have government bills been proposed, though anti-hate speech laws have yet to be passed (Krieger 2015).

On the positive side, social media serves as a connector to help save lives when phone contact is impossible. In 2011, when a 9.0-magnitude earthquake struck on the southwestern coast, devastating the country, many people turned to social media sites such as Facebook and Twitter to reach out to family and friends and to share their whereabouts. While cell phone networks were congested in the aftermath, social media allowed users to send messages and photos to other users in Japan and around the world. The Red Cross also took advantage of social media by reaching out to non-Japanese-speaking residents on Twitter to share emergency information and the locations of shelters. Many others simply used online services to share the severity of the situation with the rest of the world.

The future of the internet in Japan is bright. As the country prepares for the upcoming 2020 Tokyo Olympics, the government has pledged support for network technology regarding the launch of 5G connections by 2020. The government also has invited hackers for a special drill for the purpose of strengthening the security of government websites. Japanese transportation operators are preparing through an initiative to offer increased free Wi-Fi access in train stations and on flights. All this will make the internet in Japan faster, more accessible, safer, and, to the delight of tourists, sometimes even free (Hofilena 2014).

Crystal L. Hecht

See also: China; China: Hong Kong; North Korea; South Korea; United States

Further Reading

Aizu, Izumi. 2009. "Japan." Accessed December 20, 2015. https://www.giswatch.org/country-report/20/japan

Akamai. 2015. "State of the Internet: Asia Pacific Highlights." October 5. Accessed December 15, 2015. https://www.stateoftheinternet.com/downloads/pdfs/2015-q3-state-of-the-internet-report-infographic-asia.pdf

Bennett, Corey. 2015. "Japanese Leader Hits Chinese Hacking in Speech to Congress." April 29. Accessed December 19, 2015. http://thehill.com/policy/cybersecurity/240480-abe-to-congress-no-free-riders-on-intellectual-property

Bennett, Shea. 2015. "Twitter Japan: 26 Million Users, Rising to 30 Million by 2018." January 28. Accessed December 19, 2015. http://www.adweek.com/socialtimes/twitter-japan-users-growth/613663

eMarketer. 2015. "Smartphone App Users in Japan Focus on Communication." December 24. Accessed December 30, 2015. http://www.emarketer.com/Article/Smartphone-App-Users-Japan-Focus-on-Communication/1013383

Freedom House. 2013. "Freedom on the Net: Japan." Accessed December 20, 2015. https://freedomhouse.org/report/freedom-net/2013/japan

Ghosh, Palash. 2012. "Dire Threat to Culture? Mobile Phones, Email Destroying Penmanship." September 25. Accessed December 10, 2015. http://www.ibtimes.com/dire-threat-culture-mobile-phones-email-destroying-penmanship-795287

Hofilena, John. 2014. "Japan's Transportation Companies Beefing up Public WiFi Access Ahead of 2020 Olympics." April 28. Accessed December 15, 2015. http://japandailypress.com/japans-transportation-companies-beefing-up-public-wifi-access-ahead-of-2020-olympics-2847850/

Huffington, Arianna. 2013. "Postcard from Japan: Talking Zen, Abenomics, Social Networking, and the Constitution with Prime Minister Shinzo Abe." May 9. Accessed December 19, 2015. http://www.huffingtonpost.com/arianna-huffington/shinzo-abe-arianna-huffington_b_3245338.html

Jackson, Mark. 2009. "Top 20 Super Fast Fibre Optic Broadband Countries." February 13. Accessed December 20, 2015. http://www.ispreview.co.uk/news/EkFuVEyEEVPvohcBfQ.html

Krieger, Daniel. 2015. "Japan Combats Rise in Hate Speech." November 30. Accessed December 19, 2015. http://america.aljazeera.com/articles/2015/11/30/japan-encounters-rise-in-hate-speech.html

Miniwatts Marketing Group. 2015. "Asia Internet Use, Population Data, and Facebook Statistics." November 30. Accessed December 15, 2015. http://www.internetworldstats.com/stats3.htm#asia

Otake, Tomoko. 2015. "1.25 Million Affected By Japan Pension Service Hack." June 1. Accessed December 15. http://www.japantimes.co.jp/news/2015/06/01/national/crime-legal/japan-pension-system-hacked-1-25-million-cases-personal-data-leaked/

Ryall, Julian. 2014. "Japanese Drug Smugglers Use Online Shoppers as Mules." February 28. Accessed December 15, 2015. http://www.scmp.com/news/asia/article/1437576 /japanese-drug-smugglers-use-online-shoppers-mules

Soble, Jonathan. 2015. "Japan Investigating Hacking Attack on Shinzo Abe's Website." December 10. Accessed December 20, 2015. http://www.nytimes.com/2015/12/11 /world/asia/japan-hacking-shinzo-abe-whale-hunting.html?_r=0

Sugaya, Minoru. 2012. "Regulation and Competition in the JP Broadband Market." January 15. Accessed December 18, 2015. http://www.ptc.org/ptc12/images/papers/upload /PTC12_Broadband%20Policy%20Wkshop_Minoru%20Sugaya.pdf

KAZAKHSTAN

The Republic of Kazakhstan is a former Soviet state located in Central Asia. Kazakhstan occupies a huge territory (it is the ninth-largest country in the world) and has a relatively small population of 17.7 million people (Committee on Statistics of Kazakhstan 2016). The Kazakh government is investing in the development of internet in the country through the Kazakhtelecom national company. In 2016, the internet penetration in Kazakhstan reached 64 percent of the population aged six to seventy-four years old. Despite the state investments in the field, the authorities exercise heavy control over the internet sphere.

Many Kazakh internet users prefer to access the web via mobile phone. Currently, there are four mobile network operators in Kazakhstan, one of which has been providing 4G wireless telecommunications service since 2002, while all others announced the launch of 4G service by the end of 2016. Social networks are becoming popular among Kazakh internet users, with the Russian VKontakte holding the top position with more than 1.9 million local users.

After the dissolution of the Soviet Union in 1991, Kazakhstan experienced a major economic crisis that resulted in the decline of living standards in the country, it was not until 2004 when the country's gross domestic product (GDP) reached the level of the pre-independence era. Since the early 2000s, the country had a rapid economic growth (during 2000–2007, its GDP grew by 10 percent per year on average), mainly due to the rise in oil production and high oil prices. Such rapid economic growth even prompted some economists and political scientists to raise discussions about the "Kazakh Economic Miracle" and the birth of a new "Asian Tiger" in the Central Asian region. The drop in oil prices during the 2007–2008 global financial crisis, and later in 2015, slowed down the country's economy.

A substantial rise in state revenues in the 2000s allowed the state to increase investments in its infrastructural projects. The Kazakh government, through its national operator, Kazakhtelecom, has been consistently working on improving information and communications technologies in the country. According to official calculations, internet penetration in Kazakhstan exceeds 64 percent of the population aged six to seventy-four (Committee on Statistics of Kazakhstan 2015). Current estimates demonstrate rapid growth of internet users in the country. Just in 2000, the internet was used by 32 percent of the population in Kazakhstan.

The internet in Kazakhstan is mostly used for working with emails, searching for information, and reading news. In addition, 54 percent of users watch or download movies, while 43 percent prefer to listen to or download music. In 2015, 63.1 percent of Kazakh users accessed the internet through mobile phones, and an equal number of users used PCs, while 23.5 percent of all users accessed web resources through laptops (Committee on Statistics of Kazakhstan 2015). Google and the Russian site Yandex are the two most popular search engines employed by Kazakh users. The Kolesa.kz website, which represents an online market for Kazakh car sellers and car buyers, is the only locally developed website included in the list of the top ten most popular websites among Kazakh users. The growing number of Kazakhs who prefer to do online shopping is evidenced by the popularity of Aliexpress, an online retail service owned by China's Alibaba Holding Group, that is also among the ten most popular websites in Kazakhstan (Shigaeva 2015).

Kazakh users mainly utilize Russian (the country's official language), as well as Kazakh (the state language of Kazakhstan) on the internet. Google Translate added the Kazakh language in 2014. The inclusion of Kazakh was supported by the WikiBilim Public Foundation, a nonprofit organization that aims at promoting online educational content in the Kazakh language. The first project of WikiBilim was the promotion of Kazakh Wikipedia by improving the amount and quality of its material in the Kazakh language. The project was launched in 2011 and supported by the National Welfare Fund "Samruk-Kazyna." When the project was announced, there were 7,000 articles in Kazakh on Wikipedia; as of August 2016, this number exceeds 216,600 (WikiBilim 2016). Among the other projects sponsored by WikiBilim was the creation of the Open Library of Kazakhstan (http://kitap.kz) in 2012. Today, the online library contains more than 4,000 books and audio files in the Kazakh language.

During 2011–2014, an annual Freedom of the Net report considered Kazakhstan a "partly free" country; in 2015, however, the state for the first time was included on the list of "not free" states (Freedom House 2015). The Freedom House report put Kazakhstan in forty-eighth place out of the sixty-five countries reviewed in the report and noted that the state had blocked pages or entire websites of some news outlets. Many of those websites were blocked for reporting on the participation of Kazakh fighters in the extremist Islamic State of Iraq and Syria (also known as ISIL or Daesh), because they are considered propaganda. Among the government actions were to increase punishments for the dissemination of rumors and for libel (a restriction that also applied to online sources), and to ban tools that can hide users' internet protocol (IP) addresses.

At the end of 2015, there were more than 31 million mobile phone users in Kazakhstan (Committee on Statistics of Kazakhstan 2016). The discrepancy between this number and the country's total population can be explained by the fact that many people prefer to have more than one subscriber identity module (SIM) card to take advantage of cheaper on-net calls (within one mobile network), which also explains popularity of dual-SIM mobile phones in the country.

As of 2016, there are four licensed Global System for Mobiles (GSM) mobile network operators in Kazakhstan: Kcell, Beeline, Tele2, and Altel (the last two announced a merger of their companies in 2015). Of these four, only Altel provides 4G service, though all the other companies announced that they would launch 4G services in the near future. Kcell and Beeline each control 38 percent of the mobile network market in the country, while the share of the Altel-Tele-2 is about 24 percent (Profit 2016).

Locally developed mobile applications are becoming popular in Kazakhstan. In 2015, the mobile application of the Kolesa.kz website became the most actively used one in the country, with more than 8 million general and 74,000 unique users. Also included in the top ten ranking are Sajde KZ (an application that notifies about the time of Islamic prayers adjusted to the geographic locations of all major cities in Kazakhstan); Kiwi.kz's mobile application that is a video-sharing website; and Tengrinews.kz's mobile application that provides online news service (Bairamov 2015).

The Russian site VKontakte is the most popular social network used by Kazakh users. In 2015, there were more than 1.9 million active Kazakh users who posted at least one public post/message per month on VKontakte. Instagram is the second most popular social network in the country, with more than 1.3 active users; it is also the most rapidly growing social network in Kazakhstan. Today, Instagram is being heavily employed by small business enterprises and cottage industries that sell handmade clothing, toys, cookery, and other products, which in many cases have no official registration and, consequently, do not pay taxes. Recently, the Tax Committee of the Kazakh Finance Ministry issued a statement that it will monitor Instagram business users and take measures to register them officially and make them pay taxes (Koskina 2016). Other social networks include Moy Mir, with 155,300 users; Facebook, with 125,800 users; and Twitter, with 16,600 users. These three sites are also popular in Kazakhstan, although the number of their users is much smaller than those of Vkontakte and Instagram (Sarafannoe Radio 2015). The most popular Instagram accounts in Kazakhstan belong to local celebrities. As of June 2016, the top Instagram profile, with 1.8 million followers, is for the Kazakh producer, actor, and singer Bayan Yessentayeva (1974–). Among the most popular Facebook pages with Kazakh users is for professional boxer middleweight champion Gennady Golovkin (1982–); his page has more than 320,000 followers.

Nurlan Kabdylkhak

See also: Kyrgyzstan; Russia; United States

Further Reading

Bayramov, Badirsha, Torgunakova, Viktoria, Gorozhankin, Konstantin, Martynova, Marina, and Musabekova, Ulzhan. 2015. "Ranking Top-30 Mobile Applications of Kazakhstan." Accessed June 10, 2016. http://forbes.kz/leader/top-30_mobilnyih_prilojeniy_kazahstana_2

Committee on Statistics of Kazakhstan. 2015. "Development of Information-Communication Technologies in Kazakhstan in 2010–2014." Accessed June 10, 2016. http://www.stat.gov.kz/faces/wcnav_externalId/publicationsCompilations?_afrLoop=383056053590 87294#%40%3F_afrLoop%3D38305605359087294%26_adf.ctrl-state%3Dgti 2n9ako_74

Committee on Statistics of Kazakhstan. 2016. "Preliminary data for 2015." Accessed June 10, 2016. http://www.stat.gov.kz/faces/wcnav_externalId/publicationsCompilations?_afrLoop=38305605359087294#%40%3F_afrLoop%3D38305605359087294%26_adf.ctrl-state%3Dgti2n9ako_74

Freedom House. 2016. "Freedom on the Net 2015." Accessed June 10, 2016. https://freedomhouse.org/report/freedom-net/freedom-net-2015

Profit. 2016. "Tele-2 and Alter: Output in the First Quarter of 2016." Accessed June 10, 2016. http://profit.kz/news/30164/Obedinennaya-kompaniya-Tele2-i-Altel-itogi-I -kvartala-2016-goda/

Sarafannoe Radio. 2015. "First ranking of social networks of Kazakhstan." Accessed June 10, 2016. http://sarafannoeradio.org/analitika/297-perviy-reyting-sotsialnih-setey-kazahs tana.html

Shigaeva, Darya. 2015. "Internet-Users of Kazakhstan." Accessed June 10, 2016. http://forbes .kz/stats/internet-auditoriya_kazahstana_portret_i_predpochteniya_polzovatelya

WikiBilim. 2016. "History of WikiBilim." Accessed June 10, 2016. http://wikibilim.kz/index .php/about-foundation/timeline

KENYA

Kenya, located in East Africa, is one of the emerging economies on the continent. With a population of slightly over 46 million, Kenya covers over half a million square kilometers of land. It gained independence from British colonial rule in 1963 and has enjoyed relative peace since then. However, ethnic tensions and proximity to unstable countries such as Somalia and South Sudan threaten its stability. Kenya is one of the fastest-growing technology hubs in Africa.

Kenya has experienced one of the fastest mobile phone and internet penetration increases in sub-Saharan Africa. By the end of 2015, Kenya had the highest internet penetration rate in Africa (69.6 percent), accounting for nearly 10 percent of all internet usage on the continent (Internet World Stats 2016). A total of 99 percent of Kenyans accessing the internet do so using a mobile device (Alexander 2013). Toward the end of 2015, the mobile phone penetration rate in Kenya stood at 88.1 percent; 97.3 percent of these subscriptions were prepaid (Communications Authority of Kenya 2016). Almost half of the country's population had an active internet/data subscription. This rapid upward trend in internet penetration may be attributed to increased access to low-cost smartphones and the competitive pricing of data bundles by mobile phone subscribers in Kenya.

Structural reforms undertaken in the past 20 years have transformed the information and communication technology (ICT) infrastructure in Kenya, making it an attractive destination for high-tech companies. Global technology giants such as Facebook, IBM, Intel, Microsoft, and Nokia have all established themselves in Kenya in recent years, owing to the country's modernized ICT environment (Alexander 2013; Bright and Hruby 2015). Kenya is now popularly referred to as the "Silicon Savanah" due to its rapidly expanding ICT infrastructure, which has spurred innovations such as the M-Pesa and Ushahidi platforms (Benequista 2015). M-Pesa is a mobile money transfer service that allows people to deposit, send, and receive money using their mobile phones. M-Pesa, Swahili for "cash," was developed by the telecom firm Safaricom after noticing that people were sending each other mobile phone airtime instead of cash (Alexander 2013).

The internet-based crowdsourcing innovation Ushahidi (a Swahili word meaning "witness") combines the use of geomapping with other mobile-based technologies to allow ordinary citizens to document corruption, human rights abuses, as well as to track crises around the world (Bright and Hruby 2015). Ushahidi was developed by bloggers in Kenya and diasporic Kenyans to track the spread of violence during the 2008 unrest that followed the contested presidential election results in Kenya in 2007. Several versions of the platform have since been used in countries all over the world to follow a variety of situations, including the earthquake crisis in Haiti in 2010 and the parliamentary elections in Egypt in the same year (MIT Technology Review 2013).

While Facebook, YouTube, Twitter, and Instagram are among the top social networking sites in Kenya, various indigenous websites rank in the top ten as well (Alexa 2016). Among these outlets are *The Standard, Daily Nation,* and the *Star,* which are largely devoted to national news about politics, business, and entertainment. Nevertheless, Facebook remains the single most popular website. A report released by Facebook in 2015 indicated that there were more than 4.5 million monthly active Facebook users in Kenya, and that 95 percent of these were active on the mobile platform (World Stage 2015). The report further indicated that 2.2 million Facebook users were active on a daily basis. Kenyans, especially young people, use Facebook as a tool to connect and maintain contact with their friends, families, and colleagues, and as an online public sphere through which they discuss issues affecting them in their everyday life. Social media platforms such as Facebook and Twitter are the main sources of news for young people, who rarely pay attention to news emanating from the mainstream media outlets. However, the interface between mainstream media and social media discussion is evident in many Facebook discussion groups that have formed to provide digital public spaces where young people question issues represented in the mainstream media. Stories picked up from other media outlets and posted in Facebook chat groups often generate much online discussion among the participants, providing a space through which young people not only access and discuss events of the day, but also contest ideologies represented in the mainstream media.

As in other parts of Africa, Twitter remains an integral aspect of Kenya's sociopolitical environment and is often appropriated by young people in galvanizing and channeling public discontent, mobilizing for civic action, and creating civil awareness. A report published by Portland Communications in 2012 indicates that Kenyans were the second most active Twitter users in Africa, with close to 2.5 million tweets posted in Kenya in a span of three months; a subsequent report revealed that Kenya had fallen to fourth place (behind Egypt, Nigeria, and South Africa), though this determination was based on geolocated data, meaning that the data's country of origin had been verified (Portland Communication 2015). The analysis further indicates that 60 percent of Africa's Twitter users were younger (twenty-one to twenty-nine years old) than the global Twitter average (thirty-nine years old), and that Twitter was mainly used in Africa for social conversations, as a news source, and a growing forum for political conversation and activism.

Kenyans on Twitter, a collective group of Twitter users popularly referred to as KOT, is one of the most outspoken social media groups in Africa. KOT is renowned not only for its sharp criticism of the political class in Kenya, but also for its Twitter wars, popularly referred to as "tweefs," with social media users in other countries in the region. Kenyans have also used Twitter to challenge Western media coverage

In Kenya, in the village of Lanet Umoja, Chief Francis Kariuki uses his cell phone to tweet on crime and other local matters. The power of social media is demonstrated at all levels in Kenya, from small villages to the group KOT, Kenyans on Twitter, that reaches international audiences by challenging Western media coverage of Kenyan issues. (AP Photo/Khalil Senosi)

of Kenyan issues, particularly those that are perceived to have represented Kenya in unfavorable terms. Recent examples include KOT criticism of the CNN reference to Kenya as a "hotbed of terror" on the eve of U.S. president Barack Obama's (1961–) visit to the country. Using the hashtag #SomeoneTellCNN, Kenyans used Twitter to ridicule CNNs sensationalism in its reporting about their country. The ensuing outcry forced the CNN news channel's managing director to travel to Kenya to offer an apology (Mutiga 2015; Tharoor 2015).

While Kenya maintains a relatively free status in terms of expression via the media, recent developments have begun threatening this freedom. More than once, government officials have called for action against bloggers who post defamatory content online. In 2015, several social media users and bloggers critical of the government had been arrested under the 2013 Kenya Information and Communications Act (KICA) for the "misuse of licensed telecommunications equipment" (Freedom House 2016). A popular Kenyan blogger, Robert Alai (n.d.–), has been arrested at least four times for content published on his Twitter and Facebook pages. In December 2014, he was arrested after calling President Uhuru Kenyatta (1961–) an "adolescent president" for his handling of a terrorist attack that occurred in

Mandera County, killing twenty-eight people. Alai's Twitter account was also suspended for allegedly posting the president's private mobile phone number, which is against Twitter's user policy (Freedom House 2016).

In 2016, the Kenyan government began working on legislation that would establish how social media should be used, with the understanding that people use social media platforms to instigate hatred against their enemies. The new law would provide rules for taking action against such abuses. The government is taking these steps in preparation for Kenya's 2017 general election to help prevent the spread of hate speech during that period. Dennis Itumbi (c. 1982–), the country's director of digital communication, emphasized that the law is not meant to target bloggers, but to protect the public (Kenya Live 2016).

Abraham Mulwo and Marilyn J. Andrews

See also: Nigeria; Somalia; Tanzania

Further Reading

Alexa. 2016. "Top Sites in Kenya." Accessed August 21, 2016. http://www.alexa.com/topsites/countries/KE

Alexander, G. 2013. "Kenya's Tech Visionaries Lead the Way." *The Guardian,* March 1. Accessed August 21, 2016. https://www.theguardian.com/world/2013/sep/15/kenya-technology-visionaries

Benequista, N 2015 "Journalism from the 'Silicon Savannah': The Vexed Relationship Between Nairobi's Newsmakers and Its ICT4D Community." *Stability: International Journal of Security & Development*, 4(1): 12, 1–16, DOI: http://dx.doi.org/10.5334/sta.fc

Bright, J., and Hruby, A. 2015. "The Rise of Silicon Savannah and Africa's Tech Movement." Accessed August 21, 2016. https://techcrunch.com/2015/07/23/the-rise-of-silicon-savannah-and-africas-tech-movement/

Communications Authority of Kenya. 2016. "Kenya's Mobile Penetration Hits 88 Per Cent." Accessed August 13, 2016. http://www.ca.go.ke/index.php/what-we-do/94-news/366-kenya-s-mobile-penetration-hits-88-per-cent+&cd=6&hl=en&ct=clnk&gl=us

Freedom House. 2015. "Freedom on the Net: Kenya." Accessed August 21, 2016. https://freedomhouse.org/report/freedom-net/2015/kenya

Internet World Stats. 2016. "Internet Usage Stats in Africa." Accessed August 21, 2016. http://www.internetworldstats.com/stats1.htm

Jack, W., and Suri, T. 2010. *The Economics of M-Pesa.* Accessed April 1, 2016. http://www.mit.edu/~tavneet/M-PESA.pdf

Kenya Live. 2016. "Curbing Social Media Platforms in Kenya." Accessed August 21, 2016. https://kenyalive.co.ke/curbing-social-media-platforms/

Meier, P. 2012. "Ushahidi as Liberation Technology." In *Liberation Technology: Social Media and the Struggle for Democracy,* edited by L. Diamond, and M. Platner. Baltimore: Johns Hopkins University, 95–109.

MIT Technology Review. 2013. "Company Profile: Ushahidi." Accessed August 4, 2016. http://www2.technologyreview.com/tr50/ushahidi/

Mutiga, Murithi. 2015. "CNN Executive Flies to Kenya to Apologise for 'Hotbed of Terror' Claim." *The Guardian,* August 14. Accessed August 21, 2016. https://www.theguardian.com/world/2015/aug/14/cnn-kenya-apologise-obama

Omanga, D. (2015) "'Chieftaincy' in the Social Media Space: Community Policing in a Twit-ter Convened Baraza." *Stability: International Journal of Security & Development,* 4(1): 1–16. http://dx.doi.org/10.5334/sta.eq

Portland Communications. 2012. "New Research Reveals How Africa Tweets." Accessed Au-gust 21, 2016. http://www.portland-communications.com/2012/02/new-research-reveals-how-africa-tweets/

Portland Communications. 2015. "How Africa Tweets 2015." Accessed January 20, 2017. http://portland-communications.com/publications/how-africa-tweets-2015/

Tharoor, Ishaan. 2015. "Kenyans Ridicule CNN Report Calling Their Country a Terror Hot-bed." *Washington Post,* July 23. Accessed August 21, 2016. https://www.washingtonpost.com/news/worldviews/wp/2015/07/23/kenyans-ridicule-cnn-report-calling-their-country-a-terror-hotbed/

World Stage. 2015. "Millions in Nigeria and Kenya Embrace Facebook Mobile." Accessed August 21, 2016. http://worldstagegroup.com/index.php?active=news&newscid=24493&catid=37

KYRGYZSTAN

Kyrgyzstan, officially known as the Kyrgyz Republic, is a small, landlocked nation located in the northeastern part of Central Asia. Kyrgyzstan gained its indepen-dence from the Soviet Union in 1991, but it remains firmly in the Russian sphere of influence. Following its independence, Kyrgyzstan implemented a number of large-scale projects focused on establishing internet communications networks around the country and expanding these networks for users. Despite these efforts, internet usage in Kyrgyzstan still remains well below the world average, although the num-ber of internet users is rapidly growing. Internet use in Kyrgyzstan is mainly enter-tainment based, as evidenced by the significant use of social media in the country, with limited political and social activism taking place over the web. Partially because of this trend, internet users face few restrictions on web content in comparison to their neighbors, such as Kazakhstan and China. The future of internet in Kyrgyz-stan will focus primarily on building broadband infrastructure and expanding in-ternet usage throughout the country.

Internet use in Kyrgyzstan expanded rapidly in the early- to mid-2000s. The number of internet users, defined as the number of individuals who can access the internet at home via any device or connection, increased from approximately 51,000 in 2000 to nearly 2 million in 2015, a near-fortyfold increase (Internet Live Stats 2016). Nevertheless, internet penetration is still low by global standards, reaching just 32.5 percent in 2015 (Internet Live Stats 2016). There is a significant digital divide between urban and rural areas, and few locations outside the major cities have access to broadband internet. Two-thirds of internet users are concentrated in central cities, with 41 percent in the capital city of Bishkek alone (M-Vector Con-sulting Agency 2012).

Young people, unsurprisingly, make up the largest group of internet users. Con-sequently, internet use in Kyrgyzstan focuses heavily on entertainment, as noted previously, with social media websites being among the most widely visited in the country. Popular social media platforms include a mix of Western, Russian, and local websites; foreign social media sites have become increasingly popular in recent years. Until recently, international websites—websites that do not use the local ".kg" domain name—could be accessed only by paying an additional fee, leading to a proliferation of local social media platforms. Although fixed-line internet services still charge fees for accessing international websites, most mobile internet plans allow free access to foreign social media websites such as Facebook and Odnoklassniki (meaning "classmates" in Russian), a Russia-based social media network. As a result, Facebook and Russian social media websites such as Odnoklassniki, Moi Mir ("My World"), and VKontakte ("In Contact") have come to dominate the social media landscape.

Mobile internet has become the most widely used means of accessing the internet in Kyrgyzstan, due in part to the low broadband penetration rate and comparatively high cost of broadband internet. According to a 2012 report, 77 percent of internet users use mobile internet, compared to only 29 percent who use the internet at home (M-Vector Consulting Agency 2012). Cybercafés, historically one of the most popular means of accessing the internet, have declined in popularity as more and more individuals turn to mobile internet as their primary means to access the web. This gradual shift from cybercafés to mobile internet is possible due to the increasing spread of mobile internet networks and decreasing cost of mobile internet services.

Broadband internet is unavailable to most of Kyrgyzstan's citizens due to lack of infrastructure. As of 2016, there is a major project underway, the Trans-Asia-Europe Fiber Optic Line, with the purpose of increasing broadband internet in the country. The 27,000-kilometer-long optical fiber line began in October 1998 and will connect Shanghai, China, to Frankfurt, Germany (United Nations 2014). Kyrgyzstan completed its portion of the fiber optic line in 2010, but the project is still not complete.

Currently, broadband internet is available mainly in Bishkek, with some limited coverage in the provinces provided by the state-run Kyrgyztelecom. Mobile internet providers have been significantly more successful in their bid to expand mobile internet throughout the country, with a projected coverage of 90 percent of the populated territory. Mobile phone penetration surpassed 128 percent in 2014 (ITU 2016).

As mobile internet coverage expands throughout the country, the price of mobile internet has decreased, becoming more affordable for citizens. As of April 2016, Beeline (www.beeline.kg), one of the largest mobile internet providers in the country, provides a 2-GB plan for subscribers that costs as little as $6.50 per month, with free access to Facebook and Odnoklassniki, two of the most popular social media platforms in the country. Megacom, another large internet provider, offers an unlimited internet plan for $8.75 per month (www.megacom.kg). In contrast, broadband internet can easily range from $14–$70 per month for limited access and up to $84 for unlimited download (Freedom House 2015; United Nations 2014).

Despite the decreasing cost of mobile internet, the internet is still considered a luxury good and is consistently available primarily for wealthier citizens. As of 2015, the average monthly wage in the country was approximately $195, with salaries often significantly lower in the provinces. To combat the high cost of internet access and increase the pool of internet users in the country, Kyrgyztelecom, the largest broadband internet company in Kyrgyzstan, launched free Wi-Fi zones in Bishkek in October 2012, expanding to sixteen hotspots around the country as of 2015 (United Nations 2014; Freedom House 2015).

The telecommunications sector in Kyrgyzstan is relatively liberal compared with other countries in the region. Nevertheless, the state-run Kyrgyztelecom dominates the market for internet access; the three other top-tier internet service providers (ISPs) are privately owned. Kyrgyzstan is heavily reliant on the neighboring countries of Kazakhstan and Russia for access to international websites. Three of the four top-tier ISPs in Kyrgyzstan are linked to the internet via Kazakhstan; the fourth is linked through Russia.

Internet use in Kyrgyzstan was rated as "partly free" in 2015 according to Freedom House's Freedom on the Net. While freedom of internet use has improved since President Kurmanbek Bakiyev (1949–) was removed from power in 2010, new restrictions have raised concerns about the future of internet freedom in Kyrgyzstan. Since February 2014, subscriber identity module (SIM) cards have been required to be registered to their owners, limiting the ability of citizens to use information and communications technology anonymously. In the same year, the government instructed internet and mobile service providers to store information about their subscribers for up to three years and to grant authorities access to communications networks. These and other new laws tightening freedom of speech on the internet have worsened internet freedom in the country and led to fears of increasing prosecution and self-censorship.

ISPs in Kyrgyzstan also suffer from "upstream filtering" (the blocking of content by a third party) due to their reliance on internet bandwidth from Kazakhstan and Russia. The Kyrgyzstani government itself does not place any restrictions on accessing social media and blogging platforms. However, websites that are blocked for domestic users in Kazakhstan and Russia, such as the popular blogging platform LiveJournal, were historically also blocked for broadband users in Kyrgyzstan, although steps have been taken recently to resolve this issue with positive results (OpenNet Initiative 2010).

Despite relative freedom in social media content, the government has made an effort to block and filter some political and social content. However, the government mandate to block "extremist" content and content that incites "national hatred" has been inconsistently administered. Limited government attention to politically sensitive material on the internet may be due to Kyrgyzstan's lack of a developed blogosphere and digital activism. Television remains the primary medium through which citizens seek out information about politics, and few websites or blogs exist that are dedicated to political or social issues (Chapman and Gerber, 2016).

That is not to say, however, that individuals have not been targeted for posting incendiary information online. In a well-known case, journalist Dayirbek Orunbekov (n.d.) was charged for "knowingly disseminating false information regarding the commission of crimes" in 2012 for writing an online article accusing the government of inciting ethnic strife in the southern part of the country in 2010 (Freedom House 2015). This case was later dropped due to lack of evidence.

Hannah S. Chapman

See also: China; Germany; Kazakhstan; Russia

Further Reading

Chapman, Hannah S., and Gerber, Theodore P. "Soft Power and the Media: Russia's Media Strategy in the Near Abroad." Presented at the American Political Science Association 2016 Annual Conference, Philadelphia, PA, September 2016.

Freedom House. 2015. "Freedom on the Net: Kyrgyzstan." Freedom House. Accessed April 20, 2016. https://freedomhouse.org/report/freedom-net/2015/kyrgyzstan

International Telecommunication Union (ITU). 2016. "Mobile-Cellular Subscriptions per 100 Inhabitants." Explore Key ICT Statistics. Accessed May 1, 2016. http://www.itu.int/net4/itu-d/icteye/

Internet Live Stats. 2016. "Kyrgyzstan Internet Users." Accessed May 1, 2016. http://www.internetlivestats.com/internet-users/kyrgyzstan/

M-vector Consulting Agency. 2012. *"Issledovanie povedenija i vosprijatija media auditorii."* Исследование поведения и восприятия медиа аудитории 2012 г. (2-я волна) [Media Consumption & Consumer Perceptions Baseline Survey 2012 (2nd Wave)]."

OpenNet Initiative. 2010. "Country Profile: Kyrgyzstan." Accessed April 29, 2016. http://opennet.net/research/profiles/kyrgyzstan

United Nations. 2014. "An In-Depth Study of Broadband Infrastructure in North and Central Asia." United Nations Economic and Social Commission for Asia and the Pacific (ESCAP) Working Paper Series. Accessed May 1, 2016. http://www.unescap.org/resources/depth-study-broadband-infrastructure-north-and-central-asia

L

LIBYA

Libya is a Mediterranean country located in northern Africa that has a population of 6.33 million. It is one of the countries that experienced revolution during the so-called Arab Spring of 2011 along with its neighbors, Egypt and Tunisia. The civil war and the revolution of 2011 are important factors in the usage of internet in Libya because of the overthrow of Muammar Gaddafi's (1942–2011) authoritarian government. In 2016, Libya had 1.335 million internet users, with a penetration rate of 21.1 percent. Although the country has a relatively low penetration rate, the number of internet users increased by 9.9 percent in 2016, which ranked it as the fifteenth country in the world in terms of internet usage growth (Internet Live Stats 2016). Desktop and laptop computers were the main source of web traffic, at 89 percent of internet users, while mobile phones and tablets were 11 percent. As of 2014, there were 10.2 million active mobile subscriptions in Libya. Of these subscriptions, 96 percent were prepaid, while only 4 percent utilized a postpaid subscription. Half of mobile subscriptions were capable of using 3G connections. However, only 23 percent of these mobile subscribers use their mobile phones to connect to the internet (We Are Social 2014).

Before the civil war of 2011, the internet was extremely censored and services were quite poor, resulting in low internet speed and penetration rates. Since it was forbidden to establish new corporations without permission from the national intelligence services, the only internet service provider (ISP) before the civil war was the state-owned monopoly, Libya Telecom & Technology (LTT). After the revolution, many companies started to provide internet service (mainly satellite internet) throughout the country. According to Akamai's State of the Internet report, in the first quarter of 2014, Libya had the slowest connection speed in the world, at 0.5 Mbps. A more recent report by Akamai, covering the third quarter of 2015, stated that Libya and Yemen were still the only countries with an average connection speed below 1.0 Mbps. However, the same report also indicated that the government of Libya announced that "broadband connectivity will play an integral role in the rebuilding of its economy." In order to achieve that, Libya started building a fiber optic network to increase its service quality (Akamai 2015).

Social media platforms are commonly used in Libya. In 2014, 1.86 million people were actively using social media, and 1.44 million users were using social media on their mobile phones; however, these statistics suggest that these numbers do not reflect unique users (We Are Social 2014). Facebook is the most popular social media platform and the most visited website in Libya. Google, YouTube, and Instagram fall just behind Facebook in terms of visits. Twitter, on the other hand, is less popular in Libya; it is the seventh most visited website, just after Opensooq .com, an e-shopping website popular throughout the Arab world (SimilarWeb 2016).

Internet cafés were also very popular places in Libya. In 2002, Libya had the third-highest number of internet cafés among the Arab countries, with 700 (Warf and Vincent 2007). Internet cafés are widely used in Libya for different reasons. Teenagers, for example, commonly use internet cafés to browse pornographic content. Before the revolution of 2011, many activists and bloggers were arrested and tortured for political reasons. Therefore, dissident writers used internet cafés to publish their work anonymously (Mesrati 2013). However, the popularity and number of internet cafés decreased with the new connection alternatives that arose after the revolution.

Currently, Libya's freedom on the net is considered "partly free," in contrast with what it had been under Gaddafi's government. It is reported that internet in Libya was very open between the revolution of 2011 and the year 2015. The first political censorship after the revolution was the blocking of Alwasat, a news website, in February 2015 due to antigovernment statements (Freedom House 2015). On the other hand, the Libyan government has always tried to censor pornographic content, including during the more open period from 2011 to 2015 mentioned previously.

Social media played an important role in the organization of protests that drove the country to civil war in 2011. People commonly used social media platforms like Facebook, Twitter, and various instant messaging (IM) platforms to communicate during the protests. Viral videos and photos also caused the protests to advance and inspired more people to take to the streets, just as the photos of Mohammed Bouazizi's (1984–2011) self-immolation did in Tunisia, which showed police violence, and the famous v-log of Asmaa Mahfouz (1985–) did in Egypt, which invited people to protest. When the wave of Arab Spring revolution extended to Libya, Gaddafi's government responded harshly to the uprising. Before the protests began, Gaddafi predicted that they would spread through Libya via the internet. He also gave a speech about the internet and the Tunisian protests in which he used the term "Kleenex" (Gaddafi was known for using unique names) instead of the Wikileaks website:

> Even you, my Tunisian brothers. You may be reading this Kleenex and empty talk on the Internet. This Internet, which any demented person, any drunk can get drunk and write in, do you believe it? The Internet is like a vacuum cleaner, it can suck anything. Any useless person; any liar; any drunkard; anyone under the influence; anyone high on drugs; can talk on the internet, and you read what he writes and you believe it. This is talk which is for free. Shall we become the victims of "Facebook" and "Kleenex" and "YouTube"! Shall we become victims to tools they created so that they can laugh at our moods? (Hounshell 2011)

Social media not only helped the organization of the protests, but it also facilitated international support when people around the world used Twitter and Facebook for the purpose of solidarity with the revolutionaries in Libya during the protests. This was characteristic of all the movements that took place during the Arab Spring. The protesters wrote English words such as "Freedom" on the walls or used them as a banner in the protests, and pictures of those writings reached the outside world with the help of social media (Gilmore 2012, 8).

Ugur Dulger

See also: Egypt; Tunisia; Turkey; Yemen

Further Reading

Akamai. 2015. "State of the Internet." Accessed July 12, 2016. https://www.akamai.com/us/en/multimedia/documents/report/q3-2015-soti-connectivity-final.pdf

Castells, Manuel. 2012. *Networks of Outrage and Hope: Social Movements in the Internet Age.* Cornwall, U.K.: Polity Press.

Freedom House. 2015. "Freedom on the Net: Libya." Accessed July 12, 2016. https://freedomhouse.org/report/freedom-net/2015/libya

Gilmore, Eamon. 2012. "Democratisation and New Media." *Irish Studies in International Affairs,* 23: 5–12.

Hounshell, Blake. 2011. "Qaddafi Mourns Tunisian Dictator, Rips Wikileaks." January 17. Accessed July 12, 2016. http://foreignpolicy.com/2011/01/17/qaddafi-mourns-tunisian-dictator-rips-wikileaks/

Internet Live Stats. 2016. "Internet Users by Country." Accessed July 12, 2016. http://www.internetlivestats.com/internet-users-by-country/

Mesrati, Mohamed. 2013. "Bayou and Laila (Libya)." In *Writing Revolution: The Voices from Tunis to Damascus*, edited by Layla Al-Zubaidi, Matthew Cassel, and Nemonie Craven Roderick. Cornwall, U.K.: I. B. Tauris & Co. Ltd, 66–91.

SimilarWeb. 2016. "Website Ranking: Top 50 Sites in Libya for All Categories." Accessed July 12, 2016. https://www.similarweb.com/country/libya.

Warf, Barney, and Vincent, Peter. 2007. "Multiple Geographies of Arab Internet." *Area,* 83–96.

We Are Social. 2014. "Digital Landscape: Middle East, North Africa, & Turkey." Accessed August 5, 2016. http://www.slideshare.net/wearesocialsg/social-digital-mobile-in-the-middle-east-north-africa-turkey.

MALAYSIA

Malaysia is located in Southeast Asia, bordered by Thailand to the north and the Malacca Straits to the south. Within the region, Malaysia ranks only behind Singapore for having the highest social media penetration; its penetration rate is around 65 percent. Around 67 percent of the population is online, equating to slightly over 20 million of its 30 million people. The country's social media users are very savvy. They utilize rich images and language and are often knowledgeable about different technologies and platforms. The most accessed websites include Google, Facebook, YouTube, Yahoo, Google.co.my, Blogspot, Life.com.tw, Maybank, Wikipedia, and Lazada (Alexa 2015). The most cited reasons for Malaysians accessing the internet are communications and chatting; additional reasons include searching/researching, blogging, banking, and shopping. English is the most common social media language, with Malay, Chinese, and some Hindi; the language choice on social media corresponds to the site and user community.

Facebook is the premier social media platform for Malaysians. It is where the people go to converse and exchange views on important socioeconomic and political issues online. It is also a space where people can discuss challenging topics and engage a wide audience. Malaysia's Facebook demographics show that the platform is most popular with men aged twenty-five to thirty-four, and that local platform users are 55.8 percent men and 44.2 percent women (Allin1 Social 2015). After Facebook, Malaysians utilize a number of social media platforms, including WeChat, YouTube, WhatsApp, Google+, Facebook Messenger, and Twitter. Smaller numbers of Malaysians are on South Korea's KakaoTalk. WeChat claims to have a 95 percent mobile phone penetration in the country since mid-2015 (Digital News Asia 2015). Its user base primarily consists of youths aged sixteen to thirty-four, who say that they want to be reachable at all times. This population group makes up approximately 80 percent of all WeChat users in Malaysia (Gomez 2014). Line, a Japanese app with chat and shopping capabilities, has over 10 million registered users in Malaysia (Line 2014). These high numbers are not surprising, given that many Malaysians carry more than one mobile phone on which to conduct work and personal business, and sometimes may choose to use a phone based on its plan or its connection to a network with coverage in the area where they are calling.

In general, Malaysians prefer chat applications and playing games on their tablet and mobile devices. The top free downloads on Google Play are WhatsApp Messenger, WeChat, Facebook Messenger, Facebook, and 武林至尊（全球公测, or chen yaoyi), a kung fu/extreme martial arts gaming app. The most popular paid apps are League of Stickman-Samurai, XPERIA Minnie Mouse, Minecraft,

In Kuala Lumpur, Malaysia, during the 6th World Islamic Economic Forum (WIEF), an employee of PERNEC, a telecommunications and IT company, poses next to a new WiFi router. Malaysia has one of the highest social media and internet penetration rates in Southeast Asia, ranking behind only Singapore. (Dinictis/Dreamstime.com)

Flightradar24, and iQuran (AppAnnie 2015). For iOS devices, the data is similar, with Line taking the spot of the kung fu game in the free app category. For paid apps, the data shows that Malaysians care about the style and content of their social media messaging. For example, the top apps are Grindr Xtra, Text and Emoji for SMS+Texting, Pimp My Text PRO, Blog Touch Pro for Blogspot, and the GPS Phone Tracker for iPhones (AppAnnie 2015). Viber and Skype, other chat apps, appear in lower places on the lists.

Malaysians love to use emojis to express their opinions and feelings online. According to one recent analysis on emoji use in social media, Malaysia's preferred emojis are hand gestures, with the thumbs up sign being a particular favorite, followed by images that have to do with being female, funny, transportation, and farm animals, mainly pigs and rams (SwiftKey 2015). According to the report, Malaysia uses the most expressive emojis because its social media contains the widest variety of symbols. Some noteworthy pieces of analysis indicate that Malaysia uses sleepy faces more than two times more than other countries, and soccer balls more than three times the rate of other countries (SwiftKey 2015). These findings suggest that Malaysians discuss sleep and soccer online considerably more than any other nation.

There is a serious side to social media and internet usage in Malaysia, however. The country has laws against spreading false information and malicious rumors, particularly those that constitute libel, the making of false claims or accusations through the media. These laws extended to the internet and social media in 1998 with the Communications and Multimedia Act. Online enforcement typically includes identifying problematic messages or websites, contacting the platform or website owner and requesting removal of the information that violates the law. The big social media platforms, such as Facebook and Twitter, complied with around three-quarters of the removal requests in 2014. In 2015, the Malaysian Communications and Multimedia Commission (MCMC) met with platform representatives in order to ensure continued cooperation and adherence to Malaysian law. From the government's perspective, illegal web content must be deleted quickly, and when deemed necessary, the responsible authors could be punished through the judicial system.

Some Malaysians viewed the MCMC's efforts as trying to enforce stricter censorship on social media and the internet. They also saw it as a potential means for the government to cover up corruption, which is one of the long-standing concerns of the public. Government actions often play into this perception, as the government's response to online criticism is usually to tighten controls on internet speech. During 2015, there was increasing correlation between opinions voiced on social media and government calls to crack down on the internet. In July 2015, for example, the MCMC blocked the Sarawak Report, a local news portal hosted in Britain, for alleging that the country's prime minister, Najib Razak (1953–), received kickbacks from an investment fund. The site's reporting relied on documents obtained through the *Wall Street Journal* (Middleton 2015). Although the claims have not been substantiated, critics argue that the site did not violate the law because it was not spreading false information, but rather using accepted reporting techniques to expose possible corruption.

After the Sarawak Report incident, another controversy arose in August 2015 when the government blocked websites about the Bersih 4.0 rally, an event where Malaysians gathered to demonstrate against embedded corruption in the government. Bersih, which refers to the idea of sweeping or cleaning up, has been a multiyear effort to remove allegedly corrupt officials and graft. The movement, which relied on social media and the internet to spread information about its cause, was blocked online, reinforcing the people's frustration that internet censorship prevents political reform and anticorruption initiatives. According to the police, however, the protest was illegal. Despite the questions regarding its legality, the protesters continued to organize and prepare. The movement deemed social media of significant importance and had its own app developed, known as Prime Kini. Protesters were also considering using FireChat, a messaging app that does not rely on Wi-Fi to transmit and receive messages, which became popular during Hong Kong's Umbrella Movement in 2014. The group uses the hashtag #Bersih across other platforms as well. By August 2015, more than 27,000 people had used the hashtag as the movement geared up to protest (Malaysian Insider 2015).

In addition to using social media as a popular platform to promote domestic reforms, Malaysians have used social media to form and convey a sense of unity. This online Malaysian identity was evident throughout 2014 in the wake of national tragedy and distress. In March 2014, Malaysian Airlines Flight MH370, bound from Malaysia to China, disappeared off the radar. The missing plane and passengers created an international search and received high levels of media coverage that increased when debris from the plane and some passenger belongings washed up on eastern African shores. The search was called off in January 2017, which led to #mh370 becoming one of the top trending hashtags. The Malaysian government has offered a reward for private searches that can locate the disappeared flight, so it is likely the topic will continue to be discussed, and it will spike if any further discoveries are made. In July 2014, Malaysia Airlines suffered another devastating incident. Flight MH17 took off from Amsterdam to Kuala Lumpur on July 17, and partway into its journey, the plane broke up over the Ukraine, killing 298 people. In the wake of these tragedies, Malaysians started using the hashtags #staystrong and #prayforMH370 on Facebook, Instagram, and Twitter to convey support for the victims' families and to encourage Malaysians not to lose faith (Molander and Edenert 2014). Within the country, these hashtags went viral and showed the world that Malaysians will come together and show unity even in the most difficult situations.

Laura M. Steckman

See also: Indonesia; Japan; South Korea; Thailand; Ukraine; United Kingdom

Further Reading

Alexa. 2015. "Top Sites in Malaysia." Accessed September 5, 2015. http://www.alexa.com/topsites/countries/MY

Allin1 Social. 2015. "Facebook Statistics for Malaysia." Accessed September 5, 2015. http://www.allin1social.com/facebook/country_stats/malaysia

AppAnnie. 2015. "Google Play Top App Charts." Accessed September 5, 2015. https://www.appannie.com/apps/google-play/top/malaysia/overall/

Digital News Asia. 2015. "WeChat Hits 549mil Global Users, Claims 95% Penetration in Malaysia." May 29. Accessed September 4, 2015. https://www.digitalnewsasia.com/mobility/wechat-hits-549mil-global-users-claims-95pc-penetration-in-malaysia

Gomez, Jennifer, 2014. "Malaysian Youngsters Just Want to WeChat." November 19. Accessed September 5, 2015. http://www.themalaysianinsider.com/malaysia/article/malaysian-youngsters-just-want-to-wechat

Line Malaysia. 2014. "10 Million Users in Malaysia." March 20. Accessed September 4, 2015. https://www.facebook.com/433965876655984/posts/715158931870009/

Malaysian Insider. 2015. "27,000 Tweets on #Bersih 4 in the Past Month, Says Politweeet." August 27. Accessed September 4, 2015. http://www.themalaysianinsider.com/malaysia/article/27000-tweets-on-bersih-4-in-past-month-says-politweet

Middleton, Rachel. 2015. "Malaysia Blocks UK-Based Portal for Breaching Internet Law." July 21. Accessed September 4, 2015. http://www.ibtimes.co.uk/malaysia-blocks-uk-based-portal-breaching-internet-law-1511768

Molander, Agnes, and Nässlin Eidenert, Kim. 2014. "Radiotystnad Resulterar i Fullt Kaos En Studie om Malaysia Airlines Kriskommunikation." Uppsala: Uppsala Universitet. Accessed September 5, 2015. http://www.diva-portal.org/smash/get/diva2:786335/FULLTEXT01.pdf

SwiftKey. 2015. "SwiftKey Emoji Report." April 2015. Accessed June 5, 2015. http://www.aargauerzeitung.ch/asset_document/i/129067827/download

MEXICO

Mexico is the largest Spanish-speaking country in the world. It is located in North America, just south of the United States, with coasts on both the Pacific and the Atlantic oceans. Its population is 127 million. The internet has strong roots in Mexico, going back at least two decades. Similar to the United States, the first networks were built for academic and research institutions. The first use of a networked system in Mexico was in 1987, when the Monterrey Institute of Technology and Continuing Studies connected to the University of Texas at San Antonio. The first permanent connection was built in 1987, and by the mid-1990s, public internet service in Mexico was firmly established (Cassinelli and Fernández 2007b). Since then, the internet has spread, and today, Mexico is one of the most connected nations in Latin America.

Internet penetration in Mexico has almost doubled since 2007, when about 24 percent of the population had access (AMIPCI 2015; Cassinelli and Fernández 2007a). By 2014, there were over 50 million internet users, or approximately 44.4 percent. In 2015, the majority of these users were under thirty-five years old, with thirteen to eighteen-year-olds being the largest age group, at 26 percent of internet users (AMIPCI 2015). This statistic has not changed much in the past several years; in 2007, only 17 percent of internet users were over the age of thirty-five, and the proportion of users younger than eighteen years old was 25 percent (Cassinelli and Fernández 2007a). In general, the age distribution was about even.

While penetration across the nation was about 10 to 15 percent in each region, the southern states had slightly less penetration and the central region, due to the presence of the nation's capital, had more penetration. As more Mexicans are gaining access to the internet, they are also spending more time online. In 2015, the average time using the internet was six hours and eleven minutes a day, which was twenty-four minutes longer than the previous year. Average use of the internet was about the same regardless of the day of the week; however, Sundays saw slightly less internet use. In 2015, 84 percent of internet users said that they usually access the internet from home. Other locations, which were used about half as much, include work, school, stores, and cybercafés. Mexicans primarily connected to the internet using either private or public Wi-Fi connections; only about a quarter of the time were they using mobile data plans (AMIPCI 2015). The ten most commonly visited websites in 2015 were Google.com.mx, Facebook, YouTube, Google, Live, Yahoo, Amazon, Wikipedia, MSN, and Twitter (Alexa 2015).

The most common device used to access the internet in 2015 was the laptop. The next most common device was the smartphone, followed by the PC. Mobile phones not compatible with apps, tablets, and gaming consoles were used only a third of the time or less (AMIPCI 2015). While the majority of smartphone users owned Android devices, which are common throughout Latin America, Mexico had one of the highest rates of iPhone users in 2015. A total of 74 percent of smartphone users owned an Android and 26 percent owned an iPhone. Apple devices were even more common among tablet owners, with 37 percent owning an iPad and only 59 percent owning an Android (ComScore 2015). While Mexicans

Pokémon Go, an augmented reality game played via a smartphone app, launched in select locations in July 2016. Within two months of its release, the app, now known as the most viral one of all time, had earned more than $500 million and had more user activity than Twitter and Facebook (Perez 2016). The game's Latin America release occurred in early August, to coincide with the 2016 Summer Olympics in Brazil. Since then, it has become routine in Mexico for thousands of locals to visit Mexico City's Chapultepec Park on the weekends, bringing their entire families to capture the virtual creatures.

primarily used their smartphones to make calls, the next most common usage was to surf the web (AMIPCI 2015).

In 2015, the most common app downloaded for all smartphones was WhatsApp Messenger. The next five most popular free apps downloaded by Android users were Facebook Messenger, Facebook, Automatic Call Storer, CM Security Antivirus AppLock, and Moto Traffic Race. The next five most popular free apps downloaded by iPhone users were Snapchat, Facebook Messenger, Facebook, YouTube, and Instagram. Interestingly, iPad owners tended to have completely different preferences of the types of free apps that they downloaded to their tablet. The top free apps downloaded were all entertainment related: Disney Infinity: Toy Box 3.0, Hotel Transylvania 2, Doodle Jump SpongeBob SquarePants, FIFA 16 Ultimate Team, and Netflix (AppAnnie 2015).

Mexicans love social media. In Latin America, they are second only to Brazil in unique visitors to social media websites per month. In 2014, Mexico had nearly 25 million unique monthly visitors to social media websites, and 4 percent growth was seen between April 2013 and April 2014 (ComScore 2014). The primary reason that Mexicans use the internet is to access social media. The rate of social media penetration among Mexicans who have internet access is over 90 percent (AMIPCI 2015; ComScore 2015). Of those who did not use social media, over half of them made that decision because they wanted to protect their personal information. Nevertheless, the primary reason for being online is to use social media websites; in 2015, 85 percent of internet users were on social media, compared with 78 percent and 73 percent, respectively, of people online to look up information or check their email. Studies have shown that engagement with social media is higher when the content includes one or more photographs. Mexican social media users are primarily young; 65 percent are younger than thirty-five years old (ComScore 2014). The most common social media websites are international platforms, including Facebook, Google+, Twitter, Instagram, and Linkedin (We Are Social 2015).

Facebook is the most popular social media website in Latin America, and this is the case in Mexico as well. A total of 44.4 million Facebook users are Mexicans, which makes Mexico the fifth-leading country to utilize the website, behind the United States, India, Brazil, and Indonesia (Mexico News Daily 2015). This marks a significant increase in users over the past several years. In 2007, of the 57 percent of internet users who said they had a personal webpage, 34 percent of them said they utilized MySpace, compared with only 4 percent who used Facebook (Cassinelli and Fernández 2007a). Mexicans use Facebook primarily for accessing news and media, which comprised 43 percent of their usage pattern in 2014. Other popular categories were entertainment (17 percent), publishers (14 percent), and food/beverage (9 percent) (ComScore 2014).

Blogging is also a common activity among Mexican internet users. Most Mexican blogs are hosted on Blogspot, which was the eleventh most popular website in Mexico. In a survey conducted by the International Center for Journalists, bloggers with more professional content reported using Blogger and WordPress, while those who kept personal blogs tend to use Posterous and Tumblr (Sierra n.d.). Fashion, health, and lifestyle topics are some of the most popular subjects for blogs. In 2014, one of the most frequently visited fashion blogs was Moda Capital (http://www.modacapital-blog.com/), a blog started by Gaby Gómez (1985–) in 2009. She is from Mexico City and blogs in both Spanish and English. In 2013, she gave an interview to *Glamour* magazine, during which she stated that her favorite social media was Twitter because it lets you learn about what is happening the moment it occurs. At the time, she reported having 145,000 visitors a month (Glamour 2013).

On the other hand, blogging can be more serious—and even dangerous. In 2011, Mexican drug cartels started targeting bloggers who used this type of social media to publish journalist pieces that were critical of the cartels. Several violent murders were committed in 2011 as warnings to other bloggers who might want to speak out against them. The drug cartels have always targeted journalists, but in 2011, this focus shifted to those who wrote online. In September of that year, two bloggers were found hanging from an overpass in Nuevo Laredo. A sign that accompanied the bodies was signed with a "Z," indicating that the murders were perpetrated by the Zetas drug cartel (Flock 2011). In another incident, a woman was found decapitated next to her computer, with a note blaming her death on her blogging practices (Espatko 2011). As recently as the end of 2014, there have been numerous reports of similar murders of activist bloggers.

Marilyn J. Andrews

See also: Argentina; Brazil; Chile; Cuba; India; Indonesia; United States

Further Reading

Alexa. 2015. "Top Sites in Mexico." Accessed September 27, 2015. http://www.alexa.com /topsites/countries/MX

AppAnnie. 2015. "Google Play Top App Charts." Accessed September 27, 2015. https:// www.appannie.com/apps/google-play/top/mexico/?date=2015-09-27

Asociación Mexicana de Internet (AMIPCI). 2015. "11º Estudio Sobre los Hábitos de los Usuarios de Internet en México 2015." Accessed September 29, 2015. https://amipci .org.mx/images/AMIPCI_HABITOS_DEL_INTERNAUTA_MEXICANO_2015.pdf

Cassinelli, Andrés, and Fernández, Javier. 2007a. "Mexico: Internet Access and Usage Patterns." Accessed September 29, 2015. http://cs.stanford.edu/people/eroberts/cs181 /projects/2006-07/latin-america/mexicoAccess.html

Cassinelli, Andrés, and Fernández, Javier. 2007b. "Mexico: It's Not Too Late to Catch Up." Accessed September 29, 2015. http://cs.stanford.edu/people/eroberts/cs181/projects/2006-07/latin-america/mexicoIntro.html

ComScore. 2014. "The State of Social Media in Mexico," August 1. Accessed September 29, 2015. http://www.comscore.com/Insights/Blog/The-State-of-Social-Media-in-Mexico

ComScore. 2015. "IMS Mobile in LatAm Research Study." April 20. Accessed September 27, 2015. https://www.comscore.com/Insights/Presentations-and-Whitepapers/2015/comScore-IMS-Mobile-in-LatAm-Research-Study

Epatko, Larisa. 2011. "Mexican Drug Cartels' New Target: Bloggers." October 13. Accessed September 29, 2015. http://www.pbs.org/newshour/rundown/mexico-bloggers/

Flock, Elizabeth. 2011. "Mexican Bloggers' Bodies Disemboweled, Hung from Bridge." September 15. Accessed September 29, 2015. http://www.washingtonpost.com/blogs/blogpost/post/mexican-bloggers-bodies-disemboweled-hung-from-bridge/2011/09/15/gIQAfs6vUK_blog.html

Glamour. 2013. "Te Presentamos a Gaby Gómez, Embajadora de Glamour Girl Goes on Air by Lacoste." May 8. Accessed September 29, 2015. http://www.glamour.mx/magazine/articulos/glamour-girl-lacoste-perfumes-concurso-moda-capital/1465

Mexico Daily News. 2015. "Mexicans Are Social, Internet Study Finds." August 22. Accessed September 29, 2015. http://mexiconewsdaily.com/news/mexicans-are-social-internet-study-finds/

Sierra, Jorge Luis. n.d. "Results of a Survey of Mexican Journalists and Bloggers." Accessed September 29, 2015. https://freedomhouse.org/report/special-reports/digital-and-mobile-security-mexican-journalists-and-bloggers

We Are Social. 2015. "Digital in the Americas." Accessed September 27, 2015. http://wearesocial.net/blog/2014/06/social-digital-mobile-americas/

MONGOLIA

Mongolia is a Central Asian landlocked country bordered by China to the south and Russia to the north. Ulaanbaatar (UB), the capital and largest city, is home to about 45 percent of the country's population of about 3 million people. Almost 60 percent of the UB population is under age 35, now consisting of a mix of city and rural-bred young residents, making Mongolia one of the youngest countries in the world in terms of its citizens.

Before 1990, Mongolia was a socialist country and a satellite of the Soviet Union. Western cultural and linguistic trends were considered a cruel weapon of capitalist ideology and were strictly banned by the ruling communist party—the Mongolian People's Revolutionary Party (MPRP), which ruled from 1921 to 1990. The Russian language and culture were the only prevalent foreign elements in Mongolia. The communist authorities replaced the classic Mongolian Uyghur script with the Cyrillic alphabet in 1941, and Cyrillic Mongolian remains the official orthography (Rossabi 2005).

Following the collapse of the Soviet Union, Mongolia transformed from a socialist regime into a democracy, embracing a free-market economy in 1990. The authorities of communist Mongolia resigned without confrontation in 1990, marking the end of the seventy-year period of communist rule. This was the beginning of the new social, political, and economic order for the newly democratic nation. Mongolia opened up to the outside world, and new technology, media, computer, cable TV, urban radio stations, and the internet started circulating quickly across the nation. Mongolia witnessed major changes in its social, cultural, and linguistic lifestyle. Mongolians began getting heavily involved with internet, computer, and mobile phones in the context of their daily practices, including text messaging, chatting, surfing the web, playing video games, listening to music, and watching movies (Marsh 2009).

English and other languages, such as Japanese, Korean, Chinese, German, French, Spanish, Italian, and Turkish, have replaced Russian as Mongolians have become more connected to the world via the internet. Many of these languages are taught at all levels of Mongolian schools; at the national level, the country has been trying to attract quality English-language teachers from abroad. It is also reforming how its university teaching programs train instructors to provide exemplary teaching of all languages offered in Mongolian schools at

Buryat, an eight-year-old Mongolian boy, speaks on a mobile phone. When Mongolia's socialist regime ended in the early 1990s, the country opened itself up to the rest of the world, a move that included access to new communications technologies such as the internet. Due to the opened economy, youths such as Buryat have the ability to communicate worldwide on phones and online. (Alexander Podshivalov/Dreamstime.com)

the elementary, secondary, and tertiary levels (Munkhbat and Lkhagva 2015). The growth of English-language education, in particular, has also transferred to the professional realm, where most jobs require proficient English-language skills and frequently require job applicants to interview in English (Dovchin 2014).

Today, internet penetration is 43 percent, reaching about 1.3 million people, which is relatively high considering the small and dispersed population and the public's level of socioeconomic development. Social media such as Facebook, Twitter, and YouTube have soared in popularity, with data from late 2015 showing equal numbers of Facebook and internet users (about 1.3 million each). There are also many locally popular news sites, such as olloo.mn, sonin.mn, zaluu.com, and news.mn, for local and global news written in Cyrillic Mongolian by the sites' editors and journalists.

Popular sites such as asuult.net, hantulga.com, and tsahim.net are specifically devoted to young Mongolians, who socialize online and discuss their favorite topics, including popular music, sports, computer games, and blogging. As a result, many young Mongolians are engaging with a range of linguistic, cultural, and literacy materials that have implications for Mongolian language and culture, new postsocialist identities, and diverse literacy practices (Dovchin et al. 2016).

One of the most common digital orthographies in Mongolia is Roman Mongolian, instead of standard Cyrillic Mongolian, due to its convenience for keyboard use. Many computers in Mongolia lack Cyrillic Mongolian fonts and require specific device and technology support (Billé 2010). However, many Mongolians have expressed their strong preference to use Cyrillic Mongolian orthography, and thus, information technology (IT) support companies and retail stores have started installing Cyrillic Mongolian as a prerequisite font for computers in Mongolia. As a result, many Mongolians started using the Cyrillic Mongolian font more frequently in recent years. In 2013, Google began providing Mongolian-language support to its search engine. Basic Mongolian translation appeared in Google Translate later the same year.

Mongolia's social media users utilize Facebook, Twitter, and YouTube to express themselves. Online users have become highly skilled and digitally literate users of social media, which forms a powerful space for user access to worldwide activities. They create varied global and local discourses that are appropriate and meaningful in their particular contexts (Leppänen et al. 2015).

English and other foreign languages are also becoming significant in the digital daily lives of Mongolians. Online users seem to increasingly integrate varied linguistic and cultural resources within their daily lives online, which redefines the role of English and other foreign languages in relation to their existing online social relationships. Mongolians innovatively use English by not only mixing it with local languages, but also with other multiple texts with both local and global media content. Their digital literacy practices on social media often involve linguistic creativity, with users recontextualizing available signs and linguistic resources to create their own versions of digital literacy (Dovchin, Sultana, and Pennycook 2015). For example, the mixture of internet-related terms and Mongolian expressions is a common practice: "*Chatlii!*"—a combination of the English root word *chat* and the Mongolian suffix –*lii* [let's], has a local meaning: "Let's chat!" There are many Facebook default features such as "share," "like," and "tag," that have been deeply recontextualized and used by Mongolians as part of their local language. These Facebook linguistic features are mixed with Mongolian linguistic features to create local sayings, such as "*Minii zurgiig 'like' khiigeechee!*" ('Why don't you 'like' my photo?'); inserting the Facebook term *like* into a Mongolian sentence, such as "*Ene zurgiig shareleed uguuch!*" ('Please share this photo!'); and mixing the Facebook share feature with the Mongolian phrase "–*leed uguuch*" ("please"). Further examples can be found in Dovchin (2016a, 2016b).

Meanwhile, foreign-language adoption and language mixing have caused some Mongolian officials and social organizations to question whether foreign influences, particularly those derived from the growing influx of digital media, will lead to the deterioration of the Mongolian language. The ideology of "linguistic dystopia" has become a concern for some Mongolians, and in fact it led to the passage of the 2015 Law on the Mongolian Language. Article 21.7.9 of this legislation states that the "National Council of the Language Policy of the President shall take control over the implementation of adherence of the standards of the Mongolian language by media organizations" (Globe International 2016).

Though it remains unknown how this law may affect the internet, there is a fear that the global influences on social media use that have led to the intermixing of languages threaten Mongolian language and culture. This sentiment was addressed in a "Letter to the Committee of the Comprehensive National Development Strategy of Mongolia, Mongolian Parliament, 2007," written by a group of well-respected, highly educated Mongolians (Zuunii Medee, 2008). One of the main points of this letter was the importance of preserving "language purity," or protecting the Mongolian culture from erosion due to exposure to outside languages and influence. For the people who support "language purity," there are also many who believe in the right to express themselves and use language to redefine their identity and place in the world.

For digital users, these new ways to communicate and play with language are a way of adapting to globalization, as the world has opened up to them since the fall of the Soviet Union in 1991, an era when Mongolia had less contact with non-Russian-speaking nations. The emergence of unconventional digital language practices in Mongolia has only been presumed, rather than analytically examined by scholars, educational researchers, and language policy makers. This is a concern for Mongolia because digital media practices of online users in Mongolia are rapidly evolving as Mongolia is steadily connecting with the rest of the world through personal contacts and international business agreements.

To keep up with these changes, the country is examining and updating its focus on foreign-language acquisition and how it educates teachers. Taking advantage of digital users' language preferences and usage, as well as considering the impact that language dynamic may have on Mongolia and in other sociolinguistic ideological contexts, could profoundly and positively affect language and social media education in the country (Dovchin 2015, 2016a, 2016b).

Sender Dovchin

See also: China; North Korea; Russia; South Korea; Turkey

Further Reading

Billé, Franck. 2010. "Sounds and Scripts of Modernity: Language Ideologies and Practices in Contemporary Mongolia." *Inner Asia,* 12(2), 231–252.

Dovchin, Sender. 2014. "The Linguscape of Urban Youth Culture in Mongolia." PhD dissertation, University of Technology, Sydney.

Dovchin, Sender. 2015. "Language, Multiple Authenticities, and Social Media: The Online Language Practices of University Students in Mongolia." *Journal of Sociolinguistics* 19(4), 437–459.

Dovchin, Sender. 2016a. "The Ordinariness of Youth Linguascapes in Mongolia." *International Journal of Multilingualism.* http://dx.doi.org/10.1080/14790718.2016.1155592.

Dovchin, Sender. 2016b. "The Translocal English in the Linguascape of Popular Music in Mongolia." *World Englishes.* http://dx.doi.org/10.1111/weng.12189

Dovchin, Sender, Sultana, Shaila, and Pennycook, Alastair. 2015. "Relocalizing the Trans-lingual Practices of Young Adults in Mongolia and Bangladesh." *Translation and Trans-languaging in Multilingual Contexts,* 1(1), 4–26.

Dovchin, Sender, Sultana, Shaila, and Pennycook, Alastair. 2016. "Unequal Translingual Englishes in the Asian peripheries." *Asian Englishes.* http://dx.doi.org/10.1080/13488678.2016.1171673.

Globe International Center. 2016. Media Freedom Report 2015. Beijing: UNESCO.

Leppänen, Sirpa, Spindler Møller, Janus, Rørbeck Nørreby, Thomas, Stæhr, Andreas, and Kytölä, Samu. 2015. "Authenticity, Normativity, and Social Media." *Discourse, Context, & Media, 8,* 1–5.

Marsh, Peter. 2009. *The Horse-Head Fiddle and the Cosmopolitan Reimagination of Tradition in Mongolia.* New York: Routledge.

Munkhbat, Bulgantsetseg, and Lkhagva, Enkhbayar. 2015. "Implementation of a New Cur-riculum for the English Teacher Program at the National University of Education in Mongolia." *10th East Asia International Symposium on Teacher Education,* Nagoya, Japan.

Rossabi, Morris. 2005. *Modern Mongolia: Fromkhans to Commissars to Capitalists.* Berkeley, CA: University of California Press.

Zuunii Medee. 2008. June, number 259/2707. Accessed January 17, 2017. http://dusal.blogmn.net/14798/yndesnii-ba-tsogts-gesen-aguulga-uguilegdej-baina.html

MOROCCO

Morocco, a North African country, is made up of roughly 34.8 million people. Since 2011, internet penetration has increased by 10 percent. Compared to the rest of the African continent, Morocco is one of the most active countries on the internet. There are approximately 20 million users in the country, with an internet penetration of 57.6 percent (Internet Live Stats 2016). Social media has become increasingly popular in the country, especially with Moroccan youth. While social media is used for social communications, it also has helped further develop the job market and increase the small business and restaurant online presence to multiple audi-ences. In addition, Moroccan activists have adapted social media platforms to act as a tool to protest government censorship and question the monarchy's authority, even going as far as to declare a social media revolution in response to a ban on apps providing free voice over internet protocol (VoIP) services. Blogging has also become very popular in the country, and maintaining a well-written blog can lead to recognition. For example, Synergie Media celebrates Moroccan bloggers every year with the renowned Maroc Blog Awards.

Social media has become increasingly popular with Moroccan youth. The average age range for social media users in Morocco is fifteen to twenty-nine years old. In many countries around the world, social media has been utilized as a powerful tool to fight censorship and human rights violations. This can be seen in recent antigov-ernment protests in Tunisia and Egypt. Not only have the youths of Morocco turned to Facebook and YouTube to express themselves, but they also have become avid bloggers who share details about their personal stories, emotions, and lifestyles. Some bloggers have also shared critiques and opinions of the government and monarchy, despite the fact that they could face government persecution. It is esti-mated that there are about 30,000 blogs in Morocco. Popular Moroccan bloggers living outside the country use their expat status to speak more critically of their former home. Moroccan expats also encourage their fellow countrymen to speak openly and encourage Moroccan leaders to increase freedom of expression and freedom of press to allow the public an outlet without fear of reprisal (Haug 2011).

To celebrate the Moroccan bloggers for their creativity and ingenuity, Synergie Media, a communications agency that specializes in social media, created the Maroc Blog Awards, also known as the Morocco Blog Awards. Thousands of blogs enter each year to win the twenty coveted awards. It gives bloggers the opportunity to distinguish themselves by the quality of their blogs and the impact on their readers. The Maroc Blog Awards has the reputation of being one of the most prominent competitions in the Middle East and North Africa (MENA) region. Its primary purpose is to promote opportunities for Morocco's digital market and to support web content creators in their efforts to cultivate creative and innovative writing (Maroc Web Awards 2016).

In Morocco, the most used social media network is Facebook; around 98.6 percent of Morocco's internet users have an account, while other social media networks are utilized by fewer than 1 percent of active internet users (StatsMonkey 2015). Despite this high number, some experts believe that this statistic may not be accurate, arguing that Moroccans have multiple Facebook profiles. Taking this into account, the final calculations may be distorted, but they give researchers a good indicator of which social media platforms Moroccans use on a regular basis (Haug 2011). Aside from Facebook, LinkedIn is another popular social media network in Morocco.

Both Facebook and LinkedIn have helped advance the Moroccan job market and economy significantly. Many restaurants and small businesses use Facebook to market their services and products to an online audience. Customers and clients are able to access news and leave feedback on the site. Local Moroccan businesses and individuals also take advantage of the tools and services offered by LinkedIn. By using LinkedIn, many Moroccans have access to job opportunities posted on the social media platform, and they can develop online networks to help them professionally. Small businesses are able to build their brand and online presence on LinkedIn and gain access to a larger pool of job applicants (Elotmani 2014).

In January 2016, Morocco's three telecommunication service providers, Maroc Telecom, Meditel, and Inwi, banned any social media networks that provide free mobile internet calls. Morocco's Agence Nationale de Reglementation des Telecommunications (ANRT), or Telecommunications Regulatory National Agency, is targeting VoIP, which allows users to make free calls over wireless connections. Mobile apps, such as Viber, Facebook Messenger, WhatsApp, and Skype, are no longer available to Moroccans. According to ANRT, the mobile apps providing free mobile internet calls do not have the authorized licenses to offer these services, nor do they follow the guidelines stipulated for the country's telecommunications market and regulations. While it is not yet illegal to use VoIP services, the ANRT regulations do limit app usage in the country. Experts are unsure of the immediate impact that this ban is having on the economy. It is believed that while the three telecommunication providers are trying to be more competitive, it could tentatively be a setback for the country's movement toward becoming a modern, developed state in the twenty-first century (Egyptian Streets 2016).

The decision to ban these mobile apps was made without letting Moroccan consumers know about it first. The beauty of VoIPs was their ability to communicate with people around the world without accruing expensive charges from local telecom service providers. As global social media providers grew, consumers around the world had access to new and improved services, such as free calls over the internet. VoIP gave Moroccans and Moroccan expats the ability to communicate with friends and families around the world without the fear of having to pay an expensive bill. The sudden cutoff by the three major telecommunication service providers angered many consumers. In fact, there was a call for a social media revolution to fight the strict VoIP regulation that was implemented in January 2016. This VoIP regulation deeply affects low-income Moroccans working in the farming and construction sectors.

Social media activists and columnists have stated that this maneuver by local telecom service providers to compel consumers to pay for calls is denying Moroccans their basic right to communication (Bendraoui 2016). The founder of Th3professional, Amine Raghib (c. 1986–), condemned the actions of the telecom providers on his Th3professional blog and his official Facebook page. In a series of statements, Raghib told his followers to protest against the three telecom providers. First, he told the followers to "unlike" the telecom's social media pages. Next, they were to boycott all events hosted by the three telecom companies. This would ultimately lead to an unidentified Moroccan consumer filing a lawsuit against the ARNT for what the public viewed as an illegal action. Since these events, the telecom providers have lost tens of thousands of followers on their official social media pages (Jabrane 2016). The "unlike" campaign and the lawsuit appear to have worked; in October 2016, some users discovered that the block had been removed, though the ARNT did not make an official announcement.

Morocco has a history of persecuting journalists and activists that question the authority of the country's monarchy or its laws. The government immediately shut down any political dissidence. As the online community grew in Morocco, it was only natural for individuals to explore and use the internet as a new public forum. Activists began to use different social media channels to air their grievances or try to spark a social revolution. In 2007, for instance, YouTube was banned for two weeks for publishing videos that were critical of the Moroccan king, Mohammed IV (1963–). The following year marked the appearance of more Moroccan activists on social media.

Moroccan activists are continuing their fight for a prodemocratic government through online platforms and addressing new audiences in Morocco and around the world (Global Voices 2016). By 2012, the threats against social media activists significantly increased as more bloggers and social media users were arrested, allegedly for violating the press law and the country's sacred values. While the Moroccan government has arrested and continues to arrest online activists, the final sentences are often reduced or pardoned outright upon public outcry (York 2012).

Karen Ames

See also: Algeria; Egypt; Libya; Tunisia

Further Reading

Bendraoui, Mounia. 2016. "The VoIP Ban in Morocco and Its Impact on the Economy." April 2. Accessed April 4, 2016. http://www.howwemadeitinafrica.com/voip-ban -morocco-impact-economy/

Cashmore, Pete. 2007. "YouTube Banned in Morocco." May 26. Accessed April 4, 2016. http://mashable.com/2007/05/26/youtube-maroc/#_eM1Z6kLpsqB

Egyptian Streets. 2016. "Morocco Bans Whatsapp, Skype, Viber, and Other Services Providing Mobile Internet Calls." January 8. Accessed April 4, 2016. http://egyptianstreets .com/2016/01/08/morocco-blocks-whatsapp-skype-viber-and-other-services-providing -mobile-internet-calls/

Elotmani, Fazine Meziane. 2014. "The Positive Impact of Social Media on Morocco's Job-Market." July 4. Accessed April 3, 2016. http://www.moroccoworldnews.com/2014/07/ 133904/the-positive-impact-of-social-media-on-moroccos-job-market/

Global Voices. 2016. "[Timeline] Morocco: Political Repression in the Era of Social Media." January 26. Accessed April 5, 2016. https://advox.globalvoices.org/2016/01 /27/timeline-morocco-political-repression-in-the-era-of-social-media/

Haug, Astrid. 2011. "Moroccan Youth Take on Social Media." February 7. Accessed April 2, 2016. https://www.mediasupport.org/moroccan-youth-take-on-social-media/

Internet Live Stats. 2016. "Morocco Internet Users." Accessed April 2, 2016. www .internetlivestats.com/internet-users/morocco/

Jabrane, Ezzoubeir. 2016. "Social Media Revolution Against Moroccan Telecom Providers." February 27. Accessed April 5, 2016. http://www.moroccoworldnews.com/2016/02/ 180834/social-media-revolution-against-moroccan-telecom-providers/

Maroc Web Awards. 2016. "C'est Quoi les Maroc Web Awards?" Accessed on April 3, 2016. http://marocwebawards.com/

StatsMonkey. 2015. "Morocco Social Media Usage Statistics Using Mobile." Accessed April 3, 2016. http://www.moroccoworldnews.com/2016/01/176915/morocco-whatsapp-with -viber-facetime-skype/

York, Jillian. 2012. "Moroccan Activist's Arrest Signals Crackdown on Speech." February 15. Accessed April 5, 2016. https://www.eff.org/deeplinks/2012/02/moroccan-activists -arrest-signals-crackdown-speech

Zerhouni, Hicham. 2016. "Morocco, Whatsapp, with Viber, FaceTime, & Skype." January 6. Accessed April 4, 2016. http://www.moroccoworldnews.com/2016/01/176915/morocco -whatsapp-with-viber-facetime-skype/

MOZAMBIQUE

Mozambique, officially known as the Republic of Mozambique, is a country in Southeast Africa with a population of 26.4 million people (as of 2016), of which 51 percent are female, 55 percent are below 19 years of age, and almost 68 percent live in rural areas. Its capital is Maputo City, which is where most technological investments are based, as all three mobile telephone companies' headquarters located there. It is one of the poorest countries in the world, with one of the lowest internet penetration rates, and yet it is considered one of the fastest-growing economies in the world.

The growth of the information and communication technology (ICT) sector in Mozambique has accelerated rapidly over the past four years, driven by economic development trends (i.e. newly discovered minerals, effective government reforms, foreign aid, and capital investments). This occurrence has accelerated the growth of mobile phone usage, internet usage, and network infrastructure development. Mobile penetration in 2012 was 48 percent, and in 2014, it had already reached 77 percent. Internet penetration in 2014 was low at 5.9 percent, while Facebook usage was 4.7 percent, encompassing a little over a million people, based on the population at that time.

Mozambique has three mobile operators: Mcel (with the most subscribers), Vodacom, and Movitel. All of them offer 3G connectivity and cover most areas of the country. In 2015, Facebook and Mcel launched Internet.org (later renamed Free Basics by Facebook), an application that gives all Mcel subscribers free access to Facebook and selected internet sites. This project is accelerating the growth of mobile internet and Facebook usage. From 2010 to 2012, the number of mobile phone subscribers increased about 11 percent, while the number of fixed telephony subscribers decreased 0.08 percent.

Radio still has the best penetration of any other device/technology in Mozambique, at above 53 percent penetration. In 2012, the country had ten state-run radio stations and eighty private stations that broadcast in more than twenty languages. Because of the number of stations, radio permeates almost the entire country and can reach every audience, including the illiterate portion of the population, which is around 49 percent. For these reasons, radio became the primary medium through which the population received news. However, television is starting to replace the radio for news consumption, especially in urban areas.

Mozambique-based users rely on foreign social media, as there is no local investment in the development of locally based social media applications. They are most active on Facebook, WhatsApp, Instagram, YouTube, Google Plus, and Twitter. Facebook and WhatsApp are the fastest-growing social media platforms in Mozambique.

Facebook is also largely used for many social activities. For example, in 2014, a local rapper, Edson da Luz (1984–), known as Azagaia, took to Facebook to crowdfund for his brain tumor operation in India, where he was successful in both gathering the amount of money that he needed. Local free speech has also been affected by Facebook and social media. In one example, a local journalism website called Canal de Moçambique, which criticized government policies, had its internet protocol (IP) blocked and the website was taken offline. The site operators responded by creating a Facebook page and started using the page to disseminate their articles. In 2015 and 2016, Mozambique had been affected by a military-political tension between the ruling party, Frelimo, and the main opposition party, Renamo. Although the government was able to contain the spread of rumors via national television and radio stations, on Facebook, new anonymous profiles (such as Unay Cambuna) started reporting on the clashes between government forces and Renamo's military forces in the central regions of Mozambique. Many users took to Facebook to get real-time knowledge of the actual situation, which was viewed as an "undeclared war."

In addition, many nonprofit organizations have used Facebook to prevent the spread of rumors. In 2014, there was a rumor that Mozambique had cases of Ebola, and the United Nations International Children's Emergency Fund (UNICEF) and the Ministry of Health mainly controlled it via Facebook.

WhatsApp is widely used due to its ease of use and its group functionalities. In 2016, an anonymous source called all Mozambicans to come out and protest against the government due to the current military-political tension and international debt crisis. The message went viral in a single day, mainly via WhatsApp and Facebook, and this prompted the government to deploy antiriot police in Maputo City and many businesses closed for the day of the protest. In the end, no protest was held.

The e-Government Strategy, approved in 2006, was conceived specifically to support the second phase, from 2006 to 2011, of Mozambique's public-sector reform. The reform intended for the government to achieve decentralization and improve service delivery in addition to improving institutional and human capacity. GovNet is a key contributor to the successful implementation of the e-Government Strategy because it constitutes the technological foundation for all other components. This success has led to many government institutions offering basic services via their own websites. Specifically, all key government ministries now have websites, and all legislation is now online.

On August 28, 2015, the government of Mozambique approved a new piece of legislation (Ministerial Decree no, 18/2015) that demanded subscriber identity module (SIM) card registration for all cell phone users in Mozambique. The law was in response to a series of riots in Mozambique that the government said had been coordinated by mobile phone via short message service (SMS). In 2016, almost 1 million SIM cards were switched off by the three mobile operators in Mozambique following the failure of mobile phone owners to register their SIM cards. Currently, it is against the law to have a functional cell phone with a valid local SIM card that is not registered. The government knows the name, address, and contacts of every national cell phone user in Mozambique.

The internet still has a low penetration in Mozambique, and among the common reasons for the nonuse or infrequent use of the technology are the costs associated with using the internet, slow connectivity speeds and lack of local language content and support. Households with internet access tend to use mobile handsets to get connected, instead of a personal computer or a laptop. The share of internet users with an email address in Mozambique is above 60 percent—higher than the share of internet users who have signed up for a social networking application.

Local mobile operators have also launched the usage of mobile money, with Vodacom using the "M-pesa" since 2013, and Mcel using the "mKesh" since 2011. Mobile money programs are designed to allow people to bank and perform financial transactions online. The effort is designed to increase commerce and financial inclusion. There is an increased usage of these new paying modalities and huge marketing campaigns by the mobile operators to push the usage of their mobile money for day-to-day transactions.

Mozambique had its first Digital Marketing workshop in 2016, led by a Portuguese digital specialist, and one of the key outcomes was that Mozambique usage of digital (social networking, websites, and related digital) tools is growing at a fast rate, and many companies are now shifting from their standard one-way communication to two-way communication, as digital communications and the internet offer new tools for companies to connect with their customers.

For many youths and adolescents, the internet is Facebook in Mozambique. For many businesses, the internet offers a cost-effective solution to counter the current debt crisis, and for the government, the internet offers new possibilities to increase the qualities of their public services. The internet will definitely see a huge increase in adoption, with more initiatives like Free Basics, but the call for more local language content and support is crucial for the growth in rural areas.

Claudio Fauvrelle

See also: Botswana; Kenya; South Africa; Zimbabwe

Further Reading

AllAfrica. 2016. "Mozambique: Phone Companies Disconnect a Million Clients." Accessed July 4, 2016. http://allafrica.com/stories/201603040272.html

Business Insider. 2015. "The 13 Fastest-Growing Economies in the World." Accessed July 1, 2016. http://www.businessinsider.com/world-bank-fast-growing-global-economies-2015-6

Facebook. 2016. "Audience Insights Page Statistics." Accessed July 1, 2016. https://www.facebook.com/ads/audience-insights/interests?act=169522799804337&age=18-&country=MZ

INCM. 2015. "Boletim da Repùblica, Decreto no. 18/2015." Accessed July 4, 2016. http://www.incm.gov.mz/documents/10157/343078/decreto%20no18-2015.pdf

Instituto Nacional de Estatistica. 2014. "Estatísticas dos Transportes e Comunicações." Accessed July 1, 2016. http://www.ine.gov.mz

Internet World Stats. 2016. "Mozambique Internet Stats." Accessed July 1, 2016. http://www.internetworldstats.com/africa.htm#mz

Mabila, Francisco. 2013. "Understanding What Is Happening in ICT in Mozambique." Policy Paper 10, 2013. Accessed July 1, 2016. http://www.researchictafrica.net/publications/Evidence_for_ICT_Policy_Action/Policy_Paper_10_-_Understanding_what_is_happening_in_ICT_in_Mozambique.pdf

Socialbakers. 2016. "Mozambique Facebook Page Statistics." Accessed July 1, 2016. http://www.socialbakers.com/statistics/facebook/pages/total/mozambique/

MYANMAR

Myanmar, sometimes referred to as Burma for historical political reasons, is a Southeast Asian nation located to the west of Thailand. It has one of the lowest internet penetration rates in the world, and yet it is poised to have some of the highest internet growth rates over the coming years. In mid-2015, Myanmar's internet penetration was approximately 5 percent of its population of 54 million. The majority of the country's internet users access the web via mobile phone. Myanmar-based users rely on both foreign and domestic social media. They are present on Facebook, Twitter, and YouTube, with growing user bases on Viber, WhatsApp, Line, WeChat, BeeTalk, and Tango. MySQUAR is the country's most used domestic platform, and there are several others emerging as potential competition in Myanmar's predominantly Burmese-language environment. Exact numbers of users on these platforms are difficult to ascertain.

Myanmar was one of the most closed-off countries of the world until its transition to democracy in 2011 and 2012. As part of that transition, the government decided to modernize some of its infrastructure to include its outdated telecommunications systems. The process included reducing the costs of accessing the system, which consisted of lowered prices for mobile phones, internet access, and installation. Before the policy to lower device and access costs went into effect, internet installation cost around $500, with a $70 monthly fee. For mobile phone users, subscriber identity module (SIM) card costs dropped from $3,000 to $500 in 2011; by 2012, the costs had dropped to $200 (Aung 2013). In early 2013, the government started sponsoring SIM card lotteries. Winners could purchase SIM cards for around $2. The price reduction was one piece of the plan to upgrade Myanmar's telecommunications. The price reduction came as a relief to the average family that earned around $82 per month, and especially to families who earned less than $200 annually (Heinrich 2014).

As for infrastructure, the government issued contracts in 2013 to two foreign firms, Norwegian Telenor and Qatar's Ooredoo, with the express purpose of modernizing cellular and mobile communications within the country. Previously, the Myanmar Posts and Telecommunications monopolized the mobile industry. Because of domestic corruption and remnants of the dictatorship, the agency was ineffective. Without competition, prices remained exorbitant and service was dismal. With the influx of the new contracts, Myanmar's mobile phone access and social media usage are poised to explode. Some estimates predict that 80 percent of the population will have internet access by the end of 2016; more conservative estimates suggest that 50 percent of the population will have access by 2019 (Heinrich 2014).

The availability of devices and the development of new Burmese-language platforms will certainly lead to a growth in the number of online users and reconnect Myanmar's people to others worldwide. Viber is Myanmar's most active chat app, with approximately 5 million registered users (Kham 2015). In a 2014 survey of social media application usage, 79 percent reported using Viber, while 27 percent reported using, or also using, Facebook (OnDevice 2014).

One of the biggest obstacles for Myanmar is using their language font online. Most devices and applications accept English and other Latin alphabet–based languages with ease. However, Burmese-language fonts and keyboards are difficult to install and unwieldly to use without the requisite site and device support. There is little notable language support at present, though this dilemma is slowly being resolved. One company paying attention to Myanmar's internet needs is Google. Google started to provide Burmese-language support to its search engine in 2013. Basic Burmese translation appeared in Google Translate in late 2014, and language support to Gmail started in mid-2015. Other companies interested in capturing Myanmar's market share, which could range from online shopping to advertising, are likely to follow suit.

One entrepreneur has also decided to tackle the difficulty people in Myanmar have using their language online. Rita Nguyen (c. 1977–), a Vietnamese-Canadian woman, aims to offer Myanmar social media in its own language. The challenge with the Burmese language is that it uses a Sanskrit-derived script that requires special keyboards and keyboard layouts. Noting that a potential market existed for a Burmese-language social media platform, Nguyen worked with a team to create MySQUAR. MySQUAR is a social media platform with a built-in chat application called MyChat, which had a user base of approximately 700,000 people in mid-2015. The platform's overall success led to a $2.5 million investment and started trading shares on Britain's Alternative Investment Market in order to increase revenue (Stewart-Smith 2015). As more people in Myanmar have access to the internet and a choice of social media applications, MySQUAR has the potential to expand quickly into one of Myanmar's premier platforms.

MySQUAR was not the only local app inspired by Myanmar's lack of native-language social media outlets. Entrepreneurs have developed several additional noteworthy apps, many of which imitate the well-known foreign platforms. Doe Myanmar is a local version of Facebook, designed to connect friends, share content, and build an online community. Hush is a platform where users can share content and messages anonymously. It was likely inspired by Myanmar's decades-long censorship and fear of government reprisal for open criticism. Pyaw Kyi is another app similar to Hush. Momo Lay is a blogging site that allows users to control not only their writings, but also the advertisements and banners appearing on the page. Currently under development is Pwar, which will also share features with sites such as Doe Myanmar and Facebook (Kham 2015). As market demand grows, Myanmar is likely to see additional apps develop to meet user demand.

Myanmar's social media users utilize Facebook somewhat differently than in other parts of Southeast Asia and the world. For many years, it was one of the only platforms available in Myanmar. However, while there is some social interaction and use of its chat features, Facebook is used in Myanmar less for social engagement and community building than as a news feed (Huq 2014). In fact, the most popular pages in Myanmar are 7Day News Journal, Eleven Media Group, and the Irrawaddy's Burmese-language news page (Socialbakers 2015).

Facebook also has a dark side in Myanmar. During times of high interethnic tension between the Buddhist majority and Rakhine Muslims, Facebook has been utilized to spread rumors and hate speech. Often disguised as news, the rumors have incited conflict and violence. For example, in March 2014, Facebook users reported a nongovernmental organization (NGO) worker for throwing the Myanmar flag on the floor, out the window, or somehow otherwise disposing of it. Although the facts were unclear, rumors raced online until a mob confronted the NGO worker, causing her to flee. An eleven-year-old girl died at the scene, with at least a dozen others reported to have died later due to injury (DiCerto 2014). In the town of Sittwe in July 2014, a rumor appeared online that two Muslim men raped a Buddhist woman. Following a similar mob mentality, fighting ensued. Two died and at least a dozen more were injured (DiCerto 2014). Although these allegations were not true, the statements that appeared online caused tremendous social damage. Facebook, of course, was not the cause of these incidents; however, its use as

A Buddhist monk uses his cellphone during a meditation at Shwedagon Pagoda, the most sacred Buddhist shrine in Myanmar. Until very recently, only a small, wealthy segment of Myanmar's population could even dream of accessing the internet. Since Myanmar's move to adopt democratic governance in 2012, domestic reforms have expanded the telecommunications infrastructure, making it more common for non-elites to obtain internet access. (Antonella865/Dreamstime.com)

a conduit indicates that social media can influence people's behavior and have a negative societal impact when derogatory and inflammatory speech go unchecked.

Myanmar has a number of hacker groups that claim to profess elite cyber skills. Popular hacking groups include Myanmar Noob Hackers, Myanmar Black Hats, Blink Hacker Group, Myanmar Hacker UniTeam, and Myanmar Hacker Warriors. Most of these groups use Facebook and blogs to communicate about their achievements, which usually consist of using denial of service (DoS) and distributed denial of service (DDoS) to prevent the targeted websites from functioning. Other hacking efforts include such a group gaining access to a website and then posting a message to declare both that the site was hacked and which group hacked it. Screenshots of the hack will often appear on the group's Facebook page or blog as proof shortly thereafter.

Myanmar's hackers tend to choose the websites they target based on the sites' geographic locations and current the current sociopolitical situation within the sites' home country. It is not uncommon for Myanmar's hackers to target Bangladeshi websites or to respond to an attack that Bangladesh's hackers initiated against them. Often, the conflict is over an issue such as the Rohingya ethnic group, a Muslim minority present in Myanmar and Bangladesh. Neither country accepts the Rohingya as their own citizens, so the tensions that exist in real life sometimes bubble over onto the internet. Less frequently, Myanmar hackers enter a hacking contest with Indian users, or more rarely, enter temporary hacking alliances, such as joining the Malaysian hackers against Bangladesh.

Laura M. Steckman

See also: Bangladesh; India; Malaysia; United Kingdom

Further Reading

Aung, San Yamin. 2013. "Burma to Hand out Free Temporary SIM Cards to SEA Games Participants." October 26. Accessed August 24, 2015. http://www.irrawaddy.org/burma/burma-hand-free-temporary-sim-cards-sea-games-participants.html

DiCerto, Bridget. 2014. "In Newly Liberated Myanmar, Hatred Spreads on Facebook." August 8. Accessed August 24, 2015. http://www.globalpost.com/dispatch/news/regions/asia-pacific/myanmar/140804/newly-liberated-myanmar-hatred-spreads-facebook

Ferrie, Jared. 2013. "In Myanmar, Cheap SIM Card Draw May Herald Telecoms Revolution." April 24. Accessed August 24, 2015. http://www.reuters.com/article/2013/04/24/us-myanmar-telecoms-draw-idUSBRE93N1AX20130424

Heinrich, Erik. 2014. "Asia's least-developed telecom market will soon become the world's fastest growing." September 18. Accessed August 24, 2015. http://fortune.com/2014/09/18/asia-myanmar-burma-telecommunications-market/

Huq, Md Mushfiqul. 2014. "Rita Nguyen: The Woman Behind Myanmar's Social Media Revolution." December 9. Accessed August 20, 2015. http://sdasia.co/2014/12/09/rita-nguyen-woman-behind-myanmars-social-media-revolution/

Kham, Aung. 2015. "Rival of Social Apps in Myanmar?" June 8. Accessed August 20, 2015. http://myanmar-entrepreneur.com/2015/06/08/rival-of-social-apps-in-myanmar/

OnDevice Research. 2014. "Myanmar: The Final Frontier for the Mobile Internet." June 23. Accessed September 1, 2014. https://ondeviceresearch.com/blog/myanmar-mobile-internet-report

Socialbakers. 2015. "Myanmar Facebook Page Statistics." Accessed August 24, 2015. http://www.socialbakers.com/statistics/facebook/pages/total/myanmar/

Stewart-Smith, Hana. 2015. "MySQUAR, Orchard Funding Join AIM, Sophos Begins Full Trading (ALLIPO)." July 1. Accessed August 24, 2015. http://www.morningstar.co.uk/uk/news/AN_1435743173459573100/mysquar-orchard-funding-join-aim-sophos-begins-full-trading-(allipo).aspx

NAMIBIA

Namibia is a country on the southwestern coast of the African continent with a population of 2.1 million, according to the 2011 census (Namibia Statistics Agency 2016). Namibia is sparsely populated over a vast geographical area of 824,292 square kilometers (CIA World Factbook 2015). This vast space demands a good communication infrastructure, which is why the country has invested in high-quality infrastructure. Statista (2015) and the World Bank (2016b) rank Namibia fourth on the African continent in their ranking of countries with the best infrastructure. This infrastructural environment affects the country's networked readiness positively, allowing it to rank 99 out of 139 countries in the world, and seventh overall in Africa (WEC 2016).

In spite of its relatively good standing on the various indexes and its competitive advantage regarding the first pillar of network readiness—political and regulatory environment, where Namibia ranked thirty-first in the world and fourth in Africa (WEC 2016)—this success has not translated into actual access to and usage of the internet. The discrepancy is partly due to the high cost of digital services, which translates into limited access and low internet penetration (WEC 2016). Broadband internet has a particularly low uptake in Namibia as a result, with subscriptions standing only at 1.76 percent in 2014. The country's fixed telephone subscriptions stood at a mere 7.78 percent, a drop from 7.97 percent the year before (World Bank 2016a). In stark contrast, mobile cellular subscriptions per 100 people were at 114 around the same period. However, not all these mobile subscribers have smartphones, so they do not all have internet access. Overall, the tally of internet users in Namibia measured through number of people with access to the internet at home, at work, or in public, stood at 14.8 per 100 people in 2014 (World Bank 2016a), although other sources put internet penetration at 17.1 percent (BuddeComm 2016).

The startling contrast between access to fixed telephone and fixed broadband internet on one hand and mobile cellular subscriptions on the other hand is a trend on the African continent. For example, Pew Research Center (2015) established that two-thirds or more people in South Africa, Nigeria, Ghana, Senegal, Tanzania, Kenya, and Uganda own cell phones, although only a few of those own internet-enabled smartphones. This has led to some to claim that sub-Saharan Africa has skipped the landline development stage, jumping straight into the digital age through the proliferation of mobile phone networks (Pew Research Center 2015). Therefore, the growth of and access to internet in Africa in general and Namibia in particular has gone hand in hand with growth in cellular phone subscriptions.

Research carried out in Namibia and ten other African countries that sought to explain why internet has gone mobile established that generally, "mobile internet requires fewer ICT skills and financial resources, and does not rely on electricity at home, compared to computer or laptop and fixed-internet access" (Stork, Calandro, and Gillwald 2013, 4). Thus unlike the first wave of internet adoption that was facilitated by access to fixed telephony, the second wave that is typical of Africa and Namibia relies on the use of mobile phones. In Namibia, about half of internet users reported to have first accessed it on mobile phones (Stork, Calandro, and Gillwald 2013).

The penetration of internet in Namibia is still low by international standards. For example, Nordic countries average more than 95 percent of internet users, with Iceland topping the world with 100 percent of internet penetration—a far cry from Namibia. Nonetheless, internet use in Namibia has grown steadily over the years, with internet users more than quadrupling from an estimated 90,349 users in 2006 to 392,181 in 2016, the latter representing 15.6 percent of the population (Internet Live Stats 2016). The local media metrics, however, put (broadband) internet access in Namibia at 23 percent in 2014, which was a 6 percent increase from the previous year (IPPR 2014). In comparison, mobile internet access in the country at that time was much higher, at 30.7 percent, which was measured in terms of the number of users with mobile handsets capable of browsing the internet (Stork and Calandro 2014, 215). This is almost the same as South Africa, the highest-ranked country in Africa, where 34 percent of mobile subscribers own smartphones (Pew Research Center 2015). With initiatives such as the recent "oSmartPhona" roadshow campaign by the Namibia's biggest mobile telecommunications operator, Mobile Telecommunications Corporation (MTC), which sought to promote access to smartphones by selling them at discounted prices of roughly $30 (Nakashole 2016), the use of internet through mobile phones in Namibia is set to continue growing.

What is the relationship between this trend of internet access through mobile phones and the use of social media? Research from emerging economies says that internet usage is highly associated with social media use. According to Pew Research Center (2016), "[I]nternet users in emerging and developing countries are more likely to use social media compared with those in the developed world." This is possibly because all popular social media sites have mobile apps, so accessing them is more convenient than accessing the sites via web browsers. In Namibia, mobile network operators also offer dedicated data for specific social media apps, which could be a factor. The most popular social media site is Facebook, with a 93.23 percent share of users (StatsMonkey 2016). Other social media sites used are Instagram, Twitter, YouTube, Snapchat, LinkedIn, Pinterest, and Tumblr. According to Peters, Winschiers-Theophilus, and Mennecke (2015, 4), "Namibians also have a relatively high Facebook adoption rate in comparison to other African countries." In 2013, over 10 percent of the population was on Facebook, with a large percentage in the age group of eighteen to thirty-five years old (IPPR 2014), a figure that will have increased since the original survey. Also popular are chatting apps such as WhatsApp and Facebook Messenger due to their high social interactivity.

How do Namibians use social media? The most popular social media site, Facebook, is used for both content consumption, such as keeping up to date with news or entertainment like videos and pictures of events or personalities; and content generation, like uploading content, sharing views, and sharing content from one platform to another. Namibians, like all other social media users, also utilize the site to connect with friends and family. In addition, social media has become a prominent tool for civic engagement. In the past two years, social media has been the tool of choice for youth activism, whereby prominent youth leaders use Facebook to spark movements and galvanize and organize their followers. They share information about their activities, report back to their "constituencies," so to speak, and use it as a logistical tool support the planning of events such as protests.

An example of this phenomenon is a Namibian youth-led movement called Affirmative Repositioning that uses its Facebook page and Twitter account to plan demonstrations, communicate information on its events, and organize actions with its members/followers. One can sometimes see more activism online than offline. Political parties also use social media sites to reach out to the electorate, while citizens use them to debate contemporary issues related to elections and society (IPPR 2014). Public institutions like government offices and institutions of higher learning use them to access their audience, and political leaders also use them to communicate with the masses. For instance, the current president of Namibia, Hage Gottfried Geingob (1941–), has Facebook and Twitter accounts that he uses to communicate frequently with citizens about the activities of his office, such as his speeches and public statements or matters of policy.

In addition, social media is used for educational purposes, such as to support university students' collaborative and networked learning (Haipinge 2013). Students and teachers, especially in higher education, use platforms such as Facebook and WhatsApp to engage beyond the face-to-face contexts and support e-learning.

As for entertainment, social media is a powerful tool in the country, whereby during national entertainment events such as annual music awards, broadcasting companies stream events live on YouTube. Short videos giving sneak peeks and pictures highlighting these events are also posted on Instagram, Facebook, and Snapchat (Kawela 2016). The use of Twitter hashtags during key events or to promote discussions or causes is also common. Other examples indicative of the entertainment use of social media is the following of celebrity fan pages. Socialbakers.com (2016) ranks a celebrity Facebook page of a Namibian model, Behati Prinsloo (1989–), as the most liked page in Namibia, with 603,797 fans; and the third-most-liked page belongs to another local celebrity, Dillish Mathews (1990–), a winner of the television show *Big Brother Africa*.

Other uses of social media include promoting causes, marketing, publishing of news by media houses, and public relations activities by public institutions. For example, a cheetah conservation Facebook page is the second most popular page, followed by a marketing page for a popular Namibian beer. Newspapers, a mobile network company, and a radio station are also among the top 10 pages.

Regarding user behavior, Peters et al. (2015) found that Namibian Facebook users behaved distinctively differently online from their offline behavior. For example, they find it easier to discuss and post information on Facebook that they consider taboo or uncomfortable in their offline lives. This is corroborated by Haipinge (2013), who found that users of social media maintained two separate conceptions of reality between their online and offline activities. There is also a high degree of naïveté among Namibian social media users regarding issues of privacy, as shown through research findings that users share intimate information about themselves or falsely believe that they have substantial control over information that they share online (Haipinge 2016; Peters et al. 2015). This reporting indicates that as the use of the internet and social media increase, that does not necessarily include knowledge about how to use these tools safely.

The future of the internet and social media in Namibia is bright; access and use have continued to increase as growth rates over the last few years have shown an upward trend. Online access gives citizens more opportunities to make their voices heard, offers enhanced access to information, and assists with marketing—all with the potential for informed decision making and enhanced democratic participation. At the same time, the growth in internet and social media use also calls for more education on the safe and responsible use of these platforms. The fast pace at which social media has grown in the country has caught the legal environment and political leadership off guard. Indicative of this is the call by politicians to tighten laws in order to control social media use by the public that they perceive as abusive and careless (Shinovene 2014).

In this regard, improving the legal framework to deal with and regulate the use of the internet in ways that are on par with international standards is another priority. This concern does not mean that citizens' use of social media should be stifled, however, given the huge benefits that these sites offer for freedom of speech, government transparency, and promoting business. At the same time, when a country's laws are not responsive to the fast-paced developments in technology, challenges will naturally arise.

Erkkie Haipinge

See also: Finland; Iceland; Kenya; Nigeria; South Africa

Further Reading

BuddeComm. 2016. "Namibia: Telecoms, Mobile and Broadband—Statistics and Analyses." Accessed August 21, 2016. https://www.budde.com.au/Research/Namibia-Telecoms-Mobile-and-Broadband-Statistics-and-Analyses

CIA World Factbook. 2015. "Namibia." Accessed August 21, 2016. https://www.cia.gov/library/publications/the-world-factbook/geos/wa.html

Haipinge, Erkkie. 2013. "Conceptions of Social Media and its Role in Supporting Networked Learning: A Global South Perspective through Student Teachers in Namibia." Accessed August 21, 2016. http://jultika.oulu.fi/files/nbnfioulu-201306051490.pdf

Haipinge, Erkkie. 2016. "Social media in educational contexts: Implications for critical media literacy and ethical challenges for teachers and educational institutions in Namibia." *Namibia CPD Journal for Educators*, 1(1): 102—111.

Internet Live Stats. 2016. "Namibia Internet Users." Accessed August 21, 2016. http://www.internetlivestats.com/internet-users/namibia/

Institute for Public Policy Research (IPPR). 2014. "Social Media and Namibian Elections: Is Namibian Politics Keeping up with the Times?" Accessed August 21, 2016. http://www .ippr.org.na/sites/default/files/ElectionWatch5%20web.pdf

Kawela, M. 2016. "NAMAs 2016: The Social Media Experience." May 6. Accessed August 21, 2016. http://www.namibian.com.na/index.php?page=archive-read&id=150448

Nakashole, N. 2016. "MTC Takes Smartphones to Low-Income Areas," July 1. Accessed August 21, 2016. http://www.namibian.com.na/index.php?page=archive-read&id=152617

Namibia Statistics Agency (NSA). 2011. "Namibia 2011 Population & Housing Census Main Report." Accessed August 21, 2016. http://cms.my.na/assets/documents /p19dmn58guram30ttun89rdrp1.pdf

Peters, A. N., Winschiers-Theophilus, H., and Mennecke, B.E. (2015). "Cultural Influences on Facebook Practices: A Comparative Study of College Students in Namibia and the United States." *Computers in Human Behavior*, 49: 259–271. doi: 10.1016/j.chb .2015.02.065

Pew Research Center. 2015. "Cell Phones in Africa: Communication Lifeline—Texting Most Common Activity, but Mobile Money Popular in Several Countries." Accessed August 21, 2016. http://www.pewglobal.org/files/2015/04/Pew-Research-Center-Africa-Cell -Phone-Report-FINAL-April-15-2015.pdf

Pew Research Center. 2016. "Smartphone Ownership and Internet Usage Continues to Climb in Emerging Economies." Accessed August 21, 2016. http://www.pewglobal. org/2016/02/22/smartphone-ownership-and-internet-usage-continues-to-climb-in -emerging-economies/

Shinovene, I. 2014. "MPs Want Law to Control Social Media," May 6. Accessed August 21, 2016. http://www.namibian.com.na/index.php?id=122886&page=archive-read

Socialbakers.com. 2016. "Facebook Pages Stats in Namibia." Accessed August 21, 2016. https://www.socialbakers.com/statistics/facebook/pages/total/namibia/

Statista. 2015. "Ranking of Countries with Best Infrastructure in 2015." Accessed August 21, 2016. http://www.statista.com/statistics/264753/ranking-of-countries-according-to-the -general-quality-of-infrastructure/

StatsMonkey. 2016. "Mobile Facebook, Twitter, Social Media Usage Statistics in Namibia." Accessed August 21, 2016. https://www.statsmonkey.com/table/21423-namibia-mobile -social-media-usage-statistics-2015.php

Stork, C., and Calandro, E. 2014. "Internet Gone Mobile in Namibia." In *ICT pathways to poverty reduction: Empirical Evidence from East and Southern Africa,* edited by E. Adera, T. Waema, J. May, O. Mascarenhas, and K. Diga. Practical Action Publishing Ltd. International Development Research Centre, 205–226. http://dx.doi.org/10.3362 /9781780448152.009

Stork, C., Calandro, E., and Gillwald, A. 2013. "Internet Going Mobile: Internet Access and Usage in 11 African Countries." Accessed August 21, 2016. http://www.researchictafrica .net/publications/Evidence_for_ICT_Policy_Action/Policy_Paper_14_-_Understand ing_Internet_Going_Mobile.pdf

World Bank. 2016a. "Connectivity Indicators." Accessed August 21, 2016. http://data .worldbank.org/indicator/IT.NET.USER.P2?locations=NA

World Bank. 2016b. "International LPI Global Ranking." Accessed August 21, 2016. http:// lpi.worldbank.org/international/global?sort=asc&order=Infrastructure

World Economic Forum (WEC). 2016. "Networked Readiness Index." Accessed August 21, 2016. http://reports.weforum.org/global-information-technology-report-2016/net worked-readiness-index/

NEPAL

Nepal is a country in South Asia, bordered by India and China and geographically most well known for the Himalayan Mountains. Nepal has gone through a number of political upheavals in its recent past. In 2008, it held its first elections in nine years; however, the country's instability has continued. A new constitution was enacted in September 2015 after seven years of controversy and deliberation. The recent political situation and Nepal's particular geography have affected telecommunications infrastructure; however, Nepal has seen substantial growth in mobile usage. In addition, Nepal experienced a devastating earthquake on April 25, 2015, which became known as the Gorkha Earthquake. This catastrophe led to an unprecedented global social media response, assisting multiple efforts to recover from the disaster.

While the earthquake slowed the development of telecommunications infrastructure in the country, Nepal has managed to move forward. Nepal Telecom (NT), the state-owned telecommunications company, built the majority of the country's infrastructure and provides coverage for internet, mobile, and landline phones. The company held a monopoly until about ten years ago, when Ncell, the country's other major provider, began offering services. Currently, these two companies represent over 90 percent of the total market, with four other smaller companies representing the remaining share (Himalayan Times 2015a; 2015b). Infrastructure in Nepal is also affected by the numerous mountainous regions.

At the end of November 2015, the internet penetration rate in Nepal grew to over 44 percent (Kathmandu Post 2015). This was a 10 percent increase from the previous year, and numbers have been growing rapidly since the introduction of 3G service in 2010. Mobile usage in particular represents the fastest-growing sector. The government formally adopted a National Broadband Policy in April 2015, aiming to continue and expand on this growth. Internet penetration has been further encouraged by the dropping prices and increased availability of mobile handheld devices. Generally, internet users skew toward urban areas, such as Kathmandu. Many rural and mountainous areas in the country still have spotty, limited, or no internet connection. With a total population in Nepal of about 28 million, internet access is still unavailable to 17 million people (Kathmandu Post 2015). The Nepal Wireless Networking Project is a well-known nonprofit in the country, founded in 2003, that works to connect rural villages wirelessly because of the difficulties in building infrastructure in these regions. Over the past ten years, they have been able to connect 175 villages in Nepal to the internet (Nepal Wireless 2016).

Mobile phone costs have been decreasing, with smartphones now available for about $45. An unlimited internet subscription through Ncell will cost an additional $8 per month for the slower connection and 1 GB of data (Ncell 2016). While these are certainly very low prices by U.S. standards, they are significant in Nepal, where the average yearly income per person in 2014 was $730 (World Bank 2016). Ncell also offers a couple of unique internet packages to its customers: one is a Facebook-only subscription; the other two are for unlimited access to Twitter and Wikipedia. These two packages are free for any Ncell customer and allow subscribers to use Twitter and Wikipedia without accruing any data charges.

Given that one of the largest internet providers in Nepal takes special notice of Facebook, the fact that it is the number one website in Nepal is hardly surprising (Alexa Internet 2016). Twitter is not far behind in seventh place. Other popular websites include Google and YouTube. For Nepali-created websites, those listed in the top twenty are all news sites, the most popular being Onlinekhabar, a Nepali-language website for national news. Eight other Nepali news sites are also among the top twenty websites used in Nepal.

Many languages in South Asia face challenges because of the lack of language support for characters that are not in a Latin script. Nepali, the language of Nepal, receives some benefits from being written in the same script as Hindi, which is a major global language. This alphabet, known as Devanagari, is used by a number of languages in South Asia. For this reason, some strides have been made in making it functional for online applications. While many people in Nepal use English-language social media and web applications, Nepali-language materials are also available. There are many Nepali-language news websites, for example.

Within Facebook itself, the most popular pages are similar to rankings for websites generally. The Facebook page with the largest audience in Nepal is "Facebook for Every Phone," designed for the Facebook app. Five of the top Facebook pages are news-related, in both English and Nepali (Socialbakers 2016). eKantipur, a

Nepali Army medics transport and treat the wounded at Tribhuvan International Airport on April 29, 2015, after the country experienced a massive earthquake. People from around the world used social media to solicit money and support for the country immediately after the disaster struck, creating a virtual support network that translated into actual goods and services aimed at aiding the Nepali people. (Mumbaiphoto/Dreamstime.com)

Nepali and English news site, and BBC Nepali come in at number two and number three, with 1,666,955 fans and 1,659,769 fans, respectively as of April 2016. Other top ten Facebook pages in Nepal are pages for Nepali singers Udit Narayan (1955–), Anju Panta (1977–), and Muna Thapa Magar (n.d.–); a page for *Bollywood News*, and a page for a Nepali radio station, Radio Kantipur.

Increasingly, global events are reflected online and on social media. On April 25, 2015, Nepal experienced a devastating earthquake measuring 7.8 on the Richter scale that killed over 8,000 people and left hundreds of thousands homeless. Social media played a particularly impactful role in the aftermath, through monetary donations, crisis mapping, helping to locate missing people, and helping to get the appropriate assistance to where it was most needed. Internet was still available through mobile phones in the immediate aftermath of the earthquake, and it was used extensively to assist in recovery and aid.

When reports of the disaster began coming in through Facebook and Twitter in the immediate aftermath of the quake, both telecommunications companies attempted to reign in and direct the flood of information. Facebook activated its Safety Check tool, a tool that contacts users near affected areas during crises and asks them to send status updates to their friends and family. In addition to the official moves by Facebook, a number of new groups were created in the aftermath of the quake. Many of these were used to organize aid and monetary donations to affected areas. Twitter attempted to focus all the various threads being used by promoting #NepalQuakeRelief as the hashtag for all relevant communications. Also, numerous official and ad hoc groups sought funding through social media in the form of Facebook groups and pages, crowdfunding sources such as IndieGogo, and official humanitarian groups seeking funds through various channels. Google launched Person Finder, an open-source database available in multiple languages. It allows users to input data about missing people and to update information about someone who has been found. Google encouraged nonprofits to also use the database within their own websites in order to centralize missing persons information.

About one week after the earthquake, the Nepalese government's National Emergency Operation Center set up its own Twitter handle, @NEoCOfficial. The account gave important information and advice about accessing clean water, aid received from various humanitarian organizations and other groups, road and travel information, and official statistics about deaths and injuries. As of April 2016, the account had over 15,000 followers. Kathmandu Living Labs, a Nepali-based nonprofit organization, launched Quakemap.org immediately following the earthquake. This crisis-mapping website helped connect resources and aid with exactly what was needed in specific places. Anyone could access the site, state damages, and request specific items or aid. In turn, both ad hoc and official aid groups could report on aid that was delivered. The Nepalese Army even used this site.

OpenStreetMap events were held across the globe in response to the Nepal earthquake. OpenStreetMap is a freely available mapping tool that allows anyone to edit the maps. The events focused on filling in important geographic data in Nepal so that relief efforts would be able to get to the areas that they were needed more efficiently.

Code for Nepal was another organization that worked on crisis-mapping tools to help direct aid efforts after the earthquake. In addition, they focused on low-tech solutions by creating and sharing a Google doc on social media to provide information about food, shelter, water, and medical assistance.

Two Nepali-based groups, the Rapid Response Team and Sankalpa, created their own texting apps in response to the disaster. The Rapid Response Team used texting and phone calls to connect people with missing family and friends. Sankalpa used a technology provided by SparrowSMS, a Nepali-based company, to organize emergency relief through a texting app, and compiled their data with Twitter and Facebook messages into a relief map showing where specific aid was needed.

Karen Stoll Farrell

See also: Bangladesh; India; Pakistan

Further Reading

Alexa Internet, Inc. 2016. "Top Sites in Nepal." Accessed February 22, 2016. http://www
    .alexa.com/topsites/countries/NP
Dey, Sushmi. 2015. "Nepal Earthquake: Govt Using Social Media to Connect and Provide
    Relief." April 28. Accessed February 22, 2016. http://timesofindia.indiatimes.com/india
    /Nepal-earthquake-Govt-using-social-media-to-connect-and-provide-relief
    /articleshow/47076269.cms
Himalayan Times. 2015a. "Country's Telephone Penetration Crossed 100pc: Telecom Regu-
    lator 2015." July 29. Accessed April 19, 2016. http://thehimalayantimes.com/business/
    countrys-telephone-penetration-crossed-100pc-telecom-regulator/
Himalayan Times. 2015b. "NT Top Telecom Service Provider." June 17. Accessed April 19,
    2016. https://thehimalayantimes.com/business/nt-top-telecom-service-provider/
Kathmandu Post. 2015. "Over 11 Million Nepalis Have Access to Internet." November 29.
    Accessed February 22, 2016. http://kathmandupost.ekantipur.com/news/2015-11-29/
    over-11-million-nepalis-have-access-to-internet.html
Nepal Wireless. 2016. "Nepal Wireless." Accessed April 19, 2016. http://www.nepalwireless
    .net/
Sinha, Shreeya. 2015. "3 Ways Nepalis Are Using Crowdsourcing to Aid in Quake Relief."
    May 1. Accessed February 22, 2016. http://www.nytimes.com/2015/05/02/world/asia
    /3-ways-nepalis-are-using-crowdsourcing-to-aid-in-quake-relief.html?_r=0
Socialbakers. 2016. "Facebook Stats for Fans in Nepal." Accessed February 22, 2016. http://
    www.socialbakers.com/statistics/facebook/pages/local/nepal/
Venkatraman, Janane. 2015. "Social Media Pitches in for Nepal Quake Aftermath." April 26.
    Accessed February 22, 2016. http://www.thehindu.com/news/national/social-media
    -pitches-in-for-nepal-quake-aftermath/article7143425.ece
World Bank. 2016. "Data—Nepal—GNI." Accessed April 19, 2016. http://data.worldbank
    .org/country/nepal
Zraik, Karen. 2015. "Google and Facebook Help Nepal Earthquake Survivors and Contacts
    Connect." April 27. Accessed February 22, 2016. http://www.nytimes.com/2015/04/28/
    world/asia/google-and-facebook-help-nepal-earthquake-survivors-and-contacts
    -connect.html?_r=1

NIGERIA

Nigeria is a nation in West Africa, located along the Atlantic coast just north of the Gulf of Guinea. With 181 million people, Nigeria is the most populous country in the continent of Africa and the eighth most populous country in the world. As of late 2015, there were 97.21 million mobile internet users, equaling 53.7 percent of its population. This is a significant increase from 2013, when only 38 percent of the population used the internet. All popular social media apps in Nigeria are foreign-based. Given that English is the country's official language, foreign-based apps are well integrated into Nigerian society. Internet users in Nigeria are active, according to an assessment of app usage in November 2015, on Facebook, WhatsApp, Facebook, Facebook Messenger, Instagram, BlackBerry Messenger, Twitter, Skype, Truecaller, and Hushed. Internet users are active across Android, Windows, and iOS smartphone platforms (Matuluko 2015).

The first attempt to introduce the internet in Nigeria was made through the Regional Informatics Networks for Africa project in 1995, sponsored by the United Nations Educational, Scientific, and Cultural Organization (UNESCO). Then, the Nigeria Internet Group was formed as a nonprofit, nongovernmental organization (NGO) with the primary aim and objective of promoting and facilitating access to the internet in Nigeria (Nairaland 2015). Three years earlier, in 1992, the government of Nigeria established the Nigerian Communications Commission (NCC) to serve as the independent national regulatory authority responsible for managing telecommunications services throughout the country.

According to the NCC, in September 2015, 97.06 million mobile web users were on Global System for Mobile (GSM) networks and 151,816 users on Code Division Multiple Access (CDMA) networks (Premium Times 2015). *CDMA* and *GSM* are shorthand references to the two major radio systems used in mobile phones, with CDMA mainly used in the United States and GSM used globally (Segan 2015). Telecommunications operators that serve Nigeria are based there, including Globacom, Airtel Nigeria, Etisalat Nigeria, Multi-Links, and Visafone, with the exception of MTN Nigeria, Africa's largest mobile network, which is based in South Africa. MTN Nigeria had 41.84 million subscribers and ranked first among the telecom operators. Globacom was second, with 21.89 million subscribers on its network. Airtel Nigeria followed, with 17.73 million internet users, and Etisalat Nigeria had 15.59 million customers browsing the web. The CDMA operators were Multi-Links and Visafone. Visafone had a total of 151,530 internet users, and Multi-Links had 286 internet users in September (Premium Times 2015).

Although Nigeria has a high number of internet users, the actual penetration of individual users is less than what the numbers indicate. An average Nigerian uses around two subscriber identity module (SIM) cards to cut costs and benefit from the many low-cost deals available in the market. According to a survey by MTN in 2013, the rate of multi-SIMming in the Nigerian market is around 40 percent (Hatt et al. 2014). When a web user takes on more than two SIM cards for a single phone or has two phones with one SIM card in each phone, this counts as two internet users; so these statistics do not accurately portray the number of users in Nigeria.

Since 2011, NCC has required telecommunication companies to verify the identity of their subscribers through SIM card registration in order to identify and reduce the amount of criminal activity conducted via mobile networks. The verification process can include submitting photographs, official documents, and even biometric data. Other African countries have enacted similar policies to fight cybercrime, in particular to prevent online fraud and reduce support for terrorism. Despite the efforts to prevent criminal activities facilitated by mobile phone usage, successful implementation is a challenge for telecommunications companies. Throughout Nigeria, it is possible to obtain SIM cards from unregistered vendors who do not have the necessary instruments to record biometric information from those who purchase SIM cards. In 2015, NCC imposed penalties on telecommunications companies that operate in Nigeria for those who were not able to register all SIM cards, including MTN Nigeria, Globacom, Etisalat, and Airtel. The most extensive fine was levied on MTN Nigeria, which was ordered to pay $5.2 billion for failing to disconnect 5.1 million unregistered SIM cards before the August and September deadlines (Tshabalala 2015). Although it is not easy for telecommunications companies to be certain that all SIM card vendors register the cards, Nigeria continues to penalize telecommunications companies as per the SIM card registration law.

Affordability for internet usage continues to be a challenge for a vast majority of Nigerians. In 2014, fixed-line broadband subscriptions cost an average of 39 percent of average income, while subscriptions for mobile broadband packages were about 13 percent of average income. In 2010, about 80 percent of Nigerians earned $2 or less per day—about $730 per year—which is just over half of the 2012 national average income of $1,440 (Alliance for Affordable Internet 2014).

Population demographics in Nigeria are also linked to inequality among mobile phone users throughout the country. Nigeria's population in urban and rural areas is split roughly in half, but mobile penetration is much lower outside cities. The north-south divide within the country also presents mobile inequality. The southern regions, including major metropolises such as Lagos, are well ahead of their northern counterparts in terms of household access to mobile phones.

Expanding coverage to rural areas is also a complex issue. Geographic terrain, vast distances, lack of electricity, poor road access, and continued security threats are all challenges to investment in rural coverage. Nevertheless, NCC has set a goal of nationwide internet coverage by 2017 (Hatt et al. 2014).

To provide lower-cost internet access, several initiatives are in currently in progress. In a joint partnership to provide web access to users throughout Nigeria, Opera Software, makers of the Opera Mini mobile browser, Naij.com, a Nigerian news and entertainment portal, and MTN Nigeria created a service that provides web users with free access to the internet for 1 million days. Beginning in June 2015, MTN offered free mobile internet through Opera's Sponsored Web Pass to more than 140 million subscribers. The Web Pass was to be given to the first 40,000 users who sign up each day. Each internet user will be offered 10 MB that can be used only with the Opera Mini browser. They can return to claim another Web Pass after two days (Aderibigbe 2015). The 10 MB limit provides ample data to read news and check email. For those who live in rural areas and do not have regular access to internet-based information, this creates an opportunity for them to stay in touch with the world around them.

In late 2015, two additional low-cost internet access plans were offered through a partnership among DataWind, Airtel, Intel, and an entrepreneurial fund from Microsoft. This partnership planned to offer low-cost tablets and smartphones. DataWind's Intel-based devices were to be bundled with free internet access through its existing partnership with Airtel (Stockhouse 2014). Through an entrepreneurial fund, Microsoft was to select companies to receive cash awards, along with free technology and mentorship from Microsoft to scale solutions that help close the digital divide. The goal was to fund internet access in Nigeria by assisting companies that have a well-developed plan to expand affordable internet access (Aginam 2015). By bringing lower-cost internet access throughout Nigeria, both plans are helping to broaden usage throughout the country.

A study conducted by Ericsson (2015) identified local data usage across the web and its customers' satisfaction. The results of the study detail how Nigerian Ericsson subscribers view their access to the internet. According to these findings, 57 percent of Nigeria's eighteen- to twenty-four years old wanted to watch television and video content at their convenience, while 67 percent wanted easy access to video content across all devices. Further, 50 percent of the time that Nigerians spent watching video was on laptops and smartphones. However, only 36 percent of Nigerian consumers were satisfied with finding content online that suited their viewing habits and preferences. Streaming issues and download speeds were the two most important factors affecting the satisfaction of those watching video content online. Only 30 percent of Nigerians were satisfied with streaming over the internet. Consumer usage of on-demand streaming services affects the 27 percent of Nigerian consumers who stream videos on a weekly basis. More than 50 percent of the Nigerians interviewed for the study thought that mobile video on-demand (M-VOD) and IPTV Triple Play were attractive services and were open to subscribing. 72 percent believed that mobile operators should provide M-VOD, and mobile service providers and operators should offer IPTV services (Ericsson 2015).

Internet users in Nigeria are active on Facebook. In June 2015, Facebook recorded that Nigeria had 15 million users. Nigeria, as Africa's most populous nation and, along with Kenya, one of the most developed countries in Africa, is seen as an important entry point on a continent of nearly 1 billion people. Internet users in Nigeria use Facebook primarily to follow shopping, banking, news, and Nigerian celebrities. The ten most popular Facebook pages as of December 2015 were as follows:

1. Kaymu, an online shopping community launched in 2012 that connects buyers and sellers
2. Mayor Boss (1986–), a Nigerian rapper, singer, songwriter, and video producer based in Europe
3. GTBank, a Nigerian multinational financial institution based in Lagos that offers online/internet banking
4. MTN Nigeria
5. P-Square—The official page for a Nigerian R&B duo composed of identical twin brothers Peter Okoye and Paul Okoye (1981–).
6. Naij.com

7. Pastor E. A. Adeboye (1942–), a Nigerian pastor who is the appointed General Overseer of Redeemed Christian Church of God Pastor
8. Information Nigeria, a Nigerian information portal
9. Omotola Jalade Ekeinde, Member of the Order of the Federal Republic (MFR) (1978–), a Nigerian actress, singer, philanthropist, and former model from Lagos
10. Goodluck Jonathan (1957–), the former president and vice president of Nigeria and former governor of Bayelsa State

Nigeria is also well known for its dark side online, with email scams regularly falling into junk mail folders. In 2014, nearly one-fifth of online advance fee scams originated in Nigeria (Engber 2014). In the United States, Nigerian scams deceived Americans into handing over large sums of money in "get-rich-quick" schemes or advance-fee ploys that asked them to make an advance deposit before supposedly receiving millions of dollars in their personal accounts. Although the United States has cracked down on Nigerian scammers, in 2011, the Federal Bureau of Investigation (FBI) received nearly 30,000 complaints. In another version of the scam, a stranger contacts a potential victim, gets to know the person, and professes love. When the scammer has ensnared the victim, money is requested. In 2012, the FBI received over 4,000 of these complaints; victims lost more than $55 million. Other typical scams involve phony lottery winnings, job offers, and inheritance notices (Eichelberger 2014).

As a result, the government of Nigeria incorporated section 419 into Chapter 34 of its criminal code. Throughout the world, "Nigerian 419" is a moniker for an advance-fee scam. Penalties for data theft and online fraud include seven years imprisonment or a fine amounting to approximately $35,000. The law increased the punishment to fourteen-year prison sentences for convictions in scams that caused physical damage or harm, and the penalty became even steeper, up to life in prison, when loss of life resulted. Lesser offences, such as hacking government infrastructure, result in ten-year sentences. Nigeria's sentencing reforms show that it recognizes the severity of cybercrime and is actively working to combat it (Oxford 2014).

Nigeria's Economic and Financial Crimes Commission (EFCC) has the responsibility of tracking down scammers throughout Nigeria. The country's cybercrime law also penalizes internet café owners who knowingly allow their premises to be

Chris Messina (1981–) used the first hashtag on August 23, 2007, when he posted on Twitter, "how do you feel about using # (pound) for groups. As in #barcamp [msg]?" The idea was to implement the hashtag across the web, not just for Twitter. The pound sign made more sense for this purpose than other symbols because it was easy to type on the phones available in 2007. Twitter initially resisted using hashtags, but they eventually caught on with the users of this and other social media platforms. Messina, a former Google employee and involved in a number of start-ups and open-source projects, joined Twitter in 2006. His Twitter account has more than 77,000 followers. Today, hashtags are important for social and political activism both online and off-line around world, exemplified by 2014's #UmbrellaRevolution in Hong Kong to 2015's #NigeriaDecides to combat electoral irregularities and corruption.

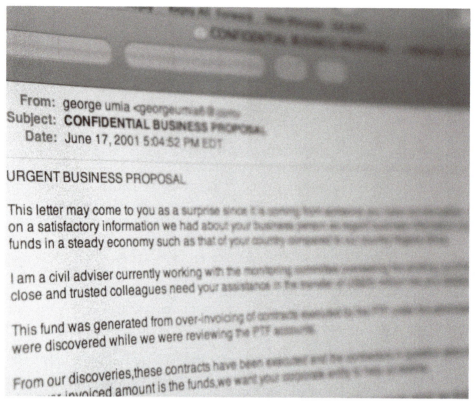

From: george umia <georgeumia@...>
Subject: **CONFIDENTIAL BUSINESS PROPOSAL**
Date: June 17, 2001 5:04:52 PM EDT

URGENT BUSINESS PROPOSAL

This letter may come to you as a surprise since it is coming from someone you don't know
on a satisfactory information we had about your business person as regard investing of
funds in a steady economy such as that of your country compared to our country Nigeria here

I am a civil adviser currently working with the monitoring committee, presently me and two of my
close and trusted colleagues need your assistance in the transfer of funds discovered by us

This fund was generated from over-invoicing of contracts executed by the PTF over the years, these
were discovered while we were reviewing The PTF accounts

From our discoveries,these contracts have been executed and the contractors fully paid, the over
... invoiced amount is the funds,we want your assistance only to help us transfer

Nigerian scammers use phishing emails to deceive unsuspecting email users to hand over their money. One of the more popular scams is to claim to be a Nigerian prince who needs international bank account access to deposit funds, promising to share a percentage with the account holder, while really siphoning existing money from the account. The Nigerian government has intensified efforts to thwart these scammers because they violate the law, and tarnish the country's reputation globally. (Just One Film/Getty Images)

used for committing a crime. To promote this law, EFCC has publicized its efforts and now bans browsing the internet at night in public spaces. The organization has been highly visible through raids on cybercafés in urban areas and has made some high-profile arrests (Wanjiku 2015). In December 2015, the EFCC teamed up with Italian police to arrest sixty-two people from Nigeria and Cameroon suspected of being members of a cybercriminal organization responsible for frauds committed against small- and medium-scale enterprises across the world. EFCC tracked down the fugitives in their homes in Lagos (AllAfrica 2015).

Anthony Ortiz

See also: Cameroon; Kenya; South Africa; United States

Further Reading

Aderibigbe, Niyi. 2015. "For the Next 9,999 Days, Nigerians Will Have Access to Free Internet." Accessed December 17, 2015. http://venturesafrica.com/for-the-next-9999-days-nigerians-will-have-access-to-free-internet/

Aginam, Emeka. 2015. "Nigeria: Firms to Get Microsoft Seed Fund to Improve Access to Affordable Internet." November 18. Accessed December 17, 2015. http://allafrica.com/stories/201511191447.html

AllAfrica. 2015. "Nigeria: EFCC, Italian Police Arrest 62 Suspected Internet Fraudsters." November 19. Accessed December 17, 2015. http://allafrica.com/stories/201511191586.html

Alliance for Affordable Internet. 2014. "Nigeria: How Africa's Largest Economy if Prioritising Affordable Internet." April. Accessed December 17, 2015. http://a4ai.org/wp-content/uploads/2014/04/Nigeria-Case-Study-Final.pdf

Eichelberger, Erika. 2014. "What I Learned Hanging out With Nigerian Email Scammers." March 20. Accessed December 17, 2015. http://www.motherjones.com/politics/2014/03/what-i-learned-from-nigerian-scammers

Engber, Daniel. 2014. "Who Made That Nigerian Scam?" January 3. Accessed December 17, 2015. http://www.nytimes.com/2014/01/05/magazine/who-made-that-nigerian-scam.html?_r=0

Ericsson. 2015. "TV and Media in Nigeria. How Changing Consumer Needs Are Creating a New Media Landscape." September. Accessed December 17, 2015. http://www.ericsson.com/res/region_RSSA/rssa_news/pdf/TV-and-Media-Report-Nigeria-2015.pdf

Hatt, Tim, Sharma, Akanksha, and Arese Lucini, Barbara. 2014. "Country Overview: Nigeria." June. Accessed December 17, 2015. http://draft-content.gsmaintelligence.com/AR/assets/4161587/GSMA_M4D_Impact_Country_Overview_Nigeria.pdf

Matuluko, Muyiwa. 2015. "The Most Popular Apps on Every Nigerian Smartphone." November 6. Accessed December 17, 2015. http://techpoint.ng/2015/11/06/popular-apps-in-nigeria-smartphone/

Nairaland. 2015. "History of the Internet in Nigeria." April 7. Accessed August 1, 2016. http://www.nairaland.com/2241209/history-internet-nigeria

Oxford, Adam. 2014. "New Nigerian Law Means Seven Years for Cybercrime." November 5. Accessed December 17, 2015. http://www.zdnet.com/article/new-nigerian-law-means-seven-years-for-cybercrime/

Premium Times. 2015. "Nigeria Internet Users Increase to 97 Million—NCC." December 1. Accessed December 17, 2015. http://www.premiumtimesng.com/news/headlines/192485-nigeria-internet-users-increase-to-97-million-ncc.html

Segan, Sascha. 2015. "CDMA vs. GSM: What's the Difference?" February 6. Accessed August 1, 2016. http://www.pcmag.com/article2/0,2817,2407896,00.asp

Stockhouse. 2015. "DataWind Partners with Intel to Launch Low-Cost Internet-Enabled Devices in Nigeria." December 14. Accessed December 17, 2015. http://www.stockhouse.com/news/press-releases/2015/12/14/datawind-partners-with-intel-to-launch-low-cost-internet-enabled-devices-in#Mh3UT8k8kAy5Zm1d.99

Strydom, TJ. 2015. "Facebook Rakes in Users in Nigeria and Kenya, Eyes Rest of Africa." September 10. Accessed December 17, 2015. http://www.reuters.com/article/us-facebook-africa-idUSKCN0RA17L20150910#StZsTQYbac4JBoyJ.97

Tshabalala, Sibusiso. 2015. "Nigeria Is Fining MTN $1,000 per Illegal SIM Card Even Though Customers Generate Just $5 a Month." October 26. Accessed December 17, 2015. http://qz.com/533041/africas-largest-mobile-network-is-being-fined-5-2-billion-for-flouting-nigerias-sim-card-rules/

Wanjiku, Rebecca. "The Story Behind the Nigerian Phishing Scam." Accessed December 17, 2015. http://www.pcworld.com/article/192664/the_story_behind_the_nigerian_phishing_scam.html

NORTH KOREA

North Korea is an Asian country that shares a border with China and South Korea in the northern portion of the Korean Peninsula. Since the end of the Korean War (1950–1953), a demilitarized zone has existed between North and South Korea; no one from either country is allowed in this area, and guards are posted to prevent travel in either direction. Internationally, North Korea is known for its totalitarian government and complete state control over most communication flowing to or from the country. Therefore, analyzing technology use in North Korea is an extremely difficult endeavor, as information that is shared by the country is most often propaganda delivered by the government and, therefore, unreliable. For this reason, the number of devices in North Korea that are connected to the internet is difficult to determine. However, considering that in 2014, there were only 1,024 internet protocol (IP) addresses in the whole country, which is composed of approximately 25 million people, it is unlikely that there are many connected devices. A national network, or intranet, is available to some citizens, but there are usually severe limitations on how it can be used. Essentially, all contact with the outside world is restricted by the North Korean government.

Cell phones are available in North Korea, but they cannot be used to access the internet. Not all North Koreans own a cell phone, but estimates fall around 2 million cell phones in the country. This number may be misleading, however, as elite North Koreans sometimes buy a new phone, which can be cheaper than adding minutes to a current cell phone (Kim and Lee 2014). Those elites who do use cell phones use them in similar ways as the rest of the world: to set up work meetings, arrange dinner plans, take pictures, and share music. They just cannot use them to access the internet (Lee 2013).

While North Koreans do not have free access to the internet, this same restriction is not true for visitors to the country. On January 18, 2013, foreigners were allowed, for the first time, to bring mobile phones into the country. They could access the internet by purchasing a local subscriber identity module (SIM) card, which was offered only to foreigners. The SIM card would work with the 3G network provided by Koryolink and was available in select North Korean cities. Koryolink is a joint venture between the Egyptian telecommunications company Orascom and the North Korean government. It is the only wireless network available in the country. Koryolink first made 3G technology available in 2008, but they did so without offering a data plan. North Koreans can only use this service to make phone calls, and all international calls are prohibited (BBC 2013; Lee 2013).

With the introduction of these foreigner-only SIM cards, visitors to the country could access the internet and make international phone calls. However, this could be a costly enterprise. According to journalist Jean Lee (c. 1973–), a foreign correspondent and expert on North Korea, in 2013 these SIM cards cost about $70 to purchase. In addition to purchasing the SIM card, there was an activation fee of $75. Moreover, international phone call rates varied wildly, depending on the country: calls to Switzerland were reasonable, at forty-three cents a minute, but phone calls to the United States cost $8 a minute. Lee broadcast what was very likely the first tweet sent over the internet from North Korea on February 25, 2013: "Hello

On the street in Pyongyang, North Korea, a young woman is holding a cell phone. While mobile penetration is on the rise in North Korea, citizens are not allowed to use the internet; therefore, the only cell phones that can connect to the web are those that use a special SIM card granted to foreigner visitors that allow access to the country's only wireless network. (Linqong /Dreamstime.com)

world from comms center in (hash)Pyongyang." Dennis Rodman also famously tweeted from North Korea during his 2013 visit: "I come in peace. I love the people of North Korea!" (Lee 2013).

As of mid-2015, foreigners were still allowed access to the internet through Koryolink. This freedom has allowed them to post to social media websites like Facebook, Twitter, and Instagram. While these posts have not been numerous, they have caused the North Korean government to suspend access to these websites occasionally. During the summer of 2015, for instance, Instagram service was interrupted by messages in English and Korean stating that the website was blocked due to harmful content. Service resumed a week later. However, the government has never officially acknowledged the ban (AP 2015). More recently, however, in April 2016, the government has changed its stance toward foreign social media platforms; it has announced an official ban of Facebook, YouTube, Twitter, and all South Korean websites. Foreigners are not exempt from the ban.

While the North Korean people do not have access to the same internet as the rest of the world, some citizens, such as elites and university professionals, do have access to a national intranet known as Kwangmyong, meaning "bright." It is a closed network only available in North Korea containing solely government-sanctioned content. People largely use this system to disseminate information, rather than to access social media sites (Kim and Lee 2014). Nevertheless, there are some instances of social networking via Kwangmyong. In 2013, Lee shared photos of an unnamed social network available on the intranet. She said the website was not much more than a message board where people posted birthday greetings (Dewey 2013).

Interestingly, North Korean leader Kim Jong-un (1983–) has identified science and technology as matters of national importance, which has resulted in broader use of laptops and tablets, even though they are devices created specifically for North Korean consumption (Lee 2013). However, owning a computer requires permission from the government and may cost as much as three months' salary (Kim and Lee 2014). In 2012, the experience of browsing the Kwangmyong was very different from surfing the internet in the rest of the world. For example, to access the national intranet, citizens used the North Korean operating system, Red Star. When users logged on at the sole cybercafé in the capital of Pyongyang, a message appeared stating the importance of the operating system aligning with the core values of the country. The computer's calendar did not display the current year, but rather the number of years since Kim Il-Sung (1912–1994) was born; the current North Korean state was founded upon his philosophies. One striking difference in the North Korean experience occurred whenever a user came across the name of North Korean leader Kim Jong-un. The programming code was written so that his name appeared slightly larger than any of the other words on the page. The websites available on Kwangmyong were limited to state-run news services, bulletin boards, and chat services (Khazan 2012; Lee 2012).

The Red Star operating system uses a web browser called Naenara, which is an adapted version of Firefox. Robert Hansen (n.d.–), vice president of WhiteHat Labs at WhiteHat Security, examined a version of this North Korean browser early in 2015 and made several interesting observations. His first impression was that the browser seemed very old. In order to navigate to a specific website, for example, the user had to type in the IP address. The browser treated the whole country like a medium-sized business network instead of a national network. All email was routed through a single Uniform Resource Locator (URL). He also found evidence that suggested that the entire network may have been run through a proxy server (Hansen 2015).

In addition, all crash reports, news feeds, and email route through a central URL. This function allows North Korea to control all the content on the national intranet tightly. In fact, every aspect of technology found in the country—right down to the hardware—is regulated by the government. This makes it very difficult for users to access anything that the authorities have not sanctioned. In 2013, Lee stated that North Koreans seem proud of their national intranet and do not display any signs of subversive use of the system (Dewey 2013).

Nevertheless, North Korean defectors have shared stories about the different ways that media have passed through North Korea's tight security. While it is very risky, some North Koreans have acquired Chinese cell phones, which are smuggled across the border. These phones generally work on the Chinese network within about six miles of the border between the two countries, allowing users access to outside internet and international calling; however, using technology in this fashion is very risky. The penalties can include forced relocation to a detention center or political prison camp. South Korean television has been smuggled across the border via universal serial bus (USB) drives attached to balloons. The balloons float across the border and despite lacking internet access, the computers that North Koreans use can play the USB drives. These drives can contain anything from South Korean soap operas to Wikipedia in Korean (Lee 2012).

While the average North Korean does not have access to the social media that is popular in much of the world, sectors of the government have utilized these media outlets. While some of the accounts have never been verified, it appears as though the country of North Korea has a webpage, a Twitter account, a Facebook page, and a YouTube account. The website www.korea-dpr.com has been around since 2000 and appears to be an official national website. Rumors have suggested that the monthly budget for the web team was only $100 a month. Therefore, it did not come as much of a surprise when a cheap blogging template was used in 2012 to redesign the website. This occurrence attracted international attention because the redesigned website template cost $15 and was cut-and-pasted wholesale, uncustomized, onto the site. In light of the attention, people reading the site uncovered the fact that it is actually published by the Korea Friendship Association. Incidentally, the group claimed this was the first North Korean website on the internet (Ackerman 2012; Khetani 2012).

The other social media attributed to the North Korean government carry some version of the name uriminzokkiri, which means "our nation," or "on our own as a nation" (Chen 2010; Williams 2010). In July 2010, @uriminzok joined Twitter. As of 2012, this account had not been officially connected with the North Korean government, but the Twitter feed, which has been updated often, is filled with positive and inspiring impressions of life and politics in North Korea. As of September 27, 2015, this Twitter account had produced 16,200 tweets, had 18,700 followers, and followed four other entities: DPRK News Service, Pyongyang DPRK, TrDPR NggP Duy, and an Austin-based investor, Jimmy Dushku (c. 1989–). On August 10, the South Korean government banned this account. South Korea has been known to block websites sympathetic to North Korea, as the two countries are still technically at war (Williams 2010).

The Facebook account with the username uriminzokkiri was the source of some ridicule when it first opened. Upon its creation, some social media consumers suggested that it appeared unprofessional and pointed out how the user had utilized some Facebook profile settings incorrectly. This account owner interacts differently with the public than the user of the Twitter account, which suggested that they are not connected (Chen 2010; Williams 2010). The YouTube channel of the same name, uriminzokkiri, has also garnered much attention. The channel was created on July 14, 2010, and, as of September 27, 2015, had 12,018 subscribers and 12,983,545 views. It was also being regularly updated. The videos posted to this YouTube channel are numerous and largely come from the Korean Central Television feed (Taylor 2012; Williams 2010).

In 2013, some of these social media accounts were hacked by Anonymous, the hacker activist group that protects its members' identity under the guise of Guy Fawkes. This group left a message on @uriminzok's Twitter feed with a pig snout and Mickey Mouse drawn on a picture of Kim Jong-un (Kim and Lee 2014). North Korea has experienced other hacking attacks that have resulted in a partial or complete shutdown of its internet. One of these instances may have occurred at the end of 2014, right after Sony Entertainment was hacked. North Korea had condemned Sony's movie, *The Interview,* a film that took a satirical view of the regime, so it seemed like a logical suspect in the attack. North Korea denied responsibility but expressed satisfaction with the hack. Shortly thereafter, North Korea's internet shut down temporarily; however, the problem was never definitively connected to a cyberattack (AP 2015).

Marilyn J. Andrews

See also: China; Egypt; South Korea; United States

Further Reading

Ackerman, Spencer. 2012. "North Korea's Cheerleaders Spent a Whole $15 Making Their Website." April 18. Accessed September 27, 2015. http://www.wired.com/2012/04/north-korea-website/

Associated Press (AP). 2015. "North Korea Instagram Access Appears to Resume." June 25. Accessed September 27, 2015. http://www.huffingtonpost.com/2015/06/25/north-korea-instagram_n_7662300.html

Chen, Adrian. 2010. "North Korea Only Has 65 Friends on Facebook and Is Gay," August 20. Accessed September 27, 2015. http://gawker.com/5617946/north-korea-has-only-65-friends-on-facebook-and-is-gay

Dewey, Caitlin. 2013. "A Rare Glimpse of North Korea's Version of Facebook," March 13. Accessed September 28, 2015. https://www.washingtonpost.com/news/worldviews/wp/2013/03/13/a-rare-glimpse-of-north-koreas-version-of-facebook/

Hansen, Robert. 2015. "North Korea's Naenara Web Browser: It's Weirder Than We Thought." Accessed September 28, 2015. https://blog.whitehatsec.com/north-koreas-naenara-web-browser-its-weirder-than-we-thought/

Khazan, Olga. 2012. "What It's Like to Use the Internet in North Korea," December 11. Accessed September 28, 2015. https://www.washingtonpost.com/news/worldviews/wp/2012/12/11/what-its-like-to-use-the-internet-in-north-korea/

Khetani, Sanya. 2012. "North Korea Spent Just $15 on the Redesign of Their Official Website," April 19. Accessed September 27, 2015. http://www.businessinsider.com/north-korea-spent-just-15-to-redesign-their-official-website-2012-4

Kim, Tong-Hyung, and Lee, Youkyung. 2014. "Look at How Bizarre North Korea's 'Internet' Is." December 23. Accessed September 27, 2015. http://www.businessinsider.com/a-look-at-north-koreas-tightly-controlled-internet-services-2014-12

Lee, David. 2012. "North Korea: On the Net in World's Most Secretive Nation." December 10. Accessed September 28, 2015. http://www.bbc.com/news/technology-20445632

Lee, Jean H. 2013. "North Korea's Twitter Activity Offers a Real-Time Glimpse into Life in Pyongyang." February 28. Accessed September 27, 2015. http://www.huffingtonpost.com/2013/02/28/north-koreas-twitter-acti_n_2781720.html

Taylor, Adam. 2012. "13 of the Strangest Videos Uploaded to North Korea's Official YouTube Channel." July 9. Accessed September 27, 2015. http://www.businessinsider.com/north-koreas-official-youtube-channel-the-13-strangest-videos-2012-7

Williams, Martyn. 2010. "South Begins Blocking North Korean Twitter Account." August 20. Accessed September 28, 2015. http://www.reuters.com/article/2010/08/20/urnidgns852573c40069388000257785000b-idUS56724690620100820

P

PAKISTAN

Pakistan is a South Asian country nestled between India to its south and Afghanistan and Iran to its north and west. It is the sixth most populous country in the world, with a population of over 200 million. Internet usage in the country is still quite low, although recent increases in availability of internet and of mobile phones have led to dramatic growth. There is still limited availability beyond urban centers. Social media usage is growing, and Urdu script fonts also have made the internet more available in the country. Censorship issues continue to affect the country, most recently through the potential approval of a cybercrime bill.

Along with many of its Asian counterparts, Pakistan initially established access to the internet in the mid-1990s. Services were offered by the government-owned Pakistan Telecommunications Company Ltd. (PTCL). Dial-up internet speed did not increase to 56 Kbps until 1998, and the first broadband connection type, digital subscriber line (DSL), became available in the country in 2001 (ISPAK 2014). By the mid-2000s, over 100 internet service providers were available to consumers. Overwhelmingly, internet access, especially broadband access, has been centered around Pakistan's three major cities: Karachi, Lahore, and Islamabad.

The overall internet penetration rate in Pakistan is still low at 15 percent in 2016. However, the rate is clearly on an upward trend, given that Pakistan's internet penetration rate in 2015 was only 3 percent (Press Information Department 2016). This increase is at least partially attributable to the launch of 3G/4G services in Pakistan's major urban centers. In addition, as with other countries in the region, mobile usage is on the rise, and the availability of wireless internet has increased the penetration rate. This trend is clear, as evidenced by the fact that the number of mobile broadband subscribers grew an estimated 100 percent from 2014 to 2015 (BuddeComm 2015). While mobile broadband is having some effect on the availability of internet in poorer and more rural communities, usage is still heavily biased toward the middle and upper classes and to major urban centers.

One other factor affecting internet usage among Pakistan's population is language. While Punjabi is the most widely spoken language in Pakistan, Urdu and English are the country's official languages. Urdu uses a Perso-Arabic script, and for many years, there was no support for this script online. For example, Google did not release its first version of an Urdu script font until late 2014 (Google 2014). The availability of online fonts for Urdu plays a role in how much of the country's population is able to access and communicate in the online world. Since Urdu and Punjabi fonts have been made available only recently, people used to transliterate their language into Roman characters to communicate online. There is no standardized transliteration system, and many sounds and letters do not have an equivalent in the Roman alphabet, which created a wide array of spelling variations in transliterated online writing.

Finally, overall low literacy rates also affect the rate at which internet usage will grow in Pakistan. United Nations International Children's Emergency Fund (UNICEF) estimates give an adult literacy rate of 55 percent as of 2013 (UNICEF 2015); however, the literacy rate for women is even lower. Men comprise approximately 69.5 percent of the literate population over the age of fifteen, with women at only 42.7 percent (UNESCO 2016). As with other accessibility issues, literacy is higher in urban centers, with the rural poor at a disadvantage. As the internet becomes available to more of the population in Pakistan, literacy will need to improve in order to make it accessible to the entire population.

The top two websites in Pakistan belong to the globally recognized Google (Alexa 2016). The first spot belongs to Pakistan's version of the site (Google.com.pk), which allows Urdu text, while the U.S.-based Google.com comes in second. Another well-known internet portal, Yahoo, is the fourth most used website in Pakistan, while Microsoft's search portal, Live.com, is in ninth place. The Pakistan-specific search portal, HamariWeb, sits in sixteenth place. News sites are a well-represented category in the top twenty websites for Pakistan. Urdupoint.com ranks sixth, Javedch.com ranks thirteenth, Express.pk ranks fourteenth, and DailyPakistan.com .pk ranks nineteenth. All four of these news sites are Urdu-language sites with country-specific news and entertainment sections. The encyclopedic information site, Wikipedia, places eighth.

A number of social media sites are also in the top twenty sites for Pakistan, with Facebook ranking as the third most used website. YouTube comes right behind, at number four. Another video publishing site, Daily Motion, is the eleventh most used website in Pakistan. LinkedIn ranks at number twelve, while the blogging sites Blogspot and Blogger are seventh and eighteenth, respectively. The social media site, Twitter, is also in the top twenty, at number nine.

Rounding out the top twenty most popular websites in Pakistan are Onclickads .net, ranking fifteenth, Alibaba.com at seventeen, and Kat.cr coming in at number twenty. Onclickads is an advertising site where users are often sent when they click an ad on a webpage. Alibaba is a site for small businesses to sell their products wholesale to other companies around the globe. Finally, Kat.cr is a peer-to-peer file sharing, or BitTorrent site, that has moved its internet service provider (ISP) location to multiple countries over time, from Somalia to Costa Rica, as many governments crack down on file sharing. Some countries, such as the United States and Portugal, have blocked or banned the site, as have Google Chrome and Mozilla Firefox.

Most of the top ten Facebook pages in Pakistan belong to broad news and entertainment sites. The most popular page, however, belongs to Atif Aslam (1983–), a contemporary Pakistani actor and musician. The second most popular page is for PTV Sports, a Pakistani channel with coverage of sports both within the country and internationally. Coming in third is *Zindagi Gulzar Hai* (Life Is a Garden), a very popular Pakistani soap opera that aired from 2012 to 2013 and achieved global recognition when it aired in India, in eleven Arab countries after being dubbed into Arabic, and in some European countries. Other popular Facebook pages are for ARY News, UrduMaza Network, Express News, Samaa TV, and Waqt News, all news sites that publish either primarily or solely in Urdu and are based in Pakistan. The rest of the top ten are Waymu.pk, an online shopping site based in Pakistan, and Allama Iqbal, a page is hosted by an organization based in Toronto, Ontario, called the International Iqbal Society. The name *Allama Iqbal* refers to Sir Muhammed Iqbal (1877–1938), a highly regarded poet and politician in India. The Facebook page posts cultural, religious, and literary material and uses English, Urdu, and Persian. Facebook's Internet.org program also became available in Pakistan in May 2015 (Dawn.com 2015), offering free limited internet access throughout the country. Facebook intends for the program to give internet connectivity to those who could not otherwise access it; however, it has been criticized because of limits to the internet sites that are available to subscribers (Khan 2015). The program does not offer full internet access, but instead allows access only to a small subset of participating websites, including Facebook, Wikipedia, Dictionary.com, ESPN, Accuweather, and a few others.

Pakistan has been criticized for a number of other censorship issues beyond the limitations of Internet.org. Over the years, via the Pakistan Telecommunications Act, the government has blocked a variety of websites, generally for either national security reasons or offensive content, although it has not always given exact reasons for doing so. YouTube, for example, was blocked for nearly four years until Google created a country-specific version of the site in early 2016 that blocks objectionable content and allows the Pakistani government to request further blocks as needed (Junaidi 2016). Finally, the Prevention of Electronic Crimes bill was passed in August 2016 (Reuters 2016). This bill has censorship critics concerned because it could criminalize speech that is critical of the government or religion, as well as other speech considered benign (Khan 2016).

Karen Stoll Farrell

See also: Bangladesh; India; Nepal; Saudi Arabia; Somalia; United States

Further Reading

Alexa Internet, Inc. 2016. "Top Sites in Pakistan." Accessed June 4, 2016. http://www.alexa.com/topsites/countries/PK

BuddeComm. 2015. "Pakistan—Broadband Market, Internet Services, and Forecasts." Accessed June 3, 2016. http://www.budde.com.au/Research/Pakistan-Broadband-Market-Internet-Services-and-Forecasts.html

Dawn.com. 2015. "Facebook Provides Free Internet Access to Pakistani Citizens." May 29. Accessed June 3, 2016. http://www.dawn.com/news/1184763

Esfahbod, Behdad. 2014. "'I Can Get Another If I Break It': Announcing Noto Nastaliq Urdu." November 10. Accessed June 5, 2016. https://developers.googleblog.com/2014/11/i-can-get-another-if-i-break-it.html?utm_campaign=noto-urdu-1114&utm_source=jabran.me

Freedom House. 2015. "Freedom on the Net: Pakistan 2015." Accessed June 3, 2016. https://freedomhouse.org/report/freedom-net/2015/pakistan

Internet Service Providers Association of Pakistan (ISPAK). 2014. "History of Internet in Pakistan." Accessed June 3, 2016. http://www.ispak.pk/internet_pakistan.php

Junaidi, Ikram. 2016. "YouTube Returns to Pakistan." January 19. Accessed June 3, 2016. http://www.dawn.com/news/1233960

Khan, Arzak. 2015. "Internet.org Risks the Web's Future in Pakistan." June 22. Accessed June 3, 2016. http://america.aljazeera.com/opinions/2015/6/internetorg-risks-the-webs -future-in-pakistan.html

Khan, Raza. 2016. "Controversial Cyber Crime Bill Approved by NA." April 13. Accessed June 3, 2016. http://www.dawn.com/news/1251853

Press Information Department. 2016. "PR No. 102: Minister for IT, Anusha Rahman Addresses the UNESCO Mobile Learning Week Policy Forum Islamabad: March 13, 2016." March 13. Accessed June 3, 2016. http://www.pid.gov.pk/?p=16345

Rafique, Jabran. 2014. "Quick Overview of Google Noto Nastaleeq Urdu Web Font." November 11. Accessed June 5, 2016. http://jabran.me/articles/google-noto-nastale eq-urdu-web-font/

Reuters. 2016. "Pakistan Passes Controversial Cyber-Crime Law." August 12. Accessed December 14, 2016. http://www.reuters.com/article/us-pakistan-internet-idUSKC N1N0ST

Socialbakers. 2016. "Pakistan Facebook Page Statistics." Accessed June 4, 2016. http://www .socialbakers.com/statistics/facebook/pages/total/pakistan

United Nations Educational, Scientific, and Cultural Organization (UNESCO). 2016. "Adult Female Functional Literacy Programme (AFFLP)." September 1. Accessed December 14, 2016. http://www.unesco.org/uil/litbase/?menu=4&programme=63

United Nations International Children's Emergency Fund (UNICEF). 2015. "The State of the World's Children—Statistical Tables." Accessed August 14, 2016. http://www.data .unicef.org/corecode/uploads/document6/uploaded_pdfs/corecode/SOWC_2015 _Summary_and_Tables-final_214.pdf

PARAGUAY

Paraguay is a small, landlocked country in the middle of South America. It is bordered to the north by Brazil and Bolivia and to the south by Argentina. In 2012, about half the population was bilingual in Spanish and the local native language, Guaraní; another 40 percent of the population was comprised of monolingual Guaraní speakers, whereas only 6 percent of the population was monolingual in Spanish (Grazzi and Vergara 2012). In 2014, the population of Paraguay was about 6.7 million. Internet penetration in the country has risen in the past few years, from about 30 percent in 2012 to just over 40 percent in 2015; however, Paraguay is still one of the least-connected nations in South America (World Bank 2015). One potential reason could be the fact that internet access in Paraguay is prohibitively expensive. In 2015, a low-range income was only about $500 a month, while a connection of 10 Mbps could cost users as much as $146 a month (Young 2015). Therefore, internet users in Paraguay tend to be members of the upper class, who can afford such an expense.

Another potential reason for low internet penetration is language. Most of the internet is in English, Spanish, and Chinese; therefore, a digital divide exists for those Paraguayans who only speak Guaraní. Studies have shown that language has played a factor in decision making about technology in Paraguay (Grazzi and Vergara 2012). However, Google does offer its search engine in Guaraní.

Of those Paraguayans who do have internet access, the number of active social media users in 2014 numbered around 2.2 million (We Are Social 2014). Twitter is a popular social media website with 400,000 users, which represents a growth rate of 13 percent over the past few years (Abad 2015). The most popular social network in the country is Facebook (Alexa 2015). From 2012 to 2015, the number of Facebook users doubled, from 1.29 million to 2.6 million (Abad 2015; Internet World Stats n.d.). A third of these users are located in the nation's capital, Asunción, and by far, the predominant age group is eighteen- to twenty-four-year-olds, with 35 percent penetration (Abad 2015). The most popular Facebook page in Paraguay is Ramón Torres Psicólogo, with 3,389,600 fans (Socialbakers 2015). Torres (n.d.–) is a relationship therapist who has published a book entitled *El Amor, Así de Simple y Así de Complicado* (Love: Simply Complex), which is available in Kindle format on Amazon. Other popular websites visited by Paraguayan internet users include Google.com.py, YouTube, Google, Amazon, Abc.com.py (news), Live, Paraguay (news), Wikipedia, Última Hora (news), Twitter, Yahoo, and Blogspot (Alexa 2015).

The number of active mobile subscriptions in 2014 was 7.3 million (We Are Social 2014). As this number exceeded the population of the country, it suggests that having multiple mobile devices was common. Android mobiles were by far the most popular type of cell phone. In 2014, they dominated the market at 74.5 percent, while iOS, the operating system on Apple devices, accounted for only 9.8 percent of the mobile device market (eMarketer 2014). The most popular free apps downloaded for Android devices were WhatsApp Messenger, Facebook, Facebook Messenger, Instagram, and CM Security Antivirus App Lock. The top free apps downloaded by iPhone users were Facebook Messenger, WhatsApp Messenger, Facebook, Snapchat, and Spotify (music) (AppAnnie 2015).

Facebook introduced Internet.org to Paraguay in an attempt to increase internet penetration in 2013 (Rey 2015). Internet.org is a comprehensive project that aims to bring internet to the whole world, aiming to collaborate with local partners to eradicate obstacles to internet access in various countries. Internet.org identifies several reasons why two-thirds of the world's population is not connected. Some of these are relevant to Paraguay, such as expensive services, lack of network infrastructure, and language barriers. Internet.org has provided a free version of the internet to certain areas of the world with low penetration rates (Internet.org 2015). Paraguay, along with Zambia, Senegal, the Philippines, and fifteen other countries, is one of these areas (Russell 2015). The free version that Internet.org offers does not require a data plan, which means that users can access it via their cell phones whether or not they have a plan. However, the number of websites that they can visit with this free access is limited to just a handful, such as Facebook, Wikipedia, and a few local news and weather websites. According to an informal paper released by Facebook in March 2014, the Internet.org initiative in Paraguay had already led to an increase of more than 50 percent in daily data usage, as well as a 50 percent increase in the number of Paraguayans using the internet (Internet.org 2014).

Emojis, wildly popular images that can be used on social media platforms, have become a communications tool for worldwide diplomacy. After the Colombian flag was used in tweets during the 2016 FIFA Football World Cup qualifiers, the country's officials started using the flag emoji in communications with other Central American nations to promote regional cooperation and security. Social media for the presidency of Paraguay almost always starts with an emoji when one of its tweets contains a link to a report, agenda, or other official political document intended for public consumption (Twidiplomacy n.d.)

However, Internet.org has also received some harsh criticism. Some have suggested that rather than broadening people's experiences, Facebook is acting as a gatekeeper, deciding for internet users what their experience should be (Young 2015; Russell 2015). In Paraguay, some users of Internet.org's free internet service have been frustrated by its limitations on internet access. One response to this has been the development of software known as the "Facebook tunnel." Matías Insaurralde (n.d.–) created the software, which can grant users the ability to access other websites through the free internet provided by Internet.org. It utilizes Facebook's chat service as a virtual tunnel that links the user to the rest of the internet. The Internet.org user connects with a friend via the chat service, and through that connection, can visit any website that is available through the friend's internet service (Young 2015).

Marilyn J. Andrews

See also: Argentina; Mexico; Philippines; Senegal

Further Reading

Abad, Daniela. 2015. "Estadísticas de Facebook y Twitter en Paraguay." March 18. Accessed September 30, 2015. https://www.latamclick.com/estadisticas-de-facebook-y-twitter-en-paraguay-2015/

Alexa. 2015. "Top Sites in Paraguay." Accessed September 28, 2015. http://www.alexa.com/topsites/countries/PY

AppAnnie. 2015. "Google Play Top App Charts." Accessed September 28, 2015 https://www.appannie.com/apps/google-play/top/paraguay

eMarketer. 2014. "Smartphone Users in Latin America Reaching for Android Devices." February 19. Accessed September 28, 2015. http://www.emarketer.com/Article/Smartphone-Users-Latin-America-Reaching-Android-Devices/1010619

Grazzi, Matteo, and Vergara, Sebastián. 2012. "ICT in Developing Countries: Are Language Barriers Relevant? Evidence from Paraguay." *Information Economics and Policy*, 24:161–171.

Internet.org. 2014. "Connecting the World from the Sky," March 28. Accessed September 28, 2015. https://press.internet.org/2014/03/28/connecting-the-world-from-the-sky

Internet.org. 2015. "Who We Are." Accessed September 28, 2015. https://internet.org/about

Internet World Stats. n.d. "South America." Accessed September 28, 2015. http://www.internetworldstats.com/south.htm

Rey, Patricia. 2015. "Internet.org Llega a Panama y Apunta a Cuba." April 10. *BN Americas*. Accessed December 14, 2016. http://www.bnamericas.com/es/news/tecnologia/internetorg-llega-a-panama-y-apunta-a-cuba

Russell, Jon. 2015. "After Internet.org Backlash, Facebook Opens Portal to Court More Operators," July 27. Accessed September 28, 2015. http://techcrunch.com/2015/07/27/facebook-internet-org-one

Socialbakers. 2015. "Paraguay Facebook Page Statistics." Accessed September 28, 2015. http://www.socialbakers.com/statistics/facebook/pages/total/paraguay

We Are Social. 2014. "Digital in the Americas," June 10. Accessed September 28, 2015. http://wearesocial.net/blog/2014/06/social-digital-mobile-americas

World Bank. 2015. "World Development Indicators." Accessed September 28, 2015. http://databank.worldbank.org/data/reports.aspx?source=2&country=PRY&series=&period=

Young, Nora. 2015. "A Tunnel to (Internet) Freedom," April 16. Accessed September 28, 2015. http://www.cbc.ca/radio/spark/282-dot-internet-dot-ethics-dot-sucks-and-more-1.3035630/a-tunnel-to-internet-freedom-1.3035651

PERU

Peru is located in the western half of Latin America, bordered by Ecuador, Chile, Bolivia, Colombia, and Brazil. As of 2014, it had the seventh largest digital population in the region. In Peru, internet users accessed news sites most frequently, but ultimately spent the majority of their time online on social media sites (García 2014). Peruvians' favorite social media platforms were Facebook, YouTube, and Twitter, with some Peruvians branching out to LinkedIn, Google+, Instagram, Pinterest, WhatsApp, and other sites. Peruvians who use social media tended to manage accounts on three to six different platforms, with few utilizing more than six.

The company GfK Perú (2015) produced a detailed report about Peru's digital population. Its research concluded that around 50 percent of urbanites and 46 percent of rural populations accessed the internet. Peru's internet users consisted of 29 percent fifteen- to twenty-four-year-olds, 28 percent twenty-five- to thirty-four-year-olds, 21 percent thirty-five- to forty-four-year-olds, 14 percent forty-five- to fifty-four-year-olds, and 8 percent fifty-five-year-olds and older (Statista 2015). Including all age ranges, Peru's top ten websites in 2015, listed in descending order, were Facebook, YouTube, Google.com.pe, Google, Live.com, Yahoo, Amazon, Wikipedia, MSN, and Elcomercio.pe (Alexa 2015). Facebook was not only the most popular website accessed, it was also the country's most popular social media platform. At the end of 2014, approximately 14 million of Peru's 31 million inhabitants, or 45 percent, had accessed the site (Trome 2015).

Peru's Facebook demographics mirrored those of people who access the internet, but on a slightly smaller scale. Of the almost 8 million men and 7 million women with Facebook accounts, 40 percent were under age twenty-four, 32 percent were between twenty-five and thirty-four, and 16 percent were between thirty-five and forty-four. A total of 56 percent of Facebook users resided in Lima, 5 percent in Arequipa, and 5 percent in Trujillo (Rodríguez 2015). Other accounts did not specify a location or provide geocoding capabilities to record the user's location.

More than half of all users accessed the platform from both a computer and a mobile phone. Most users chose Google Chrome as the best browser to access Facebook.

Twitter usage split almost evenly between Lima, other urban areas, and the provinces, meaning that it likely connected more loosely defined communities that participated in more diverse conversations than they do on Facebook. By the end of 2015, approximately 4 million Peruvians participated on Twitter, producing an average of 280,000 tweets daily, or more than 11,000 tweets per hour (Quantico Trends 2015). One of the more fun, nationalistic Twitter campaigns challenged local communities to take pride in their culture by tweeting about what makes them unique, including the hashtag #MásPeruanoQue [More Peruvian Than]. Respondents used the phrase to start their messages by describing how they were more Peruvian. Most responses included colorful images of food and locals, and often contained meme-inspired humor.

Peruvian tweeters were passionate about social and political causes. Issues taken up on Twitter and other social media had a greater chance to be addressed and resolved publicly than causes advocated in other forums. For example, a concerned public adopted the hashtag #LeyStalker (Stalker Law) in 2015 to warn other citizens about presidential decree No. 1182, permitting the police to access mobile phone global positioning system (GPS) data on any phone without a warrant. The decree went even further, requiring phone carriers and other data storage enterprises to retain data for at least three years. Police intended to utilize the decree to confront social plagues such as corruption, extortion, or blackmail—all issues endemic to Peru. Much of the public, however, believed that police would implement the new law to track and harass innocent civilians without cause. The law was taken to court for appeal.

Another notable social campaign tackled the problem of wildlife smuggling and abuse. Neotropical Primate Conservation, an organization seeking to abolish all wildlife mistreatment, launched a public education campaign on traditional and social media to inform the people about the dangers and consequences of wildlife trafficking. The campaign encouraged people to report all observed abuses and created several social media spaces where witnesses could submit information anonymously. After incidents were brought to their attention, the organization turned them over to the government for review and action. A long-term objective of this program was to map the trafficking networks across Peru, as little data exist about how local poachers operate.

While Facebook helped to promote awareness of some causes, it also had a downside. Some Peruvian communities have high crime rates and limited authorities to investigate and reduce overall crime in an area. After a neighbor's scare over a break-in that resulted in police releasing the perpetrator prior to trial, one Peruvian woman created a Facebook page called "Chapa tu choro" (Catch your thief). The page encouraged Peruvians to take matters into their own hands (i.e. vigilantism) instead of relying on the police. Peruvian Facebook users resonated with the sentiment, and more than 100 similar pages appeared (BBC 2015). However, some of the pages took a more violent approach, advocating that criminals be injured or killed immediately when apprehended, without waiting for a trial.

Promoting vigilantism and violence presents additional potential problems for Peruvian society, especially if smaller crimes lead to homicides and assaults. Facebook, while not responsible for page content, can be used to spread hate speech and extreme sentiments. The copycat sites have started posting gruesome videos and slogans, indicating that even though there are few such sites, this new phenomenon may be more serious and long lasting than the typical Facebook fad. The use of social media for malignant purposes is not limited to Peru; other nations, such as Germany, Myanmar, Ukraine, and the United States have had to deal with hate-speech incidents on the site.

Peruvians tended to prefer established international social media platforms. However, in spite of this preference, a small number of Peruvians sought to design niche networks. For example, the Council of Science, Technology, and Technological Innovation (Concytec) produced a unique new platform in 2015. ACTIvanet found a niche in connecting Peruvian scientists with investors and entrepreneurs across the globe. The platform offered users a way to learn about new technologies, create communities of interest, and investigate new potential sources of funding for scientific research. The site connected to Facebook, allowing account registration with a Facebook account and some additional personal information. A driving force behind the network's development was the desire to encourage local innovation and technological advancement. ACTIvanet was a new endeavor with no international equivalent or competition.

Peru's app preferences somewhat mirror those of other Latin American nations. On Google Play, for example, the top five free apps were WhatsApp, Facebook Messenger, Facebook, DU Battery Saver, and Hola Launcher. The top five paid apps were Minecraft: Pocket Edition, Monopoly, Quizduell Premium, Geometry Dash, and XPERIA Donald Duck. These preferences almost exactly matched those in Venezuela. In contrast, Peru's iOS preferences differed from those of its neighboring countries. The top free apps were WhatsApp, Facebook Messenger, Facebook, YouCam Perfect, and the game Apensar. The paid apps were Geometry Dash, SFind for Spotify, FACIE, Minecraft: Pocket Edition, and the Cut the Rope game (AppAnnie 2015). These app preferences suggest that Peruvians value messaging and games, adding a little diversity by choosing apps that improved their social networking experiences.

Another similarity that Peruvians have with other Latin American populations is their love for memes. In Peru, creating and sharing memes has become an important part of social media and national cyberculture. Most Peruvian memes contained humor, but also a hint of biting ruthlessness and social criticism that poked fun at celebrity scandals. Many memes poke fun at *farándula* (celebrities), such as Figueroa. A famous meme from 2014 developed when model Milett Figueroa (1992–) stole a man away from his girlfriend. Online, pictures of Figueroa gave her the sarcastic nickname "Candy," mocking her for being less than sweet in orchestrating the breakup. The name stuck, and most Peruvians now refer to her as "Candy" rather than by her real name (Fernandez Coraza 2015). The deeper the scandal, the more likely it is that related memes will flourish across Peruvian social media. Other Latin American nations, including Argentina, Ecuador, and Venezuela, also produce prolific memes.

Laura M. Steckman

See also: Argentina; Ecuador; Germany; Myanmar; Ukraine; United States; Venezuela

Further Reading

Alexa. 2015. "Top Sites in Peru." Accessed September 14, 2015. http://www.alexa.com/topsites /countries/PE

AppAnnie. 2015. "iOS Top App Charts." Accessed September 14, 2015. https://www .appannie.com/apps/ios/top/peru/?device=iphone

BBC Trending. 2015. "The Facebook Vigilantes Catching Thieves—and Punishing Them." September 14. Accessed September 15, 2015. http://www.bbc.com/news/blogs-trending -34224196

Fernandez Coraza, Arnold. 2015. "Memes and Peruvian Wit Claim Cuzco's Social Media." January 6. Accessed September 14, 2015. http://www.cuzcoeats.com/2015/01/memes -and-peruvian-wit-claim-cuzcos-social-media/

García, Marco. 2014. "Futuro Digital Perú 2015." Accessed September 14, 2015. http://www .comscore.com/lat/Prensa-y-Eventos/Presentaciones-y-libros-blancos/2015/Peru -2015-Digital-Future-in-Focus

GfK Perú. 2015. "Uso de Internet en el Perú." May 14. Accessed September 14, 2015. http://es.slideshare.net/GfKPeru/gfk-per-uso-de-internet-en-el-per-abril-2015

Quantico Trends. 2015. "Infografía de Twitter en Perú." January 31. Accessed September 14, 2015. http://www.quanticotrends.com/usuarios-de-twitter-2015/

Rodríguez, Gustavo. 2015. "Algunas Cifras de Facebook en Perú a Julio del 2015." August 31. Accessed September 14, 2015. http://www.hashtag.pe/2015/08/31/algunas-cifras -de-facebook-en-peru-a-julio-del-2015/

Statista. 2015. "Distribution of Internet Users in Peru as of 1st Quarter 2015, By Age Group." Accessed December 14, 2016. https://www.statista.com/statistics/446468/age-distribution -internet-users-peru/

Trome. 2015. "Facebook: Seis Datos sobre el Impacto de esta Red Social en Perú." January 20. Accessed September 11, 2015. http://trome.pe/redes-sociales/facebook-seis-datos -sobre-impacto-esta-red-social-peru-2036797

PHILIPPINES

The Philippines is an archipelagic nation in the Pacific Ocean consisting of more than 7,000 islands. It lies north of Indonesia and east of Vietnam. Culturally, the Philippines has integrated emerging technologies such as social media and apps into what has become a permanent, digital component to its contemporary society. This phenomenon has paralleled that which is occurring in Southeast Asian nations. Approximately 100 million people live in the Philippines, around 44 million of whom have access to the internet, with 90 percent or more having at least one active social media account (Revesencio 2015). The country used to be the short message service (SMS), or texting, capital of the world; today, it has established a reputation as an active, global social media hub.

In 2014, *Time* magazine analyzed over 400,000 Instagram photos labeled "selfies" and developed a list of the top 100 "selfie capitals of the world." The actual selfie capital, in first place, is Makati City, along with Pasig in the Philippines. The rest of the top ten, in order, were Manhattan; Miami, Florida; Anaheim and Santa Ana, California; Petaling Jaya, Malaysia; Tel Aviv, Israel; Manchester, England; Milan, Italy; Cebu City, Philippines; and George Town, Malaysia (Wilson 2014).

Filipinos use multiple social media and chatting platforms. Overwhelmingly, Filipinos have a strong presence on Facebook, where they connect to close friends and family and make new friends online. In fact, Filipinos are often ranked the number one population worldwide in terms of their commitment to using Facebook. In addition to Facebook, Filipinos engage on Google+, Pinterest, Reddit, StumbleUpon, Tumblr, Twitter, YouTube, Instagram, Viber, Skype, Yahoo Messenger, WeChat, Line, WhatsApp, Kakaotalk, and LinkedIn. Filipinos' top ten accessed sites confirmed their interest in social media; the sites were Facebook, Google, Google.com.ph, YouTube, Yahoo, abs-cbnnews.com, Gmanetwork, Amazon, Twitter, and Blogspot (Alexa 2015). Blogging was not as popular as in other Asian nations, such as Indonesia and China. However, more Filipinos have begun initiating blogs in addition to their usual social media channels.

Filipinos love Facebook to the extent that many reacted negatively to a Harvard study on usage of the site in their country. The title, "The Unintelligent Facebook Users," shocked Filipinos, who then criticized and even rejected its findings. The researcher, a Filipino-American psychology student at Harvard, concluded that Filipinos online were not as cultured as most of the rest of the world, that they were not prepared to implement the necessary legal guidelines and etiquette standards, and that most of them functioned online at a lower intellectual level, with less emotion, decency, and awareness than other nationalities (Green and Pagulayan 2014). The paper used several case studies to back up these findings. In one example, the author posted a picture of an unhealthy child with the caption "Don't Like Before You Read the Article." That picture pointed to an article, but users reportedly shared the photo 85,000 times, while the article was shared fewer than 10,000 times. The picture alone received 90 percent more "likes" than the article, presumably because users did not read it (Green and Pagulayan 2014). However, though the psychology experiments led to the study's main conclusions, they did not prove that Filipinos who used Facebook were somehow less ready to participate in a globalized social media environment. Instead, the assertions caused great discomfort domestically and challenged local Filipinos to think about how foreigners perceived their interactions online.

The study's conclusions also ignored the Philippines strict cybercrime laws. In 2012, the government proposed the Cybercrime Prevention Act. Among the outlawed offenses were identity theft, child pornography, cyberbullying, cybersquatting, illegal access to and use of online data, and defamation. The law was controversial; many Filipinos responded negatively on social media, with many users

In response to the Philippines' highly controversial Cybercrime Prevention Act, multiple activist groups, such as the Philippine Internet Freedom Alliance, Bayan Muna, Piston, and Gabriela, staged a silent protest near the Supreme Court in Manila. The public responded to the law on social media, particularly on Twitter, to express its outrage at many of the law's provisions. Worldwide governments are trying to prevent illegal activities on the internet; because the medium is still relatively new, fighting online crime has some inherent political, social, and cultural challenges. (Hrlumanog/Dreamstime.com)

utilizing the #notocybercrimelaw hashtag. As a result, the law was suspended and readdressed. It remained controversial until 2014, when the Supreme Court upheld the law, dropping only a handful of its provisions. The defamation provision remained, with the stipulation that only the original poster could receive a punishment for insulting statements, and that others who liked or otherwise supported the original post were not liable. The Filipinos responded with another hashtag, #NonLibelousTweet. The court's decision also infuriated several hacker groups, including Anonymous, which decided to deface government websites in protest.

In contrast to the Harvard study, an international, social media–focused study determined that Filipinos were extremely interactive and social online. The study, conducted by UM, a division of IPG Mediabrands, concluded that Filipinos spend an average of fifty-three hours a week socializing online, with their main goals being maintaining friendships, starting new acquaintances, entertaining themselves, trading stories, and feeling more closely connected (Locsin 2014). By engaging online, Filipinos met some of their core personal, professional, spiritual, and cultural needs, such as creating strong support networks online and offline. In addition to uncovering the reasons why Filipinos socialize online, the study offered insights on some of their main issues and concerns. Privacy protection was a major issue in the nation, with people saying that they hesitated before sharing personal information (Locsin 2014). This study showed Filipinos to be not only internet-savvy, but also caring, sharing, and concerned about the well-being of the people in their networks.

Social media demographics on the Philippines revealed that the average social media user was a young man. Younger users tended to prefer interacting and sharing online. For example, 42 percent of internet users were fifteen to twenty-four years old, 33 percent were twenty-five to thirty-four years old, 16 percent were thirty-five to forty-four, and 9 percent were over forty-five (Leet Digital 2015). Additional research on social media also showed that 56 percent of users were male (OnDevice 2014). Because internet penetration remained relatively low, the potential exists that these demographics could shift as more people gain access to the internet.

The Philippines has achieved some unique benchmarks for its digital culture. The country's social media users have earned acclaim as the population that spends the most time on the internet every day, more than in any other country. The average laptop user spent 6.3 hours online daily, whereas the mobile user was online 3.3 hours each day (Revesencio 2015). These statistics were collected based on device, and they did not determine how many people use both laptops and mobile internet access. Interestingly, it was possible that some Filipinos spent up to 9.6 hours online per day, combining their time on a laptop and a mobile device.

Social applications played an important role in the Philippines. Each month, around one-third of mobile phone users downloaded at least six apps (OnDevice 2014). In September 2015, the top free Google Play apps included Clash of Clans, Facebook Messenger, SHAREit, and Twitter. The top paid apps were League of Stickman-Samurai, Minecraft: Pocket Edition, EvoCreo, Xperia Daisy Duck, and Poweramp. The iOS top free apps were Google Street View, Facebook Messenger, Facebook, Twitter, Instagram, YouTube, Viber, Magic Piano by Smule, Spotify, and Free Music Play—Mp3 Player. The top paid apps were Emoji Keypad, NBA 2K15, Flightradar 24, Instant Street View, Afterlight, Plague Inc., Grindr Xtra, Crystal—Block Ads, Real Steel, and Cake Mania Celebrity Chef (AppAnnie 2015).

Apps also contributed to a geographically specific emerging trend in the Philippines. The country is often at risk for natural disaster, particularly typhoons due to its location in the Pacific Ocean. After Typhoon Haiyan hit the country in late 2014, technology companies rushed to create apps to help people understand weather risks. RaincheckPH and iTyphoon were two apps released around this time to ensure that Filipinos with internet access had the ability to learn quickly about potential weather threats. The Filipino government also decided to use technology to warn its people and enacted laws requiring mobile service providers to send SMS warnings of impending storms throughout the network. Typhoons and other natural disasters are impossible to prevent, but their affects can be mitigated with the proper warning systems in place. The increased availability of weather apps and legally required messaging will ensure that every person with a laptop or mobile phone can receive advance warnings and up-to-date information.

The Philippines has one of the most robust and vibrant online communities. In spite of these colorful, fast-paced channels of communication, one major obstacle has plagued the country: its internet speed is rated as one of the slowest in the world. In the country, the average speed is 3.64 Mbps. Of all Asian nations, the Philippines ranked number twenty-one of twenty-two, with the slowest spot belonging to Afghanistan, at 2.52 Mbps (Philstar 2015). To put this speed in perspective, until January 2015, the U.S. Federal Communications Commission (FCC) set the minimum recommended bandwidth for streaming audio or video was 4 Mbps, though this speed usually became problematic with high-definition video or when the system handled two or more devices. However, the FCC updated this guideline to 25 Mbps for downloads and 3 Mbps for uploads, in order to reflect changes in quality and demand for streaming audio and video. In the Philippines, the average download speed continued to reflect around half of even the outdated FCC standard.

Laura M. Steckman

See also: Afghanistan; China; Indonesia; United States

Further Reading

Alexa. 2015. "Top Sites in Philippines." Accessed September 20, 2015. http://www.alexa
.com/topsites/countries/PH

AppAnnie. 2015. "iOS Top App Charts." Accessed September 22, 2015. https://www
.appannie.com/apps/ios/top/philipines/overall/?device=iphone

Green, Rhodora, and Pagulayan, Ruel. 2014. "Harvard: Pinoy Facebook Users Are Among the Most Unintelligent People." November 11. Accessed September 21, 2015. http://
www.thephilippinepride.com/harvard-study-finds-80-13-of-filipino-facebook-users
-are-dumb/

Leet Digital. 2015. "Philippines Social Media and Internet Statistics." July 10. Accessed September 21, 2015. http://ph.leetdigital.com/marketing/philippines-social-media-and
-internet-statistics-2015

Locsin, Joel. 2014. "Pinoys Lead the World in Social Media Engagement—Study." June 30. Accessed September 21, 2015. http://www.gmanetwork.com/news/story/367983/
scitech/technology/pinoys-lead-the-world-in-social-media-engagement-study

OnDevice Research. 2014. "The Philippines Mobile Internet Crowd: Young, Affluent, and Growing." July 8. Accessed September 21, 2015. http://www.slideshare.net/OnDevice/
philippines-mobile-internet-trends

Philstar. 2015. "If Philippines Joined an Internet Speed Race." [Infographic] August 25. Accessed September 21, 2015. http://www.philstar.com/news-feature/2015/08/25/
1492054/infographic-if-philippines-joined-internet-speed-race

Revesencio, Jonha. 2015. "Philippines: A Digital Lifestyle Capital in the Making?" May 4. Accessed September 21, 2015. http://www.huffingtonpost.com/jonha-revesencio/
philippines-a-digital-lif_1_b_7199924.html

POLAND

In 2016, Poland, a country located in Eastern Europe, was ranked tenth in the world in internet connection speed, at 10.6 Mbps. This rate is not only significantly higher than the global average of 5.1 Mbps, but also a huge jump from the 2.6 Mbps that earned it the nineteenth spot in 2013. With a total net population of 38.6 million people in the country, the internet penetration is 67 percent, which is significantly higher than its social media penetration rate of 36 percent. While Poles are not avid users of social media sites, they are on an upward trend in the country. Poles have access to a variable mix of international and local social media providers, such as Facebook, Google+, Nasza-Klasa (nk.pl), Twitter, LinkedIn, GoldenLine, and Fotka. Within three years, Facebook and Google+ users have steadily grown to surpass the local favorite, Nasza-Klasa. Twitter, LinkedIn, and GoldenLine are also slowly gaining in popularity, while Blip and Fotka have closed down or lost traction within Poland. Polish entrepreneurs are also successfully integrating social media into marketing strategies for their companies. Brand24 is a prime example of a Polish start-up that heavily uses social media in its business practices while maintaining a dominant presence on the internet.

In 2006, Nasza-Klasa was founded in Wroclaw, Poland. Most people refer to Nasza-Klasa, often shortened to nk.pl, as the Polish version of Facebook, while some people call it the Polish counterpart of Classmates.com. Nasza-Klasa allows students and alumni to connect or reconnect with classmates from their alma mater. In 2009, Nasza-Klasa reached the height of its popularity in Poland, where at least 86 percent of its users lived. At the same time, the social media provider was also popular in Norway, ranking fifth in that northern European country (Mohsin 2009). As Facebook steadily grew in popularity, the number of active users on Nasza-Klasa dropped nearly 40 percent. During this decline, Forticom bought the struggling Nasza-Klasa. To maintain its relevance as a social media provider to its remaining 5 million active users, Nasza-Klasa installed a newsfeed similar to Facebook and integrated a microblog known as Sledzik. Nasza-Klasa also added a feature where companies can promote products and advertisements (Kijuc 2014).

Other local social providers that have fallen out of favor of the Polish population are Blip and Fotka, the Polish counterparts for Twitter and Hi5. Blip was founded in 2007. Its sole purpose as a social media provider was to be a user-friendly microblog. Initially, it grew in popularity with the help of famous Polish celebrity accounts, much like the instant popularity that Twitter enjoyed in the United States with Hollywood celebrities. Despite the fact that Blip transmitted 10 million messages, the number of user accounts grew slowly (Kotowski 2009). Blip ultimately closed in 2013 as Twitter became the Poles' primary choice for microblogging (Kijuc 2014). Established in 2001, Fotka is the one of the oldest social media platforms in Poland. One of the first features introduced by Fotka was uploading photos for users to "like" or leave commentary. Eventually, it branched out to include other forms of entertainment, such as online games. In 2014, Fotka recorded the lowest membership rate since its conception with 600,000 active users. It is predicted to continue on this downward trend. As more Poles create long-distance friendships, it becomes harder for them to connect with new friends in different countries on a local provider. Despite Fotka's commitment to its remaining active users, it is quickly losing its appeal and could potentially close down like Blip (Wirtualnemedia 2014).

While Poles spend approximately four hours and twenty-five minutes each day on a PC or tablet, they use their mobile phones on the internet for an average of only one hour and seventeen minutes. Since January 2015, a 6 percent growth of internet users was documented in data collected by We Are Social (Kemp 2016). Poland continues to improve internet availability and accessibility to its citizens across the country. In December 2015, Satellite Solutions Worldwide bought two additional subsidiaries in Poland—AVC Solutions and Hetan Technologies. These recent acquisitions will allow an even faster internet speed to more than 5 million homes in the country. More than a third of Poland's users get a broadband speed of only about 2 Mbps, which is significantly lower than the country's average internet connection speed of 10.6 Mbps. To bring these homes up to date, Satellite Solutions Worldwide plans to offer Poles faster service through wireless internet via satellite (Sharecast 2015).

Poles spend, on average, up to of 4.5 hours online, with about 1.2 of those hours spent on social media (Kemp 2016). The Polish population has a wide mix of social media providers to choose from, including Polish counterparts to popular international providers such as Facebook and Twitter. As global social media providers create multiple platforms for international connectivity and worldwide visibility, it has become harder for local social media providers, such as Nasza-Klasa, Blip, and Fotka, to compete with international platforms that feature global values and contemporary trends.

The primary use of social media in Poland is to stay connected with friends or to reconnect with old acquaintances. However, that is changing with its integration into the business industry. With the growing popularity of LinkedIn and GoldenLine in the country, recruiters are turning to social media to find future employees and develop company branding. Founded in 2005, GoldenLine is Poland's biggest business social networking platform that is similar to LinkedIn. Users are able to connect with business contacts, discover new clients, and search for jobs. The development and functionality of GoldenLine is changing the trend of job recruitment in Poland. Companies and employers are seeing a shift from traditional job searching to a more passive approach that allows applicants to submit their résumés online. With an analysis of the applicants' experience, skills, and references available for review, recruiters may use the resulting reports to find the right candidate. Employers also use GoldenLine to build the desired image for the company as well as market branding concepts. More than 6,000 recruiters in Poland use GoldenLine to find international and local talent for Polish companies (Viluckyte 2015).

Before the introduction of the start-up culture in Poland, small and medium-sized businesses in the restaurant and hospitality industries utilized social media to market their services to an online audience. Initially, many people did not perceive social networks as a practical method for business strategies or marketing. In recent years, tech start-ups and Polish entrepreneurs have truly taken advantage of social media marketing. After merging social media strategies into their marketing approach, many Polish tech start-ups have launched successfully. Some experts believe that social media is integral to the beginnings of any start-up. Social media allows new businesses to gain more visibility with an online community and market their products more creatively and efficiently. Through social media, start-ups can keep in touch with their clientele, as well as establish new relationships with potential clients.

One particular start-up that integrated social media into its business plan is Brand24 (Kijuc 2014), founded in Wroclaw, Poland, in 2011. The start-up takes a unique approach to social media. While some companies utilize social media in their marketing and advertising strategies, Brand24 fully incorporates the internet and social media into its business. Brand24's customers use the company's tools to monitor and analyze anything that mentions their brand, products, or services online. Brand24 collects and archives all data regarding a specific brand for review. While there are other global monitoring companies on the internet, Brand24 dominated the online space and obtained international recognition. The start-up has a second headquarters in New York City and clients in multiple countries, such as China, Indonesia, and the United Kingdom.

Not only does Brand24 monitor social networks, such as Facebook and Twitter, but it also utilizes YouTube to celebrate new clientele. For its video recognizing IKEA as a new client, Brand24 amassed 70,000 views. Some experts believe this is a distinguishing factor for the company. Creating and posting a video brought more attention to the newfound relationship than a typical press release (Kijuc 2014). Brand24 also is working with Michelle Obama on one of her educational projects targeting Generation Z, the fourteen- to nineteen-year-old age bracket. The initiative is geared toward motivating students to pursue higher education after high school. The First Lady approached the company to support the campaign and facilitate online conversations with participants (Goldman 2015).

Karen Ames

See also: China; Germany; Indonesia; United Kingdom; United States

Further Reading

Goldman, Stefania. 2015. "Brand24 to Cooperate with the White House." December 2. Accessed May 4, 2016. https://itkey.media/brand24-to-cooperate-with-the-white-house/

Kemp, Simon. 2016. "Digital in 2016: We Are Social's Compendium of Global Digital, Social, and Mobile Data, Trends, and Statistics." January 27. Accessed May 2, 2016. http://wearesocial.com/uk/special-reports/digital-in-2016

Kijuc, Dominika. 2014. "#Poland: Interview with Social Media Experts." March 24. Accessed December 26, 2015. http://news.bitspiration.com/news/people/poland-interview-with-social-media-experts-3/

Kotowski, Daniel. 2009. "Sławy na Blipie Drogą do Sukcesu?" July 17. Accessed December 27, 2015. http://di.com.pl/slawy-na-blipie-droga-do-sukcesu-27579

Mohsin, Saleha. 2009. "Hey, America, We Have Our Own Facebooks." July 15. Accessed December 27, 2015. http://www.businessweek.com/globalbiz/content/jul2009/gb20090715_921142.htm

Sharecast. 2015. "Satellite Solutions Expands Poland Broadband Business." December 21. Accessed December 26, 2015. http://www.sharecast.com/news/satellite-solutions-expands-poland-broadband-business/23753971.html

Vilukyte, Aurelija. 2015. "How Recruitment Is Adapting to Social Media in Poland." September, 18. Accessed December 28, 2015. http://www.candarine.com/blog/how -recruitment-is-adapting-to-social-media-in-poland/

Wirtualnamedia. 2014. "Fotka.pl Rekordowo Traci Użytkowników, ale Wciąż Mocno Ich Angażuje," December 18. Accessed December 27, 2015. http://www.wirtualnemedia .pl/artykul/fotka-pl-rekordowo-traci-uzytkownikow-ale-wciaz-mocno-ich-angazuje

Q

QATAR

The State of Qatar is a small country in the Arabian Peninsula, located between Africa and Asia. As of December 2014, its internet penetration rate was 91.9 percent of 2.2 million people, the third highest in the Middle East (Internet World Stats 2015). In 2013, 89 percent of the country's internet users accessed the web via smartphone (Northwestern University of Qatar 2014). Persons aged twenty-five to fifty-four years old represent 70 percent of the population, and the median age is 32.6. Non-Qataris (mainly expatriate workers) comprise 90 percent of the population. Because the overwhelming majority of these workers are men, less than one-quarter of the country's inhabitants are women (IndexMundi 2014).

After deposing his father in a bloodless coup in 1995, Qatar's then-ruler, Emir Hamad bin Khalifa Al Thani (1952–), launched the pan-Arab satellite news network Al-Jazeera in 1996. He also ushered in a new constitution in 2003, which protected privacy and guaranteed freedom of expression. In 2013, Hamad transferred power to his 33-year-old son, Tamim (1980–). The move marked the culmination of a decade of rapid development, during which Qatar, which shares a massive natural gas field with Iran, became the world's wealthiest country, with an estimated per-capita gross domestic product (GDP) of $143,400 in 2014 (CIA World Factbook 2015).

State-owned Qatar Telecom (since renamed Ooredoo) had a monopoly on the internet market in Qatar prior to 2013 (Alqudsi-ghabra et al. 2011). Nine internet service providers (ISPs) currently are licensed in the country (allISPs, 2015). Infrastructure improvements have featured prominently in Qatar's preparations for playing host to soccer's World Cup in 2022, with information technology (IT) at the forefront. The government is working to expand the country's fiber-to-the-home network, increase access to universal high-speed broadband, install two new submarine cables for enhanced international connectivity, and extend free Wi-Fi service across the country (ictQATAR 2013).

Qatar is the second most wired country in the Middle East (after Saudi Arabia), and in 2002, it became the first in the region to adopt asymmetric digital subscriber line (ADSL) internet. In 2011, 98 percent of primary- and secondary-schoolchildren in Qatar had access to the internet (The Peninsula 2011). In 2013, 67 percent of Qatar-citizen internet users utilized social networking sites, spending 3.9 hours per day on them. Among non-Qataris, 77 percent of Arab-expatriate internet users utilized social networking sites, spending 2.9 hours per day on them; the figures for Asian expatriates were even higher (80 percent and 3.2 hours per day), and for

Hamad International Airport in Doha, Qatar, offers travelers computer kiosks with amenities such as internet access. Qatar has one of the highest internet penetration rates in the Middle East, and aims to connect more people through continuous government infrastructure improvement programs. (9darek/Dreamstime.com)

Western expatriates, the usage was at 85 percent and 2.8 hours per day (ictQATAR 2013).

Regarding social media applications, 88 percent of Qatari residents—the Arab world's fourth-highest percentage—use Facebook. Among Qatari-citizen internet users, however, Facebook penetration is 65 percent. Qatari citizens are more likely to use Twitter (65 percent) and Instagram (48 percent) than are non-Qatari residents of the country (33 percent and 8 percent, respectively). Further, 94 percent of Qatari citizens use Arabic when accessing the internet, but 56 percent also are able to engage online in English, and 10 percent of residents use Google+ (Northwestern University of Qatar 2014).

Perhaps owing to its wealth, Qatar experienced little of the political upheaval that roiled the Middle East following the 2010–2012 "Arab Spring" uprisings. Nevertheless, and in spite of the 2003 constitution's guarantees, the Qatari government closely monitors residents' online and offline behaviors and punishes dissent. Organized opposition to the government is banned, and political parties are illegal.

Internet usage in Qatar is governed by Decree Law No. 34, a telecommunications law passed in 2006. It criminalizes, with fines and imprisonment for up to one year, using "a telecommunications network . . . for the purposes of disturbing, irritating or offending any persons" (Alqudsi-ghabra et al. 2011). The U.S. State

Department has criticized Qatar for restricting civil liberties and reported that the government censors the internet through a proxy server that monitors websites, email, and chat rooms. Pornography, sites containing political criticism of the Arabian Peninsula countries, and material deemed offensive to Islam are blocked. Gay and lesbian content, reproductive health information, and sites offering tools for enhancing privacy or evading controls also are filtered. Because of the strict regulations, journalists practice self-censorship when reporting on the government, the ruling family, and Islam (OpenNet Initiative 2009). Officials at the Communications Regulatory Authority and the Ministry of Information and Communications Technology have denied censoring the internet and insisted that ISPs are responsible for ensuring that subscribers comply with the 2006 law (Kovessy 2015).

In 2011, Qatari citizen Muhammad Rashid al-Ajami aka Muhammad ibn al-Dheeb (c.1985–) was arrested for criticizing the country's leaders in a poem titled "Tunisian Jasmine." It read in part, "We are all Tunisia, in the face of the repressive elite." While studying in Cairo in 2010, al-Ajami recited the poem at a private gathering with friends, one of whom uploaded a video of the recitation to YouTube. He was sentenced to life in prison in 2012 for inciting an overthrow of the regime, but his punishment was reduced on appeal to fifteen years in 2013. International human rights groups criticized the case, with Amnesty International calling for his unconditional release (Bollier 2013). In March 2016, Emir Tamim bin Hamad pardoned al-Ajami, and he was released from prison.

In 2015, Human Rights Watch urged Qatar to remove "loosely worded provisions penalizing criticism of Qatar or neighboring governments" from a draft media law that has been pending since 2012. The legislation eliminates criminal penalties for media law violations, but Article 53 prohibits publishing or broadcasting information that would "throw relations between the state and the Arab and friendly states into confusion" or "abuse the regime or offend the ruling family or cause serious harm to the national or higher interests of the state." Violators would be fined up to $275,000.

Qatar residents generally were positive about the internet's benefits, according to a 2013 survey. For 79 percent of the respondents, the internet was the first place that they went for information; 60 percent agreed with the statement "People like you can better understand the nature of powerful institutions"; 58 percent concurred that the internet enabled people to "have more influence on society"; and 57 percent believed that it allowed them to "have more say about public issues" (Northwestern University of Qatar 2014).

In light of the well-known monitoring and filtering of the internet in Qatar—illustrated by a "technical issue" at Ooredoo that resulted in the temporary blocking of several popular websites that made headline news (Kovessy 2015)—some survey responses appear naïve or reflective of a desire to avoid controversy. For example, 60 percent of respondents agreed that "it is okay for people to express their ideas on the internet even if they are unpopular"; 54 percent admitted feeling comfortable "saying what I think on public issues"; and 46 percent said it was "safe to say whatever one thinks about public issues." Only 38 percent worried that powerful institutions were checking their online activities. Tellingly, in this conservative society, 60 percent of Qatar citizens agreed that "the internet should be more tightly regulated in Qatar" (Northwestern University of Qatar 2014).

Mark A. Caudill

See also: Saudi Arabia; United Arab Emirates

Further Reading

allISPs. 2015. "List of ISPs in Qatar." Accessed November 23, 2015. http://222.allisps.com/en/offers/QATAR

Alqudsi-ghabra, Taghreed M., Al-Bannai, Talal, and Al-Bahrani, Mohammad. 2011. "The Internet in the Arab Gulf Cooperation Council (AGCC): Vehicle of Change." *International Journal of Internet Science,* 6(1): 44–67.

Arab Social Media Report. 2014. "Twitter in the Arab Region." March. Accessed October 27, 2015. http://www.arabsocialmediareport.com/Twitter/LineChart.aspx?&PriMenuID=18&CatID=25&mnu=Cat

Bollier, Sam. 2013. "Qatari Poet's Sentence Reduced to 15 Years." February 25. Al Jazeera. Accessed November 23, 2015. http://www.aljazeera.com/news/middleeast/2013/02/20132251151345579.html

CIA World Factbook. 2015. "Qatar." Accessed November 20, 2015. https://www.cia.gov/library/publications/the-world-factbook/geos/qa.html

Democracy Now. 2015. "Life Sentence for Qatari Poet Reduced to 15 Years." February 26. Accessed April 10, 2016. http://www.democracynow.org/2013/2/26/headlines#2269

Dennis, Everette E., Martin, Justin D., and Wood, Robb. 2014. "Media Use in the Middle East: An Eight-Nation Survey." Northwestern University in Qatar. Accessed November 20, 2015. https://www.scribd.com/fullscreen/148423818?access_key=key-hhrnkzjb5s6ns9vu3md&allow_share=true&view_mode=scroll

GO-Gulf. 2013. "Smartphone Usage in the Middle East—Statistics and Trends." October 2. Accessed November 20, 2015. http://www.go-gulf.ae/blog/smartphone-middle-east/

Human Rights Watch. 2016. "Qatar: Events of 2015." Accessed April 10, 2016. https://www.hrw.org/world-report/2016/country-chapters/qatar#5b498e

Hume, Tim, and Elwazer, Schams. 2016. "Qatari Poet Accused of Insulting Emir Freed After 4 Years, U.N. Says." CNN. Accessed April 10, 2016. http://edition.cnn.com/2016/03/16/middleeast/qatar-poet-released/index.html

ictQATAR. 2013. "Qatar's ICT Landscape 2013." Accessed November 20, 2015. http://www.ictqatar.qa/sites/default/files/documents/Qatar%20ICT%20Landscape_EN.pdf

Internet World Stats. 2014. "Internet Usage in the Middle East." December 31. Accessed October 27, 2015. http://www.internetworldstats.com/stats5.htm#me

Kovessy, Peter. 2015. "Several Popular Websites Temporarily Blocked in Qatar." February 11, 2015. *Doha News.* Accessed November 23, 2015. http://dohanews.co/popular-websites-temporarily-blocked-qatar/

Northwestern University of Qatar. 2013. Media Use in the Middle East, 2013. Accessed October 27, 2015. http://menamediasurvey.northwestern.edu/

OpenNet Initiative. 2009. "Internet Filtering in Qatar," August 6. Accessed October 27, 2015. https://opennet.net/sites/opennet.net/files/ONI_Qatar_2009.pdf

Reyaee, Sulaiman, and Ahmed, Aquil. 2015. "Growth Pattern of Social Media Usage in Arab Gulf States: An Analytical Study," April 14. Accessed October 27, 2015. http://dx.doi.org/10.4236/sn.2015.42003

The Peninsula. 2011. "e-class." October 15. Accessed November 20, 2015. http://thepeninsulaqatar.com/news/qatar/169224/e-class

ROMANIA

Romania is a country located in southeastern Europe, with a geographical size of 238,391 square kilometers and a population of 19.8 million in 2015. The country has been a European Union (EU) member since 2007. Set in a traditional competition with the United States and Japan in the area of information and communication technologies (ICT), the European Union has been making serious efforts to become a world leader in the field. To this end, it has sponsored multiple programs and strategies that support and stimulate investment, research and development (R&D), and innovation in ICT. Digital Agenda, one of these initiatives, aims at bringing Europe out of the economic crisis that it has recently experienced and back on track to a smart, sustainable, and inclusive society; its general objective is to bring durable economic and social benefits, thanks to the digital single market based on fast and ultra-fast internet and interoperable applications. These strategies have been adapted and adopted in the twenty-eight member states, including Romania, and have imprinted on them a clear positive trend in ICT.

Some Romanian strengths in the ICT sector include a great interest and major accomplishments in cybersecurity (e.g. Bitdefender, a "visionary" endpoint protection platform, according to Gartner; and Roboscan, an advanced truck/plane scanning system designed and developed by a team of specialists at MBTechnology), functional online procurement/payment platforms at the governmental level (e.g. www.e-licitatie.ro, www.ghiseul.ro, and www.seap.ro), the presence of strong national and multinational ICT companies, and a well-developed ICT infrastructure in major cities, which is progressing in terms of wireless internet access in public spaces. In terms of electricity consumption, Romania is truly a green country: in 2014, the electricity generated from renewable sources (like hydroenergy, biomass, solar energy, wind energy and geothermal energy) represented 41.7 percent of gross electricity consumption (Eurostat 2014).

In Romania, 52 percent of individuals aged sixteen to seventy-four years old (more than 10 million people) use the internet regularly (on a daily or weekly basis), while 32 percent of the Romanian population has never used the internet (European Commission 2015). Further, 68 percent of Romanian households have internet access. Most connections (90 percent) are broadband, but there is a significant gap between urban and rural areas. Inhabitants in rural areas do not subscribe to internet access due to the costs of the necessary service or equipment, which are perceived as high, although prices are falling, especially for 3-in-1 packages (fixed internet data connection, cable TV, and landline). They access internet using mainly mobile phones and tablets (Stanca 2015). Internet is used mainly for sending/

receiving emails, finding information about goods and services, reading newspapers online, posting messages on online social platforms, searching for tourist services, and creating a website. Only around 10 percent of individuals aged sixteen to seventy-four use the internet for ordering goods or services, interacting with public authorities, or internet banking.

More than half of Romanians today have a personal subscription to one of the main four mobile operators. One of four Romanians use a prepaid card for mobile telephony. Mobility is also encouraged by high 3G (96 percent) and 4G (25 percent) coverage, as well as the recent but rapidly developing Long-Term Evolution (LTE) coverage, which causes the migration of increasingly tech-savvy consumers from wired access to mobile phones (Digi24 2015).

The penetration rate of smartphones among the population of Romania has exploded in recent years, currently reaching about 40 percent; the average download speed has doubled in recent years. The consumption of mobile data is similar, with 32 percent of Romanians using the internet on mobile devices as soon as they wake up (Ziarul Financiar 2015). However, low purchasing power, demographic structure of the population, and the low level of digital literacy of the older population translate into a still-reduced rate of mobile internet penetration.

Regarding digital skills, 85 percent of Romanians have low or no ICT skills, compared with the European average of 47 percent. Therefore, steps have been taken to promote e-inclusion, which will allow greater penetration of ICT in the daily life of citizens. Since 2005, ICT has been compulsory for all specializations and qualifications in theoretical and vocational high schools. Also, a mandatory exam that certifies the ICT skills of Romanian high school graduates has been in place since 2009. Meanwhile, over 70 percent of universities have implemented an e-learning platform, and most libraries (especially in rural areas) are equipped with computers.

Between 2005 and 2015, Romanians have been embracing social media with enthusiasm. As a result, in July 2016, there were 8.5 million Facebook users, which represent 42.5 percent of the population and 88 percent of all internet users (Facebrands 2016). The strong attraction to Facebook has transformed it into a second internet. Other social network sites used are WhatsApp, Twitter, YouTube, Instagram, and LinkedIn (Mediafax 2016). The cohesion and empathy expressed by citizens on social media increased considerably after a fire raged through Collectiv, a nightclub in Bucharest, killing 64 people and injuring 147 (Colectiv 2015); this unfortunate event caused mass anger against the government, its poor enforcement of safety regulations, and its tolerance of corruption. Social media is now used in a more conscious manner by its aficionados, who are convinced by the real power of expression and action that this media provides, and is also used in a more professional way by public institutions and private organizations.

Even after the pressure of the civic solidarity expressed by Romanians during the last several years, the presence of public institutions on social online platforms remains unremarkable. Most institutions have developed a type of unidirectional communication that reports information but does not give citizens the capacity to provide comments and feedback in real time. The e-Government services generally satisfied users who appreciate the results of interaction with the administration, such as the usefulness of information, the ease of finding that information, ease of use, and the transparency of services provided (Georgescu and Popescul 2014). The main e-Government processes by the city halls that have moved into the virtual environment include the following: "payment transactions of debts to the local public administration; filling in forms; searching for documents (functioning certificates, of taxi companies etc.); booking city hall services (hearings, marriages etc.)"; information delivery; and discussions with citizens (Popescul 2012, 239).

Despite having a presence on Web 2.0, the city halls are not putting enough effort into using the web to interact with citizens. This shortfall most likely comes from the lack of competition in the environment where they operate, as competition often drives innovation and interaction (Popescul 2012). City hall has tried a number of initiatives, such as having a YouTube channel and creating web content accessible on mobile devices, but these strategies have not yet been integrated into a municipal strategy to improve relations with the population at large (Popescul 2012). The role of social network sites was understood better by supporters of certain poltical parties than by the parties themselves (Tasențe 2014). Interaction with candidates online in a colloquial manner is now possible, and offers the potential to provoke discussion and elicit reactions. These media have the power to attract young people to express their opinions.

Since 2004, blogging in Romania has grown in popularity, becoming one of the most common formats for self-expression and communications in social media. A series of studies show the rise of Romania's blogging phenomenon and the effects that blogs can have on their readers (Tomiuc and Stan 2015). By 2014, according to the *Zelist Monitor,* blogging activity reached over 92,000 blogs, with 18.5 million posts and over 15.7 million comments (Zelist 2016). A total of 56 percent of Romanians read blogs, which cover an extensive area of subjects: personal interest, entertainment, politics, tourism, news, sports, education, economics, new media and technology, environment, music, film, lifestyle, and other diverse subjects (Tomiuc and Stan 2015).

The main reason for students' social media use is to foster social interaction with old or new friends, as social relationships with other persons are universal human needs critical to the psychological well-being of individuals, and these needs are very well addressed by social media. Other reasons for use that students mentioned,

---

In mid-2015, Swiftkey, a social media app developer, released a two-part report on emoji use around the world. Their findings for Eastern Europe revealed that there are varying preferences concerning the most popular emoji. Croatian-, Hungarian-, and Serbian-language users utilized more hearts than any other emoji, while Czech, Romanian, and Slovakian users preferred smiley faces. Additional results showed that Hungarians tend to avoid the thumbs-up emoji, while most of the other languages steer away from sad or strange faces. Swiftkey's reports analyzed over a billion posts using emoji to produce these results (Swiftkey 2015, 11–13).

in descending order of importance, were listening to music, seeking out academic information, passing time when idle or bored, education, reading posts/comments, playing games, looking for new friends, shopping, expression of opinions, fighting loneliness, surveillance/knowledge about others, and self-promotion. Even if the entertaining aspects prevail, the academic reasons are also very well represented as many students seek out school-related information in social media frequently; also, they often use social media for educational purposes (Georgescu and Popescul 2016). There is a difference in students' preferences for one media site or another. The uncontestable Romanian students' favorite is Facebook (checked hourly), followed by YouTube (for listening to music, watching entertainment videos, and watching educational videos). Blogs and Wikipedia are read regularly, and a few students use them for posting, commenting, and editing. Instagram is used by almost half of students. A few students occasionally use podcasting, Twitter, Really Simple Syndication (RSS) feeds, Flickr, LinkedIn, or MySpace.

Three-quarters of the companies in Romania use social networks to promote business, and Facebook is the number one platform for communication in social media, as revealed in a study by EY Romania. Although the number of users is growing, the presence and actions of brands in social media has somehow stabilized since 2015. Facebook is preferred by 92 percent of companies; 44 percent have social media initiatives at the marketing level; and 21 percent have a social media strategy integrated into the global company strategy (Badea 2015). Their actions include company/brand campaigns on blogs, campaigns by video bloggers, Facebook pages/campaigns, Instagram/Twitter/LinkedIn/Google+/Snapchat/Tumblr/Flickr accounts, and the initiation or support of events in the blogosphere, especially with video bloggers. Banks are becoming more of an online presence and invest in social media, primarily to launch specific services to reach consumers.

Daniela Popescul

See also: Germany; Greece; Hungary; Italy

Further Reading

Badea, D. 2016. "74% Dintre Companiile din România Se Promovează Pe Rețele de Socializare, în Special pe Facebook (studiu)." February 24. Accessed June 2, 2016. http://www.agerpres.ro/economie/2016/02/24/74-dintre-companiile-din-romania-se-promoveaza-pe-retele-de-socializare-in-special-pe-facebook-studiu—12-14-22

Bitdefender. "Digi24. 2015. Studiu de Piață: Ce Operatori de Telefonie Mobilă Apreciază Românii în Prezent." June 9. Accessed June 23, 2016. http://www.digi24.ro/Stiri/Digi24/Actualitate/Stiri/Studiu+de+piata+Ce+operatori+de+telefonie+mobila+apreciaza+roman

European Commission. 2015. "Digital Agenda Scoreboard Key Indicators." Accessed March 22, 2016. http://digital-agenda-data.eu/datasets/digital_agenda_scoreboard_key_indicators/visualizations

Eurostat. n.d. "Your Key to European Statistics." Accessed June 9, 2016. http://ec.europa.eu/eurostat.

Facebrands. 2016. "Date demografice Facebook Romania." July 26. Accessed August 30, 2016 http://www.facebrands.ro/demografice.html.

Georgescu, M., and Popescul, D. 2014. "The Uncertainty of Using Web 2.0 Technologies in E-Government Development. Romania's Case." *Procedia Economics and Finance*, 15: 769–776.

Georgescu, M., and Popescul, D. 2016. "Students in Social Media: Behavior, Expectations, and Views." *Proceedings of the IE 2015 Conference*, Cluj Napoca, Romania.

Idea Connection. 2009. "Roboscan." Accessed August 30, 2016. https://www.ideaconnection.com/invention-success/ROBOSCAN-00062.html

Mediafax. 2016. "Peste Jumătate Dintre Liderii Companiilor, Prezenți pe Rețele de Socializare: România Este o țară Care Are o Cultură de Leadership ce Poartă încă Amprenta Tranziției." April 14. Accessed June 13, 2016. http://www.mediafax.ro/economic/peste-jumatate-dintre-liderii-companiilor-prezenti-pe-retele-de-socializare-romania-este-o-tara-care-are-o-cultura-de-leadership-ce-poarta-inca-amprenta-tranzitiei-15251951

Popescul, Daniela. 2012. "Process Innovation in the Romanian Public Administration: Observations and Recommendations," *Lucrări Științifice. Seria Agronomie*, 55: 239–243.

Stanca, A. 2015. "Peste 10 Milioane de Români au Acces la Internet. Care Este Profilul Utilizatorului Din Mediul Rural." March 5. Accessed June 22, 2016. http://www.gandul.info/it-c/peste-10-milioane-de-romani-au-acces-la-internet-care-este-profilul-utilizatorului-din-mediul-rural-13936734

Tasențe, T. 2015. "The Electoral Campaign Through Social Media. Case Study—2014 Presidential Elections in Romania." *Sfera Politicii*, XXIII, 1(183): 92–104.

Tomiuc, A., and Stan, O. 2015. "The Fashion Blogosphere in Romania. Fashionscape and Fashion Bloggers." *Postmodern Openings*, 6 (1): 161–174.

Zelist. http://www.zelist.ro/

Ziarul Financiar. 2015. "Aplicațiile Mobile au Devenit o Normalitate. Peste 30% Din Români Folosesc Internetul pe Mobil Imediat ce se Trezesc." April 22. Accessed July 27, 2016. http://www.zf.ro/business-hi-tech/aplicatiile-mobile-au-devenit-o-normalitate-peste-30-din-romani-folosesc-internetul-pe-mobil-imediat-ce-se-trezesc-14135443

RUSSIA

Russia, or the Russian Federation (RF), emerged from the ruins of the Union of Soviet Socialist Republics (USSR), or the Soviet Union. After the USSR's first and last president, Mikhail Gorbachev (1931–), publicly resigned his office on December 25, 1991, the RF ended up as the world's largest country measured by geography—just over 17 million square kilometers, stretching from Eastern Europe to the Pacific Ocean—with a population of between 144 and 146 million as of 2016 (World Bank 2015; Russian Federal Service of State Statistics n.d.). Geographically, RF is nearly twice the size of the United States. The RF also inherited the USSR's dilapidated industrial, transportation, institutional, and technological infrastructure. The development of information technology (IT), from mass media to telephony to the Russian Internet (RuNet) and social media, has shaped RF's social and political history from Boris Yeltsin (1931–2007) through Dmitry Medvedev (1965–) to current RF president Vladimir Putin (1952–). Whether the internet will provide an avenue for liberal democracy or greater centralized federal authority in Russia remains to be seen (Gorham 2014; Greene 2014).

The USSR's quiet dissolution surprised Western analysts, but nearly all of them acknowledge the importance of media. It had a vast nuclear arsenal, extensive storehouses of chemical and biological weapons with the requisite delivery mechanisms, and more than 5 million soldiers, along with additional Commissariat of State Security (KGB) and interior ministry battalions. The USSR was a near-peer to the United States in military and technological capabilities. It also had an astounding number of soldiers and police at its disposal. However, these forces never experienced a major mutiny, nor were they employed to stave off the collapse. Surprisingly, the USSR disappeared quietly (Kotkin 2001). By contrast, civil wars in the post-Soviet space occurred in nominally independent nation states after the collapse of the USSR, the dynamics of which were shaped to a large degree by the explosion of independent mass print and television media (Beissinger 2002). Since Putin's surprising ascension to the presidency on December 31, 1999, mass media in general, and the RuNet in particular, have provided avenues for dissent and repression as RF finds its way from Soviet political life (McFaul 2001). RF's IT infrastructure is new, but the visions for its use have roots in the USSR.

Like the space industry, the USSR's IT sector emerged immediately after World War II. However, unlike the space race, which the USSR won over the United States with Yuri Gagarin's (1934–1968) trip around the Earth aboard the *Vostok 1* on April 12, 1961, the Soviet Union lost the IT race. Theoretically and chronologically, both the USSR and the United States shared the same point of departure. Through economic exchanges during World War II, Soviet economic planners and mathematicians encountered the ideas of the father of modern IT, Norbert Wiener (1894–1964), the son of a Russian Jewish émigré. During World War II, Wiener developed mathematical theories and techniques for controlling chance in support of U.S. wartime military projects on fire control, cryptology, and basic military command and control. In 1948, Wiener codified them by coining the term *cyber* from the ancient Greek for "govern" because he saw the potential for digitized information structuring and dissemination to govern the world beyond the military (Weiner 1965, 11–12).

Soviet researchers and academics quietly reinvented Wiener's pioneering field after the death of Joseph Stalin (1878–1953). Privately, the Soviet/Ukrainian mathematician Viktor Glushkov (1923–1982), who became aware of Wiener's ideas while on an exchange in the United States, developed his concept of cybernetics. Then in 1962, he publicly extended cybernetics to the Soviet context in an article for the *Literary Gazette* entitled "Cybernetics, Progress, and the Future." Economic planners found the possibility of obtaining reliable and timely information about Soviet production and consumption essential for accurate planning purposes. For political reasons, however, the Soviet Union never developed any digital network comparable to the U.S. military's Advanced Research Projects Agency Network (ARPANET), whose foundational infrastructure originated in 1950s automated air defense systems designed to detect Soviet nuclear intercontinental ballistic missile (ICBM) attacks and which laid the groundwork for the modern internet through the 1960s, 1970s, and 1980s (Gerovitch 2002; Peters 2016).

The contemporary Russian government inherited the Soviet dilemma that free access to information posed: namely, the possibility that the freedom, social justice, and economic growth that it might promote could undermine political security, order, and stability in the country. Russia has unevenly pursued both strategies since

2000. Roughly speaking, President Putin, who headed the Russian government from 2000 to 2008 and, after a constitutionally mandated break, again from 2012 to the present, regards IT development as a national security concern. By contrast, President Medvedev promoted IT research and development as a source of economic development during his term in office from 2008 to 2012.

This distinction between Putin's and Medvedev's policy preferences, however, should not be drawn too sharply. Putin's now-famous lament that he "really regretted that the Soviet Union had lost its position in Eastern Europe" probably reflects the USSR's loss of legitimate authority, which could have checked much of the violence that erupted throughout the former Soviet Union in the 1990s (Putin 2000, 80). What is less well remembered is that Putin also understood, at least theoretically, that "a position built on walls and dividers," which the internet undermines, "cannot last" (Putin 2000, 80). These tensions have shaped popular Russian use of the internet.

Putin's national security staff has consistently emphasized the national security dimension of IT and information. As early as September 9, 2000, this principle was codified in the "Russian Federation's Doctrine of Information Security." The Russian government reiterated the primacy of security in the IT and information realms in the "Development Strategy of an Information Society in the Russian Federation," released on February 7, 2008. The role of information security and an emphasis on cyber-sovereignty are part of the broader "National Security Strategy of the Russian Federation to 2020," published on May 12, 2009.

The RF government has expressed its national security concerns particularly in relation to Ukraine's "Orange Revolution" in 2004. In this uprising, Ukrainian citizens used social media to organize public protests against the Party of Region's candidate, Viktor Yanukovych (1950–), who they thought had won a rigged presidential election. RF still publicly claims that foreign governments were somehow behind the protests. There are also reports that Russian security services, or people acting on behalf of the security services, launched cyberattacks against the Estonian government and financial websites in April 2007 and against Georgian websites immediately preceding the Georgian War in August 2008. Since RF's annexation of Crimea in February 2014, social media sites have been a significant component of news, propaganda, and government messaging in Kyiv (Kiev) and Moscow. Despite evidence of Russian cyber-offenses, this national security policy has tended to be conservative and defensive (Greene 2014).

Alongside the national security dimension of IT and digital media, RF has also pursued internet growth for economic development, freedom, and justice. During his presidency from 2008 to 2012, Medvedev promoted this line of development with his September 10, 2009, pamphlet "Go Russia!" This document proposed a series of measures to stimulate IT innovation, including e-governance, education programs, the development of a national grid of supercomputers, and Skolkovo, a 600-acre plot of land in Moscow's western suburbs. The innovation hub in Skolkovo was intended to be "something on the lines of Silicon Valley" (Appell 2015). Medvedev reiterated his commitment to IT and information development in his "Address to the Federal Assembly" on November 12, 2009. At first, the Skolkovo initiative was successful. The Medvedev government negotiated a partnership between Skolkovo and the Massachusetts Institute of Technology (MIT); Cisco Systems provided a $100 million investment; and the number of tech companies grew from 332 in 2011, to 793 by 2012, and more than 1,000 in 2013 (Appell 2015). Since the first months of 2014, however, growth has stalled (Appell 2015).

Between these two broad policy agendas, Russian IT and digital technologies have thrived. In the summer of 2003, Russians eighteen years of age and older who used the internet daily comprised 3 percent of the population, or 3.1 million users. As of fall 2015, 63.9 million Russians eighteen years of age or older used the internet daily (FOM 2015). The rate of growth has slowed since the summer of 2014, probably due to market saturation and infrastructure limitations in rural Russia (FOM 2015). The percentage and number of Russians eighteen years of age or older who use the internet weekly or monthly has grown even more than daily usage, to 74.4 million users and 78.2 million users, respectively (FOM 2015). The internet penetration rate as a percentage of the 78.2 million monthly users, according to region, is as follows: (1) the Central region, including Moscow—28.4 percent; (2) the Trans-Volga region—19.6 percent; (3) the Southern and North-Caucasus regions (combined)—15.8 percent; (4) the Siberian region—12.9 percent; (5) the Northwest region, including Saint Petersburg—10.8 percent; (6) the Ural region—8.3 percent; and (7) the Far Eastern region—4.2 percent (FOM 2015).

Russia's internet backbone enables these penetration rates. Data available from Russian and European internet exchange point (IXP) websites (i.e. euro-ix.net and ix.ru) as of December 2015 indicate that seventeen cities host IXPs: Moscow, Saint Petersburg, Yekaterinburg, Yaroslavl, Ulyanovsk, Krasnodar, Saratov, Perm, Omsk, Krasnoyarsk, Nizhniy Novgorod, Rostov-on-Don, Stavropol, Samara, Kazan, Novosibirsk, and Vladivostok. IXPs are centralized hubs that route data from PCs through internet service provider (ISP) networks of varying size or autonomous systems (ASs) around the world (Wilson 2015). Skolkovo's results may be mixed, but the Russian government under President Medvedev certainly expanded Russia's internet infrastructure, which was a precondition for its current internet penetration rates.

RF's internet audience consumes a vast array of foreign and domestic online resources, many of which have mobile apps. Given differences in language and alphabet, foreign social media and internet platforms have interwoven themselves with indigenous offerings. As of late 2015, Russia's top ten websites were Yandex.ru, Vkontakte (VK), Google.ru, Mail.ru, YouTube, Google.com, Odnaklassniki (OK), Facebook, Aliexpress.com, and Avito.ru (Alexa.com). Yandex is the most popular search engine in the RuNet, while VK and OK are the most popular social networking platforms. Aliexpress is an online merchant site, and Avito is a Russian equivalent of Craigslist. Russian news sites, the blogging platform LiveJournal, Instagram, eBay, and Twitter all fall somewhere in the top twenty-five websites in the RuNet (Alexa.com). Worldwide, Yandex and VK come in at twenty and twenty-one, respectively, in terms of frequency of use (Alexa.com).

These platforms and services are owned by a small number of Russian IT and media companies. According to an early 2014 ratings list, Yandex was Russia's largest internet company; as Google did until recently, the company has developed

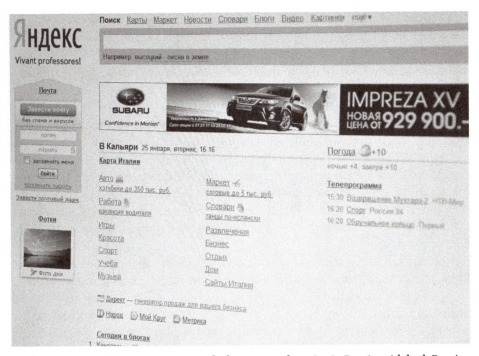

Yandex, a Russian IT company, operates the largest search engine in Russia, with both Russian and English language search capabilities. Launched in 1997 when the internet was fairly new in Russia, Yandex.ru created its domain name from the English phrase Yet Another iNDEXer. Over time, it has grown to become Russia's most popular website. (Raluca Tudor/Dreamstime.com)

numerous services and search tools under one name (Forbes Russia 2014). The Mail.ru Group, which owns VK, OK, Mail.ru, and the instant messaging services Agent Mail.ru and ICQ (an Israeli-developed messaging service later bought by the Mail.ru group), is the second largest. The fifth company on the list, Rambler&Co, owns Russia's second search engine, Rambler, the popular news sites Lenta.ru and Gazeta.ru, and the important blogging platform LiveJournal. LiveJournal, which is sometimes rendered as its Russian abbreviation ZhZh (short for *zhivoi zhurnal*), was developed in the United States and bought by Rambler&Co in 2007. Interestingly, because the servers for Facebook, Twitter, LiveJournal, and Instagram are located outside RF, the Russian government faces limits on monitoring and censoring their users, which include the opposition blogger Alexei Navalny (1976–), as well as the president of the Republic of Chechnya, Ramzan Kadyrov (1976–).

After Russia's major cities experienced social media–organized protests against President Putin's reelection in December 2011, media coverage in Europe and North America became focused on the Russian government's filtering and censoring policies. Since the February 2014 annexation of Crimea, the media have added "information warfare" to the list of the Russian government's strategies of repression and manipulation. This critical coverage is not surprising because Russia's government aspires to maintain the ability to monitor internet traffic and even shut off sections or specific pages of the RuNet (Diebert et al. 2008; Wilson 2015).

To some extent, this impulse for surveillance and control is rooted in genuine law enforcement attempts to fight cybercrime, which is a serious problem throughout the former Soviet Union. Between 2012 and 2015, this criminal activity originating in Russia, Ukraine, Kazakhstan, and some Baltic states caused financial damages that exceeded $790 million, $509 million of which was stolen outside the boundaries of the former Soviet Union (Stoyanov 2015). However real and legitimate this threat from the internet may be, the Russian government also executes politically motivated internet outages of political and social dissidents. In Europe and North America, the best-known victims of such actions are probably the nationalist Kremlin critic and blogger Alexei Navalny (1976–), and the performance art group Pussy Riot; the socialist activist Sergei Udal'tsov (1977–) is often forgotten, as he is serving a jail sentence for allegedly using violence against police during the May 2012 protests in Moscow against Putin's inauguration (Greene 2014).

Neither censorship and repression nor 2011's protests tell the whole story of the Russian internet and social media, however. One of the first political scandals to occur in Russia occurred in the midst of uncontrollable fires in the rural areas around Moscow from late July to early August 2010. As a result of a complex and poorly understood relationship between the federal and local governments, rural institutions were unprepared to fight the fires. The damage to government and private property was immense, and some regions lost significant crop harvests. In response to this emergency, a LiveJournal blogger with the user name "top-lap" posted a long, profanity-laden letter on August 1, 2010, criticizing then–Prime Minister Putin and the central government for not rooting out local corruption. Then top-lap alleged that the inability of local authorities to locate, much less use, firefighting equipment was the cause of the property damage (top-lap 2010). This post probably would have remained unread in the vast sea of LiveJournal pages had a semi-independent radio station, Echo of Moscow, not reposted the letter to its website. From then on, it drew so many views and became enough of a scandal for Putin to take the unprecedented step of answering the blogger with a letter that Echo of Moscow also posted on its site. After noting the difficulties that record-high temperatures posed to the government, Putin answered the "surprisingly open and direct" author of the post that he "on the whole agreed with [his] remarks" about local political corruption (Putin 2010).

Russia employs hundreds, perhaps even thousands, of people for "internet research," where they write a daily quota of blogs and other social media posts that attack world leaders and political ideals that do not promote pro-Russian sentiment. These troll farms, sometimes referred to in the press as "troll armies," keep blogs on LiveJournal, a site that at one time was used heavily by Western bloggers and has since become popular in Russia. They also post to forums, vKontakte, Twitter, YouTube, Facebook, and Instagram, among other platforms, to promote pro-Russian government messages (Chen 2015).

Since 2010, RF's policy toward internet-based or internet-facilitated dissent has been less conciliatory. During Putin's presidential campaign from December 2011 until his election in March 2012, the government used internet censorship and surveillance together with traditional police tactics to control the "Moscow Spring," a term that deliberately echoes the more familiar "Arab Spring" uprisings from 2010 to 2012 in the Middle East, and describes the public protests in Moscow against the results of parliamentary and presidential elections in late 2011 and mid-2012, respectively. Although its digital organizers deliberately linked their movement to the uprisings in the Middle East, the use of LiveJournal, Facebook, forums, TVRain's YouTube channel, and VK for political organization had originated with a nationalist movement protesting the portion of the state budget assigned to subsidies for the North Caucasus. The radical nature of the Moscow Spring activist political demands is still under dispute, but demographically and geographically, the movement did not extend very far outside the cities of Moscow and Saint Petersburg. The government's ability to shape media has certainly fostered vast support for Russia's annexation of Crimea. However, much of that support likely comes from Putin's rural base that, until February 2014, did not have the same digital access as the Muscovites.

Sean Gillen

See also: Estonia; Georgia; Israel; Ukraine; United States

Further Reading

Alexa.com. 2016. "Top Sites in Russia." Accessed July 30, 2016. http://www.alexa.com/topsites/countries/RU

Appell, James. 2015. "The Short Life and Speedy Death of Russia's Silicon Valley." *Foreign Policy,* May 6. Accessed July 30, 2016, http://foreignpolicy.com/2015/05/06/the-short-life-and-speedy-death-of-russias-silicon-valley-medvedev-go-russia-skolkovo/

Beissinger, Mark. 2002. *Nationalist Mobilization and the Collapse of the Soviet State.* Cambridge, U.K.: Cambridge University Press.

Deibert, Ronald, Palfrey, John, Rohozinski, Rafal, and Zittrain, Jonathan (Eds.). 2008. *Access Denied: The Practice and Policy of Global Internet Filtering.* Cambridge, MA: MIT Press.

Forbes Russia. 2014. "10 krupneishikh internet-kompanii Rossii: Reiting Forbes." *Forbes,* February 27. Accessed July 30, 2016. http://www.forbes.ru/reitingi-photogallery/251148-10-krupneishikh-internet-kompanii-rossii-reiting-forbes/photo/1

Fond Obshchestvennoe Mnenie (FOM). 2015. "Internet v Rossii: Analiticheskii Biulleten' Demo." Accessed July 3, 2016. http://fom.ru/uploads/files/Бюллетень_Интернет_в_Россиию_Выпуск_51._Осень_2015_-_демо.pdf

Gerovitch, Slava. 2002. *From Newspeak to Cyber Speak: A History of Soviet Cybernetics.* Cambridge, MA: MIT Press.

Gorham, Michael S. 2014. *After Newspeak: Language Culture and Politics in Russia from Gorbachev to Putin.* Ithaca, NY: Cornell University Press.

Greene, Samuel. 2014. *Moscow in Movement: Power and Opposition in Putin's Russia.* Stanford, CA: Stanford University Press.

Kotkin, Stephen. 2001. *Armageddon Averted: The Soviet Collapse, 1970–2000.* New York: Oxford University Press.

McFaul, Michael. 2001. *Russia's Unfinished Revolution: Political Change from Gorbachev to Putin.* Ithaca, NY: Cornell University Press.

Peters, Benjamin. 2016. *How Not to Network a Nation: The Uneasy History of the Soviet Internet.* Cambridge, MA: MIT Press.

Putin, Vladimir. 2000. *First Person: An Astonishingly Frank Self-Portrait by Russia's President.* Translated by Catherine A. Fitzpatrick. New York: Public Affairs Reports.

Putin, Vladimir. 2010. "Vladimir Putin: Otvet." *Ekho Moskvy,* August 4. Accessed July 30, 2016. http://echo.msk.ru/doc/700728-echo.html

Russian Federal Service of State Statistics. n.d. "Russian Federal Service of State Statistics." Accessed July 30, 2016. http://www.gks.ru/wps/wcm/connect/rosstat_main/rosstat/ru/statistics/population/demography/#

Stoyanov, Ruslan. 2015. "Russian Financial Cybercrime: How It Works." Kaspersky Lab's SecureList, November 19. Accessed July 20, 2016. https://securelist.com/analysis/publications/72782/russian-financial-cybercrime-how-it-works/

top-lap. 2010. "Znaete pochemu gorim?" August 1. Accessed July 30, 2016. http://top-lap.livejournal.com/1963.html

Wiener, Norbert. 1965. *Cybernetics: Or Control and Communication in the Animal and Machine Worlds,* 2nd ed. (orig. 1948). Cambridge: MIT University Press.

Wilson, Steven Lloyd. 2015. "How to Control the Internet: Comparative Political Implications of the Internet's Engineering." *First Monday,* 20(2). Accessed July 30, 2016. http://journals.uic.edu/ojs/index.php/fm/article/view/5228/4204

World Bank. 2015. "Population Total." Accessed December 21, 2015. http://data.worldbank.org/indicator/SP.POP.TOTL

S

SAUDI ARABIA

The Kingdom of Saudi Arabia is the largest country in the Arabian Peninsula, located between Africa and Asia. As of December 2014, its internet penetration rate was 65.9 percent of its population of 28 million, the eighth highest internet penetration rate in the Middle East (Internet World Stats 2015). A total of 65 percent of the country's internet users access the web via smartphone (Nielsen 2014), 70 percent of the population is under thirty (Black 2013), and 29 percent actively use social media (Fraij 2015). In 2014, Saudi Arabia had the world's highest Twitter penetration, at 60 percent of the country's internet users. It also accounted for over 40 percent of the active Twitter users among the 22 Arab countries, plus Iran, Israel, and Turkey, and the Saudis produced 47 percent of all tweets in the region (Arab Social Media Report 2014).

In a 2015 survey of approximately 3 million Saudis aged eighteen to twenty-five years old, the most popular social media platforms were WhatsApp (used by 22 percent of the respondents), Facebook (21 percent), and Twitter (19 percent). That same year, Saudi Arabia accounted for more than 90 million daily YouTube views, placing it among the world's top countries for viewership. Half of Saudi YouTube users are women (Fraij 2015). In 2009, a local company estimated that two-thirds of Saudi internet users were women (Alqudsi-ghabra, Al-Bannai, and Al-Bahrani 2011). At least 300,000 Saudis subscribe to UTURN, a domestic social media company airing a variety of shows on YouTube (The Economist 2014). A total of 90 percent of Saudi internet users interact via Arabic, 8 percent use English, and the remainder communicate in French or other languages (Fraij 2015).

Saudi Arabia's deeply conservative culture presents unique challenges in the information technology (IT) context. The government enforces an austere version of Sunni Islam, with which many citizens identify. Restrictions on Saudis' ability to communicate freely were evident in 2010, when the government blocked BlackBerry's then-popular messaging service (Ahmed 2013). Many users migrated to social media, which provided "virtual" freedom of association and assembly (Black 2013). Twitter experienced a 39.94 percent increase in its Saudi market share during 2012–2013 (Reyaee and Ahmed 2015). Perhaps not coincidentally, Saudi billionaire Prince Alwaleed bin Talal (1955–) was one of Twitter's largest investors, controlling a $300 million stake (Ahmed 2013).

Political parties are illegal in Saudi Arabia; activism can result in arrest, and self-censorship is common among journalists (OpenNet Initiative 2009). The state-owned Saudi Telecom Company, created in 1988, is the primary provider of telecommunications services. Twenty-five internet service providers operate under licenses granted by the Ministry of Culture and Information's Communications and Information Technology Commission. All of them access the internet through the same centralized state portal (Faris 2013). The King Abdulaziz City for Science and Technology (KACST) monitors internet content and imposes censorship, denying access to websites containing materials deemed objectionable on moral, political, or security grounds. In 2005, a KACST official said that the majority of blocked websites were pornographic and that over 90 percent of Saudi internet users had attempted to access blocked websites. KACST relies on citizens to help identify sites to be blocked (OpenNet Initiative 2009).

In 2008, Saudi Arabia implemented a law, Royal Decree No. M/17, governing the use of IT. It imposed fines and imprisonment for website operators who advocate or support terrorism. It also criminalized other cybercrimes, including financial fraud, invasion of privacy, and the distribution of pornography or other materials contrary to the country's laws or religious values and social standards (OpenNet Initiative 2009). A number of Saudis and foreigners have been prosecuted under this law.

In 2014, a 24-year-old gay Saudi man was arrested for tweets he sent to other gay Twitter users. Saudi Arabia's religious police, the Commission for the Promotion of Virtue and the Prevention of Vice (CPVPV), entrapped the man by arranging for an undercover investigator to meet him after agreeing on Twitter to a date with him. After arresting the man, the CPVPV reportedly found several "indecent" and "homosexual" images on his phone. The authorities sought the death penalty, but the man pleaded guilty and was sentenced to three years in prison and 450 lashes administered over fifteen sessions (Williams 2014). Also in 2014, a Saudi court, citing the 2008 law, sentenced human rights activist Fowzan al-Harbi (1978–) to seven years in prison for disseminating information "harmful to public order" (The Economist 2014).

The internet is contested space in Saudi Arabia. The government uses it to monitor public opinion and sources of dissent. In 2016, the special court for terrorism sentenced an unidentified Saudi to 10 years in prison for using Twitter to "call for protests and spread chaos to release detainees that are held for security and terrorism charges" (Agence France Presse 2016). In 2014, Human Rights Watch reported that authorities in Qatif, an eastern governorate with a high concentration of minority Shia Muslims, were employing spyware (The Economist 2014). In 2013, a Saudi judicial official warned citizens abroad that they, too, could be punished for violating the country's cyber laws (Malouf 2013). Government and clerical figures regularly complement official television broadcasts and mosque sermons with social media messaging aimed at reinforcing the country's conservative political and religious orthodoxy. Saudi Arabia's senior religious authority, Abdul Aziz al-Sheikh (1941–), described Twitter users as "a bunch of clowns" and accused them of using the application "to corrupt values and to spread lies and rumors" (Perlov and Guzansky 2014).

Liberal activists have used the internet to call for improving the status of women and minorities. While promoting greater participation by women and youth within

A woman kneels on the ground in Riyadh in March 2011, while surrounded by police who had been deployed to deal with protests calling for democratic reforms. Saudi Arabia's restrictive policies extend to the internet and, while citizens rarely use it for political purposes, it serves as a tool for the government to track public opinion and identify political dissenters. In 2016, an unidentified Saudi was sentenced to 10 years in prison for using Twitter to call for protests. (AP Photo/Hassan Ammar)

existing power structures, they have stopped short of explicitly demanding the implementation of democracy. Radical Saudi clerics, propagators of anti-Western, anti-Shia, and anti-Semitic views, are among the country's most popular internet personalities. Three leading preachers, Salman al-Ouda (1955–), Muhammad al-Arifi (1970–), and Ahmed al-Shugairi (1973–), boast 5–7 million social media followers each (Perlov and Guzansky 2014).

As for ordinary Saudis, a 2013 Saudi survey indicated that while only 8 percent used the internet for political purposes, 20 percent of citizens relied on it for reading and watching religious content (Perlov and Guzansky 2014). Matchmaking is another popular online activity in a country that maintains strict gender segregation in public spaces. In 2014, a female Saudi Twitter user described meeting her husband through the application, noting that her mother—who, along with her brothers, actively monitored her account—approved of the arrangement. Surprisingly, the CPVPV has sanctioned such practices, with its former head, Sheikh Ahmed bin Qassim al-Ghamdi (n.d.–), characterizing social media as "one of the permissible means to be used in all legitimate matters in the lives of people, whether they are men or women" (Al-Hayat 2014).

Mark A. Caudill

See also: Qatar; Syria; United Arab Emirates

Further Reading

Agence France Presse, 2016. "Saudi Gets 10 Years for Twitter 'Terrorism' Support." February 2. Accessed March 4, 2016. https://www.dailystar.com.lb/News/Middle-East/2016/Feb-02/335245-saudi-gets-10-years-for-twitter-terrorism-support.ashx

Ahmed, Qanta. 2013. "Saudi Arabia's Struggles with Social Media: Twitter Clowns and Facebook Fatwas." April 11. Accessed October 21, 2015. http://www.theblaze.com/contributions/saudi-arabias-struggles-with-social-media-twitter-clowns-and-facebook-fatwas/

Al-Hayat. 2014. "Social Media Plays a Matchmaking Role in Saudi Arabia." March 9. Accessed October 21, 2015. http://www.al-monitor.com/pulse/culture2014/03/social-media-saudi-breaking-barriers.html

Alqudsi-ghabra, Taghreed M., Al-Bannai, Talal and Al-Bahrani, Mohammad. 2011. "The Internet in the Arab Gulf Cooperation Council (AGCC): Vehicle of Change." *International Journal of Internet Science,* 6(1): 44–67.

Arab Social Media Report, 2014. "Twitter in the Arab Region." March. Accessed October 27, 2015. http://www.arabsocialmediareport.com/Twitter/LineChart.aspx?&PriMenuID=18&CatID=25&mnu=Cat

Black, Ian. 2013. "Saudi Digital Generation Takes on Twitter, YouTube . . . and Authorities." December 17. Accessed October 21, 2015. http://www.theguardian.com/world/2013/dec/17/saudi-arabia-digital-twitter-social-media-islam

Faris, David. 2013. *Dissent and Revolution in a Digital Age: Social Media, Blogging, and Activism in Egypt,* p. 17. London: Taurus.

Fraij, Ibrahim. 2015. "The State of Saudi Arabia Social Media." February 22. Accessed October 21, 2015. https://linkedin.com/pulse/state-saudi-arabia-social-media-ibrahim

Internet World Stats. 2014. "Internet Usage in the Middle East." December 31. Accessed October 27, 2015. http://www.internetworldstats.com/stats5.htm#me

Malouf, Alex. 2013. "Big Brother and Social Media in Saudi Arabia." October 24. Accessed October 21, 2015. http://www.yourmiddleeast.com/features/big-brother-and-social-media-in-saudi-arabia_18928

Nielsen. 2014. "Smartphones Dominate the Saudi Market." June 17. Accessed January 29, 2017. http://www.nielsen.com/sa/en/press-room/2014/smartphones-driving-mobile-sales-in-saudi-arabia.html

OpenNet Initiative. 2009. "Internet Filtering in Saudi Arabia." August 6. Accessed October 27, 2015. https://opennet.net/sites/opennet.net/files/ONI_SaudiArabia_2009.pdf

Perlov, Orit, and Guzansky, Yoel. 2014. "The Social Media Discourse in Saudi Arabia: The Conservative and Radical Camps Are the Dominant Voices." February 5. Accessed October 21, 2015. http://www.inss.org.il/index.aspx?id=4538&articleid=6563

Reyaee, Sulaiman, and Ahmed, Aquil. 2015. "Growth Pattern of Social Media Usage in Arab Gulf States: An Analytical Study." April 14. Accessed October 27, 2015. http://dx.doi.org/10.4236/sn.2015.42003

The Economist. 2014. "Social Media in Saudi Arabia: A Virtual Revolution." September 13. Accessed October 21, 2015. http://www.economist.com/node/21617064/print

Williams, Steve. 2014. "Saudi Arabia Is Using Social Media to Entrap LGBTs." July 31. Accessed October 21, 2015. http://www.care2.com/causes/saudi-arabia-is-using-social-media-to-entrap-lgbts.html

SENEGAL

Senegal is a West African nation bordered by Mauritania, Mali, Guinea, Guinea-Bissau, and almost bisected by the Gambia. It had approximately 14 million people at the end of 2014, with an estimated 3 million of those having regular internet access. The country also has a unique relationship with the internet, in that it ranked as the top African nation in terms of internet gross domestic product (iGDP), meaning that transactions over the internet contributed more to the country's total economy than for any other country on the continent. In Senegal, internet sales and services comprised around 3.3 percent of national GDP (McKinsey & Co. 2013). As the internet expands in Africa and Senegal, more people will conduct business, both personal and corporate, online. Senegal's current internet and social media usage will adapt as the country experiences higher levels of internet use and penetration.

In Senegal, as in many parts of the world, social media is becoming a part of everyday life, especially in urban areas. The people split their social media preferences between the well-known international platforms and other platforms built specifically for African users. Recent statistics showed Senegal's social media users visited Facebook (57 percent), YouTube (24 percent), Google+ (23 percent), Twitter (16 percent), FaceDakar (7 percent), Eskimi (5 percent), 2Go (5 percent), biNu (3 percent), Twoo (2 percent), Hi5 (2 percent), and Mxit (1 percent) (Southwood 2014). Although Facebook was the most popular site, it is clear that interest in Facebook is lower in Senegal than in other countries because there are many social media alternatives. On Facebook, the top pages belong to Seneweb, Orange, Tigo Senegal, Expresso Senegal, and Samsung Senegal (Allin1 Social 2015). Three of these sites represented telecommunications and service providers; the others are Seneweb, a news portal, and Expresso Senegal, a local business.

In June 2015, Facebook launched a new initiative designed to attract a wider market share in Senegal and provide basic internet services to areas that have remained disconnected. Senegal was one of thirteen countries throughout Africa, Asia, and Latin America where Facebook launched the new effort. The program, called Internet.org, was intended to bring free services to Tigo subscriber identity module (SIM) cardholders, and it started by offering information and apps direct from sources such as AccuWeather, BBC News, Bing, UNICEF, Wattpad, Wikipedia, Dakaractu, Senjob, and Facebook itself (Linington 2015). Support and content was available in French; English and the Wolof language were not supported. Facebook chose Senegal and the other locations because the lack of internet access corresponded to the price and amount of data and devices. Through this effort, Internet.org offered a cheaper alternative and greater accessibility to the internet.

Senegal's popular apps indicated that the people use apps for communication, for practical ways to improve their lives, and for entertainment. The top Microsoft Windows phone apps in the free category were Podcasts, Viber, Facebook, Imo.im, and Facebook Messenger; the top paid apps were Fruit Ninja, iStunt 2, SIMS Medieval, Free Music Downloader Pro, and Photo Private & Lock and Local Pro. The top free iOS apps were WhatsApp, Viber, YouTube, Facebook, and Facebook Messenger. The top paid iOS apps were Smart Alarm Clock, Salat-Comment Prier, Fitness Point Pro, Scanner Pro, and Temple Run: Oz (AppAnnie 2015). There was no data available from Google Play or the Amazon app store.

Senegal has started to develop its own platforms and forums. Mivasocial is a new, emerging social platform in Africa. Founded in 2013, the site's purpose is to promote Africans online and to connect Africans to each other via deeper, richer communication. The site differs from more traditional social media in that it combines the popular pieces of other platforms, such as photo sharing, with a meaningful way for people to collaborate. As of mid-2015, Mivasocial offered tools to chat, view African events in real time, share videos and photos, create communities within the site, access apps, coordinate real-time actions, and promote personal and corporate brands. Over time, the site has added classified ads, a directory, and an online shopping feature (Mivasocial 2015). Local forums create additional opportunities to form hobby-specific communities or create spaces for specific types of discussion. One of Senegal's more active forums belonged to Orange, the telecommunications firm based in Paris, France, with worldwide operations. The company operated forums tailored to the countries where it operates. The Senegal-specific page, for instance, allowed people in Senegal to discuss local culture and current issues in French.

Senegal's access to the World Wide Web has not yet matured enough to produce definitive data on top websites. However, the nation's most popular search terms over a one-year period functioned as a proxy measure to gauge popular websites. From September 2014 to September 2015, Senegal's most searched terms, in order, were "Facebook," "YouTube," "Google," "Senegal," "Dakar," "Telecharger," and "Seneweb" (Google Trends 2015). These terms verified the data and statistics available on Senegalese social media platforms and apps—most people who use social media used Facebook, YouTube, and Google services, though a smaller percentage of the population preferred other platforms.

Social media and the Senegalese preference for chatting online have led to an innovation in the language of communication. French, Senegal's official language, and English are common languages online. However, despite French being the official national language, there are approximately thirty-eight languages spoken in the country. The most widely spoken native language is Wolof, which arrived in Senegal during the mid-fourteenth century as the Wolof kingdom expanded and absorbed existing African communities. This expansion of Wolof domination occurred well before Europeans established colonies on the continent and led to a large Wolof-speaking majority in Senegal when the national borders were eventually established. Wolof has been difficult to use on social media because its orthography (i.e. spelling conventions) was never standardized. To use Wolof in social media, people are mixing it with French, including French-influenced spellings. It is also popular to mix in some Arabic and to use English interjections such as "cool" and "awesome." The Wolof language has taken on a new identity as a means of communicating on social media, and it will most likely develop its own shorthand forms and elite speak (e.g. leetspeak or eleet) as more people decide to use it.

Social media has helped to shape Senegalese politics. Cheikh Fall (n.d.–), a noted Senegalese blogger and social activist, designed Sunu2012 as an election monitoring site to provide coverage of the 2012 presidential elections. Fall became concerned when the incumbent president, Abdoulaye Wade (1926–), tried to reinterpret the country's constitution in order to run for a third term in office. Sunu means "our" in Wolof, and the site intended to unite the people's voices during the election.

One of the coverage efforts included the hashtag #sunu2012. Sunu2012 spread the news of Wade's controversial run, which was downplayed in the media (Napolitano 2013). In addition, other candidates utilized the site to disseminate information about their campaigns. Wade eventually lost the election. The site's success led to a broader effort to influence and improve Senegalese society.

As a spinoff, Fall designed Sunucause ("Our cause," with the primary hashtag #sunucause) to focus on the use of social media to identify and address social concerns across the country. Its core objective was to raise awareness, and funding when needed, to resolve issues that the government could not, or chose not, to address. Fall's third effort is to engage other Senegalese citizens through social media directly. This effort, called Kebetu, or "to twitter," allows people to participate in actualizing social and political change through soft power, or through peaceful means. The effort had grown on Twitter as of September 2015 and managed its own hashtag, #kebetu.

Fall and some of his followers did not limit their activism to issues affecting only Senegal. In 2014, Senegal's Twitter activists advocated for international support to halt the spread of Ebola in Africa. As the disease spread across Sierra Leone, Liberia, and Guinea, activists worried about the suffering in these neighboring countries and feared that the deadly disease would devastate their nation as well. Through hashtags such as #SenStopEbola (Senegal Stop Ebola) and #GiveUSTheSerum, Senegalese activists rallied and advocated online (Demey 2014). They also disseminated their messages on both Facebook and Twitter to reach a wider, more diverse audience. Fall's group has tackled other domestic issues, such as helping children with rare diseases and arranging charity concerts to help flood victims.

Laura M. Steckman

See also: Cameroon; France; Nigeria; Paraguay

Further Reading

Allin1 Social. 2015. "Facebook Statistics for Senegal." Accessed September 18, 2015. http://www.allin1social.com/facebook-statistics/countries/senegal

AppAnnie. 2015. "iOS Top App Charts." Accessed September 18, 2015. https://www.appannie.com/apps/ios/top/senegal/overall/?device=iphone

Demy, Juliette. 2014. "Sénégal: Les Blogueurs Mobilisés contre Ebola." September 15. Accessed September 19, 2015. http://www.lejdd.fr/International/Afrique/Senegal-les-blogueurs-mobilises-contre-Ebola-687442

Google Trends. 2015. "Explore Topics." Accessed September 19, 2015. https://www.google.com.gi/trends/explore#geo=SN&date=today%2012-m&cmpt=q&tz=Etc%2FGMT%2B4

Linington, Darryl. 2015. "Facebook Takes Internet.org to Senegal." June 5. Accessed September 16, 2015. http://www.itnewsafrica.com/2015/06/facebook-takes-internet-org-to-senegal/

McKinsey & Co. 2015. "Lions Go Digital: The Internet's Transformative Potential in Africa." November. Accessed September 18, 2015. http://www.mckinsey.com/insights/high_tech_telecoms_internet/lions_go_digital_the_internets_transformative_potential_in_africa

Mivasocial. 2015. "The Social Network of Sénégal!" Accessed September 18, 2015. http://
    senegal.mivasocial.com/
Napolitano, Antonella. 2013. "Senegal's 'Soft Revolution' Makes Change in Digital Space."
    June 25. Accessed September 18, 2015. http://techpresident.com/news/wegov/24087
    /senegals-%E2%80%9Csoft-revolution%E2%80%9D-makes-change-digital-space
Southwood, Russell. 2014. "Social Media in Africa: Who's Leading the Way?" October 4.
    Accessed December 18, 2016. https://businesstech.co.za/news/internet/70279/social
    -media-in-africa-whos-leading-the-way//

SINGAPORE

With a strong economy and a small geographic area, Singapore, a city-state on the Malay Peninsula in Southeast Asia, is also a well-connected country with one of the world's highest internet penetration rates. Singapore's diverse population allows it to host a multicultural society that serves at the forefront of technological advances and closely aligns with the latest global and regional internet trends (Tey 2015). The number of internet users in Singapore reflects this makeup. As of late 2015, Singapore had 8.2 million web subscriptions. With a population of 5.26 million, including a nonresident population—students, foreign workers, and their dependents—of 1.46 million, Singapore's internet penetration is 155 percent. In early 2015 and late 2014, Singapore had similar totals for web subscriptions with 8.1 million. Internet users in Singapore are active on foreign-based apps, including Facebook, Facebook Messenger, Google+, Instagram, LINE, LinkedIn, Pandora, Snapchat, Skype, Twitter, WeChat, and WhatsApp mainly across Android and iOS smartphone platforms (AppAnnie 2015; Statista 2015).

Singapore gained access to the internet before most countries in the region. Its debut as a full-fledged member of the international networking community came in 1987, when the National University of Singapore (NUS) established a 4.8-kbps BITNET link to the City University of New York (CUNY) (Tan 2001). BITNET, the "Because It's Time" network, began in the United States in 1981 when CUNY and Yale University started using a leased telephone circuit for communications between accounts on their mainframe computers (BITNET n.d.). Through BITNET, those who had been using the internal campus email service, including NUS, which was mainly based on the existing IBM mainframes, could realize the benefits of international email. By 1990, NUS and Princeton University established a 64-kbps internet link called NUSNET to make NUS the home of the first internet site in Singapore and the region. NUSNET was subsequently launched in 1991 by Dr. Tay Eng Soon (1940–1993), the then–senior minister of state for education, as a campuswide network with full internet connectivity. By 1994, SingTel established Singnet, Singapore's first internet access service provider (Tan 2001).

In 2014, Singapore's mobile subscriber base was split among three mobile service providers, Singtel, M1, and StarHub, for prepaid and postpaid subscription plans. For prepaid subscription plans, Singtel had 4.09 million mobile subscribers (making up 50.6 percent of the subscriber base), M1 had 1.85 mobile subscribers (22.8 percent), and StarHub had 2.15 mobile subscribers (26.6 percent).

Commuters on the Mass Rapid Transit (MRT) in Singapore use cell phones to check their email, instant message with friends, or follow celebrities on Facebook. The country boasts one of the highest internet penetration rates in the world, with one of the fastest internet speeds. (Junpinzon/Dreamstime.com)

For postpaid subscription plans, Singtel had 2.26 million mobile subscribers (48.3 percent), M1 had 1.15 mobile subscribers (24.5 percent), and StarHub had 1.28 mobile subscribers (27.3 percent) (Eu 2015). In the fourth quarter of 2014, 54 percent of multidevice users owned two devices, 29 percent owned three devices, and 17 percent owned at least four (Louisse 2015).

Singapore has one of the fastest peak internet and average internet connection speeds in the world. In the second quarter of 2015, Singapore's peak internet connection speed was 108.3 Mbps, the fastest in the world, which marked a 12 percent increase from the first quarter of 2015. For average internet connection speed, Singapore ranked fourteenth in the world, with 12.7 Mbps. Throughout the city-state, 50 percent of its population surfs the web at speeds above 10 Mbps. This data on peak connection speeds provide insight into the peak speeds that web users can likely expect from their internet connections while in Singapore (Today Online 2015).

Nationwide outdoor service coverage for 4G cellular networks in the first quarter of 2016 is 99 to 100 percent for each of the three mobile service providers: Singtel, M1, and StarHub. Outdoor 4G mobile service coverage is the ability of a cellular network to achieve a minimum signal strength of at least −109 dBm (this is an abbreviation for the power ratio in decibels of the measured power referenced to 1 mW). For Singtel, nationwide outdoor 4G mobile service coverage is 99.89 percent. For M1, coverage is 99.36 percent, and for StarHub, coverage is 99.43 percent (IDA 2016).

In the fourth quarter of 2015, 64 percent of the population had an active account with any social network. The most popular social network was WhatsApp, with a 46 percent penetration rate. Facebook had a 43 percent penetration rate, Facebook Messenger had 26 percent, Instagram had 18 percent, Google+ had 14 percent, LINE had 14 percent, LinkedIn had 14 percent, Skype had 13 percent, Twitter had 13 percent, and WeChat had a 12 percent (Statista 2015). Of these social networking apps, all originate from the United States, with the exceptions of WeChat, from China, and LINE, from South Korea and Japan.

Web users in Singapore are active on Facebook. In November 2015, Facebook recorded that it has 3.6 million users in Singapore. Web users in Singapore primarily use Facebook to follow Singaporean celebrities. The ten most popular Facebook pages, as of December 2015, are (Socialbakers 2015):

1. Joseph Prince (1963–), the senior pastor of New Creation Church in Singapore
2. Aaron Aziz (1976–), a Singaporean actor
3. Tila Tequila (1981–), an American television personality who was born in Singapore
4. Jose Cuervo, a brand of tequila
5. 林俊傑 J. J. Lin (1981–), a Singaporean singer based in Taiwan
6. Singapore Airlines, the flag carrier airline of Singapore
7. ONE Championship, a Singapore-based mixed martial arts promotion
8. Willy Foo (n.d.–), a Singaporean photographer, marketer, and technopreneur
9. Tenashar (1989–), a Singaporean DJ and model
10. VR-Zone, a news publication reporting on trends in PC and mobile gadgets

According to a 2014 survey, nearly nine out of ten Singaporeans have access to a smartphone. When asked about their smartphone use over the last seven days, email usage is more common than social networks, with 58 percent having used their phones for email and 54 percent using their phones for social networks. From 2013 to 2014, for smartphones, weekly short message service (SMS) usage declined 8 percent, while instant messaging (IM) usage increased 9 percent (Deloitte 2015).

With regular smartphone access across Singapore, the Singapore–Massachusetts Institute of Technology (MIT) Alliance for Research and Technology (SMART) runs a project called the MIT Senseable City Lab, with the goal of improving urban life in Singapore. By tapping into real-time information about the city-state, the project aims to use collected data to enable Singaporeans to "make day-to-day decisions based on their environment—creating a feedback loop between people, their actions and their city—while simultaneously contributing to the data they are collecting" (Senthilingham 2015). The project aims to use this feedback to improve factors "influencing daily life, from overcrowding and traffic jams to temperature control and taxi availability" (Senthilingham 2015). One example of how this data can improve the quality of life for Singaporeans is by tracking phones to understand, and alleviate, crowding in the city-state (Senthilingam 2015).

Singapore has strict regulations to manage media published on the internet. In a country that ranks 153 out of 180 on the Reporters Without Borders 2015 World Press Freedom Index, Singapore's internet rules follow this regulatory pattern. In 2013, the Media Development Authority (MDA) established regulations to "demand that all websites concerned with the news be licensed, and that each put down a performance-bond of 50,000 Singapore dollars. Any content deemed to be in breach of standards would have to be removed within twenty-four hours of being notified" (F.C. 2013). This licensing framework affects "everything that could be called a Singapore news program, as defined by two criteria: (1) the news program reports an average of one article or more about Singapore's news and current affairs, per week, over a period of two months; and (2) the content of the news program reaches at least 50,000 unique IP addresses from within Singapore" (F.C. 2013). According to the MDA, a Singapore news program includes "any programme to contain any news, intelligence, report of occurrence, or any matter of public interest, about any social, economic, political, cultural, artistic, sporting, scientific, or any other aspect of Singapore," though the law excludes content and programs that the government produces (F.C. 2013).

Singapore's web regulations also extend to Facebook. In September 2015, Ello Ed Mundsel Bello (c. 1988–), a 28 year-old nonresident Filipino who worked as a nurse at a local hospital, was convicted of one count under the Sedition Act for promoting feelings of ill will and hostility after making negative remarks about Singaporeans in a Facebook post. The judge in the case stated that Singapore would not condone any act that poses a threat to the city-state's social stability and security. As a result of his conviction, Bello was sentenced to four months in prison (Chelvan 2015).

Anthony Ortiz

See also: China; Japan; Malaysia; South Korea; Taiwan

Further Reading

AppAnnie. 2015. "Top App Charts." December 28. Accessed December 28, 2015. https://www.appannie.com/apps/google-play/top/singapore/

BITNET. n.d. "A Brief History of BITNET." Accessed December 28, 2015. http://bit.net

Chelvan, Vanessa Paige. 2015. "4 months' jail for Filipino Who Made Xenophobic Remarks Against Singaporeans." September 21. Accessed August 1, 2016. http://www.channelnewsasia.com/news/singapore/4-months-jail-for/2140908.html

Deloitte. 2015. "Mobile Multiplies Global Mobile Consumer Survey." Accessed December 28, 2015. http://www2.deloitte.com/content/dam/Deloitte/sg/Documents/technology-media-telecommunications/sg-tmt-global-mobile-consumer-survey-noexp.pdf

Eu, Goh Thean. 2015. "Singapore Telcos: Winners and Losers in 2014." April 9. Accessed December 28, 2015. https://www.digitalnewsasia.com/mobile-telco/singapore-telcos-winners-and-losers-in-2014

F.C. 2013. "Two Steps Back." June 5. Accessed December 28, 2015. http://www.economist.com/blogs/banyan/2013/06/regulating-singapores-internet

Info-communications Development Authority of Singapore (IDA). 2016. "4G Mobile Service Coverage in Q1 2016." Accessed August 1, 2016. https://www.ida.gov.sg/Tech

-Scene-News/Facts-and-Figures/4G-Service-Monitoring/4G-Mobile-Service-Coverage
-in-Q1-2016

Louisse, Donna. 2015. "Japan, Taiwan, and Singapore Top in Multi-Device Users in APAC:
Appier." January 29. Accessed December 28, 2015. http://e27.co/japan-taiwan-singapore
-top-multi-device-users-apac-appier-20150129/

Senthilingam, Meera. 2015. "How Smartphones Are Improving City Life in Singapore."
June 24. Accessed December 28, 2015. http://www.cnn.com/2015/06/24/tech/big-data
-urban-life-singapore/

Socialbakers. 2015. "Singapore Facebook Page Statistics." Accessed December 28, 2015.
http://www.socialbakers.com/statistics/facebook/pages/total/singapore/

Statista. 2015. "Penetration of Leading Social Networks in Singapore as of 4th Quarter 2014."
Accessed August 1, 2016. http://www.statista.com/statistics/284466/singapore-social
-network-penetration/

Tan, Bernard. 2001. "The Origins of the Internet in Singapore." Accessed December 28,
2015. http://www.physics.nus.edu.sg/~phytanb/bitnet4.htm

Tey, Yvonne. 2015. "The State of Social Media in Singapore." August 20. Accessed December
28, 2015. https://www.techinasia.com/talk/state-social-singapore

Today Online. 2015. "Singapore Tops Global Survey in Peak Internet Connection Speed."
September 27. Accessed December 28, 2015. http://www.todayonline.com/singapore/
singapore-tops-global-survey-peak-connection-speed

SOMALIA

The Federal Republic of Somalia, an African nation long steeped in conflict throughout the twentieth and twenty-first centuries, has an extremely weak infrastructure, which makes conducting field research in Somalia extremely challenging. Furthermore, there are separatists regions, such as Puntland and Somaliland, with variations in media access and restrictions that are difficult to assess. Access to complete data on all types of media usage is virtually impossible; thus, the data given here come from available sources that, as a composite, present as accurate a portrayal of the country as possible.

Somalia, despite its long-standing violence and lack of governmental authority since the collapse of the long and divisive rule of the regime of Siad Barre (1919–1995), has developed a technologically advanced and competitively priced telecommunications industry in urban areas relative to that of other African nations. The regime came to power following a military coup in 1969 and held power until Barre's forced departure in 1990. Following his ouster, clan rivalries within Somalia deteriorated to an all-out civil war that has yet to be fully quelled. Islamist militant groups and other clans still vie for power and control of various segments of the country, though a federal government has been in place since August 2012.

Mobile-phone usage has been on the rise in Somalia due to an expanding telecommunications sector, increased competition in media and telecom services, reduced cost of services, a growing private-sector wireless-based communication infrastructure, and reduced government ability to regulate the telecom industry (Immigration and Refugee Board of Canada 2015). Mobile-cellular services that provide wireless internet access, banking access, and money exchange services, which are not available in many African countries, are available in some parts of Somalia.

These services developed from entrepreneurial ventures by Somalis, with backing from large international firms, in a time of collapsed federal infrastructure and nonexistent government regulation. These market-developed services have wide adoption among Somalis and give them the possibility of utilizing communication services that might be unexpected in a nation facing so many recent setbacks and violence.

Roughly 70 percent of Somalis have access to some type of mobile device, and mobile-cellular companies in Somalia provide some of the cheapest rates for telecommunications services in all of Africa (BBG 2013b). Mobile-cellular services are used to access and share news. According to one source, "The majority of Somalis (65.6 percent) access news at least once per day" (BBG 2013b). More than a quarter of Somalis shares news with their family and social networks either daily or most days (BBG 2013b). Younger, more educated, and urban populations are more likely to access and share news more frequently than older (aged fifty-five years old and more), less educated, and rural populations. However, despite its access to mobile-cellular technology, Somalia has one of the lowest literacy rates in the world, at roughly 38 percent (UNESCO 2016). The consequences of this low literacy level shape both which media are used and how they are used, the importance of various media to Somali society, and how this expanded mobile-cellular access is being utilized by the population.

Coupled with a long-standing oral tradition of information sharing, Somalia's high illiteracy rate results in the dominance of radio in the media. Radio delivers news to the illiterate, requires little infrastructure maintenance to disseminate messages over great distances, and uses cheap receivers; in addition, mobile-cellular technologies can be used to broadcast radio channels. These factors make this medium extremely well suited to a country such as Somalia. Past survey data in Somalia demonstrates that increased access to mobile-cellular technology has strengthened radio's appeal; "The use of radio on mobile has increased dramatically with 56 percent of the respondents listening to radio on mobile in the 2011–2012 and 62 percent in 2013" (African Union/United Nations Information Support Team 2014). Television viewership is limited to urban areas and to those who can afford permanent domestic residences, electricity, satellite dishes, and cable connections. Newspapers have only limited demand due to the illiteracy rates, widespread poverty, and restrictions of newspaper transportation (African Union/United Nations Information Support Team 2014). Due to the population's economic and social makeup, radio is likely to remain the dominant media for some time (BBC World Trust 2011).

There is also a clear divide between urban and rural Somalia in media use and adoption. Specifically, the expansion of internet, mobile-cellular use for access to internet services, and the ability to access and utilize social media channels are available only to a small, albeit growing, subset of the urban Somali population. The result is an expanding digital divide between urban and rural populations in a nation whose citizens are already disparate.

The internet penetration for Somalia, as of 2015, was roughly 1.7 percent of the population, while the average penetration across the rest of Africa is 28.6 percent (Internet Live Stats 2016). The national population of just over 11 million features a staggering estimate of 10.8 million people with no access to the internet; thus, Somalia ranks among the very bottom in terms of internet usage, with only Burundi, Timor-Leste, and Eritrea having lower percentages in the world (Real Time Statistics Project 2016). However, for those populations living in urban areas, internet and social media usage are expanding in Somalia, despite the problems with infrastructure, violence, education, and access.

Although the numbers for internet access are extremely low, they are actually representative of a continual improvement in internet access for the urban population over the past decade and are likely to increase in future years. While only a small percentage of the population has access to the internet, that percentage has been steadily growing since 2012, with increasing numbers of internet cafés in urban areas.

The roughly 200,000 Somalis with direct access to the internet are mostly concentrated in large urban areas, where users have a relatively fast and inexpensive internet connection; however, survey data shows that despite not having direct access to the internet, Somalis in rural areas are able to gain access to online content through mobile device sharing (BBG 2013b). Mobile cellular subscription has risen from just over half a million people in 2010 to well over 5.5 million as of 2015 (Statista 2015). A 2006 Somali Telecommunication Association report showed 234 internet cafés in Somalia, with the total number growing at 15.6 percent per year. That number today (though unverifiable) is believed to be significantly higher, especially in urban areas such as Mogadishu. The African Union/United Nations Information Support Team (2014), conducting surveys over recent years in Mogadishu, showed that over half of the respondents claim to have used the internet in the last six months, and over 40 percent claimed to have used the internet in the past week.

Facebook is by far the most utilized social media platform in Somalia, with pages such as BBC Somalia and VOA Somalia drawing over 160,000 local fans each, and current local Somalia users at over 204,000 (Socialbakers 2016). Over 29 percent of the population reports having interacted with Facebook in the most recent survey data (BBG 2013a), while Twitter, Snapchat, and other social media platforms are expanding. The majority of Facebook users in Somalia are younger and predominantly male, though the United Nations has specifically aimed some recent efforts at correcting such gender inequality in media and in media usage (UNDP 2012). While a sizeable gender gap does exist, Somalis with access to the internet use Facebook and other social media extensively (BBC News 2014). Reports on Facebook usage show dramatic increases in the presence of Somalis on the Facebook platform over the past years. This high usage of the social media platform holds true for those Somalis living in diaspora, serving to reconnect Somalis who have been separated geographically from one another.

It is estimated that over 1 million Somalis live outside Somalia, which has created large diaspora communities (BBC World Service Trust 2011). The internet has helped unite many of the diaspora with those living in Somalia, particularly through Facebook, Twitter, Instagram, Snapchat, blogs, and targeted Somali websites. Surveys of social media usage among Somali students in Malaysia revealed that the majority of them (more than three-quarters) extensively adopted Facebook and reported it as their first preference among social media sites (Ruslan and Dhaha 2012). Twitter has also become increasingly utilized by Somalis in diaspora to link with one another, in addition to having conversations with and creating networks with others living in Somalia.

A number of activist movements, such as #Dontbuydeath, have been launched on social media platforms by Somalis in an attempt to encourage infrastructure investment and development rather than exodus among Somalis wealthy enough to make such a choice. Many news websites specifically target a diaspora audience to keep them informed of events in the country (BBC World Trust 2011). These websites attempt to dispel the notion that the only solution for young, educated Somalis is to leave their country for the West. Others living in Somali cities such as Mogadishu have taken to new social media platforms to promote Somalia as a vibrant place that is about more than just violence. For instance, after the success of Snapchat's Nairobi Day in Kenya on May 26, 2015, Somalis have utilized Snapchat to show their country in its best light. Young Somali activists post videos and pictures of beaches in Mogadishu, their family life, and the beautiful places in Somalia (Al Jazeera 2016). These activists claim that global media have portrayed Somalia negatively to the world; they are using social media to show a more human side of what it really means to be Somali.

Unfortunately, in some areas of Somalia, the violence is very real, and the expansion of internet services and mobile availability has been met with resistance by radical Islamists. These militants have attempted to ban internet access in their spheres of control and pressured providers to terminate service. Although the Somali government seems to embrace social media, as seen in the government's usage of Facebook, Twitter, and Tumblr, it still faces criticism for its inability to maintain mobile services and counter militant bans on internet access (Freedom House 2015).

Despite the continued violence, infrastructure problems, and threats of militant disruption with minimal government protections, one in four Somalis still has access to the Internet at least once a week (BBG 2013b). Furthermore, as wireless technologies increase and their costs come down, their proliferation seems inevitable in places like Somalia, whose citizens, even in the face of significant hardship, have seen an exponential increase in wireless advancement and internet connectivity compared to past years.

Skye Cooley and Emily Belle Damm

See also: Egypt; Ethiopia; Kenya; Malaysia; Timor-Leste; Yemen

Further Reading

African Union/United Nations Information Support Team. 2014. "Somalia Media Mapping Report." Accessed June 20, 2016. http://somali-media.so/wp-content/uploads/2014/05/2014_05_01_Somali-Media-Mapping-Report.pdf

Al Jazeera. 2016. "Somalia in a Snapchat, More Than Just Violence." Accessed August 1, 2016. http://www.aljazeera.com/indepth/features/2016/06/somalia-snapchat-violence-160614133403700.html

BBC. 2015. "The Somali Woman Who's Become a Global Star on Instagram." Accessed June 20, 2016. http://www.bbc.com/news/blogs-trending-31504423

BBC News. 2014. "Somalia Profile—Media." Accessed June 20, 2016. http://www.bbc.com/news/world-africa-14094550

BBC World Service Trust. 2011. "An Analysis of the Somali Media Environment." Accessed June 20, 2016. http://downloads.bbc.co.uk/rmhttp/mediaaction/pdf/AnAnalysisOfTheSomaliMediaEnvironment.pdf

Broadcasting Board of Governors (BBG). 2013a. "Global Hotspots: Media Use in Mali and Somalia." Accessed June 20, 2016. https://www.bbg.gov/wp-content/media/2013/11/Gallup-Somalia-and-Mali-briefing.pdf

Broadcasting Board of Governors (BBG). 2013b. "Media Use in Somalia 2013." Accessed June 20, 2016. https://www.bbg.gov/wp-content/media/2013/11/gallup-somalia-brief.pdf

Freedom House. 2015. "Somalia." Accessed June 20, 2016. https://freedomhouse.org/report/freedom-press/2015/somalia

Immigration and Refugee Board of Canada. 2015. "Somalia: Prevalence of Cell Phones and Internet Cafes in Mogadishu, Including the Ability to Use Cell Phones for Financial Transfers (2012–February 2015)." Accessed June 20, 2016. http://www.refworld.org/docid/550c35904.html

Internet Live Stats. 2016. "Somalia Internet Users." Accessed June 20, 2016. http://www.internetlivestats.com/internet-users/somalia/

Internet World Stats. 2015. "Africa Internet Stats Users Facebook and 2015 Population Statistics." Accessed June 20, 2016. http://www.internetworldstats.com/africa.htm

Socialbakers. 2016. "Facebook Stats for Fans in Somalia." Accessed August 1, 2016. https://www.socialbakers.com/statistics/facebook/pages/local/somalia/page-1-2

Statistia. 2015. "Number of Mobile Cellular Subscriptions in Somalia from 2000–2014." Accessed June 20, 2016. http://www.statista.com/statistics/501103/number-of-mobile-cellular-subscriptions-in-somalia/

United Nations Development Programme (UNDP). 2012. "Gender Equality and Women's Empowerment." Accessed June 20, 2016. http://www.so.undp.org/content/somalia/en/home/operations/projects/environment_and_energy/gender-equality-and-women-s-empowerment-project-.html

United Nations Educational, Scientific, and Cultural Organization (UNESCO). 2016. "Somalia Distance Education and Literacy." Accessed June 20, 2016. http://www.unesco.org/uil/litbase/?menu=13&country=SO&programme=100

SOUTH AFRICA

South Africa is Africa's largest and most developed economy. However, within the country, this dominance has not translated into mass access to cheap, accessible, and fast internet connections. Surveys show that smaller countries on the continent have faster internet speeds than South Africa. Nevertheless, mobile technology is bridging the digital divide in the country, significantly affecting the tenor and tone of political debates, but prohibitive costs (mostly for poor, black South Africans) result in very uneven access to online media. Most voters, especially the working class and the poor, still access the political realm via offline media and through forms of mass mobilization. Increasingly, the government, ruling party, and opposition groupings are cognizant of the power of social media, and disputes in the public sphere are increasingly fought over or settled online.

The most recent statistics for internet penetration (collected in 2014) show that 48.7 percent, or nearly half, of South African households had at least one member who used the internet, either at home, a workplace, a place of study, or internet cafés. About one-fifth (20.9 percent) of households owned one or more computers. Significantly, 41.3 percent access the internet using mobile devices. Broken down by access, however, only 10.9 percent access the internet at home, 15.6 percent access the internet at work, and 9.7 percent access the internet at cafés or educational facilities. Internet access at home is highest in urban or metropolitan areas, particularly in the Western Cape (23.8 percent) and Gauteng (17.3 percent) provinces, the two economic hubs; and lowest in rural provinces like Limpopo (2.3 percent) and North West (3.3 percent). An interesting statistic is that in rural areas, mobile devices have a major impact on internet access, in that only 2.4 percent, 2.8 percent, and 2.9 percent of households have access to the internet at home, at work, and elsewhere, respectively, but more than a quarter (26.8 percent) obtained access through mobile devices (South African Statistical Service 2014).

Social media use has gradually increased among South Africans. By 2015, nearly 11.8 million South Africans had accounts on Facebook (a 25 percent growth on the previous year), while the number of Twitter users have increased to 6.6 million, up 20 percent. Instagram (1.1 million) is favored by celebrities and sports stars and shows the strongest growth (65 percent). A local app, Mxit, was doing well (4.9 million users) until it was overtaken by WhatsApp (GCIS 2015), though Mxit was still predicted to fare well in India (Caulderwood 2014).

A major factor affecting social media or internet use is cost. South Africa ranks among the most expensive in the world for the cost of broadband. In an index measuring cost per Mbps, the country ranked in sixty-second place out of sixty-two nations. In the same index, South Africa was fifty-seventh out of sixty-two in terms of relative cost of broadband internet in relation to its gross domestic product (GDP) (MyBroadBand.com 2015). But mobile internet is not the panacea that it was made out to be: South Africans spend on average 24.7 percent of their income on mobile services, which is high compared to other parts of the world (Sanchez 2016).

Under apartheid (1948–1994), a political system based on forced racial segregation, telecommunications was a monopoly of a state company, Telkom, and phone access was skewed around racial lines. The first internet link via an email was connected in 1988 at Rhodes University in the Eastern Cape, and the first commercial internet service providers (ISPs) were established in 1993. Two years later, in 1995, the co.za country domain was awarded to South Africa.

In 1994, South Africa held its first democratic elections, bringing apartheid to an end officially. Around the same time, two local operators, Vodacom and MTM, established the first cellular services in South Africa in 1994. The postapartheid government pledged that by 2030, all South Africans would have access to the internet.

The South African government has been accused of using the internet to spy on its citizens. Comparisons are made to the United States, where the National Security

Agency (NSA) routinely breaks the law. Case law and legislation are too weak about what the government can and cannot do. All mobile phone users in South Africa are required to register their subscriber identity module (SIM) cards under the terms of the Regulation of Interception of Communications and Provision of Communication-Related Information (RICA) Act, enacted in 2003. While consumer groups praise RICA's role in complicating fraud and cell-phone theft, a nongovernmental organization (NGO) called Right2Know detailed case studies and examples of government spying and harassing of activists groups (Right2Know 2016).

As in other locations, public debate is driven increasingly by social media controversies on hot topics like sexism, xenophobia, and homophobia, but especially racism and hate speech. These controversies are usually characterized by emotion, outrage, and reducing complex social and political problems to individual virtues and manners. More recently, black participation on social media has increased, reversing the earlier monopolies that whites held in these media. In the United States, this phenomenon is called "black Twitter." A similar grouping is identified in South Africa.

In one of the most mediated instances, in January 2016 (which coincided with South Africa's annual summer season), a white real estate agent in Kwazulu-Natal province named Penny Sparrow (1947–) posted on Facebook that black beachgoers were "monkeys" and "littered the beaches on New Year's Eve." Sparrow was skewered by Twitter users, who not only called out her racism, but made her phone number and address public. Others created emojis and threatened to take her to the country's Human Rights Commission. Sparrow was eventually forced to retreat from social media. In other instances, racist outbursts by newspaper columnists, government employees, or television newscasters have led to them losing their jobs after they were amplified on social media.

More significantly, however, has been the use of social media and mobile phones in protests and electoral politics. The ruling African National Congress (ANC) has been slow to adapt to social media; the party leader and the country's president, Jacob Zuma (1942–), hardly tweets at all, and the party is hostile to social media when it is not treating Twitter and Facebook as bulletin boards for announcements and promotions. At one point, the ANC announced that it would submit some of

---

Social media offer an opportunity for the everyday person to become famous. One unusual star rose to prominence in South Africa in 2015. Born in 2013, Oratilwe Hlongwane (c. 2013–) learned to DJ before he could talk. His parents credited his talents to an iPad, which they used to start teaching him with educational software. Once he got ahold of his father's DJ equipment, Oratilwe started playing with the buttons until he created a rhythm, and he even figured out how to use the knobs to create sound effects. After gaining 25,000 fans from an uploaded cell phone video that his parents uploaded to Facebook, Oratilwe occasionally plays at shopping malls and small events. His music and sound effects, along with his age, astonish the crowds who gather to watch him (Macias 2015).

its public representatives to disciplinary procedures for discussing party business on Facebook. However, smaller parties have been more successful. The Democratic Alliance (DA) is the second largest party in Parliament; it controls the biggest cities. It was the first to open Facebook and Twitter accounts, and it shaped which debates trended or what formed the bases of online political disputes. However, since 2013, a smaller, insurgent party, the Economic Freedom Front has also proved to be a creative force online.

Outsider political parties, social movements, and nongovernmental organizations (NGOs) agitating around secondary education, housing, electricity, and gender rights see a social media and internet presence as central to any media or advocacy cam-

A student protester demonstrates outside Parliament in Cape Town, South Africa, on October 26, 2016. #FMF, or #FeesMustFall, originated in response to an announced tuition hike, and gained momentum over multiple online platforms, resulting in a real-world movement for free higher education and better governance. (AP Photo/Schalk van Zuydam)

paign. However, it was university students who during much of 2015 reshaped the relationship between protest and social media.

While there has been long-simmering discontent over tuition fees, insufficient government support for students, and outsourcing of services (mostly on black-majority campuses), dissatisfaction over colonial and apartheid symbols boiled over into protest in March 2015. At the University of Cape Town, a historically white campus, a group mostly comprised of black students protested the presence of a statue of Cecil John Rhodes (1853–1902), a late nineteenth century British colonial figure associated with land theft and violence against blacks. The movement against Rhodes's statue was encapsulated in the hashtag #RhodesMustFall. The hashtag translated into on-the-ground activism, and its supporters eventually identified themselves by it (this tactic was not unique, however, as they were mimicking developments elsewhere, most notably the rise of #BlackLivesMatter in the United States). When the University of Cape Town sought an interdiction against the protesting students, one of the parties identified in the suit was #FeesMustFall.

Social media accounts run by students began to set the news agenda about the protests, as mainstream media failed to keep up. Online student journalists provided updates on the protests as they developed, interviewed their colleagues (posting videos on social media like Facebook, Twitter, WhatsApp, and Instagram), and used apps like Dropbox for "alternative education," posting links there to PDFs of articles that usually were only accessible behind a paywall or through library services.

The student protests coincided with the emergence of online-only news media in South Africa. An older site, Daily Maverick, had excelled with its coverage of the police massacre of thirty-four striking mine workers in August 2012, while a second, Daily Vox, openly identified with the students. Smaller regional sites like GroundUp, which grew out of earlier social movements focusing on health and education politics, also have emerged as reliable alternatives to mainstream South African media.

Over time, the movement combined opposition to colonial symbols with curriculum reform and changes to the mostly white university faculty. Similar protests took place at other campuses and were exclusively identified with hashtags (e.g. #TransformWits [at Wits University] and #RhodesSoWhite [at Rhodes University]). Later, the movement took on the struggle for free public higher education (#FeesMustFall) and protesting the national government's performance (#NationalShutdown). President Zuma eventually met with university vice chancellors in October; eventually, he declared that there would be no public tuition hikes in 2016.

Sean Jacobs

See also: India; Kenya; Nigeria; United States

Further Reading

Caulderwood, Kathleen. 2014. "Here's Why Mxit, A New Messaging System, Will Overtake WhatsApp, BBM and WeChat in India." January 30. Accessed August 25, 2016. http://www.ibtimes.com/heres-why-mxit-new-messaging-system-will-overtake -whatsapp-bbm-wechat-india-1551941

Government Communications and Information Systems (GCIS). 2015. *South Africa Yearbook*. Government Printer. Accessed December 27, 2016. http://www.gcis.gov.za/content /resourcecentre/sa-info/yearbook2014-15

Jacobs, Sean, and Wasserman, Herman. 2015. "The Day Mainstream Media Became Old in South Africa." November 25. Accessed August 25, 2016. https://www.washingtonpost .com/news/monkey-cage/wp/2015/11/25/the-day-mainstream-media-became-old-in -south-africa/

MyBroadBand.co.za. 2014. "The History of Internet Access in South Africa." November 30. Accessed August 25, 2016. http://mybroadband.co.za/news/internet/114645-the -history-of-internet-access-in-south-africa.html

MyBroadBand.co.za. 2015. "South Africans Are Getting Nailed over Broadband Prices." March 5. Accessed August 25, 2016. http://mybroadband.co.za/news/broadband /119516-south-africans-are-getting-nailed-over-broadband-prices.html

South African Statistical Service. 2014. *General Household Survey*. Government Printer.

SOUTH KOREA

Once torn apart by the Korean War, South Korea (officially the Republic of Korea) is now a democracy with a gross domestic product (GDP) that ranks among the top countries in the Organisation for Economic Co-operation and Development (OECD). Located in East Asia, the peninsular country is bordered by the Democratic People's Republic of Korea to the north and surrounded by the Yellow Sea and the Sea of Japan (also known as the East Sea).

Home to over 50 million people, the Republic of Korea rightfully boasts that it is one of the most connected countries in the world, with over 92 percent of the population reportedly connected to the internet in the first quarter of 2015 (Miniwatts 2015). The South Korean broadband market is also one of the world's most heavily penetrated, with almost 20 million subscribing to broadband service (KISA 2015). The capital, Seoul, has often been called the bandwidth capital of the world. In addition to the highly saturated market, South Korean connection speeds are some of the fastest in the world. With an average connection speed of 20.5 Mbps and average peak connection speed of 86.6 Mbps, South Koreans enjoy connection speeds well above the global average connection speed of 5.1 Mbps and average peak connection speed of 32.2 Mbps (Akamai 2015). There are several internet service providers (ISPs) in South Korea, but the main three, KT Corp, SKBroadband, and LGU+, own most of the market share.

South Korea's high penetration rate of the internet and lightning-fast speeds can be accredited to several factors, including government planning, market competition, and population density. As the country moved from an industrial economy to an information economy, the government provided support by launching a ten-year infrastructure project starting in 1995 to make internet access available across the country. There were also no heavy regulations that would have created barriers for new ISPs entering the market. As a result, the spread and upgrade of the internet in South Korea was enhanced due to healthy competition between ISPs in the private sector. The ease of structure development also played a key role, as building the necessary infrastructure was comparatively less costly than other nations because 83 percent of South Korea's population lives in urban areas (Braun 2014).

With a highly connected population enjoying fast connection rates at low cost, it is no wonder that South Korean life is heavily mobile integrated. Home to the technology conglomerates LG and Samsung, the variety of smartphones available is wide-ranging, and young Koreans are known for changing their mobile phones on a yearly basis in an effort to keep up with the latest technology. Data service is also considerably cheaper than service in the United States. On average, Koreans pay about half the price that New Yorkers pay for 25 Mbps (Kehl 2014). South Koreans use their smartphones for daily communication, banking, payment, food delivery, and transportation. University students use their smartphones to book seats in the overcrowded study libraries or to listen to K-pop, a form of pop music that arose in South Korea in the 1990s, through one of the many streaming music subscription services.

One app that has had the most commercial success in South Korea is KakaoTalk, an online messaging application that has essentially replaced costly text messaging. While it is mainly a messaging service that serves over 100 million users (with 5 billion messages sent daily), it now has integrated social gaming, online shopping, a taxi-hailing service, video-chatting, and personal blogging into the platform

(Acuña 2013). While services like WeChat and Line dominate other Asian markets, KakaoTalk rules the South Korean market, with reported revenues of $319 million in 2014 (Mac 2015). Users can purchase emoticons or emojis to send in one-on-one or group messages, get messages from their favorite brands about sales and styles, buy coffee gift certificates or other presents to send through messages, and then pay for those services with KakaoPay, a credit card, or by adding the fee to their phone bill. Many of the games in the app have a virallike quality, as users are urged, or sometimes required, to send messages to friends in order to gain more lives or unlock levels.

Facebook has enjoyed huge success in the South Korean market. With a reported 16 million members, the site has a market penetration of 32.6 percent (Miniwatts 2015). This success came after a battle with Cyworld, which was the main social networking platform until about 2011. Run by SK, Cyworld users accepted each other's "friend" requests and could then post on each other's journal entries, photos, or "walls." Unfortunately, Cyworld failed to gain success outside of South Korea, and with the increasing need to connect with a global audience for social and business purposes, the local-oriented service could not keep up with global heavyweights such as Facebook and Twitter. Another cause for Cyworld's downfall was its failure to create a mobile-friendly platform, a certain death sentence in a society where

A smartphone features the logo of KakaoTalk, the most successful messaging app in South Korea. The app has not only replaced text messaging, but has grown to include additional services such as in-app games, shopping, blogging, and taxi-hailing. The app is starting to gain more popularity across Asia, competing with WhatsApp and Line. (AP Photo/Hye Soo Nah)

almost everything is done through a smartphone. Conversely, according to *Business Korea* (2013), 90 percent of Facebook users in South Korea connect through their mobile phones.

Although a Western company has beat out its Korean competitor in the world of social media, one area that remains almost impenetrable is web searching. While Google does have a presence in South Korea, Naver, a search engine run by NHN Corp, held 77 percent of the market in 2013. The second runner-up was Daum, holding 20 percent of the market share, while Google trailed with less than 5 percent (Return On Now 2015). Google's failure to seize market share in South Korea may be due largely to its late arrival, but it also fails to fulfill South Koreans' desire for a portal page. Naver, a portal site like Yahoo, has a landing page that highlights the latest news from outlets all over the web. It also provides services such as a multi-language dictionary, specialized message boards known as "cafés," a blogging platform, online comics called *webtoons*, and a service called Ji-shik, whose name translates as "knowledge" in Korean and is a platform where users can post any question and get responses from people all over the web.

YouTube has seen consistent growth in South Korea since its Korean-language service launched in 2008. South Korean viewers enjoy streaming videos from their mobile phones, and local users create channels for gaming, cooking, and makeup tutorials. Local users also have the chance to gain global attention, just as the musician Psy became famous worldwide when his song "Gangnam Style" reached 1 billion YouTube views in 2012 (Bahk 2015).

South Koreans also enjoy a local video-streaming service called AfreecaTV. Unlike YouTube videos that are prerecorded and often edited, AfreecaTV broadcasts live, user-streamed videos. Comments on "performances" from viewers are displayed in a live feed, with the "broadcaster" often replying to viewer comments and requests and sending thanks and shout-outs to viewers who send star balloons to pay for the pleasure of viewing the stream. These star balloons are a serious source of income for many broadcasters, with top earners making a lot of money after AfreecaTV's 40 percent cut. Some of the popular broadcasts of the several hundred thousand streams aired monthly include drinking broadcasts, study broadcasts, and dancing broadcasts. The most famous of these are *mokbang*, or eating broadcasts. A result of the South Korean communal food culture mixed with the growing number of one-person households, *mokbang* videos allow viewers to tune into a live-stream of gorging and gluttony, where people talk to the audience while enjoying tremendous amounts of food. Viewers rejoice at the sight of broadcasters scooping

South Koreans watch more videos online than any other nation. A total of 95.9 percent of South Koreans watch at least one video online per week, with many viewers watching more than one program. Following the South Koreans are the Spanish, at 92 percent digital video viewer penetration. The top five nations with the highest online video viewing percentages are rounded out by Italy, at 91 percent; Mexico, at 90.1 percent; and China, at 89.6 percent (Edelman 2015).

copious amounts of food into their mouths, describing each delicious bite over the course of the meal. Many of these videos are later posted on YouTube as well, so those who missed out on the live feed of chewing can catch up later. Top-earning *mokbang* stars have reported making up to $9,000 a month—or even £1,000 per broadcast—and many of them consider eating on camera their profession (The Observers 2014; Stanton 2015).

More successful than the social networking and user-created entertainment genres combined is gambling. Mobile gaming spending is an estimated $1.3 billion, making South Korea one of the top three spenders in Asia (Grubb 2015). Even more successful are multiplayer games, such as League of Legends. Despite the abundance of high-speed wireless internet, PC rooms and internet cafés are still quite popular. Groups of friends gather in PC rooms to connect through games and eat ramen, a major social activity for young men. The activity is so popular that internet addiction camps have been created to deal with those who cannot curb their appetite for online gaming.

Online shopping is also a major source of revenue. South Koreans regularly shop online for food products, cosmetics, shoes, and handbags. In October 2015, yearly online sales had already reached $4.24 billion, with about $2 billion of that figure specifically from purchases made through smartphones, tablets, and other mobile devices (Yonhap News 2015). Popular online shopping sites include GMarket, 11street, Emart, and Coupang.

With high connectivity must come some regulation, but South Korea has some very serious restrictions regarding their neighbors to the north. It is estimated that sixty-five sites considered sympathetic to North Korea have been banned through internet protocol (IP) blocking, and as a result, a large number of websites may be inaccessible due to shared hosting (Yoon 2010). In the realm of social media, South Korea has blocked the Twitter account operated by the North Korean government. Expressing pro–North Korean sentiment can also get South Korean citizens in serious trouble; for instance, a Democratic Labor Party activist was sentenced to one year in prison for comments regarding North Korea on the party's website (Baek 2002). In 2011, prosecutors charged a man for using social media to praise North Korea and share media clips, a serious violation of South Korea's National Security Law (Kim 2011). Police have blocked foreign sites, social media accounts, and online communities, and also have deleted material, ostensibly in the name of national security (Park 2014).

Censorship does not end with talk about North Korea. Content can be blocked if it falls into one of the categories of obscenity, gambling, illegal food and medicine, violating other's rights, and violating other laws and regulations. If a site has been blocked, users are redirected to a warning that the government has blocked the site. In 2008, the Korea Communications Standards Commission replaced the Information and Communication Ethics Committee to manage internet censorship. They are responsible for deleting 23,000 Korean webpages and blocking another 63,000 in 2013 (S.C.S. 2014). Made up of members appointed by the president, the agency has come under fire for possibly curbing free speech. Surveillance is also a point of contention in South Korea, with ISPs being accused of turning over user data without requiring warrants. The mobile messaging app KakaoTalk saw many users abandon it for foreign-based applications when they turned over user data in compliance with government orders.

The government also has created regulations for monitoring children's activity on the internet. Online gaming is banned from midnight to 6 a.m. for children under sixteen, requiring users to enter their government ID numbers to prove their age. This proved disastrous for a 15-year-old who was locked out of Starcraft II for a daytime competition in France in 2012 that ran after midnight in South Korea; when he logged back in with his parent's ID number, he had lost the match and a chance at the competition's major monetary prizes (S.C.S. 2014). In April 2015, a law was passed requiring all smartphones sold to those eighteen and under be equipped with a specific app so their parents could monitor their social media activity. The app was meant to protect children from cyberbullying, pornography, and other damaging material. However, just six months later, the app was pulled from the market after an internet watchdog group discovered serious security concerns in an analysis of the app (AP 2015).

Hacking, especially from North Korea, is a serious threat to internet security in South Korea. The National Intelligence Service has pointed the finger at its northern neighbor for attacks on banks, government agencies, military institutions, TV broadcasters, and websites. South Korea believes that North Korea is running a special cyberwarfare unit and has blamed Pyongyang for attacks on the Seoul metro system and a nuclear power plant operator, Korea Hydro and Nuclear Power (AFP 2015). It is suspected that North Korea orchestrated a 2013 cyberattack that wreaked havoc for a few hours when many automatic teller machines (ATMs), payment terminals, and mobile banking services were temporarily inaccessible after the attack took two banks and three television stations temporarily offline (Branigan 2013).

Crystal L. Hecht

See also: China; Japan; North Korea; United States

Further Reading

Acuña, Abel. 2013. "Why Is Mobile Gaming So Popular in South Korea?" October 19. Accessed December 20, 2015. http://venturebeat.com/2013/10/19/why-is-mobile-gaming-so-popular-in-south-korea/

AFP. 2015. "North Korea Suspected of Hacking Seoul Subway Operator: MP." October 5. Accessed December 15, 2015. http://www.securityweek.com/north-korea-suspected-hacking-seoul-subway-operator-mp

Akamai. 2015. "State of the Internet: Asia Pacific Highlights." October 5. Accessed December 15, 2015. https://www.stateoftheinternet.com/downloads/pdfs/2015-q3-state-of-the-internet-report-infographic-asia.pdf

Associated Press (AP). 2015. "S. Korea Pulls Plug on Child Monitoring App." November 2. Accessed December 15, 2015. http://www.koreatimesus.com/s-korea-pulls-plug-on-child-monitoring-app/

Baek, Gi-cheol. 2002. "내가 유별난가요? 사회가 이상한가요?" October 6. Accessed December 15, 2015. http://legacy.www.hani.co.kr/section-005000000/2002/10/005000000200210062252297.html

Bahk, Eun-ji. 2015. "YouTube Sees 'Tremendous' Growth in S. Korea." May 19. Accessed December 19, 2015. http://www.koreatimesus.com/youtube-sees-tremendous-growth-potential-in-s-korea/

Branigan, Tania. 2013. "South Korea on Alert for Cyber-attacks After Major Network Goes Down." March 20. Accessed December 10, 2015. http://www.theguardian.com/world /2013/mar/20/south-korea-under-cyber-attack

Braun, Andrew. 2014. "Why Does South Korea Have the Fastest Internet?" October 20. Accessed December 15, 2015. http://www.idgconnect.com/abstract/8960/why-does-south -korea-have-fastest-internet

Business Korea. 2013. "90 Percent of Facebook Users in Korea Connect on Their Mobile Phones." August 15. Accessed December 15, 2015. http://www.businesskorea.co.kr /english/news/ict/1272-connections-facebook-90-facebook-users-korea-connect-their -mobile-phones

Freedom House. 2014. "Freedom on the Net in 2014." Accessed December 10, 2015. https:// freedomhouse.org/report/freedom-net/2014/south-korea

Grubb, Jeff. 2015. "China, Japan, and South Korea Lead Massive $14B Mobile Gaming Business in Asia." November 5. Accessed December 5, 2015. http://venturebeat.com/2015 /11/05/asian-mobile-gamers-are-young-and-spending-tons-of-cash/

Kehl, Danielle. 2014. "New Yorkers Get Worse Internet Service than People in Bucharest." November 21. Accessed December 5, 2015. http://www.slate.com/blogs/future_tense /2014/11/21/cost_of_connectivity_study_2014_americans_pay_more_for_slower _internet_access.html

Kim, Eun-jung. 2011. "S. Korean Man Indicted for Pro-Pyongyang Postings on Internet, Twitter." January 10. Accessed December 5, 2015. http://english.yonhapnews.co.kr /news/2011/01/10/36/0200000000AEN20110110007200315F.HTML

KISA. 2015. "Broadband Subscribers." September. Accessed December 20, 2015. http://isis .kisa.or.kr/eng/sub01/?pageId=010400

Mac, Ryan. 2015. "How KakaoTalk's Billionare Creator Ignited a Global Messaging War." March 2. Accessed December 1, 2015. http://www.forbes.com/sites/ryanmac/2015/03 /02/kakaotalk-billionaire-brian-kim-mobile-messaging-global-competition/

Miniwatts Marketing Group. 2015. "Asia Internet Use, Population Data, and Facebook Statistics." November 30. Accessed December 15, 2015. http://www.internetworldstats .com/stats3.htm#asia

Park, Hong-du. 2014. "박근혜 정부 1년 '국가보안법 위반 사범' 대폭 증가." February 19. Accessed December 15, 2015. http://bigstory.ap.org/article/c61a87e602af44df8c98bf73 d47ac788/apnewsbreak-south-korea-pulls-plug-child-monitoring-app

Return On Now. 2015. "2015 Search Engine Market Share by Country." Accessed December 20, 2015. http://returnonnow.com/internet-marketing-resources/2015-search-engine -market-share-by-country/

S.C.S. 2014. "The Economist Explains Why South Korea Is Really An Internet Dinosaur." February 10. Accessed November 11, 2015. http://www.economist.com/blogs /economist-explains/2014/02/economist-explains-3

Stanton, Jenny. 2015. "The Skinny Korean 14-Year-Old Who Makes £1,000 a Night by Gorging on Fast Food on Webcam—While Thousands of Fans Watch." August 19. Accessed December 15, 2015. http://www.dailymail.co.uk/news/article-3203221/Skinny -Korean-14-wanted-company-ate-dinner-makes-1-000-night-gorging-fast-food -webcam-thousands-fans-watch.html

The Observers. 2014. "The New South Korean Craze for Live-stream Eating." March 14. Accessed December 10, 2015. http://observers.france24.com/en/20140314-south-korea -mokbang-trend-eating

Yonhap News. 2015. "S. Korean Retail Sales Hit 10-Month High in October." December 1. Accessed December 15, 2015. http://english.yonhapnews.co.kr/news/2015/12/01/020 0000000AEN20151201003300320.html

Yoon, Sangwon. 2010. "North Korea Uses Twitter, YouTube for Propaganda Offensive." August 17. Accessed December 5, 2015. http://www.huffingtonpost.com/2010/08/16/north-korea-twitter-propa_n_682920.html

SPAIN

Spain, located on the southwest peninsula of the European continent and north of the Mediterranean Sea, has a robust social media environment that demonstrates entrepreneurship and innovation. The Spanish enjoy utilizing foreign and domestically produced platforms and share a national pastime of watching online videos. As of January 2015, Spain had more than 23 million internet users out of a 47.7 million population, or about 48 percent. Of these internet users, at least 17 million claimed to use social media actively, with an overwhelming 90 percent having registered Facebook accounts (OBS 2015). The Spanish also have presences on Twitter, Google+, Instagram, WhatsApp, Tumblr, and Pinterest. Spain's dating population prefers using the Badoo and Lovoo apps to meet people. An examination of national social media demographics indicated that more men than women use Facebook, whereas women participate in higher numbers on Instagram and Pinterest; of all social media users, the most prevalent age group is the sixteen- to twenty-four-year-old category.

Facebook is Spain's most popular social media platform, and the app is one of the country's most downloaded programs. Facebook users reportedly access the site to "like" content such as a post or photo, read news stories, share photos, comment on their friends' posts, and chat with their contacts (OBS 2015). Professionals create profiles on LinkedIn, often in addition to their Facebook accounts, to maintain some distinction between friends and colleagues.

Facebook was not always Spain's key social media platform or networking site. In 2006, Tuenti launched in Spain in response to Facebook and Hi5, a social networking site based out of San Francisco that around 2007 was second only to MySpace. To differentiate the platform from its competitors, Tuenti promised to guarantee user privacy and not share information about individuals or aggregate data. Users could join Tuenti only through a personal invitation, and each user received only ten invitations to pass on to friends and family. However, around 2009, Facebook's global popularity and ease of sharing information caused it to overtake

Some historians are finding a new niche on Twitter. Using their knowledge of peoples' lives from major events in the past, such as the Spanish Civil War and the American Revolution, these fans tweet as if those events were unfolding in real time. One of the first historic role-play accounts is an anonymous Twitter user play acting as Antonio Hernández (1907–1992), a policeman from Murcia, Spain, who spent four and a half years as a German prisoner during World War II (1941–1945). The purpose of such accounts is to share knowledge, or in this specific case, to shed light on the fates of 9,000 Spaniards captured by German forces (Noticias Cuatro 2015).

Tuenti in popularity. In 2013 and 2014, the site still ranked in the top 25 on Spain's most accessed website list, but by 2015, it no longer appeared even within the top 200. While releases such as Tuenti Móvil (Mobile) and Tuenti Local, similar to Foursquare, a search-and-discovery app, helped modernize the brand, the app has continued to lose members. Instead of continuing to focus on the social media market, Tuenti is concentrating on its mobile phone division and looks to expand in Latin America.

In addition to Tuenti, Spain has a presence on a multitude of platforms for purposes ranging from chatting to shopping and news aggregation, where sites collect data on news pertaining to specific themes. In Spain, the main science and technology news aggregators are Divúlgame and Barrapunto; microblogging post aggregators are Divoblogger and Bitáconas; general news aggregators are Menéame and the Huffington Post in Spanish; and specialized aggregators, which allow career-specific communities and hobbyists such as automobile enthusiasts, include Feedly and Flipboard.

Spain's popular websites provide a glimpse into the preferences of the country's user base. Its top ten accessed websites were Google España, Google, Facebook, YouTube, Amazon España, Twitter, Live, Yahoo, Wikipedia, and Marca, a sports news site (Alexa 2015). These sites indicate that the people frequently use Google for online searches, prefer globally recognized social media platforms, and spend time online reading the news. The top free apps on GooglePlay confirmed that the Spanish like social media, as the top five platforms are WhatsApp, Facebook, Google Photos, Facebook Messenger, and the 360 Security Antivirus Boost, an app dedicated to removing viruses, such as those that could infect a phone through social media and internet use. The paid apps showed another dimension to the Spanish social media user. These apps included games and entertainment, such as Minecraft Pocket Edition, Geometry Dash, Runtastic Running Pro, Farming Simulator 16, and Flightradar24 to track flights (AppAnnie 2015). From these app preferences, Spain enjoyed online chatting, sharing photos, improving their math skills, and working on physical fitness.

European internet laws are often stricter than in other parts of the world, including the United States and Asia. In 2014, the Spanish government enacted a law that topped the copyright concerns of Belgium, France, and Germany against news aggregator sites. The law, which went into effect January 2015, required all Spanish publishing sites to charge aggregator sites, such as Google, for using their content. Refusing to levy the fee would result in a fine of around $750,000. Google responded by removing all Spanish publishers in its news section and dismantling its local aggregators. The result was that Spanish publishers received significantly fewer hits, with estimations that their sites received a two-thirds reduction, equating to a huge drop in online circulation. Other sites remained cautious about linking to Spanish publishers, because even the tiniest excerpt from a news story required that site to compensate the publisher.

For advocates of internet freedom, the Spanish government's decision could have additional consequences. Though Google and similar sites were not banned under the new law, their ability to point users to relevant information has diminished.

Google and other news aggregators do not receive income from pointing people to sites matched by user search queries, and they believed that it was not a good business model to capitulate to paying the fees imposed by the Spanish government. Some internet users in Spain and other parts of Europe speculated that the new trend to impose fines and other charges restricted the availability of internet content. They also speculated that the next step would be to curb content on social media sites, such as YouTube, by levying similar fee structures. Internet freedom in Spain could be in jeopardy if the law is expanded because it is likely that many sites will follow Google's example and refuse to serve the affected areas.

YouTube is the most popular video-sharing site in Spain. Because Spanish users visit the site around 20 million times a month, its popularity has transformed how users interact with it. For individuals with camera skills, the current trend is to create homemade videos and post them to an original YouTube channel. Successful channels attract subscribers, and sometimes advertisers, and once a channel has several million subscribers and views, the person controlling the channel can earn between €2000 and €3000 a month. Examples of Spain's successful YouTubers are MangelRogel, Abi Power, ElRubiusOMG, and LokoOfLucky (León 2014). This YouTube phenomenon is transforming the lives of its users and creating an industry of locally produced content that is posted on the web for mass consumption.

A second way in which Spain's YouTuber phenomenon is changing the way that people think about media is that it directly influences the country's businesses. Many people, particularly from the millennial generation and younger, no longer spend much time watching television. On the rare occasions that they do, members of this generation are more likely to watch using a smartphone rather than a television set, and might not be viewing from a traditional location like a living room. Stations such as Spain's Antena 3TV adapted to the demand for YouTube by altering their business models. Antena 3TV created its own YouTube channel, where people can access programming and now operates through traditional television and electronic platforms.

Other stations, such as Telecinco, sued YouTube for copyright infringement because users had uploaded its content. However, a judge ruled that YouTube could not be held responsible for what its users upload, at least in Spain, but the site was obligated to remove content that violated intellectual property laws. As YouTube has continued to rise in popularity, many businesses, not only television companies, have chosen to establish official channels on the site, advertise, or both in order to engage larger segments of Spain's population.

Spain does hold one unique claim to fame in the use of internet communications. In 2011, one Spanish town started relying on Twitter to enable communication between the local government and town residents. Jun, a small town located north of Granada, developed its own Twitter account because its mayor, José Antonio Rodríguez Salas, advocated that technology was the key to improving relationships between the government and the population. The town then convinced its citizens to create their own accounts and then verified that each account belonged to an actual resident. Next, the government instructed the townspeople to use Twitter to report any issues or complaints in order to resolve them. If the report required the involvement of other government employees, such as a representative from maintenance or sanitation, the government's initial response to the message included that employee's Twitter handle, along with the hashtag #JunSeMueve, or "Jun gets moving" (Powers and Roy 2015). The inclusion of the hashtag signaled that a public promise has been made to resolve the problem or address the grievance. When the required action was completed, there was a follow-up message, sometimes with photos, to confirm that the original issue had been resolved. For the town's verified 3,500 residents, the system functions well; the media even occasionally refers to the town as "Jun 2.0" for having successfully integrated Twitter and new technologies into its operations. This unique model has received attention from the news media, and there also has been academic interest in Jun's innovative use of social media.

Laura M. Steckman

See also: France; Germany; United Kingdom; Venezuela

Further Reading

Alexa. 2015. "Top Sites in Spain." Accessed August 18, 2015. http://www.alexa.com/topsites/countries/ES

AppAnnie. 2015. "GooglePlay Top App Matrix." Accessed August 18, 2015. https://www.appannie.com/apps/google-play/matrix/overall/?date=2015-08-18

León, Pablo. 2014. "The Rise of Spain's YouTubers." November 21. Accessed August 20, 2015. http://elpais.com/m/elpais/2014/11/19/inenglish/1416406298_035188.html

Martínez, Jesús. 2014. "Tuenti contra Tuenti: De 'Startup' de Nombre Filial de Telefónica." November 9. Accessed August 18, 2015. http://www.elconfidencial.com/tecnologia/2014-11-09/tuenti-contra-tuenti_440005/

NERA Economic Consulting. 2015. "Impacto del Nuevo Artículo 32.2 de la Ley de Propiedad Intelectual Informe para la Asociación Española de Editoriales de Publicaciones Periódicas (AEEPP)." July 8. Accessed August 16, 2015. http://www.aeepp.com/pdf/InformeNera.pdf

OBS Business School. 2015. "España Aumenta el Número de Usuarios Activos en Redes Sociales en 2014 y Llega a los 17 Millones." January 25. Accessed August 16, 2015. http://www.obs-edu.com/noticias/estudio-obs/espana-aumenta-el-numero-de-usuarios-activos-en-redes-sociales-en-2014-y-llega-los-17-millones/

Powers, William, and Roy, Deb. 2015. "The Incredible Jun: A Town that Runs on Social Media," April 20. Accessed August 16, 2015. http://www.huffingtonpost.com/william-powers/jun-twitter-social-media_b_7102780.html

Vara, Vauhini. 2014. "Spain versus the Internet." December 13. Accessed August 19, 2105. http://www.newyorker.com/business/currency/spain-versus-internet

SURINAME

Suriname is a country in the northeastern region of South America, surrounded by Guyana, French Guiana, Brazil, and the Atlantic Ocean. A former Dutch colony, it has an extremely diverse population due to the colonial period's reliance on importing African and Indonesian, specifically Javanese, workers to support agriculture and the extraction of mineral wealth. Technology has developed at a slower rate than in other parts of Latin America because of Suriname's small population and lack of major cities beyond the capital, Paramaribo. As technology has become more important, the Surinamese have used social media for many reasons, including communicating with friends and family. Political parties also have harnessed the power of social media during elections to address the major issues that affect their constituents.

Relatively few studies exist about Suriname's social media demographics. From the information that is available, in 2015 Suriname had around 250,000 internet users, of whom more than 200,000 had Facebook accounts and fewer than 20,000 were active on LinkedIn and Twitter (van Charante, 2015a). The country's total population hovered around 540,000 at that time, with 122 percent having mobile phones. In other words, 22 percent of people in Spain had more than one mobile phone or subscriber identity module (SIM) card; similar to other countries such as neighboring Brazil, where people maintain multiple phones, they are often split between business and personal use, or between service providers to call areas where those carriers offer coverage. With the amount of phones in use, Surinamese access the internet primarily via mobile phone.

Suriname is a Facebook country. While local users have experimented with various other platforms in the past, from 2010 onward, the Surinamese have been dedicated Facebook aficionados. Approximately 200,000 Surinamese have Facebook accounts, which was around 40 percent of the population (van Charante 2015b). The top Facebook pages belonged to Starnieuws, Dagblad Suriname, McDonald's Suriname, Surinam Airways, GFC Nieuws, Digicel Suriname, Pizza Hut Suriname, KFC, KLM Suriname, and Europcar (Allin1 Social 2015). From these favorites, the Surinamese preferred Facebook pages that supported local news coverage, transportation, and fast food brands.

According to a Surinamese social media specialist, YouTube has always been popular in the country. However, in the past, slow connection speeds prevented high levels of online video consumption. Prior to the expansion of mobile phone access and services, a five-minute video could take up to thirty minutes to buffer, which took time and accrued charges for the data connection (van Charante 2015b). With the rise of high-speed mobile access, however, more Surinamese are viewing YouTube videos. In fact, one of the top apps in 2015 is from PewDiePie, a Swedish vlogger (video blogger) whose account was the first to have more than 10 billion views. Suriname's top websites for late 2015 reflected this renewed preference for YouTube, which led the list, followed by Facebook, Google.sr, Google, Live.com, Amazon, Yahoo, Msn, Starnieuws, Xvideos, and Alibaba (SimilarWeb 2015). These sites reconfirmed the impression that Surinamese are avid online news consumers.

Twitter did not appear on the list of top sites accessed; it is not a favorite app because it has few Surinamese users. The platform has never been popular, due in part to the lack of businesses and local people who engage on the site. One of the major reasons that Surinamese use social media was to connect with friends and family, which they can already do successfully over Facebook, so switching to Twitter did not make sense culturally. In Suriname, Twitter still appears to be a primary medium to communicate internationally rather than internally. If more Surinamese choose to interact on the platform and can initiate exchanges of information in the future, more people will likely view it as another communication channel.

In contrast to Twitter, Instagram has gained ground in the country. The platform's Surinamese users were primarily youths who wanted to share images with friends and family. Instagram was the second fastest growing social media platform in 2015 after YouTube (van Charante 2015b). Sharing images through platforms like Instagram is likely to take hold within Suriname's growing tourism industry. Although the sector remains small, Suriname has started exploring its potential to expand its hospitality and services to attract more tourists. Instagram reaches a wide international audience and can allow the government and tourism companies to promote the country without investing exorbitant sums of money up front.

Other social media outlets besides Instagram have the opportunity to gain additional Surinamese users. Telesur, one of the country's primary mobile phone service providers, started offering data packages featuring connections to specific chat apps. In one of its 2015 packages, Telesur offered special data and tariffs for mobile internet plans that accessed WhatsApp, Facebook, Google+, Twitter, LinkedIn, and the Chinese platform QQ. Digicel, another Surinamese phone carrier, created a similar plan to provide free access to WhatsApp, Facebook, Instagram, and Twitter for its 4G customers. There are also small numbers of Surinamese on Pinterest, Reddit, StumbleUpon, and Tumblr—all sites that could gain a larger local user base given the right incentives.

Suriname does have a site to promote local activities to local people. Social Suriname is a Dutch-language social networking megasite, meaning that it has multiple internal means of promoting events and information to its users. The site's main purpose is to allow the Surinamese to discover more about their country and its culture; to do so, it features information on noteworthy items ranging from large events and festivals to smaller social gatherings. The site also contains a section where users can post and establish their own pages. They can also find happenings to attend in person and connect with other users. Some of the site's content is available to the public, while other portions are password protected.

App usage has started rising in Suriname because of mobile phone usage and increasingly faster connection speeds. Most popular apps have entertainment value, though chat apps have also seen a surge in popularity. For example, Suriname's top free iOS apps were iMP3 Pro, WhatsApp, FIFA 16 Ultimate Team, Facebook, and Free Music–Music Streamer. The top paid apps were PewDiePie: Legend of the Brofist, FunBridge Quiz Bundle, Grand Theft Auto III, Monopoly, and Thief Alert (AppAnnie 2015). Little data exist on popular apps for Android or other phone types. More information will become available as social media takes greater hold over the country and becomes more frequently used and easily accessible.

Social media plays a role in Surinamese politics as well. Most parties have their own pages, and social media has proved that it can influence voters, particularly new voters, to participate in the electoral process. In 2015, Suriname experienced its first social media–influenced election. Historically, voting in the country has occurred along ethnic lines, isolating younger voters who did not feel the need to participate. During the 2015 election, parties earned gains in their support base by developing colorful Facebook pages. With this strategy, they engaged younger voters to follow campaigns and learn about the major issues. Around 26,000 people used social media to engage in the election, which was more people than Suriname has on Twitter and LinkedIn. The National Democratic Party, the victorious party in the 2015 elections, gained 9,000 followers prior to the elections. Two other political contenders, the Party for Democracy and Development Through Unity and the Progressive Reform Party, received 5,000 and 4,200 people providing "likes" or other support, respectively (Times of Suriname 2015).

One issue during the elections was to elect more women to political posts, as women have traditionally not been involved in Surinamese politics. During the 2015 elections, a campaign to promote women in politics called "Ook Zij" (They Too), used all communications channels, including television, radio, social media, and billboards; these campaigns received support from Surinamese organizations, as well as the United Nations Development Programme (UNDP), the Dutch embassy, and other international supporters. As a result, women's representation increased from 9.8 percent in 2010 to 33 percent in 2015 (UNDP 2015). Social media, while not the sole cause of a higher number of women being elected to office, was an important contributing factor.

Social media's role in society has evolved in Suriname. In early September 2015, the government raised tariffs, and ultimately the prices, of basic goods. As the prices of fuel, electricity, and water rose quickly, more than 110,000 Facebook users, or about half of all Facebook users in the country, went online to express their discontent (Radio10 2015). This episode marked the first time that Surinamese had ever protested the government online, en masse. The issue sparked debate that rolled over from social media into traditional media, demonstrating how social media has gained ground in the country as important communications media.

Laura M. Steckman

See also: Brazil; Colombia; Indonesia; Venezuela

Further Reading

Allin1 Social. 2015. "Facebook Statistics for Suriname." Accessed September 24, 2015. http://www.allin1social.com/facebook-statistics/countries/suriname?page=1&period =six_months

AppAnnie. 2015. "Top iOS Charts." Accessed September 25, 2015. https://www.appannie .com/apps/ios/top/suriname/?device=iphone

Radio10. 2015. "Surinamers Uitten Kritiek op Tariefsaanpassingen Massaal op Sociale Media." September 3. Accessed September 26, 2015. http://www.radio10.sr/nieuws /surinamers-uitten-kritiek-op-tariefsaanpassingen-massaal-op-sociale-media/45572

SimilarWeb. 2015. "Website Rankings: Top 50 Sites in Suriname for All Categories." Accessed September 25, 2015. http://www.similarweb.com/country/suriname

Times of Suriname. 2015. "Belangstelling voor Politiek via Sociale Media nog Gering." February 4. Accessed September 25, 2015. http://www.surinametimes.com/belang stelling-voor-politiek-via-sociale-media-nog-gering/

United Nations Development Programme (UNDP). 2015. "Suriname: Boosting Women Participation in Parliament District by District." September 16. Accessed September 26, 2015. http://www.latinamerica.undp.org/content/rblac/en/home/ourwork/womenem

powerment/successstories/suriname—boosting-women-participation-in-parliament
-district-by.html

Van Charante, Jean-Luc. 2015a. "Interne Communicatie en Social Media in Suriname."
April 27. Accessed September 23, 2015. http://www.slideshare.net/JeanlucCharante
/spang-power-pointinterne-communicatiesocial-media-2015

Van Charante, Jean-Luc. 2015b. "Social Media in Suriname: Join the Twitter Bandwagon"
February 12. Accessed September 23, 2015. https://www.linkedin.com/pulse/social
-media-suriname-join-twitter-bandwagon-jean-luc-van-charante?trk=pulse-det-nav_art

SYRIA

Syria, a Middle Eastern nation surrounded by Turkey, Iraq, Lebanon, Israel, and Jordan, has around 18 million people, though the number is decreasing rapidly as people flee internal conflict between the government, a terrorist insurgency, and international military forces. Since 2011, the Syrian struggle has been defined as one of the most socially arbitrated in history, with online social media playing a significant part, though the multiyear conflict has real-world ramifications, both online and offline. The Syrian government, its opponents, terrorist combatants, citizens, and activists all have used the internet and social media to report on the country's sectarian and terrorism crises, underscoring the high humanitarian costs associated with the continued conflict and instability.

The Arab Spring uprisings of 2011 were a sequence of demonstrations that sprung up across the Middle East. They concentrated on unfair and cruel governments that controlled those nations and were fueled by the people's refusal to endure their governments' torments. These uprisings have openly contested numerous regimes with repressive policies, including those of Libya, Egypt, Tunisia, and, of course, Syria. Social media sites such as Skype, Facebook, Twitter, and other communications channels such as short message service (SMS) and multimedia messaging service (MMS), played an essential role in connecting and promoting these demonstrations; their use was particularly significant since Syria's president, Bashar al-Assad (1965–), maintains tight control over the internet, allowing only 30,000 people to have access when it was introduced in 2000 and keeping user numbers low, and then imposing strict censorship laws that retaliated vigorously against online policy violators (Freedom on the Net 2012). When the government eased restrictions on some websites in 2011, demonstrations were organized with the support of social interactive websites, with Twitter and Facebook being the most prevalent. These websites allowed demonstrators to spread news of uprisings at a pace that would have been unattainable without social media.

Syria, a country that had a technological development culture prior to the Arab Spring, scheduled its first major protest over Facebook. The "Day of Rage" was intended to have a similar effect as Egypt's and Iraq's "Day of Anger," but only a dozen people showed up for the first event. However, more people joined in later events; tens of thousands reportedly protested in the streets of Damascus, the capital, on March 29, 2011 (Sutter 2011). The Facebook page, also called "Days of Rage," which was an organizing force in the protests, has been removed from the site.

As of 2014, approximately 5.86 million Syrians used the internet, about 26 percent of the population at that time. For those who accessed the internet, 95 percent used computers and 5 percent used mobile phones. A total of 12.5 million mobile phone subscriptions existed, all but 9 percent prepaid (We Are Social 2014). Mobile phones with internet capabilities have reportedly been on the rise, though the exact change in penetration rates is unconfirmed. According to a 2015 survey, Syrian social media users reported 98 percent usage of WhatsApp, 97 percent of Facebook, 36 percent of Google+, 14 percent of Twitter, 12 percent of YouTube, and 1 percent of Instagram (Arab Social Media 2015). In the same survey, 49 percent reported WhatsApp as their most preferred social media app, and 47 percent reported Facebook as their favorite.

These numbers are reflective of the small number of Syrians who are already online (hence the high penetration rates). However, they do not conclusively show how many Syrians actually use these platforms. Note that there have been no updates to Syrian social media statistics in 2016 due to the ongoing sectarian conflict that intensified in 2014 when the so-called Islamic State, or Daesh, illegally usurped Syrian lands and brought in thousands of foreign fighters to fight for its radical, militant beliefs.

The use of social media in the country and throughout the conflict since 2011 has led Syria to being called the "First Social Media War" or the "First YouTube War" because it seemed like every faction, citizen, and combatant had its own social media channel, network, and followers (Van Bulck 2012). Anything that happens in Syria can be recorded or photographed and immediately uploaded online. It is not uncommon to find videos and pictures online of the conflicts' casualties or towns that have been devastated.

Daesh has used the internet and social media in Syria to promote the destruction of the Syrian state and support a violent jihad, or Islamic holy war, against the state and its people, including Muslims and non-Muslims. These terrorist insurgents in Syria and Iraq use all types of social networking sites and file-sharing platforms. The most significant of these are Tumblr, justPaste.it, kik, PalTalk, viper, Ask.fm, Instagram, and WhatsApp. Onion routers (sites that use layers of encrypted communications) like TOR are used to protect internet protocol (IP) address locations.

Twitter also plays an important role in communication, though the company actively removes accounts, once identified, that promote terrorism. Daesh's tweets often include embedded pictures, text, and links to other platforms to disperse the group's propaganda locally and abroad. Because the terrorists use so many different platforms, encrypt their messages, disguise their locations, jump from platform to platform with ease, and use cross-platform dissemination techniques, they have shown an unusual adeptness and prowess in spreading their pro-violence, intolerant philosophy online. There are presently no good estimates of how many insurgents in Syria use the internet and social media.

In response to the multiyear civil war and the ongoing terrorist threat, many Syrians are attempting to flee the country, often heading to neighboring countries or Europe. Due to the numbers of refugees trying to escape the violence, there has been an upsurge in research on how Syrian refugees use social media to flee the country. According to these reports, refugees consider their phones more valuable than food or shelter. To communicate, they use WhatsApp and Viber; for navigation, they use Google Maps. They follow unfolding events and read communiqués from friends and family on Facebook, Twitter, and specific news channel apps. When necessary, they use translation tools and buy subscriber identity module (SIM) cards for their phones (Gillespie et al. 2016). The main advantages of mobile phones are that they are compact and portable. They also have disadvantages, however, in that some of the information that they can access may not be reliable or privacy protected.

Some Syrian refugees have received assistance through social media. For example, in late 2015, a Norwegian activist posted a picture online that he took in Beirut, Lebanon. In the picture, a desolate-looking man was carrying a small child over his shoulder while selling pens on the street. Once the photo went online, people wanted to offer assistance. After a social media manhunt to identify the vendor using the hashtag #BuyPens, he was identified as Abdul (c. 1980), and within twenty-four hours, the campaign had raised $80,000 to support him and his children (Abdelaziz 2015). Some Syrian parents are trying to attract Western Pokémon Go

During an interview with The Associated Press in November 2011, Syrian refugee Abdul Halim al-Attar and his daughter, Reem, look at the picture that touched the world through the hashtag #BuyPens, and raised $80,000 to support their family. Taken in Beirut, Lebanon, the photo shows a desolate looking al-Attar with Reem asleep on his shoulder while he attempted to sell pens to passing motorists. After the 2011 Arab Spring uprisings, in which social media was instrumental, refugees have been fleeing the resulting civil war. (AP Photo/Hussein Malla)

players to assist children in war-torn areas. They take photos of their children holding pictures of Pokémon characters with slogans seeking help and then post them
online (Graham-Harrison 2016). Another effort, #IamSyrian, asks Syrians to tell
their stories online to make their plights visible outside the country. Internationally, citizens of countries in the United States, United Kingdom, Canada, and others
have started promoting the #RefugeesWelcome hashtag to indicate that they would
like to support, in some fashion, the rising numbers of Syrians fleeing the country's internal crisis.

For those Syrians who cannot escape the fighting or who promote civil action to
stop the bloodshed, social media sites continue to play a prominent role in their
activities. Online activists in Syria have asked people around the world and companies such as Google to color their avatars, profile pictures, and logos red in
May 2016. The campaign, called "Aleppo Is Burning," asked supporters to use the
color red to raise awareness about the fighting in Aleppo and show solidarity with
the Syrian people (Al Arabiya 2016). These types of social media and social networking campaigns can be expected to continue while Syrians struggle to survive
the devastation of war.

Laura M. Steckman and Susan Makosch

See also: Canada; Egypt; Iraq; Libya; Tunisia

Further Reading

Abdelaziz, Salma. 2015. "Social Media Finds Syrian Refugee, Provides Dad a New Start."
    August 31. Accessed August 27, 2016. http://www.cnn.com/2015/08/28/middleeast
    /social-media-and-syrian-refugee/
Al Arabiya. 2016. "As Aleppo Bleeds, Social Media Switches to Red Pictures." May 1. Ac-
    cessed August 27, 2016. http://english.alarabiya.net/en/media/digital/2016/05/01/As
    -Aleppo-bleeds-social-media-users-switch-to-red-pictures.html
Arab Social Media Influencers Summit. 2015. "Arab Social Media Report." Accessed Au-
    gust 27, 2016. http://dmc.ae/img/pdf/white-papers/ArabSocialMediaReport-2015.pdf
Freedom on the Net. 2012. "Syria." Accessed August 27, 2016. https://freedomhouse.org
    /report/freedom-net/2012/syria
Gillespie, Marie, et al. 2016. "Mapping Refugee Media Journeys: Smartphones and Social
    Media Networks. May 13. Accessed August 27, 2016. http://www.open.ac.uk/ccig/sites
    /www.open.ac.uk.ccig/files/Mapping%20Refugee%20Media%20Journeys%2016%20
    May%20FIN%20MG_0.pdf
Graham-Harrison, Emma. 2016. "Syrian Campaigners Use Pokémon Go to Ask World to
    Save Children." July 21. Accessed August 27, 2016. https://www.theguardian.com
    /world/2016/jul/21/campaign-pokemon-go-craze-attention-syrian-conflict
Sutter, John D. 2011. "Syria Tests Internet Freedom Theory." March 30. Accessed August 27,
    2016. http://www.cnn.com/2011/TECH/innovation/03/30/syria.internet.revolution/
Van Bulck, Magali. 2012. "Syrië is eerste 'YouTube-oorlog.'" January 20. Accessed August 27,
    2016. http://www.knack.be/nieuws/wereld/syrie-is-eerste-youtube-oorlog/article-normal
    -46906.html
We Are Social. 2014. "Digital Landscape: Middle East, North Africa, & Turkey." Accessed
    August 27, 2016. http://www.slideshare.net/wearesocialsg/social-digital-mobile-in-the
    -middle-east-north-africa-turkey

T

TAIWAN

Taiwan, or its official name, the Republic of China (ROC), is an island nation off the coast of mainland China in East Asia. Taiwan is committed to continuing its development of information technology (IT) infrastructure, positioning itself as one of the leading high-tech centers in the Asia-Pacific. It has one of the higher internet penetration rates on the globe, falling within the top thirty-five worldwide as of December 31, 2013 (Internet World Stats 2013). Over 75 percent of Taiwanese report having experience using the internet, primarily through personal computers, although mobile phone internet access has recently grown with over 63 percent of Taiwanese now utilizing mobile phones to access the internet (TWNIC 2014). Taiwanese internet users rely on both foreign and domestic search engines and social media. They are on Facebook, Google, YouTube, and Yahoo, along with Taiwanese news, entertainment, and community portals such as Ettoday, Udn, Ruten, Ltn, and Buzzhand. While Facebook was the number one website in 2011, by February 2016, Line had become the favorite app for Taiwanese smartphone users beating out not just Facebook, but WeChat and Instagram (China Post 2011; Taipei Times 2016). Line's popularity reflects the shift toward mobile phone access where users increasingly utilize phones to connect with friends, play online games, find news and information, and watch videos. Mobile phone use will likely continue growing; in 2016, nearly nine in ten Taiwanese had a smartphone (eMarketer 2016).

While only formally recognized as a country by 22 nations, Taiwan has been an independent nation, practically speaking, since 1950 following the ROC's loss of mainland China to the People's Republic of China (PRC) during the Chinese civil war. Mainland China still views Taiwan as a renegade province and tensions between the two governments exist, albeit to a lesser extent today than during the 1990s. In order to combat its diplomatic isolation, Taiwan strives to be an essential regional trading partner and a top producer of computer technology and IT. In doing so, Taiwan established the National Communications Commission (NCC) in 2006 to help Taiwan continue to modernize its economy and telecommunication infrastructure. Taiwan's commitment to developing an advanced network and information society can be seen by the 2014 Network Readiness Rating rankings placing Taiwan fourteenth worldwide (Ministry of Economic Affairs, Department of Investment Services 2016). Overall, Taiwan remains at the vanguard of both high-tech manufacturing and as a technology-driven economy.

Internet usage in Taiwan has increased significantly from 2003, when only 60 percent of the population accessed the internet, to the present, where over three-quarters of citizens reported that they did so. In fact, over 50 percent of Taiwanese children under the age of twelve already reported having used the internet (TWNIC 2014). In May 2016, top internet websites visited by Taiwanese were Pixnet.net, Google, Facebook, YouTube, Ettoday, Yahoo, and Gamer. Taiwan's indigenous social media sites have fared well with the public. Wretch.cc, a site that allows blogging and contains functionality to share photos and videos, took the second spot in the top ten; and Gamer.com.tw, which connects gamers, and Eyny .com, an online discussion forum, placed at sixth and eighth, respectively (Similarweb 2016; Alexa 2016).

While internet use is widespread, the popularity of internet gaming causes concern that it may lead to what researchers call internet addiction. In one of the most severe examples of the problem, a thirty-two-year-old gamer was found dead at a Taiwanese internet café after a three-day gaming session in 2015 (Griffiths 2015). While this type of case is rare, internet addiction has become such an issue that in one Taiwanese children's hospital, two-thirds of patients were diagnosed with "internet overdose" (Wu and Chung 2016), a condition where a person spends so much time online that he or she cannot unplug from the internet and digital devices, and it can lead to sickness, withdrawal symptoms, and even more serious conditions. To address the problem in 2015, Taiwan became the first nation in the world to create a law addressing internet addiction, banning smartphone and tablet use by children under two years of age (Crook 2016). With today's increase of smartphone use, typical internet addicts are now less likely to be teenagers binging in internet cafés and more likely to be individuals in their twenties checking their phones for social network updates (Crook 2016).

Like internet gaming, social media sites are popular in Taiwan. The Taiwanese use social media sites to keep tabs on their friends, pass the time, share news, and relax, and they use instant messaging (IM) services to text and share photos. Of those who use social networks, 30.49 percent of them reported spending between thirty minutes to an hour each day, while 25.58 percent of them spend less than half an hour on social networks (TWNIC 2014). Social media sites have become an important part of the daily lives of not just computer users, but also of smartphone owners. Popular social networking apps and websites include Facebook, Instagram, WooTalk, BeeTalk, and Line (AppAnnie 2016).

The increase of social networking platforms has prompted Taiwan's political leaders to cultivate their own web presence as well. For instance, former Taiwanese president Ma Ying-jeou (1950–) launched his official Facebook page on January 28, 2011, stating that he "should not be absent from the global trend" (China Post 2011). Since 2008, social media has helped Taiwanese politicians to disseminate information and make themselves more visible in the mass media (Chia Shin-lin 2011).

In addition to using the internet for accessing search engines, social media, and gaming sites, Taiwanese use the internet for shopping. Clothes and accessories are the most frequently purchased items through online shopping, with daily essentials taking second. Other items purchased include books, magazines, and writing tools, cross-cultural competence (3C) information tools, and food/drink (TWNIC

In parts of Asia, job seekers use Twitter in an unusual way to apply for jobs. Since 2011, certain Asian markets (most notably in China) ask applicants to submit their entire résumés in 140 characters or less using its local version of Twitter, Sina Weibo. The shortened résumé format is called a *microrésumé*. Shortly after some employers started accepting the micro format, more than 20,000 microrésumés appeared on the platform (Tan 2011). Apart from Asia, this résumé format is not yet a globally accepted way to seek or be selected for a job.

2014). The top three largest ecommerce sites are PCHome, Momoshop, and Rakuten Ichiba Taiwan. Social networks such as Facebook, Twitter, and Plurk, a Taiwanese social network platform, are used to drive ecommerce traffic by providing sharable links to ecommerce sites (Mehra 2016).

Hand in hand with the growth of smartphone use is an increase in app downloads. One study found that 62.19 percent of smartphone users have downloaded at least one app, with 24.42 percent reporting to have downloaded eleven to twenty, and 10.91 percent having downloaded between twenty-one and thirty apps (TWNIC 2014). Social media apps are the most frequently downloaded, while games are second. Although Taiwan represents a growing source of revenue for app makers, only 11.14 percent of smartphone users have paid to download apps, and 49.24 percent reported that they are unwilling to do so. Of those willing to pay for apps, approximately 30 percent are willing to pay between 1 cent and $4 (TWNIC 2014). While the popularity of apps changes frequently, as of May 2016, the most popular free app from Google Play was Tank Commando, a real-time strategy game published by Taiwan Kuro Times. The most popular free iOS app was Chronos Gate, a game combining role playing and Match-3 gaming styles. Clash of Kings, a U.S.-developed real-time strategy game, was the top paid app in for both iOS and Google Play. While gaming apps are popular, so are social networking apps, with Line and Facebook both falling within the top ten free apps for Android and iOS users alike (AppAnnie 2016). As of 2015, the majority of Taiwanese smartphones used Google's Android operating system.

Taiwan's tenuous political situation vis-à-vis the mainland Chinese government and its focus on developing a networked, technology-driven economy has raised considerable concern from Taiwanese leaders over hacking and cybersecurity. Taiwan has found itself the victim of considerable hacking, with attacks originating from mainland China and ostensibly supported by the Chinese government. According to FireEye Inc., a U.S.-based data security firm, Taiwan was the most-targeted country for hacking attempts in the Asia-Pacific region during the first half of 2014 (Gold and Wu 2015). Taiwanese vice premier Chang San-cheng (1954–) has stated that Taiwan has been the victim of cyberattacks from the Chinese government, and that "Chinese cyberattacks have not been deterred by the calming of cross-strait relations as Beijing wishes to know what we are doing and our modes of thought, especially during negotiations" (Li-hua and Chung 2016). He explained that these attacks tend to come in two ways: either through denial of service (DOS) attacks designed to crash specific websites or through backdoor hacking that enables the theft of sensitive information. Taiwanese leaders believe that China regularly uses Taiwan as a testing ground for its most advanced hacking attempts. However, the Chinese government has vehemently denied these accusations.

To help combat these attacks, the Taiwanese government has expressed interest in participating in the U.S. joint antihacking drill known as Cyber Storm. Vice Premier Chang (1954–) warned that there is the potential for hackers to use Taiwan as a back door into U.S. systems (Gold and Wu 2015). In addition to more traditional hacking, Taiwan has recently experienced a tsunami of criticism from Chinese netizens. Following the election of Taiwan's president, Tsai Ing-wen (1956–), in January 2016, Chinese mainlanders flooded the president-elect's Facebook page with tens of thousands of pro-China comments while criticizing Tsai and her party's position on Taiwanese independence (Didi 2016).

Robert Hinck

See also: China; Japan; South Korea; United States

Further Reading

Alexa. 2016. "Website Ranking." Accessed August 1, 2016. http://www.alexa.com/topsites/countries/TW

AppAnnie. 2016. "Top App Charts." Accessed July 14, 2006. https://www.appannie.com/

Bray, David. 2015. "Taiwan and the Internet of Everything." April 9. Accessed May 22, 2016. http://www.huffingtonpost.com/david-a-bray/taiwan-and-the-internet-o_b_7034420.html

Chia-shin Lin, Lue. 2011. "Social Media Politics: the Experience of Taiwanese Politicians." *The Fifth Annual PhD Student Conference in Journalism and Communication Studies,* 164–178.

China Post. 2011. "Facebook Named No. 1 in Taiwan's Top 100 Sites List." March 8. Accessed May 22, 2016. http://www.chinapost.com.tw/taiwan/national/national-news/2011/03/08/293799/Facebook-named.htm

Crook, Steven. 2016. "Too Much of a Good Thing." February 1. Accessed July 25, 2016. http://www.taiwantoday.tw/ct.asp?xItem=241186&ctNode=2235&mp=9

eMarketer. 2015. "Mobile Phones Are Top Internet Access Point in Taiwan." November 16. Accessed July 25, 2016. http://www.emarketer.com/Article/Mobile-Phones-Top-Internet-Access-Point-Taiwan/1013233

Gold, Michael, and Wu, J. R. 2015. "Taiwan Seeks Stronger Cyber Security Ties with U.S. to Counter China Threat." March 30. http://www.reuters.com/article/us-taiwan-cybersecurity-idUSKBN0MQ11V20150330

Grlffiths, Mark. 2015. "Gaming to Death: What Turns a Hobby into a Health Hazard?" January 21. Accessed July 25, 2016. http://www.cnn.com/2015/01/21/opinion/gaming-addiction-risks/

Internet World Stats. 2013. "Top 50 Countries with the Highest Internet Penetration Rates—2013." December 31. Accessed May 22, 2016. http://www.internetworldstats.com/top25.htm

Li-hua, Chung, and Chung, Jake. 2016. "Chinese Hackers Prowling Taiwan's Systems: Chang." May 15. Accessed May 22, 2016. http://www.taipeitimes.com/News/taiwan/archives/2016/05/15/2003646307

Mehra, Gagan. 2016. "Ecommerce in Taiwan: Thriving Market, Huge Mobile." May 12. Accessed July 25, 2016. http://www.practicalecommerce.com/articles/122958-Ecommerce-in-Taiwan-Thriving-Market-Huge-Mobile

Ministry of Economic Affairs, Department of Investment Services, ROC. 2016. "Why Taiwan: Complete Infrastructure." Accessed May 22, 2016. http://investtaiwan.nat.gov.tw/eng/show.jsp?ID=422

Ministry of Transportation and Communications, ROC. "Seamless Networks, Happy Life: Bringing Broadband and Modern Intelligent Life to Every Household." Accessed May 22, 2016. https://www.motc.gov.tw/en/home.jsp?id=253&parentpath=0,150,250

SimilarWeb. 2016. "Website Ranking." Accessed July 14, 2006. https://www.similarweb.com/country/taiwan

Taipei Times. 2016. "Poll Shows Line Is Most Popular Social Platform." February 4. Accessed July 25, 2016. http://www.taipeitimes.com/News/taiwan/archives/2016/02/04/2003638815

Taiwan Network Information Center. 2014. "A Survey on Broadband Internet Usage in Taiwan: A Summary Report May 2014." Accessed May 22, 2016. http://www.twnic.net.tw/download/200307/20140820d.pdf

Tang, Didi. 2016. "Chinese People Are Flooding the Internet with a Campaign Against Taiwan." January 21. Accessed May 22, 2016. http://www.businessinsider.com/chinese-people-are-flooding-the-internet-with-a-campaign-against-taiwan-2016-1

Wu, Hsin-tien, and Chung, Jake. 2016. "Internet Addiction Making Children Sick, Doctor Says." June 20. Accessed July 25, 2016. http://www.taipeitimes.com/News/taiwan/archives/2016/06/20/2003649058

TANZANIA

Tanzania is located on the east coast of Africa, along the shore of the Indian Ocean. It is officially known as the United Republic of Tanzania. Tanzania is bordered by Kenya and Uganda to the north; the Indian Ocean to the east; Zambia, Mozambique, and Malawi to the south; and the Democratic Republic of Congo, Burundi, and Rwanda to the west. Although the country has more than 120 ethnic groups with diverse languages and cultures, Swahili is spoken as the main language and is the first language of most urban and semiurban dwellers. Tanzania has a relatively low internet penetration percentage. Of a population of 55.1 million, Tanzania has only 2.9 million internet users (5.3 percent), 0.1 percent of total world internet users (3.4 billion) (Internet Live Stats 2016). Nevertheless, there has been a significant growth of internet users over the years, from 0.1 percent (2000) to 2.9 percent (2010), and from 3.2 percent (2011) to 5.3 percent (2016). This is largely due to advancement in technology and an improvement in the digital divide worldwide.

Tanzania's information and communication technology (ICT) development has gone through various phases and has faced some challenges, such as lack of communication networks, high technological illiteracy level, lack of power/energy, extreme poverty, and lack of government initiatives and support (Tedre, Ngumbuke, and Kemppainen 2010). However, lately Tanzania has experienced huge advancement in ICT in different social and economic domains, such as communication, industry, journalism and mass media, commerce, and entertainment (Kafyulilo 2011). This advancement has, in part, been made possible through China's technology industry, which provides access to affordable mobile devices.

Until the mid-2000s, the majority of Tanzanians did not own any kind of cellphones and their mobile communication depended on calling booths commonly known as *vibanda vya simu*. Mobile telephone services were available mostly in urban areas, and only a few service providers were licensed to operate in the country. To access the internet, the majority of Tanzanians depended on cybercafés, which provided poor broadband service, in most cases with a speed of only 1 mps (Biztech Africa 2012). Many Tanzanians today, particularly young people, own some form of digital device such as a smartphone, tablet, laptop, or other form of digital technology. The Chinese technology industry has given the majority of Tanzanians access to affordable, multifunctional mobile phones. To date, calling booths have become a thing of the past, while mobile communication has become the easiest and cheapest mode of communication in the country. Smartphones and other digital devices are available at affordable rates, which has increased technology literacy and bridged the digital divide in the country.

In addition to the availability of affordable mobile phones, the number of Global System for Mobile Communications (GSM) service providers have noticeably increased. Before 2003, the Tanzania Communications Commission (TCC) had the sole authority to license operations of telephone communications, including data communication services like internet bandwidth (Tanzania Ministry of Communications and Transport 2003). Due to TCC's strict and unfriendly policy, there were only about six companies that provided telecommunication services in the country until 2005. In 2003, the TCC was merged with the Tanzanian Broadcasting Commission (TBC) and the Tanzania Communications Regulatory Authority (TCRA) was formed. In 2006, TCRA amended its telecommunications policy, and more cellular phone companies were licensed to operate in Tanzania. That brought about competition for subscribers and led to improvement in and affordable accessibility of telecommunication services. Today, telephone towers can be seen everywhere, and even the most rural parts of the country have one or more service providers accessible. In 2010, the SEACOM and the EASSy undersea fiber cable system projects were implemented, which brought affordable high-speed internet (up to 4.2 mps) connectivity in the country.

To date, mobile telephone service operators are the primary providers of internet service and have contributed to revolutionizing accessibility of internet services in the country. GSM internet is now provided by all the service providers. Subscriptions are available in daily, weekly, and monthly data bundles, which give subscribers choices depending on their needs and income levels. For instance, Vodacome, the country's largest service provider, provides the internet bundles ranging from $0.50 for up to 700 Mb per day, $3.50 for up to 2.048 Mb per week, and $12 for up to 8.192 Mb per month. This has given an opportunity to the majority of Tanzanians, especially the youth, to spend much of their time switching from one online platform to another, performing a range of online multitasks like watching videos and movies, playing music, playing online games, social networking, blogging, and web browsing (Lubua 2015).

Tanzania, as in most parts of the world, has experienced a tremendous growth in online social networking. There are few Tanzanian-based social networks; therefore, the majority of Tanzanian internet users have been relying on the foreign social network sites such as Facebook, Twitter, WhatsApp, Tango, Viber, Google Plus, and YouTube. The few Tanzanian-based social networks are solely used for sharing information and discussing matters of national interest. While the foreign social media sites have been used for the same purpose, most Tanzanians use them mainly for entertainment and connecting with other global social network users.

Facebook, WhatsApp, Instagram, and Twitter are some of the most used of all foreign social networks. While most people use Instagram to share pictures and follow updates by various celebrities in the country, they also use Facebook as a source of news, as well as for other social interactions and entertainment. Likewise, they mainly use WhatsApp for multimedia messaging services (MMS), due to its easy use of local phone numbers. Since most mobile phone operators provide poor MMS services, WhatsApp fills that gap in service. WhatsApp enables people to share pictures, audios, short videos, and files. Many people also join WhatsApp groups in which they discuss many issues of common interest. Many people also use WhatsApp and other platforms to communicate with their friends and loved ones outside Tanzania.

There are several Tanzanian local online networking sites, although most of them take the form of forums. These include the country's most popular JamiiForums and MwanaHalisi Forum. On these sites, people can post new topics, and other people can comment and "like" the posts similar to Facebook. In fact, these two forums have been the country's top way to share information about matters of national interest. Most of the country's popular politicians and leaders have subscribed to these social networks; several times, they have used the platforms to communicate important information to the community and to respond to accusations directed toward them. These forums have been the main source of domestic and foreign breaking news, and they have done a better job of communicating than many formal news outlets.

During the 2015 general elections, Tanzanians used social networks to communicate results, announce events, and spread propaganda. While the news outlets were reluctant to report some incidents, people shared information about them freely online, without being censored. This created national tension, and the government worried that people might riot due to the information that they were learning.

Since these events, internet censorship in Tanzania has been on the rise. Although there is no explicit government control of access to the internet and what people can share on social media, there have been unacceptable levels of monitoring and unnecessary police interventions in what people share and publish on the internet. In 2015, the Cybercrimes Act was signed into law. This act forbids sharing online any "confidential" government document, even when the document is believed to have national interest. The law also requires that all internet service providers (ISPs) divulge the personal data of their subscribers to law enforcement on demand. Critics charged that these and other requirements of the law infringe upon freedom of speech and threaten people's freedom to use social media in Tanzania. Just a few months after the law was passed, a few individuals were arrested and charged with insulting the president and publishing false information online. This has resulted in self-censorship by the majority of Tanzanians, in that they decide to limit the amount and type of information that they publish online. Due to the effects of the Cybercrimes Act, by 2016, Tanzania's civil liberties ranking had decreased, though technically only by one point from 2015 to 2016 according to Freedom House (2015, 2016). However, observers and activists now consider Tanzania a country with high internet censorship (Freedom House 2016).

Filipo Lubua

See also: Kenya; Mozambique; Somalia; Zimbabwe

Further Reading

Ali, L. October 12, 2011. "The Digital Revolution in Sub-Saharan Africa." Al Jazeera. Accessed July 31, 2016. http://english.aljazeera.net/indepth/features/2011/10/201110108635691462.html

Communications Technologies Policy. Dar es Salaam, Tanzania: Ministry of Communications and Transport.

Custer, C. 2012. "Chinese Mobile Phones Hold a Big Chunk of East African Market, But That's Not a Good Thing." TECHNASIA. Accessed August 1, 2016. https://www.techinasia.com/chinese-mobile-phones-hold-big-chunk-east-african-market-good

Fortune of Africa. 2016. "Tanzania ICT Sector Profile." Accessed August 2, 2016. http://fortuneofafrica.com/tanzania/ict-sector-profile/

Freedom House. 2013. "Freedom of the Press: Tanzania." Accessed August 2. https://freedomhouse.org/report/freedom-press/2013/tanzania

Freedom House. 2015. "Freedom of the Press: Tanzania." Accessed January 18, 2017. https://freedomhouse.org/report/freedom-press/2015/tanzania

Freedom House. 2016. "Freedom in the World: Tanzania." Accessed August 7, 2016. https://freedomhouse.org/report/freedom-world/2016/tanzania

Hesselmark, O., and Engvall, A. 2005. "Internet for Everyone in African GSM Networks." Stockholm: Scanbi-Invest HB.

Lubua, F. 2015. "Exploring the Opportunities for Integrating New Digital Technologies in Tanzania's Classrooms." *International Journal of Learning, Teaching and Educational Research,* 14(2): 131–150.

Sambira, J. 2013. "Mobile Youth Drive Change. Africa Renewal." May. Accessed July 31, 2016. http://www.un.org/africarenewal/magazine/may-2013/africa%E2%80%99smobile-youth-drive-change

Tedre, M., Ngumbuke, F., and Kemppainen, J. 2010. "Infrastructure, Human Capacity, and High Hopes: A Decade of Development of eLearning in a Tanzanian HEI." *RUSC: Revista de Universidad y Sociedad del Conocimiento,* 7(1).

U.S. Department of State Bureau of Democracy, Human Rights and Labor. 2012. "Country Reports on Human Rights Practices for 2012: Tanzania." Accessed July 31, 2016. http://www.state.gov/j/drl/rls/hrrpt/2012humanrightsreport/index.htm?year=2012&dlid=204176#wrapper

The Citizen. 2015. "Things You Didn't Know About New Cyber Law." April 7. http://www.thecitizen.co.tz/News/national/New-cyber-law-/-/1840392/2677970/-/jb198e/-/index.html

THAILAND

More than 68 million people reside in Thailand, a country located in Southeast Asia. The country's current internet penetration is 56 percent, which is significantly higher than the 37 percent reported in 2015. In one year, the number of active internet users increased by about 19 percentage points, or a gain of approximately 12.9 million new internet users. Largely due to the rural population's lack of access to the internet, most active internet users are concentrated in urban city centers, such as Phuket, Chiang Mai, and Bangkok. To improve Thailand's internet presence, the Ministry of Information and Communication Technology (MICT) has created a series of pilot projects to develop the country's online presence. While these plans may push Thailand into becoming the digital leader in Asia, the censorship imposed on the internet and social media community by the Thai government may stop the country from having a dominant presence online.

Before 2006, Thailand had only one internet gateway to serve the country's entire population. It has since opened up considerably, with over 100 internet service providers (ISPs) providing access to the internet, with the top two internet providers operated by Chulalongkorn University and King Mongkut's Institute of Technology (Bangkok Post 2015). In May 2016, thirty-four cable TV operators from multiple provinces, such as Phetchabun, Chon Buri, and Ubon Ratchathani, submitted a joint application to provide internet services. Once approved by the Thailand's National Broadcasting and Telecommunications Commission, companies initially plan to sell to preexisting cable customers to create a new revenue stream (Hawkes 2016).

After Indonesia, Thailand's smartphone market is the second largest in Southeast Asia. In 2015, nearly 22 million units were sold within the country, and about half of the Thai population uses a smartphone (Leesa-Nguansuk 2016). Because of the economical convenience of smartphones being available to the average Thai, mobile phone users have easy access to social media apps. The regular use of social media apps has allowed global social media providers dominate the Thais' online environment. Of the social media platforms available, Thais use Facebook, Google+, Instagram, Line (also known as Line Up Messenger), and Twitter the most. Mobile users also spend about three hours and fifty-three minutes on their phone perusing the internet and social media platforms (Kemp 2016).

While Facebook and Twitter are the main choices for Thailand in terms of social networking sites, Thais use the Line app for instant messaging (IM). Created in South Korea and launched in Japan in 2011, the Line messaging app quickly grew in popularity among the Thai population, which boasts nearly 33 million monthly active users. The Line app not only allows Thais to communicate with one another, but also allows users to take photos, play games, and download third-party apps to enrich the user's experience (Orsini 2015). To remain competitive with other global social media providers in Thailand, Line opened a research and development division in Bangkok. With locally recruited engineers, Line is developing a number of products and services specifically for the Thai market in 2016 (Quigley 2015).

With a digital economy master plan and a series of pilot projects, MICT plans to develop Thailand's online presence in the hope of steering the country toward becoming the digital hub in Southeast Asia. One of MICT's pilot projects is to develop an e-commerce platform by helping small and medium-sized businesses establish an online presence. In order to sustain the e-commerce project, MICT plans to work with the Commerce Ministry to develop an e-payment platform. These two projects will allow the government to verify and monitor online businesses and merchants (Phuket Gazette 2015).

Another pilot project will be e-education. MICT plans to have high-speed internet available to every village in Thailand. This would allow remote rural areas to connect with the rest of Thailand and allow schools in the countryside to utilize the internet for educational purposes. MICT hoped to have this particular project completed by the end of 2016 (Phuket Gazette 2015). However, that goal appears to have been shifted to 2017 to ensure that 79,000 villages all receive the fast internet connectivity. Geographically, Thailand lies at the heart of the Asian market and has the eighth fastest internet speed in Asia (Bangkok Post 2015). The country is an ideal location for multinational companies to establish a central base to gain a foothold in the Asian market. MICT plans to transform the cities of Phuket and Chiang Mai into smart cities, in the hope of attracting tech start-ups. It is believed that this decision will move Thailand in the direction of becoming ASEAN's digital hub (Phuket Gazette 2015).

Prior to 2006, the Thai government monitored internet traffic on a single gateway. The primary reason for monitoring was to block pornography, pornographic-related material, and gambling, all of which are illegal in Thailand. In 2007, the Computer-related Crimes Act (CCA) was written specifically to target internet users and online freedom. The CCA allows MICT to block online content that is seen as a threat to national security, as well as that which violates public morals and order. From 2006 until 2013, the Thai state blocked tens of thousands of social media pages and websites. YouTube, a popular global video-sharing platform, was blocked multiple times during this period for airing videos that were considered offensive to the Thai state. The computer-related crime law also allowed the state to prosecute and imprison people for sentences up to five years because of their online opinions and activities that are critical or insulting of the Thai monarch and government, a direct violation of the Thai morals and public order (Talcoth 2015).

A coup d'etat in May 2014, in which the Thai military removed Prime Minister Yingluck Shinawatra (1967–), only tightened internet censorship. General Prayuth Chan-o-cha (1954–), also known as the Commander of the Royal Thai Army, declared martial law. Following this coup d'etat, fourteen progovernment and antigovernment TV networks shut down and nearly 3,000 unlicensed radio stations were closed. The military set up a special committee for the purpose of monitoring the social media community and summoning internet service providers to discuss the censorship. The CCA is currently under revision, but there is no inclination if the law will become stricter or more lax with regard to internet users critical of the Thai monarchy and government (Pitman 2014).

The Thai government continues to limit the information available to Thai internet users, even going as far as reaching out to international providers in order to remove content. The Thai government applauds as well as utilizes Twitter's country

Chiranuch Premchaiporn (n.d.), director of Prachatai's website, stands in front of the criminal court in Bangkok, Thailand, in May 2012. She had been sentenced under the 2007 Computer Crimes Act to an eight-month suspended sentence for not acting quickly enough in removing posts that were deemed insulting to Thailand's royalty. Since Premchaiporn's conviction, and the 2014 military coup d'etat, the government has only increased internet censorship, which includes reaching out to international providers, such as Twitter and Facebook, to remove content. The government also plans to consolidate all of Thailand's internet gateways into one system, which would make monitoring internet traffic easier. (AP Photo/Apichart Weerawong)

withheld content tool (Gilbert 2015: Twitter 2016). While Twitter respects users' freedom of expression, it also respects the local laws of the countries in which it operates. If content does not violate Twitter rules, it cannot be removed. However, if a tweet violates a country's content restrictions, a government/law enforcement agency representative can request that its content be withheld. If the request is approved, Twitter users in Thailand would not be able to view the tweet (Twitter 2016). A similar situation applied with Facebook: from July 2014 to December 2014, the Thai Computer Security Incident Response Team (CERT) or the Ministry of Foreign Affairs reported all content that it found offensive or in violation of national law to Facebook. In response to these requests, Facebook restricted access to thirty pieces of content that violated local laws prohibiting slander and condemnations of the Thai monarchy (Siam Voices 2015). The amount of content that officials have requested to be removed from Facebook has been increasing. In 2013, no requests were made; of the thirty requests made at the end of 2014, only five had been made prior to July.

MICT also developed a plan to revert Thailand to a one-internet gateway. Consolidating all of Thailand's internet gateways into one system would allow the government to monitor internet traffic easily. This program, approved in August 2015, was reminiscent of China's Great Firewall system. In China, the government has placed the internet under the strictest government control in the world. In October 2015, however, Deputy Prime Minister Somkid Jatusripitak (1953–) announced that the one-internet gateway plan had been halted. There were concerns that internet speeds would plummet, which not only would hurt Thai online business, but also sidetrack plans to make Thailand the digital hub of the Association of Southeast Asian Nations (ASEAN). However, some people do not believe that this announcement from the deputy prime minister means the end of the one-internet gateway. They think that the plan was originally scrapped because of the poor reception it received from the public, but once the Thai people forget about the plan, the government may execute it at a later date (Reuters 2015).

Karen Ames

See also: China; Indonesia; Myanmar; Singapore

Further Reading

Bangkok Post. 2015. "Thailand Internet 8th Fastest in Asia." May 21. Accessed December 9, 2015. http://www.bangkokpost.com/tech/local-news/568859/thailand-internet-8th-fastest-in-asia

Hawkes, Rebecca. 2016. "Thai Cable Operators Seek Broadband Internet Licenses." May 10. Accessed May 13, 2016. http://www.rapidtvnews.com/2016051042812/thai-cable-operators-seek-broadband-internet-licenses.html#axzz48m2CTzul

Kemp, Simon. 2016. "Digital in 2016: We Are Social's Compendium of Global Digital, Social, and Mobile Data, Trends, and Statistics." January 27. Accessed May 2, 2016. http://wearesocial.com/uk/special-reports/digital-in-2016

Leesa-Nguansuk, Suchit. 2016. "Smartphone Market Will Be Stale Following Rapid Growth." January 26. Accessed May 11, 2016. http://www.bangkokpost.com/tech/local-news/839728/smartphone-market-will-be-stale-following-rapid-growth

Phuket Gazette. 2015. "ICT Considers Developing 'Smart' Phuket." September 15. Accessed December 11, 2015. http://www.phuketgazette.net/phuket-news/ICT-considers-developing-smart-Phuket/61957

Pitman, Todd. 2014. "Thailand Imposes Media Censorship as Military Coup Begins." May 22. Accessed December 7, 2015. http://www.ctvnews.ca/world/thailand-imposes-media-censorship-as-military-coup-begins-1.1832916

Reuters. 2015. "Thailand Scraps Unpopular Internet 'Great Firewall' Plan." October 15. Accessed December 10, 2015. http://www.reuters.com/article/us-thailand-internet-idUSKCN0S916I20151015

Siam Voices. 2015. "Thailand: Facebook Content Removal Requests Rise Sharply Under Junta." March 16. Accessed May 6, 2016. https://asiancorrespondent.com/2015/03/thailand-facebook-content-removal-requests/

Talcoth, Roberth. 2015. "Thailand's Social Media Battleground." March 26. Accessed December 7, 2015. http://asiapacific.anu.edu.au/newmandala/2015/03/26/thailands-social-media-battleground/

Twitter. 2015. "Country Withheld Content." Accessed December 11, 2015. https://support.twitter.com/articles/20169222?lang=en#

TIMOR-LESTE (EAST TIMOR)

The territory of Timor-Leste (East Timor) occupies the eastern half of Timor, the easternmost island of the Sunda archipelago in Southeast Asia. Timor-Leste became an independent nation in 2002 after close to 500 years of Portuguese colonial administration, 24 years of Indonesian military occupation (1975–1999), and 3 years of a United Nations transitional administration.

As one of the newest nations in the world, Timor-Leste is also one of the least economically developed, and yet it has one of the fastest growing economies due to the exploitation of petroleum deposits. However, two-thirds of Timorese live a life devoted to subsistence farming (Timor-Leste, National Statistics Directorate 2010) and over a third (37.4 percent) live on less than $1.25 a day in purchasing power parity terms (World Bank 2012). The currency used is U.S. dollars, which has the positive effect of stability, but the negative effect of keeping prices much higher than in neighboring countries.

According to preliminary results of the latest national government census (Timor-Leste, National Statistics Directorate 2015) the total population is 1.17 million as of July 2015. Approximately 55 percent of the population is nineteen years old or under (Timor-Leste, National Statistics Directorate 2010, 12). Literacy rates for Timor-Leste are low (58.3 percent): Half of all women and more than 40 percent of men in Timor-Leste are illiterate (UNICEF, 2016). However, literacy rates are higher among the younger generation, reflecting stability and better education systems (UNFPA Timor-Leste 2016).

The Timorese constitution establishes two co-official languages: Tetum (the indigenous lingua franca) and Portuguese; in addition, it values the use and development of its other thirty-two national languages, including dialects (census 2010). English and Bahasa Indonesia are also recognized as working languages. Tetum is the most used language, followed by Indonesian, Portuguese, and finally English. Tetum is largely a spoken language, and attempts to standardize its written form are far from complete. While many Timorese also speak Indonesian, levels of fluency vary substantially between spoken and written ability. This language diversity affects the ability to communicate fluently in any single language and has an obvious impact on the way in which Timorese people communicate and interact through telecommunications systems, the internet, and social media platforms.

The diverse language ecology of Timor-Leste presents a challenge for internet usage for the majority of citizens. Devices do not offer the option of Tetum-language support, so most Timorese use either Indonesian, Portuguese, or English. Tetum-language support in applications is almost nonexistent, with even Google.tl using Portuguese as its search engine language (as well as Indonesian and English) rather than Tetum.

In 2002, the government of Timor-Leste granted a fifteen-year exclusive telecommunications license to Timor Telcom, a subsidiary of Portugal's Telecom. Due to this monopoly, the telecommunications sector had some of the highest service rates in the world for the first ten years of independence. This made internet access prohibitively expensive, and therefore inaccessible, for the majority of people. A network of civil society organizations set up its own independent V-Sat infrastructure to be able to access more affordable internet in the capital, Dili, but in particular also in the rural districts. Limited V-Sat internet bandwidth service fees could run to over $1,000 monthly, but they were still considerably cheaper than accessing the internet through the only service provider in the country, Timor Telcom.

In 2012, in order to liberalize the telecommunications sector in the country and make it more competitive, the Timorese government issued additional licenses to two new telecommunications operators: PT Telekomunikasi Indonesia International (Telin), from Indonesia; and Viettel Global Investment JSC (Telemor), from Vietnam. This diversification, as well as a more competitive telecommunications sector, greatly reduced the costs of mobile subscriptions and provided greater free internet access in Timor-Leste, with mobile customers increasing from 11.9 percent in 2008 to 57.4 percent in 2013 (UNESCO 2015). According to BuddeComm (2015), the mobile market in Timor-Leste has grown much more, to around 90 percent in June 2015. Quoting Timor-Leste's General Directorate of Statistics, TeleGeography (2015) reported a total of 1.25 million mobile customers in Timor-Leste in mid-2015, including 145,085 3G network users. Further, the Population and Household Census 2015 included questions related to ICT access in the country, such as the number of households having mobile cellular telephone(s), households having personal computer(s), households accessing the internet from home, households accessing the internet from elsewhere other than home, and households without access to the internet. The complete results of the official census, including the ICT indicators, have not yet been published, however. Timor Telcom reported 620,204 subscribers, including 7,592 3G users; Telemor reported 470,730 subscribers, including 87,493 3G users; and Telkomcel reported 160,000 subscribers, including 50,000 3G users (TeleGeography 2015).

Notwithstanding the high growth level of mobile phone subscriptions, data from the International Telecommunication Union (ITU), the World Bank, and the United National Population Division report that internet usage in Timor-Leste in 2016 remains very low, with only 14,030 internet users, a penetration rate of only 1.2 percent (Internet Live Stats 2016). This low number is based on the ITU definition of internet users as those who have access to the internet, with no financial barrier preventing usage. The majority of the 145,085 Timorese internet users who have 3G access, as reported by TeleGeography (2015), can access the internet only through extremely limited free services. This access is primarily from promotional mobile phone credit recharges, which include a limited amount of free data, or by using free Wi-Fi hot spots at the branch offices of telecommunications providers and other key locations such as at the shopping mall in Dili or at selected restaurants, cafés, and organizations. Despite the liberalization of the telecommunications sector in 2012 and the subsequent rapid increase of mobile phone subscriptions, with limited access to the internet, most Timorese are still unable to afford such access and rely almost exclusively on free access provided by the service providers and other third parties.

Timor-Leste social media consumers predominantly use Facebook as the network platform for social engagement and interaction, as well as for more public political and community purposes, including the sharing of news and information. Statistics for social media usage of social network platforms in Timor-Leste report that 94.91 percent of consumers use Facebook, followed by 4.04 percent using Twitter and the remaining consumers using other social media networks and platforms (StatsMonkey 2015). Some 85 percent of all Facebook users in Timor-Leste are between thirteen and twenty-nine years of age, compared to only 15 percent above thirty years of age (ASEANUp 2015).

In terms of media and information access, the main sources of information for the majority of Timorese come from analog broadcasting (radio and television), with over 95 percent penetration rates in the country, with the next most significant source of information being newspapers, and then local leaders (Niner et al. 2015). Despite a pluralistic media landscape, with more than twenty news media organizations in the country (including printed and online newspapers, state and private television stations, and community radio stations), most of these fall into the category of traditional media (newspapers, radio, and television); however, there has been a modest but growing trend in investment on online content.

The majority of the media broadcasts and publishes in the Tetum language, with limited content in Portuguese and Indonesian. There is one local online newspaper, *The Dili Weekly* (www.thediliweekly.com), which publishes news in both English and Tetum, and SAPO Timor-Leste (www.sapo.tl), a subsidiary of the Portugal Telecom Group, which provides a news service in Tetum and Portuguese.

Sara Niner and Emanuel Braz

See also: Australia; Indonesia; Vietnam

Further Reading

ASEANUp. 2015. "Southeast Asia Digital in 2015." Accessed April 27, 2016. http://aseanup .com/southeast-asia-digital-social-mobile-2015/

BuddeComm. 2015. "Timor Leste (East Timor)—Telecoms, Mobile, and Internet." Accessed April 27, 2016. http://www.budde.com.au/Research/Timor-Leste-East-Timor-Telecoms -Mobile-and-Internet.html

Internet Live Stats. 2016. "Internet Users by Country (2016)." Accessed April 27, 2016. http://www.internetlivestats.com/internet-users-by-country/

Niner, Sara, Wigglesworth, Ann, dos Santos, Abel Boavida, Tilman, Mateus, and Arunacha-lam, Dharmalingam. 2013. "REPORT: PyD Baseline Study—Attitudes & perceptions of gender and masculinities of youth in Timor-Leste." Dili, Timor-Leste: Paz y Desarrollo (PyD).

StatsMonkey. 2015. "Mobile Facebook, Twitter, Social Media Usage Statistics in Timor-Leste." Accessed April 27, 2016. https://www.statsmonkey.com/packedcircle/21483-timor-leste -mobile-social-media-usage-statistics-2015.php

TeleGeography. 2015. "Mobile Subscribers Reach 1.25m at Mid-2015." September 2015. Accessed April 27, 2016. https://www.telegeography.com/products/commsupdate /articles/2015/09/10/mobile-subscribers-reach-1-25m-at-mid-2015/

Timor-Leste, National Statistics Directorate. 2010. *Population and Housing Census 2010.* Dili:, Ministry of Finance.

Timor-Leste, National Statistics Directorate. 2015. *Population and Housing Census 2015: Preliminary Results.* Dili: Ministry of Finance.

UNFPA Timor-Leste. 2016. "Timor-Leste." Accessed May 3, 2016. http://countryoffice.unfpa .org/timor-leste/2009/11/02/1482/timor-leste_democratic_republic_of/

United Nations International Children's Emergency Fund (UNICEF). 2016. "At a Glance: Timor-Leste-Statistics." Accessed May 3, 2016. http://www.unicef.org/infobycountry /Timorleste_statistics.html

United Nations Educational, Scientific, and Cultural Organization (UNESCO). 2015. UNESCO Science Report: Towards 2030. New York: UNESCO Publishing.

UNMIT United Nations Integrated Mission in Timor-Leste. 2011. *Timor-Leste Communication and Media Survey*. Accessed January 18, 2017. https://unmit.unmissions.org/Portals /UNMIT/Media_Survey_Report_CPIO_FINAL_ENG.pdf

World Bank. 2012. *World Development Indicators 2012*. Accessed February 2014. http://data .worldbank.org/data-catalog/world-development-indicators

TUNISIA

Tunisia is a small country in North Africa located southwest of Italy, along the coast between Libya and Algeria. Much of the early development of the internet in Tunisia occurred under the regime of Zine el Abidine Ben Ali (1936–), which the Tunisian military removed on January 14, 2011, as part of a series of popular uprisings throughout the Middle East commonly known as the "Arab Spring." As such, the internet in Tunisia is widely considered to be going through a transformational process after democratization. Many legal reforms have been undertaken since the overthrow of the earlier authoritative regime to increase citizens' abilities to communicate freely and privately. However, as much of Tunisia's infrastructure was originally designed to help monitor political opposition, existing laws and structures (alongside a need to protect against attacks on officials) continue to affect the development of the internet in Tunisia.

Historically, the Tunisian government and its various agencies have been wary of the internet as a means of communication. While Ben Ali recognized the importance of developing telecommunications infrastructure to strengthen Tunisia's economy, there were efforts to limit the social and political ramifications of new media. At the time of the country's official public opening to the internet in 1996, the ICT sector was seen as a potentially powerful mechanism for economic development. Both national policy makers and international development agencies saw Tunisia's relatively small and young population as being particularly able to profit from a stronger telecommunications sector and encouraged investment into building stronger infrastructures to develop businesses that worked within the digital economy, as well as providing projects that could attract investment in Tunisia. In addition, many summits on the issue of the global digital divide between the developed and developing worlds have been held in Tunis, such as the World Summit on the Information Society in 2005 and the African Internet Summit in 2015.

Much of Tunisia's planned growth in ICT infrastructure came under fire for excluding voices from the political opposition, especially in the early phases of development. The regime would frequently remove or censor bloggers suspected of supporting the opposition to Ben Ali, and the government blocked video-sharing sites entirely. There were also incidents of Facebook users' login information being intercepted and organizations' internal emails scrambled.

Until recently, the state-owned Tunisie Télécom held an effective monopoly over internet service in Tunisia due to its ownership of the ICT infrastructure and the need for broadband subscribers to use the company's phone lines to transfer data. Authorities largely used Tunisie Télécom's monopoly as a means of restricting citizens' access to the internet through high prices for international communications. Access has been especially restricted in the countryside, which depends on 3G mobile phone networks for internet access, given the lack of extensive broadband infrastructure.

Private companies and state-led initiatives have provided more diverse ways for Tunisians to access the internet. Currently, there are three major telecommunications providers in Tunisia: Tunisie Telecom, Ooredoo, and Orange Tunisie. Recent infrastructural projects have focused on laying cables between Tunisia and Europe to increase connection speeds from Tunisia and position the country as a hub for telecommunications across Africa. Plans to reform investment laws in the hope of financing more infrastructural projects in the interior had been prepared prior to the 2011 Arab Spring revolution, but since then, other political concerns, such as the reconciliation process, have sidetracked them.

In spite of efforts to bring other telecommunications companies such as Orange Tunisie and Ooredoo into competition for customers, Tunisie Télécom continues to provide coverage to most of the country. Because of this dominant position and previous ties to state censorship, many Tunisians suspect that the company may help the country's rulers filter information coming into the country and disrupt political communications, either now or in the future. Furthermore, the scarcity of infrastructure to provide high-speed internet to many parts of Tunisia has encouraged corporations to limit access to bandwidth-intensive activities, such as voice over internet protocol (VoIP) services (e.g. programs that allow users to make telephone calls). According to companies and state officials, the restrictions are intended to build the demand for telephone use, but the restrictions are reminiscent of censorship for many, as civic groups often use these technologies to coordinate activities, advocate for policies, and especially target international communications.

Following the removal of Ben Ali, subsequent elected governments have made efforts to create legal protections for free access to information online. Article 24 of Tunisia's new constitution, adopted in January 2014, gives the state the responsibility of protecting rights to private communication and personal data except in cases of national security or criminal activity.

Nevertheless, there are still two major ways in which government agencies can use existing laws to monitor internet activity. First, laws originating under Ben Ali's regime continue to be in effect, allowing government agencies to monitor private communications. Many of the laws governing internet service providers (ISPs) were originally developed to assist in the monitoring of preinternet telecommunications infrastructure. Notable among these laws is a prohibition on the encryption of telecommunications and the requirement of purchasers of subscriber identity module (SIM) cards to provide identification that can be used to monitor their online activities. As SIM cards are increasingly the predominant means by which many

Just days before voting in a new constitution, members of Tunisia's Constitutional Assembly are seen here attending a session on January 23, 2014. Article 24 of the new constitution provides for the right to private communication and personal data. However, pre-existing laws allow government agencies to monitor internet activity, such as activity associated with media publications, which are not covered by privacy protections, and can undermine a blogger's ability to post freely. (AP Photo/Hassene Dridi)

Tunisians access the internet, these requirements maintain the infrastructure that the regime previously used to monitor activities.

In addition, the structure of ISP regulation and security laws may be used to encourage nongovernmental censorship of content and allow covert surveillance. Laws originating in the late 1990s placed legal liabilities on ISPs that distributed illicit content, including criticism of the regime as well as pornography and criminal activity. As a result, the structure of Tunisian ISPs allowed for blocking content critical of the government without direct involvement by the authorities. Security regulations also played a role in allowing Ben Ali's government to monitor online activity. As data provided to the state and media publications were not covered by privacy protections, private messages and posts could potentially be visible to officials, undermining bloggers' ability to post freely.

Second, intermittent states of emergency have made it difficult to create consistent legal protections for internet use. In a state of emergency, authorities are permitted to monitor activity suspected of undermining national security, including online communications. In addition, protections on privacy can be lifted temporarily to monitor potential terrorist activity. Attacks on government ministers, tourists, and members of the military have caused several declarations of a state of emergency, making it difficult for users to be aware of their rights on a continual basis and allowing officials to remove Facebook pages that may fuel violence or extreme views.

These attacks and efforts to prevent them have also incentivized the creation of bodies such as the Technical Agency for Telecommunications (ATT, or l'Agence Tunisienne des Télécommunications) in 2013. This agency, sometimes referred to as the Tunisian equivalent of the National Security Agency (NSA), provides technical support for investigations, but it also has often been criticized for preventing suspects from receiving a fair trial because defendants are not allowed to challenge any evidence that ATT provides at trial. Many similar internet governance agencies formed under Ben Ali's regime continue to wield significant power over the operation and regulation of the internet. However, this control has been gradually reduced in recent years. One example of this development is the Tunisian Internet Agency (ATI, Agence Tunisienne d'Internet), which used to filter most internet traffic in Tunisia through its servers to monitor communications and censor material. The ATI's mission changed drastically following the revolution, but the institution continued to be seen as a tool of censorship by many bloggers. In 2013, the ATI began to allow ISPs to circumvent state-run servers.

As of 2014, approximately 46 percent of Tunisians used the internet (World Bank 2014). Internet usage grew rapidly during the first decade of the twenty-first century and has continued to rise steadily since that time. Although a minority subscribe to fixed broadband lines, most Tunisian internet subscriptions are purchased through mobile plans or universal serial bus (USB) keys that can be reloaded as needed. At the time of writing, a typical 3G USB key cost the equivalent of $20.50, with a monthly subscription cost of approximately $13 (Freedom House 2015). These subscriptions are often criticized for being beyond the financial means of most Tunisians and with antiquated or nonexistent infrastructure in many rural areas of the country's interior. High-speed mobile internet access is concentrated primarily in major coastal cities, while major interior areas frequently have significantly slower connections.

As for social media use, the majority of Tunisians online choose Facebook over other major platforms. This preference largely follows the pattern of social media elsewhere in the Arab world, but there is considerably less penetration of Twitter in Tunisia than elsewhere in the region. With about half of the country able to access the internet in 2014, 42.1 percent of the internet-using population had Facebook accounts, compared to just 4.2 percent with LinkedIn accounts and 0.3 percent with Twitter accounts (Arab Social Media Report 2014).

The vast majority of social media activity in which Tunisians take part is held in French, with a smaller amount in English and Arabic. There are several reasons for this situation. A major reason is the heavy influence of the French language on Tunisian society, especially among more affluent members of society who are more likely to have access to the internet. Although no longer an official language of government as it was prior to independence, French remains an important language in the country, both due to continuing trade ties with French-speaking countries and the social prestige associated with the language.

Furthermore, many content creators may choose to use French for search engine optimization (SEO). Search engines often fail to return accurate results from text in the Arabic script, as short vowels and other diacritical marks are often used inconsistently in practice. The right-to-left reading of Arabic also can create challenges for coding content on websites using programming languages or interfaces that were designed only for a left-to-right alphabet. In addition, using French can allow these individuals to engage a broader audience and gather a wider following among Tunisians living abroad, who often follow Tunisian politics and news very closely.

As Tunisia was one of the first countries to erupt into protests during the Arab Spring of 2011, there is significant debate about the role of the internet in promoting the unrest there. Those who see the internet as an integral factor in the fall of the regime point to the abundance of political writers online, the ability of gathering protests to organize and spread through social media, and the fact that the regime shut down the internet shortly after the protests started as a means to prevent the use of the internet in the protests. These factors hint at the potential for the internet to galvanize protest activity. However, many other observers writing in the years following 2011 have argued that the internet at best took a back seat to offline civil society organizations, and that the protests themselves were not as clearly planned as they might have been if they really had been coordinated online.

Tyler Overfelt

See also: Algeria; France; Libya

Further Reading

Arab Social Media Report. 2014. "Citizen Engagement and Public Services in the Arab World: The Potential of Social Media." Accessed May 19, 2016. http://www.mbrsg.ae/getatt achment/e9ea2ac8-13dd-4cd7-9104-b8f1f405cab3/Citizen-Engagement-and-Public -Services-in-the-Arab.aspx

Freedom House. 2015. "Freedom on the Net: Tunisia." Accessed April 16, 2016. https:// freedomhouse.org/report/freedom-net/2015/tunisia

Privacy International. 2016. "State of Surveillance in Tunisia." Accessed April 16, 2016. https://www.privacyinternational.org/node/743

Scola, Nancy. 2011. "Why Tunisia Is Not a Social-Media Revolution." *The American Prospect*, January 21. Accessed May 19, 2016. http://prospect.org/article/why-tunisia-not-social -media-revolution-0

Shirayanagi, Kouichi. 2015. "Five Years After Revolution, Internet Censorship Is Creeping Back into Tunisia." December 1. Accessed May 19, 2016. http://motherboard.vice.com /read/five-years-after-revolution-internet-censorship-is-creeping-back-into-tunisia

World Bank. 2014. "World Development Indicators—Tunisia—Internet Users per 100 people." Accessed May 19, 2016. http://data.worldbank.org/indicator/IT.NET.USER.P2 ?locations=TN&name_desc=true

Wagner, Ben. 2012. "Push-Button Autocracy in Tunisia: Analysing the Role of Internet Infrastructure, Institutions, and International Markets in Creating a Tunisian Censorship Regime." *Telecommunications Policy*. 36, 6.

TURKEY

Turkey is a geographically, culturally, and politically diverse nation located at the crossroads between the European and Asian continents. While the nation is predominantly Muslim, it is also home to a sizeable Orthodox Christian and Jewish community. Though the population is often tense and fragmented, a religious and ethnic diversity remains somewhat in Turkey from the ruins of the Ottoman Empire. Turkey straddles several important waterways and is neighbor to Bulgaria, Greece, Georgia, Armenia, Azerbaijan, Iraq, Iran, and Syria, making it a geopolitically important and strategic country. Its internet culture is vibrant and thriving, with numerous local online endeavors. Turkey's online environment, while prosperous, is also a highly politicized and contested space. In recent years, it has become a hotbed of contention between the ruling Justice and Development Party; the Gülen Movement, a global religious organization led by Fetullah Gülen (1941–), and leftist oppositional factions. Social networking sites, such as Twitter and Facebook, now figure prominently in debates on national security, unity, and oppositional political activism. Censorship and an air of suspicion of social networking sites feature prominently in official national rhetoric amid domestic and international political challenges to the ruling government.

As of 2016, Turkey's internet penetration rate is 58 percent of the total population. This is a marked increase from the past decade (Turkey's internet penetration rate was a mere 18 percent in 2006). In 2013, the Turkish Statistical Institiute reported that approximately 53.9 percent of all internet users in Turkey were male and 39.8 percent female. In addition to the prevalence of internet cafés, which line the busy streets of Istanbul and the city centers of most cities and towns in Turkey, the International Telecommunications Union reports that 49.1 of Turkey's population have access to the internet at home.

A variety of localized online and digital services remain popular among the urban and semiurban populations of Turkey. For example, Yemeksepeti.com, meaning "food basket" is an immensely popular one-stop shop for ordering food online. This service is completely free to users and provides a centralized platform for ordering food from nearby restaurants. In addition, the smartphone application BiTaksi, meaning "a taxi," allows users to request a taxicab using geolocating software, as the Uber app does.

With just over half of Turkey's population using the internet, Turkey's social media environment continues to grow and remain active (Doğramaci and Radcliffe 2015). There are 39 million monthly Facebook users in Turkey, and of that number, 85 percent access Facebook through mobile devices (Daily Sabah 2015). Globally, Turkey ranks fourth in percentage of the population with Facebook accounts. However, while data may indicate a high penetration of Facebook usage through mobile devices, only 34 percent of the population has access to mobile broadband service (Doğramaci and Radcliffe 2015). Private businesses, sports teams, and politicians all maintain active and public social media profiles. The Facebook pages of Turkish Airlines, the Galatasary soccer team, and President Recep Tayyip Erdoğan (1954–) are among the most followed in the country (Socialbakers 2015).

In Istanbul, thousands of people protested the government's decision to censor the internet in May 2011. Many governments in and around the Middle East feared the wave of government-toppling protests that occurred in early 2011, later known as the Arab Spring. (Evren Kalinbacak/Dreamstime.com)

Social media usage in Turkey has often been characterized as an outlet for sharing and finding news, particularly in light of a general mistrust of mainstream media platforms. Given the historically tight controls that both the military and government have had on media outlets amid military coups and during an aggressive neoliberalization process beginning in the 1990s, this distrust is not surprising. During the Gezi Park protests of 2013, for example, Facebook and Twitter became key outlets for distributing videos, photos, and updates of the unfolding events. These platforms not only served as a way to quickly coordinate gatherings and medical attention for activists affected by police intervention, but also countered mainstream media blackouts and government-imposed censorship. Indeed, social media in Turkey has become a source for bypassing traditional sources of news and highlighting the growing dissatisfaction with Erdoğan and his ruling Justice and Development Party.

Beginning in December 2013, for example, recordings and other official documents began emerging that linked prominent government ministers, businesspeople, and Erdoğan, who was then prime minister, to money transfers to an Iranian businessman, Reza Zarrab (1983–). In addition to exposing Turkey's violation of international sanctions against Iran, the documents and recordings posted by the Twitter handle @HARAMZADELER333 exposed other instances of nepotism, cronyism, and money laundering between government officials and progovernment businessmen (Pamuk and Tattersall 2014). While the corruption scandal was purported by official channels to be an attempt by the Gülen religious organization (led by a former Justice and Development Party ally) to undermine the government, what followed was a massive Twitter battle between online activists and the government of Turkey. Following the corruption scandal, Twitter was blocked across the country, purportedly due to privacy and security concerns. The largest blackout came in early 2014, when access to Twitter was blocked for about two weeks (Özbilen and Coşkun 2014).

The government frequently cites concerns for security and national unity as a main reason for blocking access to social media platforms. YouTube, for example, was blocked for several years after a video emerged that apparently insulted the founder of the modern Turkish Republic, Mustafa Kemal Atatürk (1881–1938). Reflecting the constitutional protection of the image and likeness of Atatürk, Internet Law 5651 (originally passed in 2007) gives the state the authority to shut down websites for distributing material related to child abuse, obscenity, or insulting Atatürk. In 2014, amid the corruption scandal and Twitter battles, the law was expanded to allow the state to shut down websites without a court order if the website was seen a threat to national security or privacy (Human Rights Watch 2014). Though this portion of the law was subsequently overturned by the European Digital Rights Initiative (Jarvinen 2014), other recent court rulings potentially limit the types of personal material that can be circulated and possessed online.

While there remains debate over the authority and legality of the state to shut down websites, the official discourse of the Turkish government frames the online world as a threat to the integrity and unity of the entire country. President Erdoğan and the ruling Justice and Development Party, though perhaps legally constrained for the time being, frame online media as a forum in which foreign and domestic conspirators are attempting to weaken Turkey's economic and political growth (Tüfekçi 2014). This discourse of fear and mistrust marks attempts by the ruling government to contend and stifle a growing opposition that views Erdoğan and his cabinet as antidemocratic.

The ongoing crisis in Syria, political instability in the southeastern corners of the country, and the presence of Islamic State of Iraq and Syria (ISIS), also known as Daesh, cells within the country further underscore a tenuous political environment that sees online media, such as Facebook and Twitter, as active threats to the sovereignty and stability of Turkey. Thus, while Turkey's online presence remains vibrant in both the business and political world, looming and ongoing crises have created a highly politicized and unstable online culture.

Caitlin Miles

See also: Georgia; Greece; Iran; Iraq; Syria

Further Reading

Corke, Susan, Finkel, Andrew, Kramer, David J., Robbins, Carla Anne, and Schenkkan, Nate. 2014. "Democracy in Crisis: Corruption, Media, and Power in Turkey." Freedom House. Accessed January 19, 2017. https://freedomhouse.org/sites/default/files/Turkey%20Report%20-%20Feb%203,%202014.pdf

Daily Sabah. 2015. "39 Million Log on to Facebook in Turkey Daily." September 7. Accessed March 30, 2016. http://www.dailysabah.com/life/2015/09/07/39-million-log-on-to-face book-in-turkey-daily

Doğramacı, Esra, and Radcliffe, Damian. 2015. "How Turkey Uses Social Media." Digital News Report—Reuters Institute for the Study of Journalism. Accessed March 30, 2016. http://www.digitalnewsreport.org/essays/2015/how-turkey-uses-social-media/

Human Rights Watch. 2014. "Turkey: Internet Freedom, Rights in Sharp Decline." September 2. Accessed April 18, 2015. http://www.hrw.org/news/2014/09/02/turkey-internet -freedom-rights-sharp-decline

Jarvinen, H. 2014. "Turkey: Constitutional Court Overturns Internet Law Amendment." European Digital Rights Initiative. Accessed January 2015. https://edri.org/turkey constitutional court-overturns-internet-law-amendment/

Özbilen, Özge, and Coşkun, Orhan. 2014. "Turkey Lifts Twitter Ban After Court Ruling." *Reuters*, April 3. Accessed May 15, 2015. http://www.reuters.com/article/2014/04/03 /us-turkey-twitter-idUSBREA320E120140403

Pamuk, Humeyra, and Tattersall, Nick. 2014. "Leaked Documents Purport to Reveal Turk-ish Graft Allegations." *Reuters*, March 14. Accessed March 30, 2016. http://www.reuters .com/article/us-turkey-corruption-idUSBREA2D1F420140314

Socialbakers. 2015. "Turkey Facebook Page Statistics." Accessed March 30, 2016. http:// www.socialbakers.com/statistics/facebook/pages/total/turkey/

Tüfekçi, Zeynep. 2014. "Everyone Is Getting Turkey's Twitter Block Wrong." *Medium*, March 23. Accessed April 18, 2015. https://medium.com/message/everyone-is-getting -turkeys-twitter-block-wrong-cb596ce5f27

Turkish Statistical Institute. 2013. "Information And Communication Technology (ict) Usage Survey on Households And Individuals, 2013." Accessed January 19, 2017. http://www .turkstat.gov.tr/PreHaberBultenleri.do?id=13569

Turkish Telecommunications Authority. 2013. "Information about the Regulations of the Content of the Internet." Accessed January 2015. http://www.tib.gov.tr/en/en-menu -47information_about_the_regulations_of_the_content_of_the_internet.html

World Bank. 2014. "Internet Users." Accessed January 6, 2015. http://data.worldbank .org/indicator/IT.NET.USER.P2

Yeşil, Bilge. 2014. "Press Censorship in Turkey: Networks of State Power, Commercial Pres-sures, and Self-censorship." *Communication, Culture, and Critique*, 7: 154–173.

U

UKRAINE

Ukraine is a former Soviet state located in Eastern Europe. It is bordered in the north by Russia, Belarus, and Poland, and in the south by the Black Sea, Moldova, and Romania. Internet usage in Ukraine has grown exponentially over the last decade, from only a few percent of the population to roughly half today. Having been largely built in the last decade, its internet infrastructure is modern, and access is affordable, but it is much more readily available in urban areas than rural ones. Despite some troubling individual incidents, the internet in Ukraine is largely free of censorship and state control. In 2013 and 2014, the Euromaidan protests showcased an impressive use of social media, while the ensuing conflict with breakaway Eastern regions, the Crimea, and Russia has resulted in a simmering information war between Russia and the new Ukrainian regime.

Over 400 internet service providers (ISPs) are currently operating in Ukraine, which is a relatively high number compared to other similar states, indicating a low level of centralized control of the country's networks (Freedom House 2015). The largest ISP is Ukrtelecom, which was controlled by the state until it was privatized in 2011. Ukrtelecom controls most of the physical infrastructure of the internet in Ukraine, but the company leases it openly to other telecommunications companies, with no evidence of the economic strong-arming that might follow in the wake of infrastructural monopoly. The connection of the Ukrainian internet to international networks is not centralized either, with large ISPs maintaining their own international lines. Most domestic ISPs interconnect at the Ukrainian Internet Exchange (UA-IX), while at least eight regional internet exchanges provide substantial interconnection at the subnational level. Internet exchanges are large facilities full of servers where different ISPs connect their networks. The quantity of Ukraine's internet exchanges, along with the multiple redundant trunk lines out of the country, indicate a robust and fault-tolerant network infrastructure.

Typical of the post-Soviet world, internet connectivity skipped several generations of technology because the rollout of services took place later than in the early-adopting Western countries. Wired broadband subscriptions connect only 15 percent of households, while most connectivity is accomplished with mobile broadband connections via both smartphones and wireless universal serial bus (USB) modems that connect directly to the cellular networks (ITU 2015). Unlike the Western world, in which service is generally purchased on a monthly basis with caps that incur overage charges, these services in Ukraine are usually paid for up front, per byte. While the urban parts of Ukraine have very high levels of internet access through cellular networks, rural areas lag significantly, representing a substantial digital divide.

A number of different polls have sought to assess the proportion of Ukrainians who regularly use the internet, with a general consensus of about 50 percent of the population. While this is lower than the near universal use of the internet in the western world, it represents a substantial increase from the approximately 5 percent who regularly used the internet a decade ago (ITU 2015).

Social media is widely used in Ukraine. The most heavily trafficked website, other than Google, is Vkontakte (known colloquially as VK and translates literally from Russian as "in contact"), a Russian-owned and -operated social media site that is very similar to Facebook, even to the extent that it originally had almost identically formatted pages. The primary blogging platform is LiveJournal, which although originating in the United States, was wholly bought out by SuP Media, a Russian media company, in 2012.

VK and LiveJournal are also the primary social media and blogging platforms in Russia and several other post-Soviet countries. Both companies are controlled by individuals with extensive ties to the regime of Russian president Vladimir Putin, suggesting to many critics that there is a troubling potential for Russian censorship, both of their own social media sites and those in neighboring countries like Ukraine.

In terms of Western social media outlets, both Facebook and Twitter enjoy growing popularity in Ukraine. Facebook use has grown quickly, to some 3 million users by 2013, but this is only a fraction of the Ukrainian user base boasted by VK. Twitter has a smaller but faster-growing footprint, with a few hundred thousand users, though its user base in Ukraine has been growing exponentially over the last several years (Freedom House 2015).

The use of the internet for political organization in Ukraine first gained prominence during the Orange Revolution of 2005, when it was used extensively by the leadership of the opposition. The explosive growth of internet usage in the ensuing decade led to its widespread use during the Euromaidan protests of 2013 to 2014, especially in the context of social media. Usage of Twitter increased steadily over the several months of mass protest and, during the three-day climax of violence when the regime fell in February 2014, a steady stream of firsthand video and imagery from the center of Kiev flooded social media sites (Onuch 2014).

While international monitors characterize Ukraine as being near the top of the list of countries in terms of internet freedom, some troublesome incidents have occurred in recent years. Several citizens and journalists have been arrested for purportedly committing anti-Ukrainian activities online, and some received prison terms. In April 2015, state security seized the servers of about 30,000 websites after an ISP refused to comply with orders to shut down certain pro-Russian sites (Poludenko-Young 2015). An uproar ensued, however, and in response, state security quickly backtracked and services were restored. While the incident itself was discouraging, it did reveal that even if Ukraine's regime had the desire to do so, it lacked the technical expertise to censor the internet with anything approaching a malicious subtlety.

In the wake of the Ukrainian Civil War, the breakaway eastern republics have experienced similarly clumsy attempts at censorship and regulation by the nominal power holders in those territories. For instance, in Donetsk, rebel forces announced that ISPs would be required to turn over user data, although none did, and there is no evidence that the militants have the technical capacity to force them to comply (Freedom House 2015). The breakaway republics have also seen a sharp rise in intimidation of online communities and posters. Violence, kidnapping, and indefinite detention have become common tools used in the breakaway regions against those who post anti-Russian materials online.

This is partial evidence of a simmering information war that has developed between Ukraine and Russia. Russia has deployed a number of dedicated "astroturfing" companies assigned to produce content online in support of pro-Russian groups in Ukraine, making them appear to be grass-roots organizations and masking their connection to Russia. Some have gone as far as masquerading as native Ukrainian sites (Euromaidan Press 2014). In addition, cyberattacks between groups have increased in number, with several reported instances of website hijackings (including of a Ukrainian television channel's YouTube feed) and of propaganda targeted explicitly at the cell phones of Ukrainian soldiers on the battlefield (Freedom House 2015).

Of particular note is the situation in the Crimea. When seized by Russian forces in March 2014, the internet was quickly shut down in the region as the Russians seized local Ukrtelecom buildings (Miller 2014). This was possible because while the region had very high levels of internet access, its connection to the web was entirely routed through the Simferopol internet exchange point (IXP). Seizure of the facilities controlling that line was accomplished in the first twenty-four hours, along with the seizure of television and radio stations (Pizzi 2015). While the outage lasted only a few hours initially, evidence has mounted that Russia has significantly censored political information available online to the Crimea (Giles 2016).

Ukraine as a whole has a comparatively modern and growing internet infrastructure, with widespread usage of social media and blogging platforms by the public. In relative terms, the country exhibits minimal levels of mass censorship, surveillance, and actions taken against normal citizens using the internet. However, the political fallout of revolution, invasion, and civil war has taken its toll, with growing repercussions for those who post materials unfriendly toward the ruling powers of their particular region of the country. Despite the great deal of positive news about the internet in Ukraine, the dark side may perhaps be an ominous warning of the vulnerability of internet users to state repression during times of crisis.

Steven Lloyd Wilson

See also: Georgia; Kazakhstan; Romania; Russia

Further Reading

Euromaidan Press. 2014. "Kremlin's Propagandists Created a Network of Pseudo-Ukrainian Internet News Portals." November 13. Accessed July 19, 2016. http://euromaidanpress .com/2014/11/13/kremlins-propagandists-created-a-network-of-pseudo-ukrainian -internet-news-portals-in-russia/

Freedom House. 2015. "Freedom on the Net: Ukraine Country Report." Accessed July 19, 2016. https://freedomhouse.org/report/freedom-net/2015/ukraine

Giles, Keir. 2016. "Russia's 'New' Tools for Confronting the West: Continuity and Innovation in Moscow's Exercise of Power." Chatham House, March. Accessed July 19, 2016. https://www.chathamhouse.org/sites/files/chathamhouse/publications/research/2016-03-21-russias-new-tools-giles.pdf

International Telecommunication Union (ITU). 2015. "ICT Facts and Figures 2015." Accessed July 19, 2016. https://www.itu.int/en/ITU-D/Statistics/Documents/facts/ICTFactsFigures2015.pdf

Miller, Christopher. 2014. "Ukrtelekom Says Internet, Phone Connections Interrupted Between Crimea and Rest of Ukraine." March 1. Accessed July 19, 2016. http://www.kyivpost.com/article/content/ukraine/ukrtelekom-says-internet-phone-connections-cut-between-crimea-and-rest-of-ukraine-337992.html

Onuch, Olga. 2014. "Social Networks and Social Media in Ukrainian 'Euromaidan' Protests." *Washington Post*, January 2. Accessed July 19, 2016. https://www.washingtonpost.com/news/monkey-cage/wp/2014/01/02/social-networks-and-social-media-in-ukrainian-euromaidan-protests-2/

Pizzi, Michael. 2015. "Could Russia Really Cut the Internet?" *Al Jazeera America*, October 26. Accessed July 19, 2016. http://america.aljazeera.com/articles/2015/10/26/could-russia-really-cut-the-internet.html

Poludenko-Young, Anna. 2015. "Ukraine's Security Service Takes Down 30,000 Websites to Fight 'Pro-Russian Propaganda.'" *Global Voices*, April 28. Accessed July 19, 2016. https://globalvoices.org/2015/04/28/ukraine-censorship-russia-propaganda-hosting/

UNITED ARAB EMIRATES

The United Arab Emirates (UAE) is a federation of seven quasi-independent emirates in the Arabian Peninsula, located between Africa and Asia. Its broadband market is the most advanced globally, and the country's fiber-to-the-home (FttH) penetration rate—over 70 percent—is the world's highest (BuddeComm 2015). The UAE's internet penetration rate was 93.2 percent, the second highest in the Middle East (Internet World Stats 2015), and 50 percent of UAE residents access the internet via mobile phones. The average per-capita daily internet use is over 5 hours for PC and tablet users and 3.75 hours for mobile phone users (Global Media Insight 2015). Persons aged twenty-five to fifty-four years old represent 61 percent of the population, and the median age is 30.3. Non-Emiratis (mainly expatriate workers) comprise 85 percent of the 9.45 million total population, 85.5 percent of which reside in urban areas (CIA World Factbook 2015).

When Friendster turned down Google's offer to purchase the social media site in 2003, Google decided to create its own platform. The site opened on January 24, 2004, and was named Orkut after its project manager, Orkut Büyükkökten (1975–), who was of Turkish decent. After the platform's initial launch, three countries banned it outright. The governments of Iran, Saudi Arabia, and the United Arab Emirates (UAE) refused to allow access to the site on the grounds that the new platform was unethical and contained security flaws (Echovme 2014).

The UAE's mobile penetration rate reached 200 percent in 2010 (Belic 2010), and there are 7,270,000 active mobile internet users (DubaiMonsters 2015). In 2013, 73.8 percent of the population had smartphones (Wissing 2014). The country's FttH providers are Emirates Integrated Telecommunications Company, also known as du (headquartered in Dubai) and Emirates Telecommunications Corporation, also known as Etisalat (in Abu Dhabi, the capital and largest emirate). Etisalat, 60 percent of which is owned by the UAE government, is the seventeenth largest mobile network operator in the world, with more than 150 million customers in Africa, Asia, and the Middle East (George-Cosh 2011). Residents use their mobile devices to watch videos (45 percent), play games (33 percent), and use banking services (31 percent). Internet access also is available via asymmetric digital subscriber line (ADSL), cable modem, and mobile broadband (BuddeComm 2015).

Social media use is pervasive in the UAE. In 2014, Facebook (93 percent), Google+ (80 percent), YouTube (77 percent), and Twitter (70 percent) were the most popular applications among residents (Wissing 2014). Facebook's dominance is reflected by a 2015 survey in which 74 percent of respondents identified the application as their social media channel of choice (Kapur 2014). In 2011, more than 85 percent of Facebook users in the UAE interacted in English, approximately 10 percent in Arabic, and the remainder in other languages (Alqudsi-ghabra et al. 2011). In 2012, 44 percent of UAE users accessed social networks during work hours, with access peaking after work between 6 p.m. and 8 p.m. (Interactive Middle East 2012).

Tech-savvy Emiratis and expatriates were early and enthusiastic adopters of new social media platforms. In 2007, rapidly growing Twitter use in the UAE prompted the government to block the site on the grounds that it constituted a gender-mixing dating site. At the same time, ham-fisted censors blocked searches for chicken breast recipes and Middlesex University (Wissing 2014). Federal authorities since have taken a more sophisticated approach to information technology (IT). Following the success of Dubai's pioneering Internet City (1999) and Media City (2000), the UAE launched a Smart Government initiative encompassing nationwide infrastructure improvements to enable smart grids and smart homes (Wissing 2014). Masdar City, a new smart city currently under construction near Abu Dhabi, is emblematic of the authorities' current embrace of IT (BuddeComm 2015).

Nevertheless, the government continues to filter the internet and to monitor and occasionally block other media. A website creator convicted in 2007 by a court in Ras al-Khayma emirate of defaming a public official was fined and sentenced to a year in prison (Alqudsi-ghabra et al. 2011). In 2012, the UAE strengthened its 2006 cybercrime law, giving the Telecommunications Regulatory Authority (TRA), an executive agency reporting to the Cabinet, broad power to monitor internet use in the country. The TRA compiles lists of websites to be blocked and relies on Etisalat and du to ensure that subscribers comply with the Authority's Internet Access Management (IAM) policy. Per the IAM policy, sites are blocked if they contain material that (1) conflicts with UAE ethics and morals, including nudity and dating; (2) expresses hatred of religions; (3) conflicts with UAE laws; (4) allows or helps users to access blocked content; (5) directly or indirectly poses

The Emirates Telecommunications Corporation, also called Etisalat, has its main tower in Dubai. The company is the seventeenth largest mobile phone operator in the world, allowing people in Africa, Asia, and the Middle East to operate their smartphones and access the internet. (AP Photo/Kamran Jebreili)

a risk to users (e.g. phishing, malware); (6) is related to gambling; or (7) promotes the manufacture, production, or use of illegal drugs (Reporters Without Borders 2014).

Voice over internet protocol (VoIP) services were blocked by the TRA prior to 2010, when the government allowed Etisalat and du to provide internet telephony (Namatalla 2010). Skype, Viber, and FaceTime, though blocked, intermittently are available. In a 2014 statement, the TRA confirmed the illegality of VoIP services other than those provided by Etisalat and du. Facetime is not included in Apple devices sold in the UAE. Following WhatsApp's 2014 announcement that it planned to launch a VoIP service, users of the channel in the Emirates reportedly experienced uneven functionality (Bui 2014).

UAE leaders see the Muslim Brotherhood (MB) as the country's biggest threat, and they argue that internet regulation is necessary to combat Islamic extremism. In 2013, the government arrested, tried, and imprisoned ninety-four members of al-Islah—an UAE-based MB affiliate—on terrorism charges (Wissing 2014). Human rights activists, however, contend that the restrictions unduly restrict civil rights, including free speech (Freedom House 2015). Also in 2013, an Asian man got arrested and jailed on defamation charges after posting cell-phone video of a UAE motorist using his headdress cord to beat an Indian van driver following a minor traffic altercation in Dubai (Al Arabiya 2013). In another widely publicized case from the same year, a U.S. citizen was convicted of defamation for posting a YouTube video depicting a fictitious Dubai military school that trained students to throw sandals, use headdress cords as weapons, and tweet their fellow fighters. Sentenced to one year in prison, the American served approximately nine months before he was released (Wissing 2014).

The UAE likely will exhibit continued ambivalence concerning the promise and perils of IT. Senior leaders, including Abu Dhabi Crown Prince Mohammad bin Zayid (1961–), regularly use Twitter to promote their initiatives and share information. Foreign Minister Abdullah bin Zayid (1972–), who reportedly has 800,000 followers, sparked a spontaneous national event in November 2013 after tweeting his desire to see "the UAE flag on top of every house." Meanwhile, the director general of the Emirates Center for Strategic Studies and Research has expressed concern about the internet's effect on national identity. "We see more time being spent on social media, weakening family ties," he told a journalist in 2014, adding he feared that it "may change social norms and values" (Wissing 2014).

Mark A. Caudill

See also: Iraq; Qatar; Saudi Arabia

Further Reading

Al Arabiya. 2013. "UAE Man Beating Driver Makes Headlines in Gulf." July 18. Accessed November 30, 2015. http://english.alarabiya.net/en/media/2013/07/18/UAE-man-beating-driver-makes-headlines-in-Gulf.html

Alqudsi-ghabra, Taghreed M., Al-Bannai, Talal, and Al-Bahrani, Mohammad. 2011. "The Internet in the Arab Gulf Cooperation Council (AGCC): Vehicle of Change." *International Journal of Internet Science,* 6(1), 44–67.

Arab Social Media Report. 2014. "Twitter in the Arab Region." March. Accessed October 27, 2015. http://www.arabsocialmediareport.com/Twitter/LineChart.aspx?&PriMenuID=18&CatID=25&mnu=Cat

Belic, Dusan. 2010. "UAE Hits 200% Mobile Penetration Rate." January 3. *Cellular News.* Accessed November 30, 2015. http://www.intomobile.com/2010/01/03/uae-hits-200-mobile-penetration-rate/

BuddeComm. 2015. "United Arab Emirates—Broadband, Internet, and Digital Media Market—Overview, Statistics, and Forecasts." Accessed November 30, 2015. http://budde.com.au/Research/United-Arab-Emirates-Broadband-Internet-and-Digital-Media-Market-Overview-Statistics-and-Forecasts.html

Bui, Joey. 2015. "Skype Ban Tightens in the UAE." February 7. *The Gazelle.* Accessed November 30, 2015. http://www.thegazelle.org/issue/55/news/skype/

CIA World Factbook. 2015. "United Arab Emirates." Accessed November 30, 2015. https://www.cia.gov/library/publications/the-world-factbook/geos/ae.html

DubaiMonsters. 2015. "United Arab Emirates Internet, Social, and Mobile Stats 2015." Accessed November 30, 2015. http://www.slideshare.net/lukhachdem1/united-arab-emirates-internet-social-and-mobile-statistics-2015

Freedom House. 2015. "United Arab Emirates: Stop the Charade and Release Activists Convicted at the Mass UAE 94 Trial." March 3. Accessed November 30, 2015. https://freedomhouse.org/article/united-arab-emirates-stop-charade-and-release-activists-convicted-mass-uae-94-trial

George-Cosh, David. 2011. "Foreign Interests Help Etisalat Beat Forecasts." February 10. *The National*. Accessed November 30, 2015. http://www.thenational.ae/business /telecoms/foreign-interests-help-etisalat-beat-forecasts

Internet World Stats. 2014. "Internet Usage in the Middle East." December 31. Accessed October 27, 2015. http://www.internetworldstats.com/stats5.htm#me

Kapur, Shuchita. 2014. "Social Media: What UAE Residents Like and Why." March 19. Accessed January 19, 2017. http://www.emirates247.com/business/social-media-what-uae -residents-like-and-why-2014-03-19-1.542154

Namatalla, Ahmed. 2010. "Four Companies to Get VoIP in the UAE." March 15. *Gulf News*. Accessed November 30, 2015. http://gulfnews.com/business/sectors/telecoms/four -companies-to-get-voip-in-the-uae-1.597911

OpenNet Initiative. 2009. "Internet Filtering in the United Arab Emirates." August 6. Accessed November 30, 2015. https://opennet.net/sites/opennet.net/files/ONI_UAE _2009.pdf

Reporters Without Borders. 2014. "Enemies of the Internet: United Arab Emirates: Tracking 'Cyber-criminals.'" March 11. Accessed November 30, 2015. http://12mars.rsf.org /2014-en/2014/03/11united-arab-emirates-tracking-cyber-criminals/

Seksek, Tariq. 2012. "White Paper: Social Media in #KSA and the #UAE." February 21. Interactive Middle East. Accessed November 30, 2015. http://interactiveme.com/2012 /02/white-paper-social-media-in-ksa-and-the-uae/

Wissing, Douglas. 2014. "Genie Uncorked: Social Media in the UAE." August 5. *Huffington Post*. Accessed November 30, 2015. http://www.huffingtonpost.com/douglas-a-wissing /genie-uncorked-social-med_b_5440160.html

UNITED KINGDOM

The United Kingdom of Great Britain and Northern Ireland includes England, Wales, Scotland, and Northern Ireland. An island nation in northwestern Europe, it is commonly known as the United Kingdom (U.K.) or Britain. With a total population of 64.91 million people living within its borders, the internet penetration accounts for 92 percent of the population. A major contributing factor for the United Kingdom's high internet penetration is the development of its superfast 4G network. Everything Everywhere (EE) Limited, a mobile network operator company and internet service provider (ISP), increased its broadband speed to give the United Kingdom one of the fastest and most reliable networks in the world. The superfast broadband is easily accessible to populations in the more rural and remote areas of the country, thus boosting the internet penetration. With the arrival of a superfast 4G network, internet users have increased their online activity on various social media platforms. The United Kingdom also capitalized on the fast broadband by creating apps specifically for tourism in the country.

For U.K. internet users, social media is not the most popular internet activity. On a daily basis, internet users spend an average of three hours and forty-seven minutes on the internet using a PC or tablet. About one hour and twenty-nine minutes of that time is used for social media (Kemp 2016). Internet users are able to do their everyday internet activities on their mobile phones, such as navigate the internet faster, read daily news, watch high-definition (HD) videos without buffering, and share large files and high-resolution pictures in seconds (Woollaston 2013).

OMG, the social media abbreviation for the explicative "Oh my God," did not originate on social media. The first known use of OMG was in a letter from Lord John Arbuthnot Fisher (1841–1920), a British navy admiral, to Winston Churchill (1874–1965), British statesman and later prime minister. The letter, written at the end of Fisher's naval career, expressed his distaste over newspaper headlines that bolstered the German fleet's reputation. He believed that Britain had a far superior fleet and could defeat the Germans in less than a few minutes. The OMG appeared at the end of the letter when mentioning the creation of a new order of knighthood. Fisher's OMG differed from the OMG in social media just slightly, however. His version, which is spelled out after the acronym in the letter, stood for "Oh! My God!" (Nuwer 2012).

Widely used by the United Kingdom's active internet population, Google's websites and apps averaged about 46 million unique visitors in March 2015. Google's search engine dominates the market, with 39.6 million unique visitors compared to other global search engines, such as Bing (17.3 million visitors) and Yahoo Search (14 million visitors). During the same monthlong period, Google was followed closely by Facebook's websites and apps, with 41 million visitors, and the BBC's websites and apps, with 40 million visitors. Even though Google has more users, people tend to spend more time using Facebook (Edge 2015).

Nearly 90 percent of the teen and young adult generation, specifically users in the sixteen- to twenty-four-year-old age bracket, own a smartphone. While older generations, such as the fifty-five- to sixty-four-year-old bracket, are slowly joining the smartphone revolution, the younger generations are more apt to use their mobile phone instead of a computer to access the internet and social media platforms (Ofcom 2015). Social media penetration accounts for 59 percent of the country's population. Over a three-year period, there has been a 66 percent increase of internet users on social media platforms (Kemp 2016). Teens and millennials are diversifying their use of social media platforms in the United Kingdom. The most used social media platform is Facebook, followed by the instant messenger apps Facebook Messenger and WhatsApp (Edge 2015). About 47 percent of U.K. active users have accounts on Facebook, which is nearly twice that of other global social media platforms, such as Twitter, LinkedIn, and Google+. Facebook users that fall within the twenty-four to thirty-four-year-old age range account for nearly 30 percent of traffic on the social media network. About half of people in the eighteen- to twenty-four-year-old demographic have incorporated Facebook into their morning routine (White 2016).

While Facebook is widely used in the country, other global social media networks are slowly growing in popularity (eMarketer 2015). Twitter has about 13 million active users in the United Kingdom. Nearly half of Twitter's users access the microblogging network once a day using mobile phones rather than computers or tablets. About 37 percent of the United Kingdom's youth, specifically in the eleven- to sixteen-year-old age bracket, use Twitter once a week. In the United Kingdom, Instagram has about 14 million users, and nearly 90 percent of its active users are under the age of thirty-five, while 38.4 percent fall into the sixteen- to twenty-four-year-old age range. Many businesses and brands utilize the social media platform to create interest for new products in different target demographics (White 2016).

In efforts to be more competitive in the United Kingdom and around the world, EE increased its mobile web speed from 3G and 4G speeds to a superfast 4G broadband, which is double the average networking speed. Initially, the company offered the superfast 4G service to twelve cities in various parts of the United Kingdom before expanding to more areas of the country. At the time of its release in 2013, EE boasted faster networking speeds than the United States and Japan, as well as a broadband speed comparable to the most wired nation in the world, South Korea (Woollaston 2013). In early 2016, the United Kingdom ranked fourth globally in connection speed. The country has a current connection speed of 13 Mbps, which is significantly higher than the global average of only 5.1 Mbps. However, it still falls behind Japan (15 Mbps), Hong Kong (15.8 Mbps), and South Korea (20.5 Mbps). The United Kingdom has become one of the world's fastest countries online since the installation of the superfast 4G network (Kemp 2016).

Communication network companies like EE, O2, Three, and Vodafone continue to expand the 4G networks and service to make the internet more accessible and easy for everyone in the United Kingdom. As much as 87 percent of the U.K population is currently covered by the 4G network. By 2017, the development of the superfast 4G technology will be accessible by 98 percent of the country. This percentage is a target goal not only for mobile phones, but also fixed-line broadband for homes and businesses. Based on the 2017 target goal for communication network companies, it is predicted that the time spent online will double for British internet users within the next decade. As more people spend more time online, many active users believe that the benefits will increase as well. Six of every ten adults in the country use the internet to keep in touch with friends and family. Nearly two-thirds of online adult users utilize the internet to read the news, research products and services, and keep themselves informed about a variety of issues (Ofcom 2015).

The United Kingdom uses superfast 4G to its advantage for tourism. EE brokered many international 4G contracts with overseas networks to give tourists and business travelers easy access to broadband connections while in the United Kingdom. The first overseas network to sign with EE was AT&T, a U.S.-based service provider. EE provided AT&T customers with an easy way to stay connected in the United Kingdom. EE representatives explained that tourists want a fast, easy, and reliable experience while abroad. Whether it is finding local landmarks, navigating through unfamiliar cities, or purchasing tickets to local tours or events, EE provides tourists with a mobile network that makes traveling easier throughout the United Kingdom. Several cities in the United Kingdom have taken advantage of the superfast network by creating apps specifically for tourism. On the City of London's website, for instance, there is a webpage for tourists to find and use apps and social media pages created specifically for London tourism. Tourists may download apps that provide audio-guided tours, locate city services or rest stops, create itineraries for easy planning, or enhance their experience with an interactive game that shares information about different exhibits, such as the Tower of London (City of London 2016).

Access to high-speed broadband has been transformational, both socially and economically, for people living in rural areas of the United Kingdom. The implementation of Wi-Fi kiosks in select locations has been widely popular with both locals and tourists. To help grow tourism, some local governments have created broadband projects to deliver world-class connectivity to more remote areas. Many hope that these plans will entice tourists to explore more off-the-map places and, consequently, develop economic growth in these areas from the tourism boost. By 2020, the Scottish Rural Development Programme (SRDP) aims to make every rural community digitally connected, as well as help businesses meet their needs and the needs of their customers. Some believe that SRDP's plans will result in community solutions to establish and develop tourism and tourist-friendly businesses throughout rural Scotland (Highlands and Islands Enterprise 2015).

Karen Ames

See also: China: Hong Kong; Japan; South Korea; United States

Further Reading

City of London. 2016. "Social Media and Apps." Accessed March 11, 2016. https://www.cityoflondon.gov.uk/about-the-city/about-us/Pages/social-media.aspx

Edge, Abigail. 2015. "Digital Habits in the UK: Social Media, Mobile Apps, and Online News." August 12. Accessed March 5, 2016. www.themediabriefing.com/article/digital-habits-in-the-uk-social-media-mobile-apps-and-online-news

eMarketer. 2015. "Young Mobile Users Drive UK Social Media Usage: Social Networking Popular Among Teens and Millennials." October 29. Accessed March 7, 2016. http://www.emarketer.com/Article/Young-Mobile-Users-Drive-UK-Social-Media-Usage/1013163

Highlands and Islands Enterprise. 2015. "Broadband Help for Rural Communities." August 24. Accessed March 11, 2016. http://www.hie.co.uk/about-hie/news-and-media/archive/broadband-help-for-rural-communities.html#sthash.V3PQH8zl.dpbs

Kemp, Simon. 2016. "Digital in 2016: We Are Social's Compendium of Global Digital, Social, and Mobile Data, Trends, and Statistics." January 27. Accessed March 7, 2016. http://wearesocial.com/uk/special-reports/digital-in-2016

Newton, Thomas. 2013. "EE's Special Relationship with AT&T Will See UK Travellers 4G Speeds Abroad." December 17. Accessed March 11, 2016. https://recombu.com/mobile/article/ees-special-relationship-with-att-will-see-uk-travellers-getting-4g-speeds-abroad_M19876.html

Ofcom. 2015. "The UK Is Now a Smartphone Society." August 6. Accessed March 8, 2016. http://media.ofcom.org.uk/news/2015/cmr-uk-2015/

Thomas, Daniel. 2015. "EE's Superfast 4G Data Traffic Surpasses 3G for The First Time." May 13. Accessed March 10, 2016. http://www.ft.com/cms/s/0/f158b2c2-f96f-11e4-ae65-00144feab7de.html#axzz43azUOcrv

White, Jules. 2016. "Social Media Statistics in the UK." February 2. Accessed May 12, 2016. http://www.thelasthurdle.co.uk/social-media-statistics-in-the-uk-for-2015/

Woollaston, Victoria. 2013. "Superfast 4G Broadband That Will Double Web Speeds to Hit 12 UK Cities Including London and Glasgow Tomorrow." July 3. Accessed March 10, 2016. www.dailymail.co.uk/sciencetech/article-2354694/Superfast-4G-broadband-DOUBLE-web-speeds-hit-12-UK-cities-including-London-Glasgow-tomorrow.html

UNITED STATES

The United States of America, the largest country in North America in terms of both land area and population, is nestled between Canada and Mexico and has a population of about 323 million people. While Asia has nearly half the world's internet users, North America's internet penetration rate is higher than anywhere else (89 percent) (Internet World Stats 2016). Despite this fact, the United States barely ranks in the top thirty countries in terms of penetration. In 2013, while more than 84 percent of the U.S. population had available connections to the internet, the penetration rate of the world's top four countries or territories for internet penetration (namely, the Falkland Islands, Iceland, Bermuda, and Norway) all had rates of 95 percent or higher (Internet World Stats 2013). Much of the world's popular social media websites originate from and became popular in the United States. As a result, social media use and popularity has shifted quite a bit since the inception of many of these sites. In addition, one of the youngest generations, today's teenagers, have grown up never knowing a time without social media; therefore, their interactions with social media sometimes differ greatly from those of the older generations.

In January 2016, over 282 million people in the United States were active internet users, which indicated some growth over the past few years (We Are Social 2016). Historically, the most active user demographic has been eighteen- to

Two young American girls sit together viewing social media. In the United States, the younger generations are among the most tech-savvy and spend significant amounts of time online. (Rui Matos/Dreamstime.com)

twenty-nine-year-olds, with 70 percent penetration in 2000 and 96 percent in 2015 (Perrin and Duggan 2015). Nevertheless, adoption rates have been faster among adults aged sixty-five and older; in 2000, internet penetration among this group was only 14 percent, but in 2015, well over half (58 percent) were using the internet (Perrin and Duggan 2015). While 85 percent of adults in the United States in 2016 owned some type of cell phone, smartphone penetration among adults was only 57 percent. Meanwhile, 72 percent of adults owned desktop or laptop computers, and 35 percent owned a tablet. Most web traffic in the United States occurs via computer, averaging daily use of just over four hours; time spent on the internet with a mobile phone was on average slightly less than two hours, accounting for just 27 percent of web traffic (We Are Social 2016).

A common perception in the United States is that the iPhone is more ubiquitous than the Android phone; however, a recent Nielsen study showed that in the third quarter of 2015, more than half of smartphone users were using Google's Android operating system (52.6 percent), while only 42.7 percent of smartphones were utilizing Apple's iOS (Beres 2015). This inaccurate perception may be derived in part from the fact that the average number of iPhone and Android users is much closer in the United States than it is worldwide: in 2015, 82.8 percent of the total smartphone market share was dominated by Android, while iOS controlled only 13.9 percent (Beres 2015).

Google Chromebooks have recently outpaced Apple iPads in schools (Taylor 2015); however, over the course of several years, millions of iPads have been integrated into the U.S. classroom, which may account for some of the choices made when downloading free iOS apps. In August 2016, the top free app for iOS was the mobile game Flip Diving; however, iTunes U (an app featuring free educational videos) and the free version of Pages (a word-processing app), Numbers (a spreadsheet app), and Keynote (a presentation app) were also among the top ten most downloaded apps (AppAnnie 2016b). Remind, a free messaging app that allows convenient communication among teachers, students, and parents, was the sixteenth most commonly downloaded free app. Snapchat ranked ninth, Google Maps ranked tenth, Facebook Messenger ranked eleventh, Facebook ranked twelfth, Instagram ranked thirteenth, and YouTube ranked fourteenth (AppAnnie 2016b). In contrast, Facebook Messenger, Facebook, and Snapchat were the top three downloaded free apps for the Android platform (AppAnnie 2016a). Other free apps in the top ten for Android included Pokémon Go (a location-based augmented reality game introduced in mid-2016), Instagram, Pandora Radio, WhatsApp Messenger, and Remind (AppAnnie 2016a).

The websites with the most traffic from the United States are Google, Facebook, YouTube, Amazon, Yahoo, Wikipedia, eBay, Twitter, Reddit, and Netflix (Alexa 2016). Facebook has a 72 percent penetration rate of adult internet users in the United States, a rate that has not changed much in the past few years. The biggest demographic group of Facebook users consists of adults aged eighteen to twenty-nine years old; 89 percent of internet users in this category use Facebook, which comprises 29 percent total active users (Duggan 2015; We Are Social 2016). In every age group, there were more female users than male (We Are Social 2016). Of

Pinterest analyzed its users' pins to see if there were any patterns emerging throughout the week. The results showed that each day had a most popular category. On Monday, users pinned items about fitness and health. On Tuesdays, they preferred images related to technology. On Wednesday, users tended to cite inspirational quotes. Thursdays saw more pins in the fashion category. Fridays ended the workweek with humor and comics. On Saturday, users pinned more travel-related images than on any other day. Finally, on Sundays, users preferred images related to food and crafts (Pinterest 2014).

current social media in the United States, Facebook has one of the largest proportions of users over the age of forty-five (Williams 2014). While the majority of internet users in the United States have Facebook profiles, less than half consider themselves to be active users (Williams 2014), which suggests a downward trend for the social media site. However, the way that Facebook is used has changed over the years to a more passive rather than an active tool. Many users simply browse through friends' posts for news, photos, or entertainment. The social media website is often used as a tool for organizing groups or outings, such as recreation-league sports and family reunions. Lately, many small businesses are utilizing Facebook to advertise their stores rather than creating their own websites.

YouTube, the second most popular social media website in the United States, has evolved over the years as well. Launched in 2005, it was originally a video-sharing website, which provided a platform for users to share homemade videos over the internet. As it grew in popularity and methods of monetizing usage became more viable, corporate media, especially the music industry, began adding content. YouTube still has a large user base that utilizes the website to watch music videos. In December 2015, the two most visited channels had been Justin Bieber, with over 650 million views from the previous month, and Adele, with over 540 million views from the previous month (Statista 2016b). Private citizens began to produce videos specifically for public consumption, many of whom were able to make a living as YouTube stars.

In the United States, possibly the most famous of these YouTube personalities is Felix Arvid Ulf Kjellberg (1989–), better known as PewDiePie. He is a Swedish video gamer who has more subscribers than some mainstream pop stars and has made $12 million by posting videos of him playing video games (Forbes 2015). His YouTube account, @pewdiepie, is the first channel ever to have more than 10 billion views and also has the most subscribers, at 39 million and counting. The statistic of 10 billion views is significant because there are fewer than 8 billion people on the planet; these numbers mean that the equivalent of 120 percent of the planet's population watches the videos that he posts to the channel. Some successful YouTube channels offer original programming with serialized stories. YouTube Red is a channel that highlights some of the content produced by its top creators. As of July 2015, video was being uploaded at rates of over 400 hours a minute (Statista 2016b). Because of his online popularity, PewDiePie has his own YouTube Red series called *Scare PewDiePie* and may star in a television show based on his success, but details of the project are scarce (TMZ 2015).

A website similar to some of the newer uses of YouTube was launched in 2011. Called Twitch.tv, it is part of the growing e-sports phenomenon of watching video-gamers compete. The website offered a live-streaming service that provided a platform for internet viewing of video game competitions, as well as individual gaming. Twitch offered the monetization of content in much the same way as YouTube. In 2014, the website had about 45 million monthly viewers and, while not ranked highly in terms of internet traffic, if measured in terms of bandwidth use, Twitch's traffic was higher than Netflix, Hulu, and Facebook (Bean 2014). One unfortunate aspect of Twitch streaming, however, is the sexism found in the chat windows. As some female gamers broadcast their streams, the chat logs fill up with sexist comments. In general, female gamers face gender discrimination on a regular basis, and Twitch has proved to reflect this reality on occasion. Some female gamers take advantage of this sexism by adjusting the camera angles to show off more of their bodies to encourage bigger monetary donations, but in most other instances, despite moderating tools, the problem of chat toxicity persists.

Twitter is the third most popular social networking website in the United States. It is more popular there than in most other places in the world (22 percent of visitors originate in the United States) (Alexa 2016). While Twitter has recently been struggling to maintain growth and the company saw its stock price fall in early 2016, Twitter nonetheless continues to be an important social networking site in the United States. The hashtag (#), a method of organizing similar topics and themes, was popularized by Twitter. The unique 140-character limit made succinct messaging essential. The hashtag has since spread to other social media and is even used to identify various social movements, such as #BlackLivesMatter, a group dedicated to erasing racial discrimination and inequality; and #IceBucketChallenge, which raises awareness of and money to fight the disease amyotrophic lateral sclerosis (ALS).

Twitter is used in a variety of ways by corporations and individuals alike. Personal and company branding is a popular use of the site. People use Twitter to follow opinions in real time during live coverage of political or sporting events, as exemplified during the 2016 presidential election, where media outlets, supporters, and opposition actively consumed and engaged with candidate Donald Trump's account. Often, breaking news hits Twitter first and is later disseminated to other media outlets. Remarkably, while most people in the United States are aware of Twitter, less than a quarter of adult internet users have Twitter accounts. Interestingly, while the digital divide in the United States usually means that African-Americans are less likely to use the internet than white Americans, the opposite is true of Twitter. In 2013, only 16 percent of white internet users utilized Twitter, whereas 22 percent of African-American internet users utilized Twitter (Smith 2014). While this gap is shrinking, African-Americans continue to lead in proportional use of the social media website.

The younger generations in the United States tend to have higher social media penetration rates; however, one of the youngest demographic categories, teenagers, often illustrates unique social media usage. According to a 2015 Pew Research study (Lenhart 2015), of internet users aged thirteen to seventeen years old, 92 percent say that they used the internet every day. While over half of these teens went online several times a day, and 12 percent went online at least once a day,

24 percent of teens who used the internet every day said that they are "almost constantly" online. African-American and Hispanic teens reported being online "almost constantly" at higher rates (34 and 32 percent, respectively) than white teens (19 percent). This phenomenon is largely due to the prevalence of smartphones and other mobile devices, which were once possessed mainly by adults but are now ubiquitous even among young teenagers. A total of 75 percent of teens either own or have access to a smartphone; 85 percent of African-American teens own a smartphone, while only 71 percent white and Hispanic teens have one. Access to desktops and laptops is more common among teenagers from the United States (87 percent) than many other countries in the world. More than half of teens have access to a tablet; while girls and teens from wealthier families are more likely to have tablets, there seemed to be no difference in tablet usage along racial or ethnic lines.

Additional findings of the 2015 Pew Research Study (Lenhart 2015) suggest that girls are more likely to use social media sites than boys, who prefer playing videogames (which may not have as many social networking functions, but nevertheless encourage online social interaction). Boys are more likely to use Facebook, while girls are more likely to use Instagram. The study also found a correlation between social media use and socioeconomic status: teens from lower socioeconomic households (less than $50,000 a year) are more likely to use Facebook, and teens from more affluent households are more likely to use Snapchat.

However, use of social media does not necessarily correlate to popularity or perceived importance. While 71 percent of teens reported using Facebook in 2015 (Lenhart 2015), the world's most popular social media site has rapidly been falling out of favor with this age group. In 2012, 42 percent of teens using social media identified Facebook as their preferred social media, with 27 percent identifying Twitter and 12 percent identifying Instagram as most important (Statista 2016a). Snapchat did not even rank in that list; however, just three years later, in the spring of 2016, more than a quarter of all teenage social media users (28 percent) identified it as their most preferred social media site. Almost the same percentage of users preferred Instagram (27 percent), and the tally for both Twitter and Facebook users had fallen (18 percent and 17 percent, respectively) (Statista 2016a). In January 2016, only 8 percent of all Facebook users in the United States were younger than nineteen (We Are Social 2016).

Marilyn J. Andrews

See also: Canada; Iceland; Mexico; United Kingdom

Further Reading

Alexa. 2016. "Top Sites in United States." Accessed August 27, 2016. http://www.alexa.com/topsites/countries/US

AppAnnie. 2016a. "Top Apps on Google Play, United States, Overall, Aug 26, 2016." Accessed August 27, 2016. https://www.appannie.com/apps/google-play/top/united-states/overall/

AppAnnie. 2016b. "Top Apps on iOS Store, United States, Overall, Aug 26, 2016." Accessed August 27, 2016. https://www.appannie.com/apps/ios/top/united-states/overall/iphone/

Bean, Daniel. 2014. "What the Heck Is Twitch, and Why Does Google Think It's Worth $1 Billion? [UPDATE: Amazon Bought Twitch]." *Yahoo! Tech.* Accessed August 27, 2016. https://www.yahoo.com/tech/what-the-heck-is-twitch-and-why-does-google-think-its -86224528319.html

Beres, Damon. 2015. "Sorry, Fanboys: Android Still More Popular Than iOS in U.S." *Huffington Post.* Accessed August 27, 2016. http://www.huffingtonpost.com/entry/android -more-popular-than-ios_us_5678203be4b06fa6887de2e7

Duggan, Maeve. 2015. "The Demographics of Social Media Users." *Pew Research Center.* Accessed August 27, 2016. http://www.pewinternet.org/2015/08/19/the-demographics -of-social-media-users/

Forbes. 2015. "The World's Top-Earning YouTube Stars 2015." Accessed August 27, 2016. http://www.forbes.com/pictures/geeg45egklg/1-pewdiepie-12-millio/#29ff88871c96

Internet World Stats. 2013. "Top 50 Countries with the Highest Internet Penetration Rates— 2013." Accessed August 27, 2016. http://www.internetworldstats.com/top25.htm

Internet World Stats. 2016. "Internet Usage Statistics: The Internet Big Picture." http://www .internetworldstats.com/stats.htm

Lenhart, Amanda. 2015. "Teen, Social Media, and Technology Overview 2015." Pew Research Center. Accessed August 27, 2016. http://www.pewinternet.org/files/2015/04/PI _TeensandTech_Update2015_0409151.pdf

Perrin, Andrew, and Duggan, Maeve. 2015. "Americans' Internet Access: 2000–2015." http:// www.pewinternet.org/2015/06/26/americans-internet-access-2000-2015/

Smith, Aaron. 2014. "African Americans and Technology Use: A Demographic Portrait." *Pew Research Center.* http://www.pewinternet.org/2014/01/06/african-americans-and -technology-use/

Statista. 2016a. "Most Popular Social Networks of Teenagers in the United States from Fall 2012 to Spring 2016." Accessed August 27, 2016. http://www.statista.com/statistics /250172/social-network-usage-of-us-teens-and-young-adults/

Statista. 2016b. "Statistics and Facts About YouTube." Accessed August 27, 2016. https:// www.statista.com/topics/2019/youtube/

Taylor, Harriet. 2015. "Google's Chromebooks Make up Half of US Classroom Devices Sold." *CNBC.* Accessed August 27, 2016. http://www.cnbc.com/2015/12/03/googles-chrome books-make-up-half-of-us-classroom-devices.html

TMZ. 2015. "Hey Bros—I'm in Hollywood." September 10. Accessed September 10, 2015. http://www.tmz.com/2015/09/10/pewdiepie-cutie-marzia-youtube-gamer-arrives-lax/

We Are Social. 2016. "Special Reports: Digital in 2016." Accessed August 27, 2016. http:// wearesocial.com/uk/special-reports/digital-in-2016

Williams, Rhiannon. 2014. "Facebook Isn't Dying. It's Just Changing." *The Telegraph.* http:// www.telegraph.co.uk/technology/facebook/11252782/Facebook-isnt-dying.-Its-just -changing.html

UZBEKISTAN

Uzbekistan, a former Soviet state, is located in central Asia. It is bordered in the north by Kazakhstan, in the west by Kyrgyzstan and Tajikistan, and in the south by Turkmenistan and Afghanistan. It is one of the poorest countries in the world, ruled by one of the more brutal regimes—a dictatorship that has not seen a change of power since the Soviet Union fell twenty-five years ago. Its record of internet use is one of rapid growth over the last several years, driven by foreign investment in its telecommunications sector, but it nevertheless remains one of the least-connected countries in the world. In addition, the regime has implemented a wide variety of repressive measures, ensuring that freedom of speech is nearly as hampered on the internet as it is in the physical reality of Uzbekistan. Prospects for improvement on that front are dim, even as the infrastructure of connectivity has gradually improved.

The telecommunications infrastructure of Uzbekistan is the largely ancient and deteriorating technology constructed during the Soviet era. Despite being the most populous country in Central Asia, with a population of some 30 million people, there are fewer than 2 million landlines in the entire country (ITU 2015). Wired broadband connections number only around 200,000 and are almost entirely concentrated in the capital, Tashkent (ITU 2015). On the other hand, like many other developing states, Uzbekistan's infrastructure has skipped several generations of technology and gone straight to mobile devices. As of 2014, there were nearly eighty mobile phone subscriptions in the country per hundred people (ITU 2015). While mobile phones are nearly ubiquitous, that has not translated into mobile internet access, as there are only about a half million mobile broadband subscriptions in the country (ITU 2015).

That does not mean, however, that Uzbeks do not use the internet. On the contrary, some 43 percent of the Uzbek population reports regularly using the internet (ITU 2015). But connectivity is very slow and limited on average. For example, as recently as 2012, even at the most expensive hotels catering to foreigners visiting Tashkent, internet speeds topped out at around 15 Kbps, which is on par with dial-up modem connections from a generation ago in the developed world (Silk 2012).

The situation has improved gradually over the last few years, though, with subscriptions promising 256 KBps bandwidth costing $4 per month, and megabit access starting at $27.50 per month (Ruddy 2014). However, these services are available only in limited urban areas and are simply far beyond the budget of most of the population. When last measured in 2003, two-thirds of Uzbeks lived on less than $2 per day. This represents a significant digital divide, with internet users largely concentrated in cities. Rural areas are among the poorest in the world, and many areas cannot even count on reliable electricity service, let alone wireless service.

In addition, Uzbekistan suffers from major bottlenecks in accessing the internet outside its borders. It is one of only two doubly landlocked countries in the world (that is, it is entirely surrounded by countries that are themselves landlocked). This geographic isolation from the spider web of suboceanic fiber-optic lines, in conjunction with the generally mountainous terrain, has contributed to the paucity of connections to the outside world. Microwave transmission systems connect the Uzbek network to other central Asian states, including Kyrgyzstan and Tajikistan. However, international network traffic relies on Uzbekistan's connection to the Trans-Asia-Europe Line, which was completed in 1998 and has maximum bandwidth that is dwarfed by modern standards. That traffic routes through a single internet exchange in Moscow. In total, Uzbekistan has only 7.8 Gbps of international bandwidth available, which is roughly equivalent to that of a large apartment complex in the developed world (Ruddy 2014).

The government announced plans in 2013 to borrow $108 million in order to expand the country's network infrastructure and greatly increase the available international bandwidth to some 100 Gbps, but little progress has been evident (Ruddy 2014). The network inside Uzbekistan has been given much higher development priority, with investment from Russian, Japanese, and South Korean telecommunications companies, which has led to a major expansion in the connectivity of universities and government facilities (Ruddy 2014). These form the backbone of the Uzbek domestic intranet (called ZiyoNET), which by law requires all internet connections to route through ZiyoNET rather than connecting directly to the international internet (Freedom House 2015). This makes censorship and control of the internet a much easier technical task for the government than it is for many countries.

While technically several dozen private internet service providers (ISPs) operate in Uzbekistan, in reality the state-controlled Uztelecom has a virtual monopoly. It has state-granted exclusivity over all international gateways. Although in 2003, the regime announced its intent to privatize Uztelecom, that process has not yet materialized (Ruddy 2014).

Government censorship is widespread, with a large number of international sites blocked entirely by Uztelecom. These sites include Amnesty International, Freedom House, Human Rights Watch, Radio Free Europe, the BBC, and Voice of America. In addition, there is systematic blocking of the websites of any international groups of Uzbek expatriates living abroad that focus on human rights or political opposition (Freedom House 2015).

Domestically, censorship is even more heavy-handed, with strict laws regulating what individuals can post online. The government passed its most recent antiblogger law in September 2014, in advance of elections held in December and March, as part of its general strategy of rigging those elections. In a pattern of repressing any dissidence, whether in traditional media or on the internet, arrests of bloggers has become a common occurrence (CPJ 2015). The new law requires ordinary Uzbeks to remove "untrue posts" if requested to do so by the government and uses intentionally vague verbiage such that the government has carte blanche to censor and arrest at will (Kendzior 2015). This has resulted in high levels of self-censorship in Uzbekistan as people aim to steer clear of any dangerous topics.

Interestingly, Uzbekistan has not yet blocked access to most Western social media, including Twitter and Facebook, although it has done so with the Russian-controlled Live Journal blogging platform. Twitter popularity in the country spiked in 2013 when Gulnara Karimova (1972–), the eldest daughter of President Islam Karimov (1938–2016), used it to post details about the presidential family and corruption in the security services (Freedom House 2015). Google is the most visited site from within Uzbekistan, followed by Mail.ru (a Russian-based email and portal), Facebook, and YouTube (Alexa 2015). The Russian social networking sites of Odnoklassniki and Vkontakte are also among the most visited sites in Uzbekistan. However, Uzbekistan has begun introducing domestically created alternatives to many of these websites, such as the YouTube clone Utube.uz and a Twitter clone called Bamboo. These developments may hint at long-term plans to introduce alternatives that can be domestically controlled before blocking the foreign services altogether (Freedom House 2015).

In 2014, opposition groups launched a new Facebook page called Qorqmaymiz, which means "We are not afraid" in Uzbek. The group had accumulated some 12,000 members by the summer of 2015, which is an impressive number by Uzbek standards given the dangerous levels of repression and the low engagement of the population in general with social media. The site contains discussion and criticism of the government and encourages users to post pictures of themselves with signs saying "Qorqmayman," meaning "I am not afraid," which is itself a prosecutable offense (Kendzior 2015). The fact remains, however, that Uzbekistan's regime could easily block access to Facebook but it has not, which suggests that the regime values being able to monitor the activities of dissident groups like Qorqmaymiz more than it feels any threat from their existence.

The story of the internet in Uzbekistan is one of censorship and violent repression, mirroring the nonelectronic experience of the country. Its infrastructure is wholly inadequate compared to modern network standards. Even what does exist is largely unaffordable to the bulk of its population, especially those living outside urban centers. While signs of infrastructural progress have advanced over the last few years, the possibility of modern and universal access to the internet remains at best years in the future. Despite the encouraging emergence of grass-roots movements like Qorqmaymiz, these initiatives remain small, rare, and vulnerable to blocking by the government.

Steven Lloyd Wilson

See also: Afghanistan; Kazakhstan; Kyrgyzstan; Russia

Further Reading

Alexa. 2016. "Top Sites in Uzbekistan." Accessed March 25, 2016. http://www.alexa.com/topsites/countries/UZ

Committee to Protect Journalists. 2015. "Attacks on the Press." Accessed March 25, 2016. https://www.cpj.org/2015/04/attacks-on-the-press.php

Freedom House. 2015. "Freedom on the Net 2015: Uzbekistan Country Report." Accessed March 25, 2016. https://freedomhouse.org/report/freedom-net/2015/uzbekistan

International Telecommunications Union (ITU). 2015. "World Telecommunications Indicators Database." Accessed March 25, 2016. http://www.itu.int/en/ITU-D/Statistics/Pages/stat/default.aspx

Kendzior, Sarah. 2015. "'We Are Not Afraid': Inside an Uzbek Internet rebellion." *Foreign Policy*, July 14. Accessed March 25, 2016. http://foreignpolicy.com/2015/07/14/we-are-not-afraid-uzbekistan-qorqmaymiz/

Ruddy, Michael, and Ozdemir, Esra. 2014. "Broadband Infrastructure in North and Central Asia," United Nations Economic and Social Commission for Asia and the Pacific. http://www.unescap.org/sites/default/files/Broadband%20Infrastructure%20in%20North%20and%20Central%20Asia%20FINAL%20_English.pdf

Silk Project. 2012. "The Condition of the Internet in Uzbekistan." Accessed March 25, 2016. http://www.silkproject.org/internetinuzbekistan.htm

V

VENEZUELA

Venezuela lies in northern South America. Colombia borders the country to the west, with Guyana to the east and Brazil to the south. In terms of social media, Venezuela's involvement is on the rise. Estimates place the social media–using population at around 10 million, which is about a third of the 31 million people residing in the country. Regionally, Venezuela comprises about 6 percent of Latin America's online population. The largest population segment online is men, primarily those aged twenty-five years old and younger; this national statistic mirrors that of Facebook's Venezuelan demographic, where the largest population is men aged eighteen to twenty-four years old. With these age groups and other demographics, Venezuela's social media has come to play a growing role in national politics and domestic economic problems.

Venezuela's social media users primarily congregate on Facebook. Fewer are present on Twitter, though those who use Twitter tend to use the service to exchange messages about politics more so than on any other platform. Approximately 68 percent of the country's social media population who use mobile phones for access have Facebook accounts, in contrast to the 30 percent who are on Twitter. Other platforms attracted less than 1 percent of the population, though there were Venezuelans on Pinterest, Google+, YouTube, Tumblr, and StumbleUpon (StatsMonkey 2015b). The numbers are slightly different for desktop users, though the same trends are apparent. Desktop users overwhelmingly access Facebook, at 85 percent, while 13 percent use Twitter and 1 percent or less use Pinterest, Tumblr, Reddit, Google+, StumbleUpon, YouTube, LinkedIn, and Russia's vKontakte (StatsMonkey 2015a). Tuenti, a platform that originated in Spain and then expanded to the Latin American market, arrived in Venezuela in 2013. Venezuelan users have also appeared on Taringa!, Ask.fm, Scribd, and Badoo.

Venezuelans use a variety of apps, typically those that facilitate ease of using social media or permit gaming. On Google Play, the top free apps are WhatsApp Messenger, BBM, Facebook Messenger, Hola Laucher, and Instagram. The top paid apps are Minecraft: Pocket Edition, Monopoly, Quizduell Premium, Xperia Donald Duck, and Geometry Dash. For iOS systems, the top free apps are WhatsApp Messenger, Movistar, Instagram, Free Music Mp3 Player, and Facebook Messenger. Popular paid apps included 1Passe Pro Password Manager, *Quien Quiere Ser Millonario?*, Where's My Water?, Angry Birds, and Fruit Ninja (AppAnnie 2015). Facebook has enabled Venezuelans to access its chat app more easily; for the United States, Canada, Peru, and Venezuela, it is no longer necessary to have a Facebook account to use the chat function.

The country's most accessed websites differed slightly from the preferred list of apps. The top websites were Google.co.ve, Facebook, YouTube, Google, Mercadolibre, Live.com, Twitter, Lapatilla, Amazon, and Banesconline (Alexa 2015). From these sites, Venezuelans liked to search on Google; log in to Facebook, Twitter, and YouTube; and shop on Mercadolibre and Amazon.

Venezuela has several unique social media platforms and apps. Poderopedia, for instance, is a "who's who" of Latin America. On the site, profiles are created on people, companies, and others that influence the lives of everyday people, and links them together to show their relationships. The site is similar to Wikipedia, in that registered users can contribute and help build the profiles. It also emulates aspects of Klout, in that it tries to determine how influential these profiles are and how they are interrelated. Poderopedia is available in Venezuela, Chile, and Colombia (Poderopedia 2015). Another local app is *Dilo Aquí*, created by Transparencía Venezuela, which allows people to report incidents of corruption, whether witnessed or experienced, through their mobile phones. An additional social media tool of note is Octopusocial, a tool developed in Venezuela to assist with consolidating an identity or brand across social media platforms. The tool also incorporates Bit.ly, Klout, and Really Simple Syndication (RSS) feeds. Venezuelans have created social media tools that meet their social and cultural needs, and as those needs become clearer, entrepreneurs will produce apps with a specific Venezuelan flavor.

Social media in Venezuela often challenges the relationship between the government and the people. The government wants to maintain control over social media, limiting the extent to which people can voice their thoughts openly online. Indeed, the government already has implemented censorship policies with other media and legally can arrest people who do not comply. At the same time, though, the people want to have an outlet where they can voice their concerns and discuss areas where the country needs to improve. These contradictory visions for social media create tensions that worsen when the government further restricts social media speech. At the same time, those who are brave speak out more when the government emphasizes the need for control and censorship. While Venezuelans do not have the right to free speech, they do want to preserve their human rights, which include exposing corruption and human rights abuses.

The Venezuelan government has always been skeptical of social networking platforms and websites. To curb unwanted outside influences, the government blocks controversial websites and has even gone so far as to arrest people who post criticisms of the government. Suspicions that users of Facebook and Twitter are spying on the country's domestic activities are not new. In fact, the Venezuelan government decried Cuba's U.S.-designed social media system, ZunZuneo, for being illegal and immoral. Venezuela pushed member states of the Bolivarian Alliance for the Peoples of Our America to denounce the Cuban version of Twitter and promoted the idea that the United States was using it to collect data about Cuban citizens (BBC Monitoring 2014). The outcry intensified against U.S. access to Cuban digital media because Venezuela keeps all of its sensitive big data, including government databases and intelligence records, on servers located outside Havana

(Forelle et al. 2015). In addition to concerns over potential foreign access, the fact that the platforms themselves sell data on their users and the content that they post have further worried the national government because it believes that these platforms have compromised Venezuela's privacy.

The government has worked to counter these websites and platforms by offering new, government-sponsored social media tools. The new system, named "Red Patria" or "Homeland Network," launched in May 2015. The system contained six tools intended to replace the popular international platforms growing in prominence domestically. The tools include *Nido* (which means "nest" in Spanish), a Facebook-like platform; *Colibrí* ("hummingbird") to replace WhatsApp, Condor to replace Tweetdeck; *Cardenalita* ("little cardinal") to replace Twitter; *Golondrinas* ("swallows") to function as national cloud storage; and *Mochuelo* ("small owl"). Mochuelo has two functions: one to allow citizens to report events as they unfold, and the other to work as a sort of crowd-mapping or crowd-sourcing application so that responders can visualize and assess reported incidents (Gobierno 2015).

While the government has developed alternatives to popular international platforms, it has also used the same platforms that it seeks to restrict to further its own goals. Venezuelan politicians, for example, have some of the most followed Twitter accounts in the world. Venezuelan president Nicolás Maduro (1962–) actively uses Twitter to promote his political agenda. He retweets, and sometimes posts, messages that promote Venezuelan nationalism and attack U.S. policies. Hashtags that he has used include #VivaVenezuela, #ObamaYankeeGoHome, and #ObamaAppeal-TheExecutiveOrder. At the same time as some politicians utilize social media to propel their agendas, the Venezuelan attorney general announced that a bill was forthcoming in 2015 to regulate social media. Censorship laws already exist, and police have the authority to arrest social media users who speak out against the government, but the forthcoming law intends to impose additional restrictions on social media speech.

Twitter is another platform used to shape the people's political opinions. In a 2015 study examining sampled Twitter accounts from Venezuelan politicians, political opposition members, and regular citizens, the researchers found conclusive evidence that Venezuelans from all categories used bots to increase their social media clout and reach. The content from Venezuelan politicians was retweeted, though bots accounted for only 10 percent of the politicians' retweeted messages (Forelle et al. 2015). That 10 percent, while relatively small, did account for some increased nationalistic political sentiment and for the spreading of information regarding political events. Although bots were not pervasive across the country's social media, the small number of bots could be significant with the country's low levels of Twitter penetration.

The other side to the challenges of media repression and access to social media is that the people want to have a voice, and they try to exercise that voice over social media. The Venezuelan economy, despite having the largest oil reserves in the world, has not been kind to the people. They suffer from the energy crisis that has wrecked Venezuela since 2013 and has created shortages of food and medicine. To

Venezuelan president Nicolás Maduro (1962–) speaks before the Non-Aligned Movement's opening ceremony in September 2016. Maduro uses social media, especially Twitter, to promote his anti-United States policies and Venezuelan nationalism. His messages and anti-U.S. hashtags have gained a large audience, making him one of the most followed people in the world. (Ognjen Stevanovic/Dreamstime.com)

connect the people with the places that have the resources they seek, some people are turning to social media. When a pharmacy is found to have supplies in stock, customers go to Twitter to spread the word. When people lack food, they also use Twitter to indicate where the need exists. In this way, Twitter connects people who have resources with those who do not. There are also sites for bartering items that are in surplus for those that are scarce. There are two apps where people turn to get supplies to meet their needs. *Akiztá* contains information about the availability of medical supplies, while *Abastéceme* advertises where to find staple foods.

While the people strive to make ends meet, Venezuela's opposition uses social media to promote political change. Videos and messages showing violence against the opposition in 2014 have spread on Facebook and Twitter. The hashtag #SOS-Venezuela, created internally, implored the world to intervene and stop the oppression. However, while the violence was real, many of the images and videos used in the campaign were faked. The Venezuelan and international media exacerbated the issue by using the fake pictures in news coverage without checking them for reliability and authenticity. The discovery of fake material weakened the online campaigns to curb the violence. In addition, it provided the government with renewed justifications to censor social media. Although the fake pictures damaged the opposition's position, the group continues organizing protests nonetheless. It utilizes hashtags to announce participation or advertise for additional support as it stages protests. For example, the February 2015 protests used #12F, #YoSalgoEl12F ("I go on 12 F"), and #YoSalgoPor ("I go on behalf of").

One issue unites the masses and Venezuelan politicians on social media: foreign interference. While Venezuelans use social media to speak out against some government-crafted economic and political policies, they are also wary of foreigners meddling in their affairs. Venezuelans have accused some news outlets via Twitter (in particular, the U.S. network CNN) of purposefully trying to destabilize the nation. To bring CNN's flawed reports to light, social media users created the hashtags #CNNmiente ("CNN lies") and #CNNVzSeLaRespeta ("CNN must respect Venezuela"). These hashtags united the people's voices as they criticized CNN and other news organizations for spreading falsified, unverified information about the country's state of affairs. In August 2015, CNN reported widespread looting in the city of Maracay. According to subsequent news, there was actually no looting in the area. CNN admitted that the report was incorrect and apologized to the Venezuelan people for the error that led to inaccurate reporting. President Maduro and Tareck El Aissami (1974–), governor of the state of Aragua (which contains Maracay), avidly rebuffed CNN (Telesurtv 2015). The people, also outraged that international media had contributed to diminishing Venezuela's reputation, turned to social media to redress the damage to Venezuela's reputation and show that the reports did not reflect reality.

Laura M. Steckman

See also: Argentina; Chile; Colombia; Cuba; Peru; Spain; United States

Further Reading

Alexa. 2015. "Top Sites in Venezuela." Accessed September 12, 2015. http://www.alexa.com /topsites/countries/VE

AppAnnie. 2015. "iOS Top App Charts." Accessed September 12, 2015. https://www .appannie.com/apps/ios/top/venezuela/?device=iphone

BBC Monitoring. 2014. "Venezuela Slams US-Developed Cuban Social Media Network." April 21. Accessed September 13, 2015. http://search.proquest.com.ezproxy.library.wisc .edu/docview/1517903269?accountid=465

Forelle, Michelle, Howard, Phil, Monroy-Hernández, Andrés, and Savage, Saiph. 2015. "Political Bots and the Manipulation of Public Opinion in Venezuela." July 25. Accessed September 12, 2015. http://arxiv.org/ftp/arxiv/papers/1507/1507.07109.pdf

Gobierno Bolivariano de Venezuela. 2015. "RedPatria" [Blog]. Accessed September 10, 2015. http://redpatria.cenditel.gob.ve/

Poderopedia. 2015. "Poderopedia." Accessed September 10, 2015. http://www.poderopedia .org/ve/

StatsMonkey. 2015a. "Desktop Facebook, Twitter, Social Network Usage Statistics in Venezuela." Accessed September 10, 2015. https://www.statsmonkey.com/table/21733 -venezuela-desktop-social-network-usage-statistics-2015.php

StatsMonkey. 2015b. "Mobile Facebook, Twitter, Social Network Usage Statistics in Venezuela." Accessed September 10, 2015. https://www.statsmonkey.com/table/21499
-venezuela-mobile-social-media-usage-statistics-2015.php

Telesurtv. 2015. "#CNNmiente es Tendencia Tras Campaña Criminal Contra Venezuel." August 8. Accessed September 13, 2015. http://www.telesurtv.net/news/CNNmiente-es
-tendencia-tras-campana-criminal-contra-Venezuela-20150808-0002.html.

VIETNAM

Vietnam is a Southeast Asian country on the Indochina Peninsula with a population of 93.5 million. It is bordered to the north by China, to the east by the South
China Sea, and to the west by Laos and Cambodia. At the end of the Vietnam War
in 1975, North and South Vietnam unified under a communist government. Impoverished and politically isolated for many years afterward, Vietnam has since improved its situation by developing global political relations and now also enjoys a
thriving economy with a robust technology sector.

Even though two-thirds of the population still lives in rural areas, by the end of
2013, internet penetration had reached 39 percent, with over 36 million internet
users. The average time spent online using a computer or a laptop was four hours
and thirty-seven minutes (We Are Social 2014). By 2015, 81 percent of internet
users between the ages of sixteen and twenty-four were going online every day.
However, older age groups were less likely to use the internet as much, with only
about two-thirds of twenty-five- to forty-four-year-olds and about half of forty-five
and older connecting every day (Statista 2015). When these users were online,
they were most likely to visit Cốc Cốc, Google, Facebook, YouTube, Google.com.
vn, Webretho, Vnexpress, Zing, 24h.com.vn, and Doisongphapluat (Alexa 2015).
Examining this data makes it very clear that while Facebook may be the most
popular social networking platform, Vietnamese internet users value their regional
websites.

As of 2013, there were 134,066,000 active mobile subscriptions in the country.
This statistic suggests that mobile phone users had multiple subscriptions. A total
of 89 percent of mobile phone subscribers had prepaid accounts. While the average user spent four hours and thirty-seven minutes accessing the internet from a
desktop or laptop, cell phone owners were spending only one hour and forty-three
minutes a day accessing the internet with their mobile devices. It is not surprising,
then that smartphone penetration was only 20 percent. Overwhelmingly, those who
had smartphones were using them to look up local information or research a product. At the end of 2013, 60 percent of smartphone users reported having made a
purchase with their phone (We Are Social 2014). In 2015, there was so much competition to provide users with smartphones that data was really cheap; the best data
plans offered one gigabyte for just $3 (Hookway 2015). The apps that smartphone
and tablet users downloaded most often demonstrate their preference for locally
developed platforms. The top five most downloaded free apps for the iPhone in 2015
were Zalo (a chat app), Zing MP3 (a music app), Facebook Messenger, Facebook, and
Than Bài Online (a gaming app). The top five most downloaded free apps for the

Well-known dissident and blogger, Le Quoc Quan (1971–), works in his office in Hanoi, Vietnam, on September 28, 2012. Historically, Vietnam has practiced censorship of content that was deemed critical of the government and included arrests, intimidation, and prison sentences for bloggers like Quan. However, Vietnam's government does not possess the resources necessary to control the flow of internet traffic and, in recent years, has become more supportive of social media, even encouraging its use in ways that benefit economic growth. (AP Photo/ Na Son Nguyen)

Android were similar: Facebook, Zalo, Facebook Messenger, Zing, and WiFi Mast Key (AppAnnie 2015).

The average time that internet users spent on social media each day was two hours and twenty-three minutes at the end of 2013. At that time, Facebook, the most popular social networking website, had a penetration rate of 22 percent, with over 20 million active users (We Are Social 2014). By 2015, the number was closer to 30 million and was rivaled only by Zalo, a messaging app that is used primarily in Vietnam. Other social media websites that were popular in Vietnam included Google+ and Twitter.

In the past, Vietnam has practiced censorship with those websites that it deems critical of the government, which has included some international human rights organizations and, at times, has briefly blocked Facebook in an attempt to control social media messages critical of the government. A few years ago, in an attempt to divert people's attention from the unrestricted atmosphere that existed on Facebook, the government introduced Go.vn. This was supposed to be the Vietnamese version of Facebook; however, the requirements for registering proved to be too restrictive. Users had to provide their real names and their government-issued identification number. Preference for Facebook rapidly overtook the Vietnamese version. However, the government has softened its stance on Facebook; in January 2015, Prime Minister Nguyen Tan Dung (1949–) publically mentioned how Facebook can benefit small business owners. Since that announcement, even some cabinet members have become Facebook members (Hookway 2015). The Vietnamese government does not have the resources to control the flow of internet traffic in the same way as other countries that practice censorship, like China. Of late, the government has been changing its opinion on social media and has been encouraging its use because of the potential benefits it has for economic growth (Economist 2015).

This change of opinion regarding social media has allowed Vietnamese users to do more than just chat or catch up with friends; in 2015, they used social media sites to protest the removal of very old trees that decorated Vietnam's capital, Hanoi. In March, at least 500 of these trees were removed as part of the first stage of a city project to replace the trees with younger varieties. Online speculation suggested that those trees were going to be sold off for timber. The project was begun without consulting the public, who turned to Facebook to express their displeasure. A campaign to stop the felling began, and within twenty-four hours, it had garnered 20,000 supporters. Hanoi's city leaders halted the project and later called for an investigation into it (Economist 2015).

Vietnam has a booming internet start-up scene, with many companies hoping to compete with the likes of Google and Facebook. While many of these endeavors inevitably lead to failure, some of them have become incredibly successful and are proving to be fierce competition for their Western counterparts. This success is reflected in the patterns of Vietnamese social media users. An example of one of these successful stories is Cốc Cốc, a Russian-Vietnamese endeavor that launched at the end of 2012 in direct competition with Google. Its name means "knock knock" in Vietnamese, and its creators believed that the search engine stood a competitive chance against Google for language reasons. While Google is very popular in Vietnam, it does not have a local office, which puts it at a disadvantage. Cốc Cốc, on the other hand, has on-site employees who understand the local language and culture. They also have the benefit of local knowledge, which can be used to make their search results more relevant to users. These advantages have paid off; toward the end of 2015,Cốc Cốc ranked as the most visited website in the country (Alexa 2015). Some Vietnamese voice concerns over whether Cốc Cốc would censor search results on behalf of the government if requested; however, this fear does not seem to have affected the website's popularity (AP 2013).

Zalo is a relatively new app introduced to the market in 2012. It is a chat app that competes with other popular messaging apps in the region, such as Line, We-Chat, Viber, and WhatsApp. After the app launched, it took off a little slowly but quickly gained momentum. By January 2013, it had fewer than 1 million users; however, by the end of that year, it was already the second most popular chat app in the country, with 7 million users. By May 2015, Zalo had 30 million registered users, and 400 million messages were being exchanged every day. At that time, the app was the most downloaded messaging app for both iPhone and Android users. As of late 2015, Zalo was the only Southeast Asia–based chat app to gain such popularity in its home market (Do 2015).

Another successful start-up, Webtretho, ranked as the sixth most popular website in Vietnam toward the end of 2015. A forum with a bulletin board format, it is the largest online community of women in Vietnam. The name means "youthful web" in Vietnamese and it had more than 1 million users at the end of 2013. The success of the website has led to the launch of several other websites that have an e-commerce focus and target specific products: LamDieu is a cosmetics website, Beyeu sells baby products, and Foreva focuses on fashion and intimates (Do 2013b). An interesting note regarding Webtretho is that one of the most sought-after topics on the forum is adultery (Do 2013a).

While the tech industry and internet users in Vietnam may be growing quickly, the country's infrastructure for providing internet services has had problems during 2015 due to several breakdowns. While the country is working on improving this system, Vietnam's only way to access the internet at the moment is through the Asia America Gateway, a cable built in 2009 to connect Southeast Asia with the United States. It is a 20,000-kilometer-long cable that is submerged in the ocean. This cable system provides Vietnamese internet users with average speeds ranging from 160 Kbps to 3.2 Mbps, which makes it the slowest internet in the Southeast Asia and Oceania region (Thanh Nien News 2015; Vietnam Briefing 2015). However, the rapidly growing number of internet users taxes this single entry point. In addition, 2015 saw three instances of breakdowns in the cables. While these breakdowns were being repaired, the whole country experienced an enormous slowdown of internet service (Vietnam Briefing 2015). An interesting note is that some of these breakages have been due to sharks mistaking the cables for food and biting them (Hooton 2015).

Marilyn J. Andrews

See also: China; Malaysia; Philippines; Thailand; United States

Further Reading

Alexa. 2015. "Top Sites in Vietnam." Accessed September 29, 2015. http://www.alexa.com/topsites/countries/VN

AppAnnie. 2015. "IOS Top App Charts." Accessed September 29, 2015. https://www.appannie.com/apps/ios/top/vietnam/?device=iphone

Associated Press (AP). 2013. "Knock Knock, We're Coming to Get You, Google! Vietnam Start-up Challenges Global Search Giant Because It Doesn't Get Nuances of Local Language." May 15, 2013. Accessed September 29, 2015. http://www.dailymail.co.uk/news/article-2324775/Coc-Coc-Vietnam-start-challenges-Google-doesnt-nuances-local-language.html

Do, Anh-Minh. 2013a. "10 Startups in Vietnam That Have Reached over 1 Million Users." November 29. Accessed September 29, 2015. https://www.techinasia.com/million-user-startups-vietnam/

Do, Anh-Minh. 2013b. "Project Lana Sheds Jobs, Refocuses on Beyeu and Webtretho." December 16. Accessed September 29, 2015. https://www.techinasia.com/project-lana-a-new-beginning/

Do, Anh-Minh. 2015. "Vietnam's Chat App Zalo Challenges Facebook with 30 Million Registered Users." May 20. Accessed September 29, 2015. https://www.techinasia.com/zalo-30-million-registered-users-vietnam/

Hookway, James. 2015. "Five Things About the Internet in Vietnam." June 12. Accessed September 29, 2015. http://blogs.wsj.com/briefly/2015/06/12/5-five-things-about-the-internet-in-vietnam/

Hooton, Christopher. 2015. "Sharks Are Eating the Internet in Vietnam." January 2015. Accessed September 29, 2015. http://www.independent.co.uk/life-style/gadgets-and -tech/news/sharks-are-eating-the-internet-in-vietnam-9962747.html

Statista. 2015. "Daily Internet Usage Rate in Vietnam in 2014, By Age Group." Accessed September 29, 2015. http://www.statista.com/statistics/348252/daily-internet-usage-age -group-vietnam/

Thanh Nien News. 2015. "Vietnam's Mobile Internet Speed Ranks Lowest in Southeast Asia: Report." September 3. Accessed September 30, 2015. http://www.thanhniennews.com /tech/vietnams-mobile-internet-speed-ranks-lowest-in-southeast-asia-report-50938 .html

The Economist. 2015. "If a Tree Falls . . . Online, Will the Communist Party Hear Anything?" April 18. Accessed September 29, 2015. http://www.economist.com/news/asia/21648 706-online-will-communist-party-hear-anything-if-tree-falls

Vietnam Briefing. 2015. "Vietnam's Internet Infrastructure Improving." July 1. Accessed September 29, 2015. http://www.vietnam-briefing.com/news/vietnam-internet-infrastruc ture.html/

We Are Social. 2014. "2014 Asia-Pacific Digital Overview." January 23. Accessed September 29, 2015. http://wearesocial.net/blog/2014/01/social-digital-mobile-apac-2014/

Y

YEMEN

Yemen, located on the southernmost coast of the Arabian Peninsula, has the lowest internet penetration in the Middle East. As of January 2015, the country had approximately 5 million internet users out of a population of over 25 million. Approximately 1.5 million had registered on Facebook and 128,000 on Twitter (BBC 2015). In a late 2014 telephone survey of social media users, Yemenis' preferred social media platforms were Facebook, WhatsApp, YouTube, Google+, and Instagram. Less than 1 percent of the people surveyed indicated that they used LinkedIn, making it one of the country's least popular platforms; despite the lack in popularity, however, Yemenis are the second most active Middle Eastern users on the site, right behind Lebanon (TNS 2015). Yemenis have social presences in smaller numbers on other platforms; the platforms that they frequent depend on the issues being discussed and the community that has formed there. There is no evidence that Yemenis have produced apps aimed solely at their domestic market; instead, they prefer to communicate through established platforms.

Yemen's low percentages of social media users correlate with high levels of poverty and lack of telecommunications infrastructure in the country, which are in part a result of internal conflict. Since 2001, Yemen's government has dealt with a number of groups and issues that threaten stability. For example, after the terrorist attacks of September 2001 shook the world, the Yemeni government cracked down on Muslim clerics promoting radical ideologies, particularly those attached to Al Qaeda. The government also was contending with a Houthi-backed, or Zaydi Shia Muslim–backed, insurgency in the north; that group, though it faced setbacks for over a decade, seized the capital in September 2014. After forcing President Abdrabbuh Mansour Hadi (1945–) to flee, the Houthis have slowly expanded their territorial control. Yemen's precarious political situation, along with severe resource shortages (with the lack of access to water being the most extreme), contributes to the non-existent state-sponsored efforts to improve the telecommunications infrastructure. Thus, Yemen's social media user base is not expected to increase dramatically and likely will increase only if private companies develop the requisite infrastructure.

However, despite the difficulties inherent in Yemenis getting online, data on users who can access the internet showed that they sought to improve their devices' performance and security through downloads. For users preferring to browse, the highest level of interest was in reading the news and looking at social media. For example, Yemenis' top ten free apps downloaded through GooglePlay were the WhatsApp Messenger, LEO Privacy Guard, SOMA Messenger, Opera Mini web browser, Facebook, DU Speed Booster, 360 Security, Clean Master (Boost and

AppLock), and the UC Browser. The top paid app was Minecraft: Pocket Edition (AppAnnie 2015). The top ten websites accessed in Yemen were Google, Facebook, YouTube, Twitter, Blogspot, Muhitelyemen, Yahoo, Sabanews, Yemen.net.ye, and Kooora (Alexa 2015).

Yemen first gained a presence on the Middle East's social media scene during the Arab Spring uprisings throughout the region. In December 2010, Tunisia's Mohamed Bouazizi (1984–2011) immolated himself in protest after Tunisian government officials confiscated his street vendor wares and subjected him to harassment and humiliation. His plight, and subsequent suicide, prompted people across the Arab world to protest repressive governments. The activity in Tunisia inspired many Yemenis to protest in solidarity. Although Yemen's domestic efforts differed from those in Tunisia, citizens in both countries wanted to stop government oppression. Yemen's protests initially revolved around poor economic conditions, corruption, and high unemployment rates. Later, as the peaceful protests incurred violent responses from the government and security forces, Yemenis demonstrated for their rights and against institutions that inflicted violence on the people. By November 2011, President Ali Abdullah Saleh (1942–) stepped down, to be succeeded by Hadi. Popular protests, with social media playing a small role, contributed to the people's victory.

Social media played a small but important role in Yemen's Arab Spring. During the protests, the Yemenis were the first Arab group to rely on hashtags as part of the wider protest movement. The hashtag #Yemen became one of the important channels of communication, allowing protestors to publicize events and solicit mass participation. In February 2011, Yemenis arranged the first major protest against the Saleh regime. #Feb3, a date predating the protests that led to the overthrow of Egyptian president Hosni Mubarak, informed the Yemeni people about the plans attached to the protests. Subsequent protests often had hashtags reflecting the date of the protest, particularly when organizers wanted to ensure high numbers of participants. Protests occurred continually, and only major events received dated hashtags.

The phenomenon of Yemenis utilizing hashtags is significant because they developed campaigns that attracted popular attention despite having a negligible percentage of people with access to the internet. In 2011, 2 percent or less of the population had access. What this meant was that, while creative and resourceful, social media did not account for the level of popular support. While it helped publicize events, other communications channels must have spread the messages appearing on social media platforms. Specifically, several more traditional techniques were used to communicate about the protests, including painting signs and distributing printed materials. One creative outlet was the baking of words such as *irhal*, the Arabic word for "go," into loaves of bread to signal that protesting was imminent.

Yemen's most popular social media user is Tawakkol Karman (1979–), a journalist, politician, and women's rights activist. On Twitter, Karman has more than double the amount of followers as the *Mareb Press*, a Yemeni media outlet that held the account with the second highest number of subscribers in 2015. On Facebook, she had almost 3 million followers, with the second most popular account belonging to Dr. Hassan Al Amri (1970–), a Saudi Arabian native with 1.7 million followers (Socialbakers Statistics, 2015). She also has her own website to share news and speeches in both Arabic and English at www.tawakkolkarman.net.

In 2011, Karman was the first Arab woman to receive the Nobel Peace Prize. Her work focused on fighting violence toward women and the promotion of women's rights more generally in the wake of Yemeni's political turmoil and instability. In 2005, she founded Women Journalists Without Chains, an organization focused on documenting and reporting cases of human rights abuses against journalists. By 2007, Karman organized weekly protests in Sana'a, Yemen's capital, to promote government reform and advocate for more human rights protections. Once the Arab Spring started in 2011, Karman effectively rolled her protests into the activities of the Arab Spring protestors who were calling for political reform and the ousting of Yemen's ineffective leadership. With her name attached to these protests, Karman met with United Nations (UN) representatives to support a UN resolution against Saleh and the violence that his regime was inflicting against the protestors.

Karman's activism has not always been without cost, as she has been jailed multiple times for speaking out against the regime and for supporting women's rights. Despite the challenges levied against her, she continues to advocate for nonviolent resistance and human rights in Yemen through speeches that reach a wider audience and through social media outlets. Because of her prominence on social media and her activism, Karman continues to be one of the most powerful women in Yemen today.

In addition to individual efforts such as Karman's, Yemen boasts a grass-roots independent media collective called Support Yemen. The group's goals are to use media, especially video, to highlight and publicize the struggles occurring within Yemen. Using videos and social media, especially on YouTube, Facebook, Twitter, and the official website at supportyemen.org, the group allows volunteers across the country to document life in Yemen and immediately share their productions worldwide. One of Support Yemen's most popular videos, entitled "Happy Yemen," featured Yemeni youth lip-syncing to Pharrell William's song "Happy." The official version earned almost a million views through mid-2015 and seemed to be contagious, as more Yemenis in different cities created and uploaded similar content. The purpose of the videos was to show that Yemenis, despite sometimes-dire circumstances and great adversity, remained optimistic about their future. The "Happy Yemen" official video appeared on the group's website alongside other videos, many of which take a more serious approach to publicizing Yemen's internal conditions.

Support Yemen's most important internet dissemination strategy relied, and continues to rely, on the use of a primary hashtag. #SupportYemen, and the shortened version #SY, spread news and content, such as the aforementioned videos, as an alternative to mainstream news. Yemen's news often focuses on major political events and does not cover the attitudes and actions of local people, nor does it always expose the social and economic realities in the country. The hashtags allow people to follow these issues in one place, as well as across social media platforms. In order to reach beyond Yemen's population, Support Yemen and its hashtags can be found on additional platforms, including Vimeo, LinkedIn, and ScoopNest, and are further shared by international blogs that promote peace.

Laura M. Steckman

See also: Egypt; Iran; Saudi Arabia; Tunisia; United Arab Emirates

Z

ZIMBABWE

Zimbabwe, a landlocked country located in southern Africa bordering South Africa, Botswana, Zambia, and Mozambique, attained its independence from British colonial rule in 1980. Half of the country's 12 million people have internet access, according to the Postal and Telecommunications Regulatory Authority of Zimbabwe (POTRAZ 2015), a remarkable feat given that the country was home to just 50,000 internet users fifteen years ago. The ubiquity of mobile phones has been a welcome development among the nation's citizens, the majority of whom access the internet through cell phones. Both urban and rural dwellers use social media religiously, with WhatsApp, Facebook, and increasingly YouTube the most commonly used platforms. Twitter is a favorite among the nation's leading political figures, journalists, and political activists.

Zimbabwe's social media use is not restricted to the elite (Mutsvairo 2016). Some domestic workers and cattle herders earning as little as $100 a month have been known to post pictures or publishing political commentary on WhatsApp or Facebook. With English as its official language in addition to the local Shona and Ndebele vernaculars, along with several other local languages, it is not surprising that much digital communication in the country is conducted in English. On a comparative basis, digital illiteracy, which is considered a hindrance to technological development in some parts of Africa, is not as big a concern in Zimbabwe, which is believed to have the highest literacy rate in Africa (Makaripe 2015).

It is partly thanks to Zimbabweans fleeing the economic and political crisis, which began in 2000 when the country launched its agrarian reforms, that the country's digital transformation is gaining ground. Take, for example, the country's innovative mobile payment system, EcoCash, which is the brainchild of Zimbabwean telecom mogul Strive Masiyiwa (1961–), whose Econet Wireless Zimbabwe is the country's largest provider of telecommunications services. With Zimbabwe's economy virtually on its knees, the country is dependent on millions of its citizens living abroad to inject remittances into the economy. Close to $1 billion is estimated to have been poured into the economy courtesy of diaspora remittances in 2015 alone (Makaranyika 2015). Given the hardships facing the country, EcoCash, with over 4 million subscribers, has come in handy for needy Zimbabweans seeking financial assistance from relatives abroad. But Masiyiwa's road to recognition has been rocky. The government of President Robert Mugabe (1924–) was initially reluctant to give Masiyiwa a license until the Supreme Court made a timely intervention in 1995, paving the way for the businessman to set up his business empire. To this day, Masiyiwa has just a handful of friends within the country's political establishment because many perceive him to be the main face behind the Movement for Democratic Change, which since its formation in 1999 has provided formidable opposition to President Mugabe's 36-year rule. Masiyiwa has been accused of funding the opposition, much to the bemusement of officials from Mugabe's ZANU PF party. Like other Zimbabwean businesspeople, he is active on social media, even though he avoids direct attacks against the country's ruling elite.

In a country where dissenting voices have traditionally been silenced, the presence of social media is giving a new voice to activists opposed to Mugabe's rule. Its growing influence has become a potent symbol of citizen empowerment and political participation among the nation's citizens. It also is fast becoming an alternative source of news, which essentially is a huge paradigm shift given Zimbabweans' reliance on state media, whose reporting is often biased against Mugabe's opponents. It was almost unthinkable to accept, and impossible to foresee, that the *Herald*, a newspaper owned and regulated by the state that has previously demonstrated unwavering support for President Mugabe, now publishes public criticism of the government and its leader. The newspaper's Facebook page also attracts both positive and negative comments about the 93-year-old leader. While it's not clear whether Mugabe follows social media, many Zimbabweans will agree that social media, for the first time, has given them the ability to appeal directly to politicians, which may include Mugabe himself. Perhaps one piece of evidence that the president knows what is being said on social media is the fact that he has publicly threatened to use Chinese-made technology to block social media in Zimbabwe, further underlining the perceived threat posed by digital activism to his 36-year uninterrupted rule. Social media bans have already been imposed in a few African countries, including the Democratic Republic of Congo and Uganda.

Even though Zimbabwe has a stand-alone information and communications technology (ICT) ministry, it must be noted that the ruling party is very skeptical about technology. The infamous "Baba Jukwa" Facebook page, which gained notoriety for sharing state secrets on social media Wikileaks-style, left ZANU PF officials bewildered going into the 2013 elections. In a social media first for Zimbabwe, the Facebook blogger published mobile phone numbers of government officials, encouraging their constituents to call them and demand good governance. More recently, digitally based citizen movements, including the #ThisFlag campaign, which was initiated by a local clergyman, have also gathered momentum. The #ThisFlag movement calls upon Zimbabweans to use Facebook and other social media platforms to vent their anger against misrule and to speak out against corruption while draping the Zimbabwean flag around their necks. The government has responded by accusing the European Union and the United States of financially supporting the campaign in a bid to destabilize the country.

Better still, the social media explosion has undoubtedly created a market for citizen journalists in the country by providing indispensible platforms for technology-savvy journalists and activists to report eyewitness accounts and share stories with remarkable speed, further bypassing media restrictions. One must agree with the argument of Norwegian scholar Terje Skjerdal (2015) that the potential impact of citizen journalism and social media depends on its sociopolitical context within a

Zimbabweans in Harare gather around a cell phone to read about the famous Facebook blogger, Baba Jukwa, in July 2013. In a country where dissident voices have historically been silenced, social media users find hope in messages posted by critics of the country's political climate. Baba Jukwa first gained notoriety when he posted state secrets just before the 2013 elections. (AP Photo/Tsvangirayi Mukwzhi)

country or community. While Zimbabwe has a history of restricting media coverage, especially of its elections, a social media ban appears highly unlikely.

While the political use of social media is widespread in Zimbabwe, other people choose to stay out of politics for fear of reprisals. On a few occasions, people directly criticizing President Mugabe on social media have been arrested, even though it is fair to point out that the courts have freed them. Memes and political jokes easily find their way into the Zimbabwean blogosphere. It is the free Wi-Fi zones offered by internet providers such as Zolspot (which offers thirty minutes of free internet surfing) that make it possible for the majority of Zimbabwe's urban dwellers to gain internet access. You only need to pay the equivalent of 50 U.S. cents for just 8 MB of data. Masiyiwa's Econet charges exorbitant $1, $2, and $6 for its daily, weekly, and monthly social bundles, respectively. In a struggling economy such as Zimbabwe's, these prices are quite high.

Long-suffering Zimbabweans see hope in social media. They believe that online movements are giving them a voice for the first time, even though so far, it appears as if no one is listening. Mugabe, who has been criticized on social media for his love for foreign travel, is not paying heed as he continues to globetrot across the world. But unlike fifteen or twenty years ago, precise details of his trips are easily shared on social media, thanks to flight-tracking websites such as Flight Radar 24.

The presence of online activism is certainly encouraging, even though there is a danger of efforts being superseded by "clicktivism," as more and more activists turn to the internet in the hope of achieving political and social change. Social media sites are not just giving the activists a voice, as Mugabe's supporters frequently chime in during online conversations. While the president's opponents clearly feel empowered, Mugabe also sees social media as a powerful tool to galvanize support since anyone can use such platforms.

The true potential of social media will be realized only once the country's economy starts to perform. With the current economic upheavals hogging the limelight, it is very much a luxury to be on social media. Being there is not everyone's top priority, however, as hunger and starvation are tormenting rural Zimbabwe. In spite of these quagmires, though, social media has refreshingly transformed people's lives in Zimbabwe, making it possible for them to communicate and seek alternative sources of news easily. Several formal and informal businesses are now resorting to social media to market their products and services.

Bruce Mutsvairo

See also: Botswana; Mozambique; South Africa; United States

Further Reading

Gambanga, Nigel. 2015. "Looking for Wifi Locally, Here Are the Hotspots." Accessed June 2, 2016. http://www.techzim.co.zw/2015/06/looking-for-wifi-locally-here-are-the-hotspots/#.V1XicecrLnU

Makaranyika, Memory. 2016. "Nearly Half of Foreign Inflows into Zimbabwe Are Remittances." Accessed June 3, 2016. http://www.bdlive.co.za/africa/africanbusiness/2016/02/04/nearly-half-of-foreign-inflows-into-zimbabwe-are-remittances

Makaripe, Tendai. 2015. "Zim Teachers Among Lowest Paid In the Region." Accessed June 3, 2016. http://www.financialgazette.co.zw/zim-teachers-among-lowest-paid-in-the-region/

Mutsvairo, Bruce (Ed.). 2016. *Digital Activism in the Social Media Era: Critical Reflections on Emerging Trends in Sub-Saharan Africa.* London: Palgrave Macmillan.

POTRAZ Quarterly Reports. 2005. Accessed May 25, 2016. http://www.potraz.gov.zw/index.php/categorylinks/120-quarterly-reports

Skjerdal, Terje. 2015. "Why the Arab Spring Never Came to Ethiopia." In *Perspectives on Participatory Politics and Citizen Journalism in a Networked Africa: A Connected Continent* edited by B. Mutsvairo. London: Palgrave Macmillan.

Bibliography

Aamoth, Doug. 2011. "A Brief History of Skype." May 10. Accessed September 26, 2015 http://techland.time.com/2011/05/10/a-brief-history-of-skype/

Ahmad, Irfan. 2014. "Bye, Bye, Orkut: Google Says Goodbye to Its 1st Social Network on 30th September #Infographic." September 27. Accessed August 25, 2015. http://www.digitalinformationworld.com/2014/09/Google-Says-Goodbye-to-its-1st-Social-Network-infographic.html

Akamai. 2016. State of the Internet: Q1 2016 Report. Accessed July 5, 2016. https://www.akamai.com/us/en/multimedia/documents/state-of-the-internet/akamai-state-of-the-internet-report-q1-2016.pdf

Alexa.com. 2015. "Site Overview: Bbm.com." Accessed September 26, 2015. http://www.alexa.com/siteinfo/bbm.com

Alexa.com. 2015. "Site Overview: Blogspot.com." Accessed September 27, 2015. http://www.alexa.com/siteinfo/blogspot.com

Alexa.com. 2015. "Site Overview: Draugiem.lv." Accessed September 26, 2015. http://www.alexa.com/siteinfo/draugiem.lv

Alexa.com. 2015. "Site Overview: Frype.com." Accessed September 26, 2015. http://www.alexa.com/siteinfo/frype.com

Alexa.com. 2015. "Site Overview: Line.me." Accessed September 26, 2015. http://www.alexa.com/siteinfo/line.me

Alexa.com. 2015. "Site Overview: Messenger.com." Accessed September 26, 2015. http://www.alexa.com/siteinfo/messenger.com

Alexa.com. 2015. "Site Overview: Mixi.jp." Accessed September 26, 2015. http://www.alexa.com/siteinfo/mixi.jp

Alexa.com. 2015. "Site Overview: Pinterest.com." Accessed September 27, 2015. http://www.alexa.com/siteinfo/pinterest.com

Alexa.com. 2015. "Site Overview: Reddit.com." Accessed September 27, 2015. http://www.alexa.com/siteinfo/reddit.com

Alexa.com. 2015. "Site Overview: Renren.com." Accessed September 19, 2015. http://www.alexa.com/siteinfo/renren.com

Alexa.com. 2015. "Site Overview: Skype.com." Accessed September 26, 2015. http://www.alexa.com/siteinfo/skype.com

Alexa.com. 2015. "Site Overview: Twitter.com." Accessed September 27, 2015. http://www.alexa.com/siteinfo/twitter.com

Alexa.com. 2015. "Site Overview: Viadeo.com." Accessed September 26, 2015. http://www.alexa.com/siteinfo/viadeo.com

Alexa.com. 2015. "Site Overview: Viber.com." Accessed September 26, 2015. http://www.alexa.com/siteinfo/viber.com

Alexa.com. 2015. "Site Overview: WhatsApp.com." Accessed September 26, 2015. http://www.alexa.com/siteinfo/whatsapp.com

Alexa.com. 2015. "Site Overview: WordPress.com." Accessed September 27, 2015. http:// www.alexa.com/siteinfo/wordpress.com

Alexa.com. 2015. "Site Overview: Xing.com." Accessed September 26, 2015. http://www .alexa.com/siteinfo/xing.com

Alexa.com. 2015. "The Top 500 Sites on the Web." Accessed September 20, 2015. http://www .alexa.com/topsites

Amberber, Emmanuel. 2013. "[Infographic] The Internet of Africa." December 2. Accessed July 31, 2016. https://yourstory.com/2013/12/internet-africa/

Angwin, Julia. 2009. *Stealing MySpace: The Battle to Control the Most Popular Website in America.* New York: Random House.

Ariss, Jesse. 2015. "10 Years of BBM." July 27. Accessed September 26, 2015. http://blogs .blackberry.com/2015/07/10-years-of-bbm/

Arrington, Michael. 2006. "Odeo Releases Twttr." July 15. Accessed September 20, 2015. http://techcrunch.com/2006/07/15/is-twttr-interesting

Arrington, Michael. 2009. "Friendster Valued at Just $26.4 Million in Sale." December 15. Accessed September 20, 2015. http://techcrunch.com/2009/12/15/friendster-valued-at -just-26-4-million-in-sale

Associated Press. 2016. "Google Buys YouTube for $1.65 Billion." October 10. *NBC News.* Accessed September 17, 2016. http://www.nbcnews.com/id/15196982/ns/business-us _business/t/google-buys-youtube-billion/#.V92BGJgrLIU

AT&T. 1999. "SBC and Prodigy Announce Alliance." November 22. Accessed August 27, 2015. http://www.att.com/gen/press-room?pid=4800&cdvn=news&newsarticleid =7168&mapcode=corporate

Avalaunch Media. 2013. "The Complete History of Social Media." April 15. Accessed August 13, 2015. http://avalaunchmedia.com/history-of-social-media/Main.html

Banks, Michael. 2008. *On the Way to the Web: The Secret History of the Internet and Its Founders.* New York: Springer.

Barboza, David. 2014. "A Popular Chinese Social Networking App Blazes Its Own Path." January 21. Accessed September 19, 2015. http://www.cnbc.com/2014/01/21/a-popular -chinese-social-networking-app-blazes-its-own-path.html

Barnett, Brian. 2015. "Flickr Pro is Back, and It's Selling Quality over Quantity." July 25. Accessed September 27, 2015. http://www.wired.com/2015/07/flickr-pro-1-tb-full -resolution-photo-storage/

Bartle, Richard. 1990. "Early MUD History." November 15. Accessed August 26, 2015. http://mud.co.uk/richard/mudhist.htm

BBC. 2011. "Facebook Inspires Israeli Couple to Name Baby 'Like.'" May 16. Accessed August 30, 2015. http://www.bbc.com/news/world-middle-east-13417930

Berners-Lee, Tim, and Mark Fischetti. 1999. *Weaving the Web: The Original Design and Ultimate Destiny of the World Wide Web by Its Inventor.* San Francisco: HarperCollins Publishers.

BBC News. 2005. "EBay to Buy Skype in $2.6bn deal." September 12. Accessed September 26, 2015. http://news.bbc.co.uk/2/hi/business/4237338.stm

Becker, Tyler. 2015. "The 9 Major Social Networks Broken Down by Age." April 9. Accessed July 6, 2016. https://socialmediaweek.org/blog/2015/04/9-major-social-networks-age/

Berners-Lee, Tim and Fischetti, Mark. 1999. *Weaving the Web: The Original Design and Ultimate Destiny of the World Wide Web by Its Inventor.* San Francisco: HarperCollins Publishers.

Blogger. n.d. "The Story of Blogger." Accessed September 27, 2015. https://www.blogger.com /about

Bloomberg. 2014. "WhatsApp's Founder Goes from Food Stamps to Billionaire." February 20. Accessed September 24, 2015. http://www.bloomberg.com/news/articles/2014-02 -20/whatsapp-s-founder-goes-from-food-stamps-to-billionaire

Bloomberg. 2015. "Executive Profile: Xueling Li." Accessed September 19, 2015. http://www .bloomberg.com/research/stocks/people/person.asp?personId=222965354&ticker=YY

Boyd, Danah M. 2008. "Social Network Sites: Definition, History, and Scholarship." *Journal of Computer-Mediated Communication*, 13(1): 210–230.

Budiu, Raluca. 2013. "Mobile: Native Apps, Web Apps, and Hybrid Apps." Accessed July 4, 2016. https://www.nngroup.com/articles/mobile-native-apps/

Campbell, Todd. 2002. "The First E-mail Message: Who Sent It and What It Said." Accessed August 20, 2015. http://www.cs.umd.edu/class/spring2002/cmsc434-0101/MUIseum /applications/firstemail.html

Carlson, David. 2009. "The Online Timeline: CompuServe." Accessed August 26, 2015. http://iml.jou.ufl.edu/carlson/history/compuserve.htm

Carlson, Nicholas. 2012. "Inside Pinterest: An Overnight Success Four Years in the Making." May 1. Accessed September 27, 2015. http://www.businessinsider.com/inside -pinterest-an-overnight-success-four-years-in-the-making-2012-4

Cavender, Sasha. 1998. "Legends." October 5. Accessed August 20, 2015. http://www.forbes .com/asap/1998/1005/126.html

Chao, Loretta. 2011. "Renren Changes Key User Figure Before IPO." April 29. Accessed September 19, 2015. http://www.wsj.com/articles/SB10001424052748704729304576 286903217555660#ixzz1KqsoJPb8

Chase, Chris. 2014. "Germany's World Cup Rout of Brazil Was the Most Tweeted Event in History." July 9. Accessed September 8, 2016. https://ftw.usatoday.com/2014/07 /germany-brazil-most-tweeted-event-history-miley-cyrus

Chen, Adrian. 2015. "The Agency." June 2. Accessed September 8, 2015. http://www.nytimes .com/2015/06/07/magazine/the-agency.html?_r=0

Chernova, Yuliya. 2015. "The Back Story of Meerkat: A Side Project That Took Off." March 4. Accessed September 13, 2016. http://blogs.wsj.com/venturecapital/2015/03/04/the -back-story-of-meerkat-a-side-project-that-took-off/

Choudhury, Saheli Roy. 2016. "Tencent Overtakes Alibaba as China's Most Valuable Tech Company." August 18. *CNBC*. Accessed September 17, 2016. http://www.cnbc.com/ 2016/08/17/tencent-overtakes-alibaba-as-chinas-most-valuable-tech-company-as- wechat-owner-posts-strong-results.html

Classmates.com. 2015. "About Classmates." Accessed August 20, 2015. http://www .classmates.com/siteui/about

Colao, J. J. 2014. "The Inside Story of Snapchat: The World's Hottest App or a $3 Billion Disappearing Act?" January 6. Accessed September 19, 2015. http://www.forbes.com /sites/jjcolao/2014/01/06/the-inside-story-of-snapchat-the-worlds-hottest-app-or-a-3 -billion-disappearing-act

CompuServ. 2015. "About CompuServ." Accessed August 20, 2015. http://webcenters .netscape.compuserve.com/menu/about.jsp

Cosenza, Vincenzo. 2016. "World Map of Social Networks." Accessed July 6, 2016. http:// vincos.it/world-map-of-social-networks/

CrunchBase. 2015. "Draugiem." Accessed September 26, 2015. https://www.crunchbase .com/organization/draugiem

CrunchBase. 2015. "Victor Koo." Accessed September 19, 2015. https://www.crunchbase .com/person/victor-koo

CrunchBase. 2015. "Xing." Accessed September 26, 2015. https://www.crunchbase.com
/organization/xing

Cutler, Kim-Mai. 2013. "Facebook Barely Poked Snapchat, Active Usage Data Shows."
March 8. Accessed September 19, 2015. https://techcrunch.com/2013/03/08/facebook
-snapchat/

Devereux, Eoin, ed. 2007. *Media Studies: Key Issues and Debates*. Los Angeles: Sage
Publications.

DiChristopher, Tom. 2015. "Verizon Closes AOL Acquisition." June 23. Accessed August 30,
2015. http://www.cnbc.com/2015/06/23/verizon-closes-aol-acquisition.html

Doll, Jen. 2011. "R.I.P., Friendster, the Social Media Site of Our Relative Youth." April 26.
Accessed August 12, 2015. http://www.villagevoice.com/news/rip-friendster-the-social
-media-site-of-our-relative-youth-6666000

Echovme. 2014. "5 Quick Facts About Orkut You Never Knew!" September 30. Accessed
September 8, 2015. http://www.echovme.in/blog/5-quick-facts-about-orkut-you-never
-knew/

Edelman. 2015. "Social Media Use in South Korea 2015." June 29. Accessed September 5,
2015. http://www.slideshare.net/EdelmanAPAC/social-media-in-south-korea-2015

Edwards, Benj. 2014. "Where Online Services Go When They Die: Rebuilding Prodigy, One
Screen at a Time." July 12. Accessed August 27, 2015. http://www.theatlantic.com
/technology/archive/2014/07/where-online-services-go-when-they-die/374099

Eordogh, Fruzsina. 2016. "Making Sense of YouTube's Great Demonetization Controversy
of 2016." September 2. *Forbes*. Accessed September 17, 2016. http://www.forbes.com
/sites/fruzsinaeordogh/2016/09/02/making-sense-of-youtubes-great-demonetization
-controversy-of-2016/#4207505b5b0d

Erdbrink, Thomas. 2013. "Iran Bars Social Media Again After a Day." September 17. Accessed September 26, 2015. http://www.nytimes.com/2013/09/18/world/middleeast
/facebook-and-twitter-blocked-again-in-iran-after-respite.html

Facebook. 2016. "Thank You! Messenger." July 20. *Facebook Newsroom*. Accessed September 17, 2016. http://newsroom.fb.com/news/2016/07/thank-you-messenger/

Fitzpatrick, Laura. 2010. "Brief History of YouTube." May 31. Accessed September 27, 2015.
http://content.time.com/time/magazine/article/0,9171,1990787,00.html

Flannery, Russell. 2012. "Land of the Large: Youku, Tudou Merger Latest in China's Web
Consolidation." March 12. Accessed September 19, 2015. http://www.forbes.com/sites
/russellflannery/2012/03/12/land-of-the-large-youku-tudou-merger-latest-in-chinas
-web-consolidation

Gagne, Ken, and Matt Lake. 2009. "CompuServe, Prodigy et al.: What Web 2.0 Can Learn
from Online 1.0." July 15. Accessed August 27, 2015. http://www.computerworld.com
/article/2526547/networking/compuserve—prodigy-et-al—what-web-2-0-can-learn
-from-online-1-0.html

Geron, Tomio. 2012. "YY.com: China's Unique Real-Time Voice and Video Service with a
Virtual Goods Twist." June 11. Accessed September 19, 2015. http://www.forbes.com
/sites/tomiogeron/2012/06/11/yy-com-chinas-unique-real-time-voice-and-video
-service-with-a-virtual-goods-twist

Giampietro, Marina. 2013. "Twenty Years of a Free, Open Web." April 30. Accessed August 27,
2015. http://home.web.cern.ch/about/updates/2013/04/twenty-years-free-open-web

Global Web Index. 2015. "Social Networking Motivations." Accessed July 31, 2016. http://
www.globalwebindex.net/blog/top-10-reasons-for-using-social-media

Gonzalez, Sean. 1995. "Prodigy." February 21. *PC Magazine.14*(4), 152, 175.

Google Blog. 2011. "Introducing the Google+ Project: Real-life Sharing, Rethought for the Web." June 28. Accessed September 26, 2015. http://googleblog.blogspot.com/2011 /06/introducing-google-project-real-life.html

Green, Hank. 2015. "A Decade Later, YouTube Remains a Mystery, Especially to Itself." February 23. Accessed September 27, 2015. https://medium.com/@hankgreen/a-decade -later-youtube-remains-a-mystery-especially-to-itself-80a1c38feeaf

Griffin, Andrew. 2015. "China's great firewall gets higher: Tools to evade surveillance and site bans are blocked as Chinese internet censors tighten grip." January 30. Accessed August 30, 2015. http://www.independent.co.uk/life-style/gadgets-and-tech/news /chinas-great-firewall-gets-higher-tools-to-evade-surveillance-and-site-bans-are -blocked-as-chinese-internet-censors-tighten-grip-10013537.html

Griggs, Brandon, and Heather Kelly. 2013. "23 Key Moments from Twitter History." September 16. Accessed September 20, 2015. http://www.cnn.com/2013/09/13/tech /social-media/twitter-key-moments

Hafner, Katie, and Matthew Lyon. 1996. *Where Wizards Stay Up Late: The Origins of the Internet.* New York: Simon & Schuster.

Hamburger, Ellis. 2014. "Facebook Poke Is Dead . . . and So Is Facebook Camera." May 9. Accessed September 19, 2015. http://www.theverge.com/2014/5/9/5700732/facebook -poke-is-dead-and-so-is-facebook-camera

Hoffman, Reid. 2013. "LinkedIn Turns 10: Celebrating 10 Years of Relationships That Matter." May 5. Accessed September 26, 2015. http://blog.linkedin.com/2013/05/05 /linkedin-turns-10/

Huffington Post. 2012. "World's First Website, Created by Tim Berners-Lee in 1991, Is Still Up and Running on 21st Birthday." August 6. Accessed August 27, 2015. http://www .huffingtonpost.com/2012/08/06/worlds-first-website_n_1747476.html

Indexmundi. 2014. "Telephones—Mobile Cellular Per Capita." Accessed July 4, 2016. http:// www.indexmundi.com/g/r.aspx?v=4010

Interlat. 2015. "6 Redes Sociales que No Imaginabas que Existían." June 23. Accessed September 8, 2015. http://www.interlat.co/redes-sociales-curiosas/

Jacob, Mark, and Stephan Benzkofer. 2013. "10 Things You Might Not Know About Social Media." September 29. Accessed August 26, 2015. http://articles.chicagotribune.com /2013-09-29/opinion/ct-perspec-0929-things-20130929_1_social-media-10-things -boston-bombing-suspect

Jones, Steve. 2003. *Encyclopedia of New Media: An Essential Reference to Communication and Technology.* Thousand Oaks, CA: Sage Publications.

Jucha, Nicolas. 2012. "QQ—China's Instant Messenger." January 9. Accessed September 24, 2015. http://gbtimes.com/world/qq-chinas-instant-messenger

Kan, Michael. 2010. "China's Top IM Client QQ Goes International." December 15. Accessed September 24, 2015. http://www.pcworld.com/article/213750/article.html

Kaplan, Andreas M., and Michael Haenlein. 2010. "Users of the World, Unite? The Challenges and Opportunities of Social Media." *Business Horizons* 53: 59–68.

Kelly, Heather. 2015. "Google+ Is Getting Dismantled." July 27. Accessed September 26, 2015. http://money.cnn.com/2015/07/27/technology/google-plus-youtube/

Kemp, Simon. 2015. "Global Digital Statshot: August 2015." August 3. Accessed August 12, 2015. http://wearesocial.net/blog/2015/08/global-statshot-august-2015

Kim, Chang-Ran. 2014. "Japan's Rakuten Buys Chat App Viber for $900 Million to Expand Digital Empire." February 14. Accessed September 19, 2015. http://www.reuters.com /article/2014/02/14/us-rakuten-viber-idUSBREA1D07M20140214

Koh, Yoree. 2015. "Twitter Mulls Expanding Size of Tweets Past 140 Characters." September 29. Accessed September 30, 2015. http://www.wsj.com/articles/twitter-mulls-expanding-size-of-tweets-1443554128

Konrad, Alex. 2014. "Snapchat Billionaires Protect Their Stakes by Settling with Ousted Cofounder Reggie Brown." September 9. Accessed September 19, 2015. http://www.forbes.com/sites/alexkonrad/2014/09/09/snapchat-settles-cofounder-lawsuit

Lee, Emma. 2013. "Baidu's BBS Service Tieba Shifts to Mobile Social Networking After Announcing 1 Billion Users on 10th Anniversary." December 4. Accessed September 19, 2015. http://technode.com/2013/12/04/baidus-bbs-service-tieba-shifts-to-mobile-social-networking-after-announcing-1-billion-users-on-10th-anniversary

Leung, Rebecca. 2003. "In Search of the Past: Reunion Web Site Brings Together Schoolmates." May 5. Accessed August 20, 2015. http://www.cbsnews.com/news/in-search-of-the-past-05-05-2003

Levy, Oren. 2015. "All the World Loves Social Media But Many Love Their Local Networks Best." May 28. Accessed August 24, 2015. http://www.entrepreneur.com/article/246485

Lievrouw, Leah A., and Sonia Livingston, eds. 2006. *Handbook of New Media: Social Shaping and Social Consequences of ICT, Updated Student Edition.* London: Sage Publications.

LinkedIn. 2015. "About LinkedIn." Accessed September 26, 2015. https://press.linkedin.com/about-linkedin

Loyd, Beth. 2009. "At Least 140 Killed in Uighur Riots in China." July 6. Accessed September 19, 2015. http://abcnews.go.com/International/story?id=8010018&page=1

Lubov. 2015. "Top Social Networks in Russia: Latest Numbers and Trends." January 20. Accessed September 26, 2015. http://www.russiansearchtips.com/2015/01/top-social-networks-russia-latest-numbers-trends/

Lynch, Kevin. 2015. "Asian Star LUHAN Sets the Record for the Most Comments on a Weibo Post." January 15. Accessed August 30, 2015. http://www.guinnessworldrecords.com/news/2014/8/chinese-star-luhan-sets-the-record-for-most-comments-on-a-weibo-post-59720/

Macale, Sherilynn. 2011. "A Rundown of Reddit's History and Community [Infographic]." October 14. Accessed September 25, 2015. http://thenextweb.com/socialmedia/2011/10/14/a-rundown-of-reddits-history-and-community-infographic/

Macias, Amanda. 2015. "This 2-Year-Old Disc Jockey Is Being a Phenomenon in South Africa." April 11. Accessed September 8, 2015. http://www.businessinsider.com/this-2-year-old-disc-jockey-is-becoming-a-phenomenon-in-south-africa-2015-4

MacMillian, Douglas. 2015. "Snapchat Charges 99 Cents to Replay Disappearing Messages." September 15. Accessed October 25, 2016. http://blogs.wsj.com/digits/2015/09/15/snapchat-charges-99-cents-to-replay-disappearing-messages

Madrigal, Alexis C. 2014. "AMA: How a Weird Internet Thing Became a Mainstream Delight." January 4. Accessed September 24, 2015. http://www.theatlantic.com/technology/archive/2014/01/ama-how-a-weird-internet-thing-became-a-mainstream-delight/282860

Matchar, Emily. 2015. "The Tweeting Potholes of Panama." June 12. Accessed September 8, 2015. http://www.smithsonianmag.com/innovation/tweeting-potholes-panama-180955507/?no-ist

McIntosh, Neil. 2003. "Google Buys Blogger Webservice." February 18. Accessed September 27, 2015. http://www.theguardian.com/business/2003/feb/18/digitalmedia.citynews

Metz, Cade. 2016. "Forget Apple vs. the FBI: WhatsApp Just Switched on Encryption for a Billion People." April 5. *Wired.* Accessed September 17, 2016. https://www.wired.com/2016/04/forget-apple-vs-fbi-whatsapp-just-switched-encryption-billion-people/

Meeuf, Kate. 2014. "Regional Use of Social Networking Tools." Accessed September 26, 2015. http://repository.cmu.edu/cgi/viewcontent.cgi?article=1818&context=sei

Miller, Stephen. 2010. "Founding Prodigy Chief Created Online Services for Consumers." January 13. Accessed August 27, 2015. http://www.wsj.com/articles/SB1263351188 60527243

Mitchell, Dan. 2011. "Skype's Long History of Owners and Also-Rans: At an End?" May 11. Accessed September 26, 2015. http://fortune.com/2011/05/11/skypes-long-history-of -owners-and-also-rans-at-an-end/

Moore, Malcolm. 2010. "Youku Founder Victor Koo Believes Only China Can Help His Company Grow, Make Money, and Even Beat Piracy." December 14. Accessed September 19, 2015. http://www.telegraph.co.uk/finance/china-business/8201764/Youku -founder-Victor-Koo-believes-only-China-can-help-his-company-grow-make-money -and-even-beat-piracy.html

Moreau, Elise. n.d. "10 Internationally Popular Social Networks You've Never Heard of Before." Accessed September 19, 2015. http://webtrends.about.com/od/Social-Networking /tp/10-Internationally-Popular-Social-Networks-Youve-Never-Heard-of-Before.htm

Mouton, Andre. 2013. "Don't Blame Yahoo for Flickr's Decline." May 27. Accessed September 27, 2015. http://www.usatoday.com/story/tech/2013/05/27/minyanville-flickr -yahoo/2363285/

Nabila. 2014. "Infographic: 25 Fast Facts on Big Data." October 29. Accessed September 3, 2015. http://bigdataanalytics.my/infographic-25-fast-facts-on-big-data-2/

New York Times. 2001. "Company News; SBC Communications Seeks to Acquire All of Prodigy." September 22. Accessed August 27, 2015. http://www.nytimes.com/2001/09/22 /business/company-news-sbc-communications-seeks-to-acquire-all-of-prodigy.html

Noticias Cuatro. 2015. "Un Prisionero Español Tuitea 'en Tiempo Real' desde el Campo de Concentracion Nazi." January 24. Accessed September 8, 2015. http://www.cuatro.com /noticias/sociedad/Twitter-prisionero-campo_de_concentracion-Mauthausen-nazismo _0_1929375106.html

Nusca, Andrew. 2015. "Pinterest CEO Ben Silbermann: We're Not a Social Network." July 13. Accessed September 24, 2015. http://fortune.com/2015/07/13/pinterest-ceo -ben-silbermann

Nuwer, Rachel. 2012. "The First Use of OMG was a 1917 Letter to Winston Churchill." November 27. Accessed September 5, 2015. http://www.smithsonianmag.com/smart -news/the-first-use-of-omg-was-in-a-1917-letter-to-winston-churchill-145636383/ ?no-ist

O'Luanaigh, Cian. 2014. "World Wide Web Born at CERN 25 years ago." March 12. Accessed August 27, 2015. http://home.web.cern.ch/about/updates/2014/03/world-wide -web-born-cern-25-years-ago

Olson, Parmy. 2013. "Free-Calling App Viber Jumps to Desktop, Hits 200 Million Users." May 7. Accessed September 19, 2015. http://www.forbes.com/sites/parmyolson/2013 /05/07/free-calling-app-viber-jumps-to-desktop-hits-200-million-users

Oregon State University. "Randal Conrads: Academy of Distinguished Engineers—2003." Accessed August 20, 2015. http://engineering.oregonstate.edu/randal-conrads-2003 -academy-distinguished-engineers

Oshkalo, Anna. 2014. "Odnoklassniki.ru Rebrands into OK." August 17. Accessed September 26, 2015. http://www.russiansearchtips.com/2014/08/odnoklassniki-ru-rebrands-ok /#more-2619

Oxford Dictionaries. 2015. "Selfie Stick, Concern Troll, and Bae: New Words Added to OxfordDictionaries.com." May 29. Accessed September 5, 2015. http://blog

.oxforddictionaries.com/2015/05/selfie-stick-concern-troll-bae-new-words-oxford dictionaries-com/

Parfeni, Lucian. 2011. "Flickr Boasts 6 Billion Photo Uploads." August 5. Accessed September 27, 2015. http://news.softpedia.com/news/Flickr-Boasts-6-Billion-Photo -Uploads-215380.shtml

Parke, Phoebe. 2016. "How Many People Use Social Media in Africa?" January 14. Accessed July 13, 2016. http://www.cnn.com/2016/01/13/africa/africa-social-media -consumption/

Perez, Sarah. 2016. "Pokémon Go Becomes the Fastest Game Ever to Hit $500 Million in Revenue." September 8. Accessed September 9, 2016. https://techcrunch.com/2016/09 /08/pokemon-go-becomes-the-fastest-game-to-ever-hit-500-million-in-revenue/

Peterson, Kim. 2004. "No. 2 Internet Service Provider Buying Classmates Online." October 26. Accessed August 20, 2015. http://community.seattletimes.nwsource.com/archive/ ?date=20041026&slug=classmates26

Phillips, Sarah. 2007. "A Brief History of Facebook." July 25. Accessed September 20, 2015. http://www.theguardian.com/technology/2007/jul/25/media.newmedia

Pierce, David. 2015. "Twitter's Periscope App Lets You Livestream Your World." March 26. Accessed September 13, 2016. https://www.wired.com/2015/03/periscope/

Pinterest. 2014. "Pin Trends of the Week." May 15. Accessed September 6, 2015. https:// blog.pinterest.com/en/pin-trends-week

Raice, Shayndi, and Spencer E. Ante. 2012. "Insta-Rich: $1 Billion for Instagram." April 10. Accessed September 27, 2015. http://www.wsj.com/articles/SB100014240527023038 15404577333840377381670

Rapoza, Kenneth. 2011. "China's Weibos vs US's Twitter: And the Winner Is?" May 17. Accessed August 31, 2015. http://www.forbes.com/sites/kenrapoza/2011/05/17/chinas -weibos-vs-uss-twitter-and-the-winner-is

Reid, T. R., and Brit Hume. 1991. "Censorship Issue Logs on at Online Bulletin Board." November 3. Accessed August 27, 2015. http://articles.chicagotribune.com/1991-11-03 /business/9104080864_1_semitic-anti-semitic-prodigy

Reuters. 2014. "Russia's Mail.Ru Buys Remaining Stake in Vkontakte for $1.5 Bln." September 16. Accessed September 26, 2015. http://www.reuters.com/article/2014/09/16/russia -mailru-group-vkontakte-idUSL6N0RH28K20140916

Rivlin, Gary. 2006. "Wallflower at the Web Party." October 15. Accessed September 20, 2015. http://www.nytimes.com/2006/10/15/business/yourmoney/15friend.html?_r =2&

Rosenberg, P. J. 2012. "Sparks Has Been Removed from the List of Search Entries on Google+." November 27. Accessed September 26, 2015. https://plus.google.com/+GooglePlus Daily/posts/5z7VHfKng2b

Rothman, Lily. 2015. "A Brief Guide to the Tumultuous 30-Year History of AOL." May 22. Accessed August 30, 2015. http://time.com/3857628/aol-1985-history

Russell, Jon. 2014. "Twitter Is Opening an Office in Indonesia, One of the World's Top Social Media Hotspots." August 29. Accessed August 24, 2015. http://thenextweb.com /twitter/2014/08/29/twitter-opening-office-indonesia-one-worlds-top-social-media -hotspots/

Sabrina. 2013. "The Story of China's Biggest Social Network: Qzone." September 13. Accessed September 19, 2015. http://www.chinainternetwatch.com/3346/tencent-qzone

Saito, Mari. 2012. "Born from Japan Disasters, Line App Sets Sights on U.S., China." August 16. Accessed September 24, 2015. http://www.reuters.com/article/2012/08/16/japan -app-line-idUSL2E8JD0PZ20120816

Sakawee, Saiyai. 2013. "Can Viber Really Compete with Asia Apps Like Line and Kakao Talk?" November 22. Accessed September 19, 2015. https://www.techinasia.com /founding-story-viber-live-blog

Sandberg, Jared. 1997. "WorldCom Agrees to Acquire CompuServe for $1.2 Billion." September 8. Accessed August 26, 2015. http://www.wsj.com/articles/SB873513339130 901000

Satariano, Adam. 2014. "WhatsApp's Founder Goes from Food Stamps to Billionaire." February 20. Accessed September 20, 2015. http://www.bloomberg.com/news/articles /2014-02-20/whatsapp-s-founder-goes-from-food-stamps-to-billionaire

Savitz, Eric. 2012. "5 Things You Need to Know About Chinese Social Media." October 25. Accessed September 19. http://www.forbes.com/sites/ciocentral/2012/10/25/5-things -you-need-to-know-about-chinese-social-media

Sengupta, Somini, Nicole Perlroth, and Jenna Wortham. 2012. "Behind Instagram's Success, Networking the Old Way." April 12. Accessed September 24, 2015, http://www .nytimes.com/2012/04/14/technology/instagram-founders-were-helped-by-bay-area -connections.html

Shah, Saqib. "Twitter for Sale? Upcoming Board Meeting Will Shape Company's Future." September 6. Accessed September 14, 2016. http://www.digitaltrends.com/social-media /twitter-board-meeting-takeover/

Sheikh, Mahnoor. 2015. "The Cost of Internet: How Does Pakistan Compare to the Rest of the World?" October 1. Accessed July 5, 2016. https://propakistani.pk/2015/10/01/the -cost-of-internet-how-does-pakistan-compare-to-the-rest-of-the-world

Shontell, Alyson. 2014. "How a Guy Who Had His Jaw Wired Shut for Three Months Used the Experience to Build a $200 Million Startup." March 27. Accessed August 20, 2015. http://www.businessinsider.com/how-klout-sold-for-200-million-2014-3

Shu, Catherine. 2014. "Japanese Internet Giant Rakuten Acquires Viber for $900M." February 13. Accessed September 19, 2015. http://techcrunch.com/2014/02/13/japanese -internet-giant-rakuten-acquires-viber-for-900m

Smith, Craig. 2015. "By the Numbers: 40 Amazing Weibo Statistics." August 14. Accessed October 25, 2016. http://expandedramblings.com/index.php/weibo-user-statistics

Soble, Jonathan. 2014. "'Monster Strike' Gives Former Social Media Giant Mixi a Second Act." December 28. Accessed September 26, 2015. http://www.nytimes.com/2014/12 /29/business/international/monster-strike-gives-former-social-media-giant-mixi-a -second-act.html

Solomon, Brian. 2014. "From Alibaba to Weibo: Your A-Z Guide to China's Hottest Internet IPOs." March 19. Accessed August 31, 2015. http://www.forbes.com/sites/briansolomon /2014/03/19/from-alibaba-to-weibo-your-a-z-guide-to-chinas-hottest-internet-ipos

Sorensen, Chris. 2009. "Canada Pension Plan Buys Skype Stake." September 2. Accessed September 26, 2015. http://www.thestar.com/business/2009/09/02/canada_pension _plan_buys_skype_stake.html

Specktor, Brandon. 2014. "Weird Facts About 5 Punctuation Marks You See Everywhere." December 18. Accessed September 6, 2015. http://www.rd.com/slideshows/punctuation -mark-facts/view-all/

Statista. 2016. "Leading Social Networks Worldwide as of September 2016, Ranked by Number of Active Users (in Millions)." Accessed September 17, 2016. http://www.statista .com/statistics/272014/global-social-networks-ranked-by-number-of-users/

Statista. 2016. "Number of Apps Available in Leading App Stores as of June 2016." Accessed July 4, 2016. http://www.statista.com/statistics/276623/number-of-apps-available-in -leading-app-stores/

Statista. 2016. "Number of Available Apps in the Apple App Store from July 2008 to June 2016." Accessed July 4, 2016. http://www.statista.com/statistics/263795/number -of-available-apps-in-the-apple-app-store/

Stempel, Jonathan. 2014. "Google, Viacom Settle Landmark YouTube Lawsuit." March 18. *Reuters*. Accessed September 17, 2016. http://www.reuters.com/article/us-google -viacom-lawsuit-idUSBREA2H11220140318

StumbleUpon. 2015. "What Is StumbleUpon?" Accessed September 26, 2015. http://www .stumbleupon.com/about/

Swiftkey. 2015. "Emoji Report Part II: Which Is the Most Popular Emoji in Your Language." June 4. Accessed July 20, 2015. http://swiftkey.com/en/blog/emoji-report-part-ii-which -is-most-popular-emoji-in-your-language/

Tabuchi, Hiroko. 2011. "Facebook Wins Relatively Few Friends in Japan." January 9. Accessed September 26, 2015. http://www.nytimes.com/2011/01/10/technology /10facebook.html?scp=1&sq=Mixi&st=cse

Takada, Kazunori. 2012. "Youku to Buy Tudou, Creating China Online Video Giant." March 12. Accessed September 19, 2015. http://www.reuters.com/article/2012/03/12 /youku-tudou-idUSL4E8EC3K320120312

Talbot, Matthew. 2015. "Looking Back at the Last 10 Years of BBM." July 30. Accessed September 26, 2015. http://blogs.blackberry.com/2015/07/looking-back-at-the-last-10 -years-of-bbm/

Tan, Francis. 2011. "140-Character Micro-resumes Change Job Applications in China." May 26. Accessed September 8, 2015. http://thenextweb.com/asia/2011/05/26/140 -character-micro-resumes-change-job-applications-in-china/

Tass. 2013. "Publishing Private Information of Russian Social Network Users Without Consent to Become Illegal." August 20. Accessed August 24, 2015. http://tass.ru/en/russia /699283

Terdiman, Daniel. 2004. "Photo Site a Hit with Bloggers." December 9. Accessed September 26, 2015. http://archive.wired.com/culture/lifestyle/news/2004/12/65958

Thompson, Cadie. 2015. "The 15 Defining Tech Moments of 2015." December 18. Accessed September 13, 2016. http://www.techinsider.io/the-15-biggest-tech-events-of-2015-12 /#apple-gets-into-wearables-with-its-watch-1

Timmer, John. 2010. "Classmates.com Settles Suit over Misleading Emails." March 15. Accessed August 20, 2015. http://arstechnica.com/tech-policy/2010/03/classmatescom -settles-suit-over-misleading-e-mails

TMZ. 2015. "Hey Bros—I'm in Hollywood." September 10. Accessed September 10, 2015. http://www.tmz.com/2015/09/10/pewdiepie-cutie-marzia-youtube-gamer-arrives-lax/

Tomlinson, Ray. "Frequently Made Mistakes." Accessed August 20, 2015. http://openmap .bbn.com/~tomlinso/ray/mistakes.html

Tomlinson, Ray. "The First Network Email." Accessed August 20, 2015. http://openmap.bbn .com/~tomlinso/ray/firstemailframe.html

Toor, Amar. 2014. "Pavel Durov Claims He Was Abruptly Fired from the Site He Created, Rising Fears of a Crackdown on Dissent." April 22. Accessed September 26, 2015. http://www.theverge.com/2014/4/22/5638980/russias-largest-social-network-is -under-the-control-of-putins-allies

Truong, Alice. 2015. "The State of Internet Connectivity Around the World." December 7. Accessed July 5, 2016. https://www.weforum.org/agenda/2015/12/the-state-of-internet -connectivity-around-the-world/

Tuchinsky, Peter. 2016. "Social Media App Usage Down Across the Globe." June 2. Accessed July 31, 2016. https://www.similarweb.com/blog/social-media-usage

Tweney, Dylan. 2009. "Sept. 24, 1979: First Online Service for Consumers Debuts." September 24. Accessed August 24, 2015. http://www.wired.com/2009/09/0924compuserve-launches

Twidiplomacy. n.d. "Emoji Diplomacy—A New Diplomatic Sign Language." Accessed September 10, 2016. http://twiplomacy.com/blog/emoji-diplomacy-a-new-diplomatic-sign-language/

Twitter.com. 2015. "Twitter Usage/Company Facts." Accessed September 27, 2015. https://about.twitter.com/company

United Nations. 2015. "World Population Prospects: Key Findings and Advance Tables." Accessed September 26, 2015. http://esa.un.org/unpd/wpp/Publications/Files/Key_Findings_WPP_2015.pdf

Viadeo. 2014. "Notre histoire." Accessed September 26, 2015. http://corporate.viadeo.com/qui-sommes-nous/historique/

Viadeo. 2014. "Viadeo Continues to Grow: 65 Million Members Worldwide, including 25 Million in China." December 11. Accessed September 26, 2015. http://corporate.viadeo.com/en/2014/12/11/viadeo-continues-to-grow-65-million-members-worldwide-including-25-million-in-china/

Wagstaff, Keith. 2014. "The Internet and the World Wide Web Are Not the Same Thing." March 12. Accessed August 20, 2015. http://www.nbcnews.com/tech/internet/internet-world-wide-web-are-not-same-thing-n51011

Ward, George F. 2016. "Social Media in Africa—A Growing Force." *Institute for Defense Analysis Africa Watch* 11: 5–7.

Washington Post. 2010. "25 Years of AOL: A Timeline." May 23. Accessed August 30, 2015. http://www.washingtonpost.com/wp-dyn/content/article/2010/05/23/AR2010052303551.html

Waters, Darren. 2007. "Web 2.0 Wonders: StumbleUpon." March 29. Accessed September 26, 2015. http://news.bbc.co.uk/2/hi/technology/6506055.stm

Waters, John. 2010. *The Everything Guide to Social Media: All You Need to Know About Participating in Today's Most Popular Online Communities.* Avon, MA: F+W Media, Inc.

Wauters, Robin. 2009. "China Blocks Access to Twitter, Facebook After Riots." July 7. Accessed September 19, 2015. http://techcrunch.com/2009/07/07/china-blocks-access-to-twitter-facebook-after-riots

Web3Schools.com. 2015. "HTML Introduction." Accessed August 27, 2015. http://www.w3schools.com/html/html_intro.asp

Weiser, Mark. 1991. "The Computer for the 21st Century." Accessed July 5, 2016. http://www.ubiq.com/hypertext/weiser/SciAmDraft3.html

Welch, Chris. 2016. "Twitter's New, Longer Tweets Are Coming September 19th." September 12. *The Verge.* Accessed September 17, 2016. http://www.theverge.com/2016/9/12/12891562/twitter-tweets-140-characters-expand-photos

Wilson, Chris. 2014. "The Selfiest Cities in the World: TIME's Definitive Ranking. March 10. Accessed September 21, 2015. http://time.com/selfies-cities-world-rankings/

Wong, Julia. 2015. "Selfie-Incrimination: SFPD Surveils Instagram." August 5. Accessed September 9, 2015. http://www.sfweekly.com/sanfrancisco/san-francisco-news-instagram-facebook-sfpd-surveillance/Content?oid=3906541

Woolley, Suzanne. 2015. "How Did Everyone Get Addicted to Trivia Crack?" January 12. Accessed August 30, 2015. http://www.bloomberg.com/news/articles/2015-01-12/trivia-crack-game-hooks-millions-of-american-players

WordPress.com. 2015. "A Live Look at Activity Across WordPress.com." Accessed September 27, 2015. https://wordpress.com/activity/

Xing. 2015. "Xing Is the Social Network for Business Professionals." Accessed September 26, 2015. https://corporate.xing.com/english/company/

Young, Doug. 2015. "Youku Tudou Eyes Overhaul in Pursuit of Respect." August 10. Accessed September 27, 2015. http://www.forbes.com/sites/dougyoung/2015/08/10/youku-tudou-eyes-overhaul-in-pursuit-of-respect/

Zijlma, Anouk. 2015. "Facts About Africa." August 5. Accessed July 13, 2016. http://goafrica.about.com/od/africatraveltips/a/africafacts.htm

About the Editors and Contributors

Editors

Laura M. Steckman has multiple master's degrees and earned her PhD in history from the University of Wisconsin–Madison. She has received Fulbright-Hays and Foreign Area and Language Studies funding to study Asian and Latin American languages abroad and conducted research in Asia on a Boren fellowship. She has worked as an Asia-focused media analyst and currently is a social scientist with research interests in digital humanities, including how people, culture, and language intersect to influence the development and use of emerging technologies. She currently holds an honorary fellowship with the University of Wisconsin–Madison's Center for Southeast Asian Studies.

Marilyn J. Andrews earned her PhD in mass communication from the University of Wisconsin–Madison. She also holds master's degrees in cultural anthropology and education. Her previous research includes examining how the Mapuche people of South America utilize the internet as a tool of self-representation to foster national identity, and how that usage reflects traditional understandings of socioterritorial identity. In addition to exploring indigenous identity expressions online, her current research interests include social media around the world and organic multicultural communities that form in online environments. She is a member of the National Coalition of Interdependent Scholars.

Contributors

Nadia Ali was born in London and now resides in the Caribbean. She is a freelance writer who has contributed to other encyclopedias published by ABC-CLIO. She is also an author in her own right. You may contact her at nadiafreelancewriter@yahoo.com.

Karen Ames received a master of arts degree in Southeast Asian studies from the University of Wisconsin–Madison. She currently utilizes various social media platforms to develop and maintain an online community for refugee youths wanting to make connections and pursue higher education in the United States.

Zoe-Charis Belenioti is a PhD candidate at Aristotle University in Thessaloniki, Greece. Her research interests include social media marketing and branding within nonprofit organizations (NPOs), social media users' behavior, and the marketing and management of NPOs.

Lina Benabdallah is a doctoral candidate of international relations at the University of Florida. Her current research focuses on China's rise and China-Africa relations. She is a Fulbright alumna and holds an MA in political science from the University of Florida.

Emanuel Braz, PhD, is an honorary fellow at Victoria University in Melbourne, Australia. He is an expert in media and strategic communications and a senior consultant in the sector in Timor-Leste. He is also the publisher of *The Dili Weekly* newspaper.

Mark A. Caudill was a U.S. Central Intelligence Agency (CIA) Analyst during 1990–2004 and a U.S. Department of State Foreign Service Officer during 2005–2014. Author of *Twilight in the Kingdom: Understanding the Saudis*, he speaks Arabic and holds an MA in Islamic Studies from Middlesex University, London.

Hannah S. Chapman is a PhD candidate in political science at the University of Wisconsin–Madison. Her dissertation, "Technologies of Participatory Governance in Putin's Russia," examines the role of information communication technology (ICT) in bolstering support for authoritarian regimes. Chapman's fields of interest include comparative politics, post-Soviet politics, and political behavior.

Wallace Chuma, PhD, a senior lecturer at the Centre for Film and Media Studies, University of Cape Town, South Africa, has worked as a journalist and editor in Zimbabwe, Botswana, and as an Alfred Friendly Press Fellow in the United States. His research areas include media policy, media and transition in southern Africa, and political economy of the media.

Francesca Comunello, PhD, is an associate professor in the Department of Humanities, Lumsa University, Rome. Her research and publications focus on the intersections between digital technology and society. This includes research on network theories, digitally mediated social relations, social media, mobile communication, digital communication and natural disasters, and digital inequalities.

Skye Cooley, PhD, is a professor in the Department of Communication at Mississippi State University. He holds degrees in international relations, international studies, political communication, and earned his doctorate in mass communication at the University of Alabama. He holds certifications of accreditation in public relations and peacekeeping operations.

Emily Belle Damm is a Shackouls Honor College undergraduate student researcher at Mississippi State University, where she studies communication (focusing on public relations and broadcasting) and math. Her research interests include public relations crisis communication, narrative analysis, and international studies.

Jonathan Dixon is a WSD-Handa fellow at the CSIS Pacific Forum, where he focuses on Chinese nationalism and irredentism. His research has been published in *Nationalities Papers* and *Comparative Strategy*. He received a BA in political science and Asian studies from Furman University in Greenville, South Carolina, and an MA in international politics from American University.

Sender Dovchin is a postdoctoral associate at the University of Technology, Sydney. She has completed her PhD degree in language education and her MA degree in TESOL. Her research interests include the sociolinguistics of media globalization in Asia. She has authored articles in multiple international peer-reviewed journals, such as the *Journal of Sociolinguistics* and *International Journal of Multilingualism*.

Ugur Dulger graduated from international relations at Ege University in 2016 and is currently attending an MS program in sociology at Middle East Technical University. His field of study focuses mainly on digital culture, social media, and the internet.

Sarah El-Shaarawi is the managing editor of the journal *Arab Media & Society*. She has written for *Foreign Policy, Newsweek*, and the *Cairo Review of Global Affairs*, among other publications. She holds an MA in international affairs from the New School.

Karen Stoll Farrell is head of the Area Studies Department at Indiana University Libraries, Bloomington, and Librarian for South and Southeast Asian Studies. She is a coeditor of *Heading East: Security, Trade, and Environment Between India and Southeast Asia*. Her research interests include web archives and access issues for international materials.

Claudio Fauvrelle, having a degree in ICT engineering, has ten years of experience working in the area of digital communications in Mozambique as a webmaster, online editor, graphic designer, social media expert, video and multimedia editor, and external communications and public advocate. Currently, he is the digital communication officer at UNICEF Mozambique.

Solen Feyissa is a PhD candidate in learning technologies in the Department of Curriculum and Instruction at the University of Minnesota. His research and teaching interests focus on the affordances of information communication technology (ICT) for education in low-income communities and the use and nonuse of technology in learning spaces.

Jeff Gagnon is a PhD candidate at the University of Toronto's Centre for Drama, Theatre, and Performance Studies and is a former digital literacy educator. He is currently writing his dissertation, which deals with the performance of protest, public space, and the use of digital communications technologies.

Sean Gillen received a doctorate in modern Russian and European social, intellectual, and political history at the University of Wisconsin–Madison in 2012. Since then, he has worked as a media analyst. He publishes on Russian affairs and is currently a visiting researcher at Georgetown University's Department of History.

Erkkie Haipinge is a lecturer of educational technology and project-based learning at the University of Namibia. He holds a master of education degree from the University of Oulu, Finland. His research focus is social media and its role in teaching and learning. He has published articles on students' social media use.

John G. Hall is a writer and researcher living and working in western North Carolina. His professional interests include African American history and culture. Hall is also director of FamilyTree Alternative Family Services, Inc., a human services agency that supports individuals with disabilities and their families.

Crystal L. Hecht, a New Jersey native, has spent the last eight years living and working abroad. While obtaining her master's degree in applied linguistics from Korea University, her research focused on language on the internet, mobile learning, and phonological acquisition. She is currently assistant professor of creative integrated studies at Daegu University, South Korea.

Robert Hinck is a doctoral candidate and Vision 2020 fellow at Texas A&M University. His research explores the role of media, narratives, and argumentation strategies in international relations. He has interned in the Technology and Public Policy program at the Center for Strategic and International Studies and the Carter Center's China program.

Sean Jacobs is associate professor of international affairs at The New School in New York City. He has held fellowships at Harvard, The New School, and New York University and is a former Commonwealth and Fulbright scholar. Sean is the founder and editor of the website Africa Is a Country (http://africasacountry.com/).

Nurlan Kabdylkhak has a master's degree in international area studies and political science. He is an independent scholar working on the issues of politics and history of Central Asia.

Filipo Lubua hails from Tanzania. He is a novelist, poet, foreign language instructor, and instructional technologist. He is currently a doctoral student in the instructional technology program at Ohio University, and his main research interests are academic entrepreneurship in instructional technology, computer-assisted language learning, and learning management systems (LMSs).

Susan Makosch was born in Lebanon and now lives in Florida. She has worked as a senior cyber Arabic linguistic specialist and as an Arabic media analyst. She is pursuing a master's degree in mental health counseling and forensic counseling from Walden University and plans to complete her doctorate.

Anamika Megwalu, PhD, MLIS, BS, BA, is a Senior Assistant Librarian/Science and Engineering Librarian at San Jose State University. She earned her doctoral degree in Information Studies from Long Island University, specializing in online scholarly communication. She has published articles in *Advances in Librarianship, The Reference Librarian, DisCover,* and *Science & Children.*

Caitlin Miles is a doctoral student in the Department of Communication at Texas A&M University. Her research focuses on ethnography, visual culture, and the rhetorics of power and resistance. Inspired by the five years she lived in Istanbul, Turkey, she became fascinated with activist and alternative media practices in Turkey.

Simone Mulargia, PhD, postdoctoral research fellow in the Department of Communication and Social Research (CORIS), Sapienza University of Rome since 2010, has taught new media, online journalism, and sociology of communication. His research and publications examine the role of digital media in the relationship between culture, technology, and society.

Abraham Mulwo teaches development communication at Moi University and is a visiting scholar at United States International University–Africa. He holds a PhD in communication and media studies (University of KwaZulu–Natal, South Africa). He has worked with UNESCO/Ministry of Education (Kenya) and the Centre for AIDS Development, Research, and Evaluation (South Africa).

Bruce Mutsvairo studies social media activism in sub-Saharan Africa at Northumbria University at Newcastle, United Kingdom, where he is also a senior lecturer in journalism. He completed his PhD at Leiden University, the Netherlands.

Sara Niner, PhD, is a transdisciplinary researcher and lecturer with the School of Social Sciences at Monash University in Australia. She is an expert in the field of gender and development with a long-term interest in those issues in the postconflict environment of Timor-Leste and is widely published in this field.

Anthony Ortiz is a Southeast Asia specialist with experience in West Africa, where he worked as an economist with projects for the European Union, the United Nations Development Programme, the United States Agency for International Development, and the World Bank. He holds a BA in political science and an MA in economics from the University of Connecticut, and an MPIA from the University of California, San Diego.

Tyler Overfelt is a graduate student and assistant at George Washington University's Institute for Middle East Studies. In addition to freelance writing and editing, he focuses his studies primarily on the political economy of North African countries.

Daniela Popescul, PhD, is an associate professor at the Universitatea Alexandru Ioan Cuza Iași, on the faculty of economics and business administration in the Business Information System Department. She is the author of two books on mobile commerce and business, one book on information security, and more than 52 scientific papers.

Troels Runge holds an MA in media science from the University of Copenhagen. He has worked as a digital consultant and as a commentator for Danish television and news media on the use of social media in politics. Currently, he is a PhD student at the IT University of Copenhagen.

Susana Salgado, PhD, is a researcher and professor of political communication at the Institute of Social Sciences, University of Lisbon. Her main research interests include political communication, comparative media studies, media and democratization, and Internet studies. Her research is currently sponsored by the Portuguese Foundation for Science and Technology (IF/01451/2014/CP1239/CT0004).

Hong-Chi Shiau is professor of communications management and gender studies institute at Shih-Hsin University in Taiwan. He is interested in building a more nuanced understanding of how larger economic and cultural structures reshape East Asian identities. His research explores how using new media transforms the global youth culture.

Andra Siibak, who holds a PhD in media and communication, is professor of media studies in the Institute of Social Studies at the University of Tartu, Estonia. Her current research interests include internet usage practices, new media audiences, and privacy issues. Her research project PUT44 is financed by the Estonian Research Council.

Jeffrey M. Skrysak began his career at the U.S. Department of Defense and Naval Research Lab before becoming a private consultant. He lived in Ecuador for five years before returning to the United States. He is now a practicing attorney in the state of Oregon and treasurer for the Oregon chapter of Mensa.

Georges Bertrand Tamokwé Piaptie is an associate professor of economics at the University of Douala in Cameroon. His research interests focus on digital economics, new institutional economics, and globalization. He is the executive director of the Group for Studies and Research in Theoretical and Applied Economics (GRETA)

Rhon Teruelle is a postdoctoral research scholar at the University of Calgary's Department of Communication, Media, and Film. He researches the tactics used by grass-roots organizations, with a particular focus on social media. Rhon holds a PhD from the Faculty of Information at the University of Toronto.

Tamás Tófalvy, PhD, is an assistant professor at Budapest University of Technology and Economics, and secretary general of the Association of Hungarian Content Providers. In 2012–2013, he was a Fulbright fellow at Columbia University's Graduate School of Journalism, and in 2011–2012, he was a Rezler fellow at Indiana University Bloomington.

Ruth Tsuria is a PhD candidate in the Department of Communication at Texas A&M University, researching digital religion and gender/sexuality. She received her MA from Copenhagen University and her BA from Hebrew University in Jerusalem.

Steven Lloyd Wilson is an assistant professor of political science at the University of Nevada–Reno and served as a research fellow at the Varieties of Democracy project. His research focuses on comparative democratization and how the internet affects authoritarian regimes, particularly in the post-Soviet sphere.

Aya Yadlin-Segal is a PhD candidate in the Department of Communication at Texas A&M University. Her research interests include online culture, identity construction, the flow of culture across globalized mediascapes, and Israel-Iran relations.

Index

Page numbers in **bold** indicate the location of main entries.

Abastéceme, 340
ABC Group, 30
Abe, 158
Abrams, Jonathan, xl
Access to Information (a2i) Programme, 22
Access to internet. *See* Internet penetration
ACORN (Australian Cybercrime Online Reporting Network), 18
Activism. *See* Social activism
Acton, Brian, xxv, xli, 154
Adbulmunem, Salam, 139
Addiction to internet, 294
Adeboye, E. A., 211
Adidas, 30
Adkimvn, 43
Adler family naming daughter Like, 146
Adobe, Ireland as hub for, 143
ADSL (asymmetric digital subscriber line) internet, 239
Advanced Research Project Agency (ARPA), xi, xxxvii
Advanced Research Projects Agency Network (ARPANET), xxxvii, 248
Advance-fee scams, 211
Affirmative Repositioning, 201
Afghan Wireless, 1
Afghanistan, **1–5**
"Afghanistan Needs You" campaign, 4
Afghanistan Social Media Awards, 3
AfreecaTV, 277
Africa, xviii–xix. *See also specific countries*
African Coast to Europe (ACE) cable, 34
African National Congress (ANC), 272
Agence Nationale de Reglementation des Telecommunications (ANRT,

Telecommunications Regulatory National Agency), 189
L'Agence Tunisienne des Télécommunications (Technical Agency for Telecommunications), 311
Agence Tunisienne d'Internet (Tunisian Internet Agency), 311
Agenda Digitale Italiana, 149
AGID (Agenzia per l'Italia Digitale), 149
Airbnb, Ireland as hub for, 143
Airtel, 210
Airtel Nigeria, 208
al-Ajami, Muhammad Rashid, 241
Akiztá, 340
Alai, Robert, 166–167
Alemarah, 2
Aleph, xlii
"Aleppo Is Burning" campaign, 291
Algeria, **5–8**
Algeria to the Core, 7
Al-Harbi, Fowzan, 256
Ali, Denny Januar, 131
Alibaba.com, xlii, 222
Aliexpress, 162, 250
Al-Jazeera, 239
Allama Iqbal, 223
Altai Technologies, 153
Altel, 162
Alwasat, 174
Amazon, 127
Amazon App Store, xvii
America Online (AOL), xxxviii, xxxix, xl, xlii, 145
Amharic language, 88
ANC (African National Congress), 272
AndroDumper, 137
Angola, **8–11**
Angola Media Libraries Network, 9

Anonymous Algeria, 7
ANRT (Agence Nationale de Reglemen-
    tation des Telecommunications),
    189
Antihacking drills, 296
AOL, xxxviii, xxxix, xl, xlii, 145
Aparat, 136
Apartheid, 271
App development, 27, 30–31, 132, 196
App Store, xvii
App usage
    Afghanistan, 2
    Argentina, 12
    Brazil, 28–29
    Chile, 43
    Finland, 93
    freemium apps, 31
    Germany, 105–106
    Ghana, 108
    Honduras, 131–132
    Iceland, 123
    Iran, 137
    Italy, 149
    Kazakhstan, 163
    Malaysia, 177–178
    overview of, xvi–xviii
    Paraguay, 225
    Peru, 229
    Philippines, 233
    Senegal, 259
    South Korea, 275–276
    Spain, 282
    Suriname, 286
    Taiwan, 295
    Trivia Crack, 14
    United States, 329
    Venezuela, 337, 338
    Vietnam, 342–343
    Yemen, 347–348
    See also specific apps
Apple, 143, 329
Apps, types of, xvi
Arab Spring
    Algeria, 6
    Angola, 10
    Egypt, 76
    Iraq, 140
    Libya, 173, 174
    Syria, 288

    Tunisia, 308, 312
    Yemen, 348
Aranis, Germán Alejandro Garmendia, 43
Argentina, **11–15**, 42
Argentine Digital Law (Law 27.078), 13
Argueta, Manlio, 80
al-Arifi, Muhammad, 257
ARPA (Advanced Research Project
    Agency), xi, xxxvii
ARPANET (Advanced Research Projects
    Agency Network), xxxvii, 248
Article 11 (Ecuador), 73
Article 24 of Tunisian constitution, 309,
    310
Article 56 (Colombia), 58
Article 474 (Ecuador), 73
a2i (Access to Information) Programme,
    22
Árukeresö.hu, 119
Aryeetey, Ernest, 109
Asia, xix–xx. See also specific countries
Asia America Gateway cable, 345
"Ask Guðmundur," 124
Aslam, Atif, 222
Association of Afghan Blog Writers, 2
"Astroturfing" companies, 319
Asymmetric digital subscriber line
    (ADSL) internet, 239
"At sign," 150
AT&T, 326
Atatürk, Mustafa Kemal, 315
al-Attar, Abdul Halim, 290
Australia, **16–19**
Australian Cybercrime Online Reporting
    Network (ACORN), 18
La Autoridad Federal de Tecnologías de la
    Información y las Comunicaciones
    (AFTIC; the Federal Authority of
    Information and Communication
    Technologies), 13
Average internet speed, xii. See also
    Connection speeds
Average peak connection speed, xii, 16
Aziz, Aaron, 250, 264
Azul, 7

Baba Jukwa page, 352, 353
Baidu Tieba, xxxiii
Balloons floating into North Korea, 217

Bamboo, 335
Bangladesh, **21–24**, 198
Bangladesh Telecommunication Regulatory Commission (BTRC), 21
Banking, 127, 193
Bar, defined, xxxiii
Barakat, 6–7
Baratikor, xxix
Barbie Beauty Center, 140
Barre, Siad, regime, 266–267
Barrio 18, 116, 117
Bartle, Richard, xxxviii
BBC, 325
BBC Nepali, 206
BBM, xxv
BBN (Bolt Beranek and Newman), xxxvii
Be Mobile, 25
Beck, Ingrid, 14
Beeline, 162, 169
Bello, Ello Ed Mundsel, 265
Ben Ali, Zine el Abidine, 308
Bendelladj, Hamza, 7
Bengali blog blackout of 2013, 23
Berlusconi, Silvio, 150
Berners-Lee, Tim, xxxix
#Bersih, 179
Bersih 4.0 rally Prime Kini, 179
Beyeu, 345
BharatNet, 126
Big data, xiv
Bild-Zeitung, 106
bin Talal, Alwaleed, 255
bin Zayid, Abdullah, 323
bin Zayid, Mohammad, 323
Bing, 325
BiTaksi, 313
Bitcoin, 13–14
Bitdefender, 243
BITNET, 262
Black Lives Matter Toronto (#BLM TO), 40
"Black Twitter," 272
BlackBerry Messenger, xxv
BlackBerry World, xvii
#BlackLivesMatter, 331
Blackmail, 154
Le Blanc d'Eyenga (Eyenga's White Man), 36
Blindworlds, xxiii

Blip, 235
#BLM TO (Black Lives Matter Toronto), 40
Blogger, xxxiv, 183, 222
Blogger killings, 23
Blog.hu, 119
Blogs
  Algeria, 7
  Angola, 10
  Bangladesh, 23
  Cuba, 62
  Iraq, 139
  Kenya, 166–167
  Mexico, 183
  Morocco, 188–189
  overview of, xxiv, xxxiv–xxxv
  Philippines, 231
  Romania, 245, 246
  Russia, 252
  Uzbekistan, 335
  See also Social media/social networking
Blogspot, xxxiv, 222
Blue Coat, 77
Bolt Beranek and Newman (BBN), xxxvii
Bookmarking forums, xxxii–xxxiv
Botbol, Hernán, xxxiii, 13–14
Botbol, Matías, xxxiii, 13–14
Botswana, **24–27**
Botswana Telecommunications Authority, 26
Botswana Telecommunications Company (BTC), 25
Bouazizi self-immolation, 174, 348
Bouteflika, Abdelaziz, 6
Brand24, 235, 237
Brazil, **27–32**
Britain. *See* United Kingdom
British Cable & Wireless Communication Group, 153–154
Bronze soldier relocation, 85–86
Brown, Reggie, xxxii
BTC (Botswana Telecommunications Company), 25
BTRC (Bangladesh Telecommunication Regulatory Commission), 21
Burma. *See* Myanmar
#BuyPens, 290
Büyükkökten, Orkut, 320

CAB (Central African Backbone) project, 34

Calderon, Juan Carlos, 73

Calling booths (*vibanda vya simu*), 298

Cameroon, **33–37**

Camp, Garrett, xxxiii

CAMTEL, 33–34

Canada, **37–41**

Canal de Moçambique, 192

"Candy" nickname, 229

Cardenalita, 339

Cardiopads, 35

La Casa del Encuentro, 14

Case, Stephen, xxxix

Casseroles, 39

Castro, Raúl, 59

Catch your thief ("Chapa tu choro") Facebook page, 228

Catster, xxiii

Cavazanni, Maximo, 14

CB Simulator, xxiii, xxxviii

CBBS (Computer Bulletin Board System), xxxvii–xxxviii

CBS, xxxviii

CCA (Computer-related Crimes Act, 2017, Thailand), 302

CCP (Chinese Communist Party). *See* China

CDMA (Code Division Multiple Access) networks, 208

Cecil John Rhodes statue, 273

Cellcom, 146

CEMAC (Central African Economic and Monetary Community), 33

CEMP, 119

Censorship, self, 115, 116, 300, 335. *See also* Government control of Internet

Central African Backbone (CAB) project, 34

Central African Economic and Monetary Community (CEMAC), 33

Los Centros Tecnológicos Comunitarios (community centers), 56

CERN (European Organization for Nuclear Research), xi, xxxix

CERT (Computer Security Incident Response Team), 303

CFCU (Communication, Forensic, and Cybercrimes Unit), 154

Chang San-cheng, 295

"Chapa tu choro" (Catch your thief) Facebook page, 228

Charlie Hebdo shootings, 98

Check Point, 146

Chen, Joseph, xxix

Children, efforts to protect
    Finland, 95
    France, 97–98
    Ireland, 143
    South Korea, 279
    Taiwan, 294

The Children That No One Wanted (Los hijos que nadie quiso), 62

Chile, **41–45**

China, xxix, xli, **45–51**, 295–296

China: Hong Kong, xii, xvii, **51–54**, 326

China: Macau, xvii, **54–55**

Chinese Communist Party (CCP). *See* China

Chowdhury, Salahuddin Quader, 23

Christensen, Ward, xxxvii

Chromebooks, 329

Chronology of history, xxxvii–xliii

Chronos Gate, 295

Churchill, Winston, letter, 325

Circles, xxvii–xxviii

CIRT (Cyber Incidents Response Team), 154

CIS (CompuServe Information Service), xxxviii

Citizen journalism, 352–353

Citizen Lab, 90

Claro, 42, 72, 115

Clash of Kings, 295

Classmates.com, xl, xli

Climaco, Cesar Vladimir Montoya, 116

Cloob, 136

CNN, 166, 341

#CNNmiente, 341

#CNNVzSeLaRespeta, 341

Cốc Cốc, 344

Code Division Multiple Access (CDMA) networks, 208

Code for Nepal, 207

Coins for Prita, 133

Colibrí, 339

Collectiv fire, 244

Colombia, **55–59**

Columbus Communications International, 153

Commentary forums, xxxii–xxxiv

Commentators, paid, 48

Commission for the Promotion of Virtue and the Prevention of Vice (CPVPV), 256, 257

Communication, Forensic, and Cyber-crimes Unit (CFCU), 154

Communications and Multimedia Act (1998, Malaysia), 179

Communications Regulation Council (Ecuador), 73

Communications Regulatory Authority (Qatar), 241

Community centers (Los Centros Tecnológicos Comunitarios), 56

CompuServe, xxiii, xxxvii, xxxviii, xl

CompuServe Information Service (CIS), xxxviii

Computer Bulletin Board System (CBBS), xxxvii–xxxviii

Computer Security Incident Response Team (CERT), 303

Computer-related Crimes Act (CCA, 2017, Thailand), 302

Concytec (Council of Science, Technology, and Technological Innovation), 229

Condor, 339

Connection speeds
    defined, xxii
    Ghana, 108
    Iran, 135
    Israel, 145
    Italy, 149
    Japan, 155, 156
    Libya, 173
    Pakistan, 221
    Philippines, 233–234
    Poland, 234, 236
    Singapore, 263
    South Korea, xii, 275
    United Kingdom, 326

Conrads, Randal, xl

Constitution, crowdsourcing of, 124–125

Cooperative Cyber Defence Centre of Excellence, 86

Copains d'avant, 98

Copyright Amendment (Online Infringement) Bill 2015 (Australia), 17–18

Copyright issues, xlii, 282–283

Correa, Rafael, 73

Corruption, 179, 314–315

Cosenza, Vincenzo, xv

"The cosmopolitans," 145

Costs, difficulty of measuring, xii–xiii

Council of Science, Technology, and Technological Innovation (Concytec), 229

CPVPV (Commission for the Promotion of Virtue and the Prevention of Vice), 256, 257

Cricket, 23

Crime
    El Salvador, 81
    Honduras, 116
    Malaysia, 179
    Saudi Arabia, 256
    on social media, 128
    wildlife smuggling, 228
    See also Cybercrime

Crimea, 319

Crowdsourcing, Iceland constitution, 124–125

Cuba, **59–63**

Cyber Incidents Response Team (CIRT), 154

"Cyber Security Strategy," 95

Cyber Storm, 296

Cyberbullying, 143

Cybercafés. *See* Internet cafés

Cybercrime
    Australia, 18
    Cameroon, 36
    China, 50
    Jamaica, 154
    Japan, 157–158
    Myanmar, 197–198
    Nigeria, 211–212
    North Korea, 218
    Philippines, 231–232
    Russia, 249, 252
    Saudi Arabia, 256
    South Korea, 279

Cybercrime (*continued*)
    Ukraine, 319
    United Arab Emirates, 321
    web defacement, 7
    See also Hacking; Piracy
Cybercrime Prevention Act (2012, Philippines), 231–232
Cybercrimes Act (2010, Jamaica), 154
Cybercrimes Act (2015, Tanzania), 299–300
"Cybernetics, Progress, and the Future" (Glushkov), 248
Cybersecurity
    Estonia, 85–86
    Finland, 95
    Georgia, 104
    Ghana, 109
    Israel, 146
    Romania, 243
    Taiwan, 295–296
Cypher (Fernando Sanz), xxxiii, 13–14
Cyrillic Mongolian font, 186
Cyworld, 276–277
Czech Republic, xii

DA (Democratic Alliance), 273
Da Silva Santos, Neymar, Jr., 30
Daesh (Islamic State), 2, 140–141, 162, 289
Daily Maverick, 274
Daily Motion, 222
Daily Vox, 274
DailyPakistan, 222
Dark web, 141
DARPA (Defense Advanced Research Project Agency), xxxvii
Das, Ananta Bijoy, 23
Data hub, Ireland as, 143
Data plans, xii–xiii
DataWind, 210
Date My Single Kid, xxiii
Daum, 277
"Day of Anger," 140
"Day of Rage," 288
dBm, 263–264
DC Hayes Associates, xxxvii
DDoS (distributed denial of service) attacks, 85
Decree Law No. 34 (2006, Qatar), 240

Deep web, 141
Deep-packet inspection (DPI), 89
Defamation
    Colombia, 58
    Ecuador, 73
    Indonesia, 133
    Malaysia, 178–179
    Philippines, 231–232
    United Arab Emirates, 322–323
Defense Advanced Research Project Agency (DARPA), xxxvii
Dekal Wireless, 153
Delfi v. Estonia, 120
Dell, Ireland as hub for, 143
Demilitarized zone, 214
Democratic Alliance (DA), 273
Demographics generally, xv. *See also specific countries, websites, and apps*
Denmark, **65–69**
Devanagari, 205
"Development Strategy of an Information Society in the Russian Federation," 249
DIGI, 118
Digicel, 153, 286
Digikala, 136
Digital Agenda, 243
"Digital Bangladesh by 2021" project, 22
"Digital India," 126
Digital literacy
    Greece, 111
    Mongolia, 186
    Namibia, 202
    Romania, 244
    Zimbabwe, 351
Digital Scoreboard by European Commission, Directorate General for Communications Networks, Content & Technology (DG CONNECT), 111
Digital Signature Act (2000, Estonia), 83
Digitization of Denmark agreement, 65–66
Dili Weekly, 307
Dilo Aquí, 338
Directive 5105.15, xxxvii
Distance-learning, 94, 109, 201, 302

Distributed denial of service (DDoS) attacks, 85
Djezzy (Orascom Telecom Algerie), 5–6
Doe Myanmar, 196
Dogster, xxiii
Domestic terrorism, 23
Donetsk, 318–319
#Dontbuydeath, 269
Dorsey, Jack, xxxi, xxxiv, xli
Douban, 49
Doubly landlocked countries, 334
Download speeds, xxii, 76
DPI (deep-packet inspection), 89
Draugiem, xxviii–xxix
Drug cartels, 183
Drug smuggling, 157
du (Emirates Integrated Telecommunications Company), 321
Durov, Pavel, xxix

Earthquakes, safety check after, 43–44, 206
EASSy cable, 298
    East Timor (Timor-Leste), 305–307
Eating broadcasts (mokbang), 277–278
eBay, xxv, 250
Ebola, 192–193, 261
e-Cabinet, 84
EcoCash, 351
e-commerce, 49, 278, 294–295, 302
Econet Wireless Zimbabwe, 351, 353
Economic and Financial Crimes Commission (EFCC), 211–212
Economic Freedom Front, 273
Ecuador, 42, **71–74**
EcuMobil, 61
EcuRed, 61
Education
    e-education, 94, 109, 201, 302
    Ghana, 109
    Jamaica, 154
    Mongolia, 187
    Namibia, 201
    Romania, 244
    South Africa, 273–274
    Thailand, 302
    United States, 329
e-education, 94, 109, 201, 302
EE (Everything Everywhere) Limited, 324–326

e-Estonia, 83
EFCC (Economic and Financial Crimes Commission), 211–212
e-government, 84, 244–245
e-Government Strategy (Mozambique), 193
Egypt, **74–79**, 173, 174
Egyptian Revolution, 75
Ehara, Masahiro, 157
eKantipur, 205–206
Ekeinde, Omotola Jalade, 211
El Aissami, Tareck, 341
"El Hueco Twitter," 117
El Salvador, **79–82**
e-learning, 109, 201, 302
Elections
    Honduras, 131
    Iran, 137
    Japan, 157
    Kenya, 167
    Russia, 253
    South Africa, 272
    Tanzania, 299
    Uzbekistan, 335
    See also Politicians/politics
Electronic Communications Act (2008, Ghana), 109
Electronic identity (ID) cards, 83
Electronic Transactions Act (2008, Ghana), 109
El-Sisi, Abdel Fatah, 77
Email, Ecuador, 72
Emirates Integrated Telecommunications Company (du), 321
Emirates Telecommunications Corporation (Etisalat), 1, 321, 322
emojis, 156, 178, 226, 245
EMOS-1 (underwater cable), 145
Employment Law Review (2012, France), 97
Entel, 42
EPRDF (Ethiopian People's Revolutionary Democratic Front), 87–88
Erdogan, Recep Tayyip, 313, 315
e-residency program, 85
Eritrea, xvii
ESAT (Ethiopian Satellite Television Service), 89–90
Escobar, Reinaldo, 62

Espncricinfo.com, 23
Estonia, **82–87**
Ethio Telecom, 88
Ethiopia, **87–91**
Ethiopia Higher Education Entrance
    Exam leak, 90
Ethiopian People's Revolutionary Demo-
    cratic Front (EPRDF), 87–88
Ethiopian Satellite Television Service
    (ESAT), 89–90
Etisalat (Emirates Telecommunications
    Corporation), 1, 321, 322
Etisalat Nigeria, 208
Euromaidan protests, 317
Europe, xx. *See also specific countries*
European Digital Rights Initiative, 315
European Organization for Nuclear
    Research (CERN), xi, xxxix
European Union, 243
"European Union's Delaware," 84
"Europe's Silicon Valley," 84
Everything Everywhere (EE) Limited,
    324–326
e-voting, 84
Expresso Senegal, 259
Express.pk, 222
Eyenga's White Man (*Le Blanc d'Eyenga*),
    36
Eyny.com, 294

Facebook
    Africa, xviii–xix
    Bangladesh, 23
    Brazil, 29
    Canada, 38
    Chile, 43
    Colombia, 56, 57
    Cuba, 60
    Daesh (Islamic State) using, 141
    Denmark, 66–67, 68
    drones and, xiii
    Ecuador, 72
    Egypt, 77
    El Salvador, 81
    Estonia, 85
    Ethiopia, 89
    Finland, 94
    France, 98
    Georgia, 102
    Germany, 104
    Ghana, 109
    Greece, 112
    Hong Kong, 52
    Hungary, 119–120
    Iceland, 123
    India, 128, 129
    Indonesia, 130
    Instagram bought by, xxxi
    Iran, 136–137
    Ireland as hub for, 143
    Ireland privacy issues, 143–144
    Islamic State (Daesh) using, 141
    Italy, 149–150
    Japan, 158
    Kazakhstan, 163
    Kenya, 165
    Kyrgyzstan, 169
    Libya, 173
    Malaysia, 177, 179
    Maple Spring in Canada, 39–40
    Mexico, 182–183
    Mongolia, 186
    Morocco, 189
    Mozambique, 192–193
    Myanmar, 195, 196
    Namibia, 200–202
    Nepal, 204, 205
    Nigeria, 210–211
    North Korea, 215, 217–218
    overview of, xxvii, xli
    Pakistan, 222–223
    Paraguay, 225, 226
    Peru, 227–229
    Philippines, 231
    Poland, 235
    popularity of, xv
    Qatar, 240
    regional social media and, xxix
    Romania, 244, 246
    Safety Check, 43–44, 99, 206
    Saudi Arabia, 255
    Senegal, 259
    Singapore, 264
    as social media and social networking,
        xiv
    Somalia, 268–269
    South Africa, 271
    South Korea, 276–277

Spain, 281–282
Suriname, 285
Syria, 125, 289
Taiwan, 293, 295
Tanzania, 299
Thailand, 301, 303
Timor-Leste, 306–307
as top site, 22
Tunisia, 311
Turkey, 313, 314
Ukraine, 318
United Arab Emirates, 321, 322
United Kingdom, 325
United States, 329–330, 332
Uzbekistan, 335, 336
Venezuela, 337, 338
Vietnam, 343–344
WhatsApp purchased by, xxvi
Yemen, 347
See also Internet.org; Social media/
    social networking
Facebook Messenger
Italy, 150
Namibia, 200
overview of, xxv, xlii
Singapore, 264
United Kingdom, 325
"Facebook tunnel," 226
Facenama, xv, xxvii, xxix–xxx, 135
FaceTime, xxv
Fake pictures, 340
Fall, Cheikh, 260–261
Farghadani, Atena, 135
FCC (U.S. Federal Communications
    Commission), 234
#Feb3, 348
#FeesMustFall, 273
Federal Authority of Information and
    Communication Technologies (La
    Autoridad Federal de Tecnologías de
    la Información y las Comunicacio-
    nes), 13
Fernandez, Joe, xxviii
Fernández de Kirchner, Christina, 14–15
Fiber to the home (FTTH), 320
Fiber to the node (FTTN), 16
Fiber to the premises (FTTP), 16
"50 Cent Party," 48
Figueroa, Milett, 229–230

File sharing, 222
Filters, 135, 157, 170
Fines
    Ecuador, 73
    Google, 97, 104–105
    Indonesia, 133
    Nigeria, 209
    Qatar, 241
    Saudi Arabia, 256
Finland, xii, **93–96**
FireChat, 141, 179
"First Social Media War," 289
Fisher, John Arbuthnot, letter, 325
Fixed-line telephony, 80, 199
Flickr, xxx
Flipkart, 127
Flow, 153
Foo, Willy, 264
Foreva, 345
Forticom, 235
Forum Research, 38
Forums, overview of, xxiv, xxxii–xxxiv
Fotka, 235
14ymedio, 62
France, **96–100**
Free Basics by Facebook. See Internet.org
Free Internet Initiative (Egypt), 75
Freedom 251 smartphone, 126
Freedom of expression, xxii, 29, 58. See
    also Government control of Internet
Freedom of the Net report. See Govern-
    ment control of Internet
Freemium apps, 31
Frelimo, 192
French, 311
Friendster, xxiii, xl, xli, xlii, 320
Friis, Janus, xxv
Frype, xxix
FTTH (fiber to the home), 320
FTTN (fiber to the node), 16
FTTP (fiber to the premises), 16

G8, 37
Gaddafi, Muammar, 173, 174
Gambling, 54, 278
Gamer.com, 294
Gangs, 81, 116–117
Gauthier, Laurent, 39
GDP (gross domestic product), xix

Geingob, Hage Gottfried, 201
Gemius data, 118
Gender differences in use
    Afghanistan, 2
    France, 98
    generally, xiii
    Indonesia, 130
    Malaysia, 177
    Pakistan, 222
    Philippines, 233
    Spain, 281
    Turkey, 313
    United States, 332
    Venezuela, 337
    video gamers, watching, 331
Génération Numérique survey, 98
Generation Y blog, 62
Generation Z initiative, 237
Georgia, **101–103**
Germany, **103–107**, 229
Gezi Park protests, 314
GfK Perú report, 227
Ghana, **107–110**
Ghazali, Ali, 135
Ghonim, Weal, 76
GiftedMom, 35
#GiveUSTheSerum, 261
Globacom, 208
Global Information Technology report,
    66
Global System for Mobile (GSM) net-
    works, 208
Glushkov, Viktor, 248
Gmail, xix–xx, 72, 195
"Go Russia!" pamphlet, 249
GoldenLine, 235, 236
Golden Shield Project, 47
Golden Tweet, 131
Golondrinas, 339
Golovkin, Gennady, 163
Gómez, Gaby, 183
Google
    Chromebooks, 329
    Côc Côc compared, 344
    Ecuador, 71
    fines against, 97, 104–105
    Hungary, 119
    India, 127, 128
    Ireland as hub for, 143

Kazakhstan, 162
Libya, 173
Mongolia, 186
Myanmar, 195
Pakistan, 222, 223
Paraguay, 225
Person Finder, 206
personal data usage, xx
Project Loon, xiii
purchase of YouTube, xli
South Korea, 277
Spain, 282
as top site, 22
United Kingdom, 325
Uzbekistan, 335
Viacom copyright lawsuit, xlii
See also Orkut
Google Play, xvii
Google Translate, 162, 186, 195
Google+
    El Salvador, 81
    France, 98
    Italy, 150
    overview of, xxvii, xxvii–xxviii
    Qatar, 240
    Singapore, 264
Gorbachev, Mikhail, 247
Gorkha Earthquake, 204, 205, 206–207
Gotze, Mario, 104
Government control of Internet
    Algeria, 7
    Angola, 10
    Bangladesh, 23–24
    Botswana, 26
    Brazil, 29
    China, 47–49
    Colombia, 57
    Cuba, 59–60
    Ecuador, 72, 73
    Egypt, 76
    Ethiopia, 89–90
    Finland, 95
    France, 97
    Georgia, 102
    Hungary, 120
    Iceland, 123
    India, 128–129
    Iran, 134–135
    Iraq, 139, 140–141

Japan, 156–157, 158
Kazakhstan, 162
Kenya, 166–167
Kyrgyzstan, 170
Libya, 173, 174
Morocco, 188, 189–190
Mozambique, 192
North Korea, 214–218
Pakistan, 223
Qatar, 240–241
Russia, 251–252, 253, 319
Saudi Arabia, 255–256
Singapore, 265
South Africa, 271–272
South Korea, 278
Syria, 288
Tanzania, 299–300
Thailand, 302–303
Tunisia, 308, 309–310
Turkey, 313, 314–315
Ukraine, 318, 319
United Arab Emirates, 321–323
Uzbekistan, 334, 335
Venezuela, 338–341
Vietnam, 343–344
Zimbabwe, 353
Go.vn, 343
Grameenphone, 23
"Great Firewall of China," 47, 51–52, 304
Great Recession, 49
Greece, **110–114**
Gross, Alan, 59
Gross domestic product (GDP), xix
GroundUp, 274
Grupo TV Cable accounting, 72
GSM (Global System for Mobile) net-
    works, 208
GTBank, 210
Guaraní, 224
Guayaquil, Ecuador, 72
Guðmundur, 124
Guillermo Martinez (n.d) v. Google (2013),
    57
Guinness Book of World Records, 104

H&R Block, xxxviii
Hacking, 7, 157–158. *See also*
    Cybercrime
Hacking Team, 90

Haenlein, Michael, xiv
HamariWeb, 222
Hamnster, xxiii
Han, Lu, 49
Hangouts, xxv, xxviii
Hansen, Robert, 216
"Happy Yemen," 349
Harim, 3
Hashtags, 211, 331. *See also specific
    hashtages*
Hate speech, 158, 196, 229
Hebrew University, 145
Herald (Zimbabwe), 352
Hernández, Antonio, 281
Hernández, Rodil, 81
Hi5, 281
Los hijos que nadie quiso (The Children
    That No One Wanted, blog), 62
Hindi, 127
Historic role plays, 281
Historical chronology, xxxvii–xliii
Hlongwane, Oratilwe, 272
Hoffman, Reid, xxvii
HolaSoyGerman, 43
"Homeland Network," 339
Honduras, **115–118**
Hondutel, 115–116
Hong Kong. *See* China: Hong Kong
Hotmail, Ecuador, 72
Houthis, 347
HTML (Hypertext Markup Language),
    xxxix
Huffman, Steve, xxxiii
Human search engines in Iceland, 124
Humor, 77–78
Hungarian National Media and Infocom-
    munications Authority (Nemzeti
    Média-és Hírközlési Hatóság,
    NMHH), 118
Hungary, **118–121**
Hush, 196
Hutchison Telecommunications Ltd., 17
Hybrid apps, xvi
Hypertext Markup Language (HTML),
    xxxix

#IAmCharlie, 98
IAM (Internet Access Management)
    policy, 321–322

#IamSyrian, 291
IBM, xxxviii
ICE Addis, 88
#IceBucketChallenge, 331
Iceland, **123–126**
ICQ, xxiv, 145
iGDP (internet gross domestic product), xix, 259
IM. *See* Instant messaging (IM)
IMO, 137
Improving Access to Employment Program, 81
INCB (Israeli National Cyber Bureau), 146
Index.hu, 119
India, **126–130**
Indian Language Internet Alliance, 127
Indian Railway website, 127
Indiatimes.com, 127
Indonesia, **130–134**
INEC, Ecuador (Instituto Nacional de Estadisticas y Censos), 72
Infolady, xiii, 22
Information Authority (Ecuador), 73
Information hub, Ghana as, 109
Information Network Security Agency (INSA, Ethiopia), 89–90
Information Nigeria, 211
Insaurralde, Matías, 226
Instagram
  Africa, xix
  Canada, 39
  Facebook purchase of, xxxi
  France, 98
  history of, xxxi
  Honduras, 116
  Iran, 136, 137
  Italy, 150
  Kazakhstan, 163
  Kenya, 165
  Libya, 173
  North Korea, 215
  popularity of, xv
  Qatar, 240
  Russia, 250
  Singapore, 264
  South Africa, 271
  Suriname, 286
  United Kingdom, 325–326

Instant messaging (IM)
  Finland, 93
  history of, xxiii
  Israel, 145
  overview of, xxiv–xxvi
  See also specific apps
Instituto Nacional de Estadisticas y Censos (INEC, Ecuador), 72
Intel, 210
Intelligent Apps, 106
International Iqbal Society, 223
International Telecommunications Union (ITU), xiii
Internet, defined, xi
Internet Access Management (IAM) policy, 321–322
Internet addiction, 294
Internet cafés
  China, 46
  Cuba, 60
  India, 126–127
  Kyrgyzstan, 169
  Libya, 174
  Nigeria, 211–212
  Somalia, 268
  South Korea, 278
  Tanzania, 298
  Turkey, 313
Internet chronology, xxxvii–xliii
Internet exchange points (IXP), 250, 319
Internet gross domestic product (iGDP), xix, 259
Internet Law 5651 (2007, Turkey), 315
Internet neutrality, 58, 128
Internet of Things (IoT), xiii
Internet penetration
  Afghanistan, 1
  Algeria, 5
  Angola, 9–10
  Argentina, 11–12
  Australia, 16
  Bangladesh, 21–22
  Botswana, 25
  Brazil, 27
  Cameroon, 33, 36
  Canada, 37–38
  Chile, 41–42
  China, 46–47
  Colombia, 55–56

Cuba, 59–60
defined, xiii
Denmark, 65–66
Ecuador, 71
Egypt, 75–76
El Salvador, 80
Estonia, 83–84
Ethiopia, 87
factors affecting, xiii
Finland, 93
France, 96
Georgia, 101
Germany, 103
Ghana, 107–108
Greece, 111, 112
Honduras, 115
Hong Kong, 51
Hungary, 118
Iceland, 123
India, 126
Indonesia, 130
Iran, 134
Iraq, 139
Israel, 144–145, 147
Italy, 148–149
Jamaica, 153
Japan, 155–156
Kazakhstan, 161
Kenya, 164
Kyrgyzstan, 168
Libya, 173
Malaysia, 177
Mexico, 181
Mongolia, 185
Morocco, 188
Mozambique, 192, 193
Myanmar, 194
Namibia, 199, 200
Nepal, 204
Nigeria, 208
North Korea, 214
Pakistan, 221
Paraguay, 224, 225
Philippines, 233
Poland, 235, 236
Qatar, 239, 240
Romania, 243
Russia, 250
Saudi Arabia, 255
Senegal, 259
Singapore, 262
Somalia, 267–268
South Africa, 271
South Korea, 275
Spain, 281
Suriname, 285
Syria, 289
Taiwan, 293–294
Tanzania, 297
Thailand, 301
Timor-Leste, 305
Tunisia, 311
Turkey, 313
Ukraine, 317
United Kingdom, 324
United States, 328, 329
Uzbekistan, 334
Vietnam, 342
Yemen, 347
Zimbabwe, 351
"Internet research," 252
Internet safety, Ireland, 143
Internet Society, 76
Internet speed, xxii. *See also* Connection
    speeds
Internet.org
    Argentina, 9
    Colombia, 56
    India, 128
    Mozambique, 192
    Pakistan, 223
    Paraguay, 225–226
    Senegal, 259
INTERRED–CETCOL (Red Nacional de
    Ciencia, Educación y Tecnología), 55
The Interview, 218
Invitel, 118
Inwi, 189
IoT (Internet of Things), xiii
iPads, 329
IPTV Triple Play, 210
Iran, **134–138**, 320
Irancell, 137
Iraq, **139–142**
Ireland, **142–144**
al-Islah, 322
Islamic State (Daesh), 2, 140–141, 162,
    289

Israel, **144–148**
Israeli National Cyber Bureau (INCB), 146
Italy, **148–151**
ITU (International Telecommunications Union), xiii
iTyphoon, 233
iWiW, 119
IXP (internet exchange points), 250, 319

Jakarta, Indonesia, 131
Jamaica, **153–155**
JamiiForums, 299
Japan, xii, xxviii, **155–160**, 326
Javedch, 222
#JeSuisCharlie, 98
#JeSuisNice, 99
Jewish immigrants, 145
Jiayuan, 49
Job recruitment, 236, 295
Jokowi, 131
Jonathan, Goodluck, 211
Jose Cuervo, 264
Journalists
    Botswana, 26
    citizen journalism, 352–353
    Honduras, 117
    Mexico, 183
    Morocco, 190
    Qatar, 241
    Yemen, 349
JuegaGerman, 43
#JunSeMueve, 284
Jun, Spain, 283–284
Jun Lei, xxxi

KACST (King Abdulaziz City for Science and Technology), 256
Kadyrov, Ramzan, 251
KakaoTalk, 177, 276, 278–279
Kalla, Jusuf, 131
Kanji language, 155, 158
Kaplan, Andreas, xiv
Karimova, Gulnara, 335
Kariuki, Francis, 166
Karman, Tawakkol, 348–349
Karta, 88–89
Kaskus, 132
Kat.cr, 222
Kaymu, 210

Kazakhstan, **161–164**
Kazakhtelecom, 161
Kcell, 162
#kebetu, 261
Kebetu, 261
Kenya, **164–168**
Kenya Information and Communications Act (KICA), 166
Kenyans on Twitter (KOT), 165–166
Khama, Seretse Ian, 26
KICA (Kenya Information and Communications Act), 166
Kim Jong-un, 216
Kimeltuwe, Materiales de Mapudungun, 43
King Abdulaziz City for Science and Technology (KACST), 256
Kiosks, Colombia, 56
Kiwi.kz, 163
Kjellberg, Felix Arvid Ulf, 330
Klassen, Gary, xxv
"Kleenex" (Wikileaks), 174
Klout, xxviii
Kolesa.kz, 162, 163
Koo, Victor, xxxi
Korea Communications Standards Commission, 278
Korea Friendship Association, 217
Korea-dpr.com, 217
Koryolink, 214
Kosher devices, 147
KOT (Kenyans on Twitter), 165–166
Koum, Jan, xxv, xli
K-pop, 275
Krieger, Mike, xxxi
Kulena Khaled Said Facebook page, 76
Kwangmyong intranet, 216
Kwikku, 132
Kyrgyzstan, **168–171**
Kyrgyztelecom, 170

Lafrenière, Nadia, 39
LamDieu, 345
Language use and issues
    Asia, xix–xx
    Australia, 18
    Ethiopia, 88
    India, 126, 127
    Israel, 147–148

Japan, 155, 158
Malaysia, 177
Mongolia, 184–187
Myanmar, 195–196
Nepal, 205
Pakistan, 221
Paraguay, 224–225
Senegal, 260
Timor-Leste, 305
Tunisia, 311–312
Lapsiporno.info, 95
Latin America, xx. *See also specific countries*
Latvia, xii
Law 27.078 (Argentine Digital Law), 13
Law on the Mongolian Language (2015, Mongolia), 187
Le Quoc Quan, 343
Lee, Jean, 214–215, 216
Legacy media. *See* Traditional media
Legal right to services, Finland, 93
Lemma, Markos, 88
"Letter to the Committee of the Comprehensive National Development Strategy of Mongolia, Mongolian Parliament, 2007," 187
#LeyStalker, 228
Li, David, xxxi
Liability, Hungary, 120
Libel. *See* Defamation
Libya, **173–175**
Libya Telecom & Technology (LTT), 173
Life Is a Garden (*Zindagi Gulzar Hai*), 222–223
Like as given name, 146
LIME, 153
Lin, J. J., 264
Line
  Hong Kong, 53
  Malaysia, 177
  overview of, xxvi
  Singapore, 264
  Taiwan, 293, 295
  Thailand, 301
LinkedIn
  Canada, 38
  France, 98
  Hong Kong, 53
  Ireland as hub for, 143

Microsoft purchase of, xlii
  Morocco, 189
  overview of, xxvii
  Pakistan, 222
  Poland, 235, 236
  popularity of, xv
  Singapore, 264
  Spain, 281
  Tunisia, 311
  Yemen, 347
LinX, 146
Literacy, 222, 267, 305. *See also* Digital literacy
Live, 222
Live Digital (Vive Digital), 55–56
LiveJournal, 250, 251, 318
Loaf and arm memes, 157
LOCALLY, 88
Look@World Project, 83–84
LostZombie, xxiii
LTT (Libya Telecom & Technology), 173
Lunati, Thierry, xxvii

M1, 262–263, 264
Ma Ying-jeou, 294
Macau. *See* China: Macau
Maduro, Nicolás, 339, 340, 341
Magyar Telekom, 118, 119
Mail-order brides, 35–36
Mail.ru Group, 251, 335
Malaysia, **177–180**
Malaysian Airlines Flight MH17, 180
Malaysian Airlines Flight MH370, 179–180
Malaysian Communications and Multimedia Commission (MCMC), 179
Malikzada, Farkhunda, 3
Mansour, Adly, 77
MaNUfestation, 39
Maple Spring, 39–40
Mapuche peoples, 43, 44
Marco, Talmon, xxvi
Marco Civil da Internet (MCI), 29
Maroc Blog Awards, 188, 189
Maroc Telecom, 189
Marshall Islands, xvii
Mascom, 25, 26
Masiyiwa, Strive, 351
#MásPeruanoQue, 228

Mathews, Dillish, 201
Mayor Boss, 210
MB (Muslim Brotherhood), 322
Mbps, xii
Mcel, 192
MCI, 137
MCI (Marco Civil da Internet), 29
MCMC (Malaysian Communications and
    Multimedia Commission), 179
MDA (Media Development Authority),
    265
Media bubble, 66
Media Development Authority (MDA),
    265
Media libraries, 9
Media-sharing, xxiv, xxx–xxxii. *See also*
    specific apps
Medicine via Internet, 26, 35
Meditel, 189
Medvedev, Dmitry, 249
Meerkat, xlii
Megacom, 169
Memes, 78, 157, 229–230
Messi, Lionel "Leo," 12, 13, 15
Messina, Chris, 211
Mexico, 61, **181–184**
#mh370, 180
Microblogs, xxxiv–xxxv. *See also* Sina
    Weibo; Twitter
MicroNET, xxxviii
Microrésumés, 295
Microsoft Corporation
    Amharic in Ethiopia, 88
    Ireland as hub for, 143
    LinkedIn purchased by, xlii
    Nigeria, 210
    Outlook in Ecuador, 72
    Skype purchase, xxv
"Microwave Ring," 1
MICT (Ministry of Information and
    Communication Technology), 301
Middle East Media Survey, 77
Ministerial Decree no, 18/2015 (Mozam-
    bique), 193
Ministry of Communications and
    Information Technology (Egypt), 75
Ministry of Foreign Affairs (Thailand), 303
Ministry of Industry and Information
    Technology (China), 47, 49–50

Ministry of Information and Communica-
    tion Technology (MICT, Thailand),
    301
Ministry of Information and Communica-
    tions Technology (Qatar), 241
Ministry of Post and Information and
    Communications Technologies
    (MPTIC), 5
Ministry of Public Security (China), 47
Minitel, 96, 97
Mirabilis, 145
MIT Senseable City Lab, 264
Mivasocial, 259–260
Mixi, xxviii
mKesh, 193
Mobile device use
    Afghanistan, 1
    Algeria, 5
    Angola, 9
    Australia, 17
    Bangladesh, 21
    Botswana, 25, 26
    Brazil, 30
    Cameroon, 34
    Chile, 42
    China, 49
    Colombia, 56–57
    Denmark, 66
    Ecuador, 71–72
    El Salvador, 80–81
    Ethiopia, 88
    Finland, 93
    Georgia, 101
    Ghana, 108
    Greece, 111
    Honduras, 115–116
    Hong Kong, 51
    Hungary, 118–119
    India, 126
    Indonesia, 130
    Iran, 137
    Iraq, 139
    Israel, 146–147
    Italy, 148, 149
    Japan, 156
    Kazakhstan, 161, 162
    Kenya, 164
    Kyrgyzstan, 169
    Libya, 173

Macau, 54–55
Malaysia, 177–178
Mexico, 181
Mozambique, 192, 193
Myanmar, 194
Namibia, 199–200
Nepal, 204
North Korea, 214
number per person, xvii
Pakistan, 221
Paraguay, 225
Poland, 235–236
Romania, 244
Saudi Arabia, 255
Singapore, 262–263, 264
Somalia, 266–267, 268
South Africa, 271
South Korea, 275
Suriname, 285
Syria, 289–290
Taiwan, 293
Tanzania, 298
Thailand, 301
Timor-Leste, 306
Tunisia, 311
United Arab Emirates, 321
United Kingdom, 324–325
United States, 329, 332
Uzbekistan, 334
Vietnam, 342
Zimbabwe, 351
Mobile money, 164, 193
Mobile Telecommunications Corporation
(MTC), 200
Mobile video on demand (M-VOD),
210
Mobilis Algerie Telecom, 6
"Mochi mochi," 157
Mochuelo, 339
Moda Capital, 183
Modems, xxxvii
Mohiuddin, Asif, 23
Moi Mir, 169
Mokbang (eating broadcasts), 277–278
MOL Global, xli
Momo Lay, 196
Mongolia, xii, **184–188**
Mongolian People's Revolutionary Party
(MPRP), 184

Moodle, 94
Moreno, Paul, 73
Morocco, **188–191**
Morsi, Mohamed, 76
Moscow Spring, 253
Movement for Democratic Change, 352
Movicel, 9
Movistar, 42, 72
Movitel, 192
Moy Mir, 163
Mozambique, **191–194**
M-Pesa, 164, 193
MPLA (Popular Movement for the
Liberation of Angola), 8
MPRP (Mongolian People's Revolutionary
Party), 184
MPTIC (Ministry of Post and
Information and Communications
Technologies), 5
MS-13, 117
MTC (Mobile Telecommunications
Corporation), 200
MTE and Index.hu Zrt v. Hungary, 120
MTM, 271
MTN Cameroon, 34
MTN Nigeria, 208, 209, 210
Mubarak, Hosni, 75, 76
MUD (multiuser dungeon), xxxviii
Mugabe, Robert, 351, 353–354
Mujahid, Ali Ahsan Mohammad, 23
Multi-Links, 208
Multiuser dungeon (MUD), xxxviii
Mulyasari, Prita, 133
Murphy, Bobby, xxxi–xxxii
Muslim Brotherhood (MB), 322
M-VOD (mobile video on demand), 210
MwanaHalisi Forum, 299
Mxit, 271
Myanmar, **194–198**, 229
Myanmar Posts and Telecommunications,
195
MySpace, xxiii, xl–xli, 130, 183
MySQUAR, 194–195, 196
Mytaxi app, 106

Naij.com, 209, 210
Nakayama, Alberto, 13–14
Namibia, **199–203**
Nasza-Klasa, 235

National Broadband Network (NBN), 16
National Broadband Policy (Nepal), 204
National Bureau of Investigation (NBI, Finland), 95
National Communications Commission (NCC), 293
National Cyber Security Strategy (Jamaica), 154
National Democratic Party (Suriname), 287
National Development Plan (Plan Nacional de Desarrollo, or PND), 58
National Information Technology Agency (Ghana), 109
National Plan to Combat Cybercrime (2015, Australia), 18
"National Security Strategy of the Russian Federation to 2020," 249
Native apps, xvi
Navalny, Alexei, 251, 252
Naver, xxvi, 277
NBI (National Bureau of Investigation), 95
NBN (National Broadband Network), 16
NBN Co Limited, 16–17
NCC (National Communications Commission), 293
NCC (Nigerian Communications Commission), 208
Nemzeti Média-és Hírközlési Hatóság, NMHH (Hungarian National Media and Infocommunications Authority), 118
@NEoCOfficial, 206
Neotropical Primate Conservation, 228
Nepal, **204–207**
Nepal Telecom (NT), 204
Nepal Wireless Networking Project, 204
Nepali, 205
#NepalQuakeRelief, 206
Net neutrality, 58, 128
Netflix, 128
Netherlands, the, xii
Networked Readiness Index, 33
"New Revolution of the Angolan people," 10
News aggregation in Spain, 282–283
News regulations in Singapore, 265
Newspapers. *See* Journalists; Traditional media

Nexttel Cameroon, 34
N'gola Digital program, 9
Nguyen, Rita, 195–196
Nguyen Tan Dung, 343
NHN Corp, xxvi
Niantic. *See* Pokémon Go
Nido, 339
Nigeria, **208–213**
Nigeria Internet Group, 208
Nigerian 419, 211
Nigerian Communications Commission (NCC), 208
Nigerian Ericsson, 210
Nikki, Matti, 95
Nippon Telegraph (NTT), 155, 156
#NiUnaMenos, 14–15
nk.pl, 235
No. 2010/012 (2010, Cameroon), 36
No! anti-fracking movement, 7
#NonLibelousTweet, 232
North America, xx. *See also specific countries*
North Korea, **214–219**
Norway, xii
Norwegian Telenor, 195
#notocybercrimelaw, 232
NRC Market Research Ltd. study, 119
NSA (U.S. National Security Agency), 271–272
NT (Nepal Telecom), 204
Ntamack, Thierry Roland, 36
Nteff, Alain, 35
NTT (Nippon Telegraph), 155, 156
Nuro, 156
NUSNET, 262

Obama, Barack, 59
Obama, Michelle, 237
O'Brien, Denis, 153
Ochoa, Eduard, 128
Octopusocial, 338
Odnoklassniki, xv, xxix, 169, 250, 335
Ohanian, Alexis, xxxiii
OICQ, xxiv
OK.RU, xxix
OMG, 325
Onclickads.net, 222
ONE Championship, 264
One-internet gateway (Thailand), 304

Onion routers, 289
Online learning, 94, 109, 201, 302
Online piracy, 17–18, 157
Onlinekhabar, 205
"Ook Zij" (They Too), 287
Ooredoo, 6, 195, 239, 241, 309
Open Business Club (later Xing), xxvii, 105
"Open Homes, Open Hearts" Facebook page, 125
Open Library of Kazakhstan, 162
Opensooq.com, 174
OpenStreetMap, 206
Opera Software, 209
Orange, 259, 260
Orange Botswana (Vista), 25, 26
Orange Cameroon, 34
Orange Revolution, 249, 318
Orange Tunisie, 309
Orascom Telecom Algerie (Djezzy), 5–6
Orkut, xxvii, 28, 29, 320
Orunbekov, Dayirbek, 170–171
"OSmartPhona" campaign, 200
al-Ouda, Salman, 257
OuedKniss, 6
Outdoor 4G mobile service, 263–264
Over-the-top (OTT) services, 35
Oxford English Dictionary, 18
Ozone, xv

Pacific Caribbean Cable System (PCCS), 72
Paéz, Chiara, 14
Paid commentators, 48
Pakistan, **221–224**
Pakistan Telecommunications Act, 223
Pakistan Telecommunications Company Ltd. (PTCL), 221
Palacio, Emilio, 73
Panama, 117
Paraguay, **224–227**
Partner, 146
Party for Democracy and Development Through Unity (Suriname), 287
PCCS (Pacific Caribbean Cable System), 72
Peak connection speed, xii, 16
Pelephone, 146
People's Liberation Army (PLA), 50

People's Republic of China (PRC). *See* China
Periscope, xlii
Persia. *See* Iran
Person Finder, 206
Peru, **227–230**
Petzoldt, Oliver, 88
PewDiePie, 330
Philippines, **230–234**
Picaboo. *See* Snapchat
Pinterist, xxxiii, 98, 116, 330. *See also* Social media/social networking
Piracy, 17–18, 157
PLA (People's Liberation Army), 50
Plan Nacional de Desarrollo (National Development Plan), 58
Plurk, 295
Poderopedia, 338
Poke, xxxii
Pokémon Go, xlii, 52, 182, 290–291
Poland, **234–238**
Police department social media officers, 128
Politicians/politics
    Hungary, 120
    Iceland, 125
    Italy, 150–151
    Malaysia, 179
    Namibia, 201
    Romania, 245
    Senegal, 260–261
    social media use by, 68, 94–95
    Suriname, 286–287
    Taiwan, 294, 296
    Tanzania, 299
    Turkey, 315
    Venezuela, 339
    See also Elections
Popular Movement for the Liberation of Angola (MPLA), 8
#PorteOuverte, 99
Portland Communications report, 165
Posterous, 183
Pothole tweets, 117
#prayforMH370, 180
Prayuth Chan-o-cha, 302
Premchaiporn, Chiranuch, 303
Prenatal care, 35
La Prensa, 116

Presidential decree No. 1182 (Peru), 228

Prevention of Electronic Crimes (2016, Pakistan), 223

Pricing, difficulty of measuring, xii–xiii

Prince, Joseph, 264

Prinsloo, Behati, 201

Prison, El Salvador, 81

Privacy
France, 97–98
Germany, 104–105
Google and, xx
Hungary, 120
Ireland, 142, 143–144
Libya, 174
Namibia, 202
Peru, 228
Philippines, 232–233

Prodigy, xxxviii–xxxix, xxxix, xl

Produsers, 110

Progressive Reform Party (Suriname), 287

Project Loon, xiii

Protests, media used for
Egypt, 77
Euromaidan protests, 317
Gezi Park protests, 314
Iraq, 140
Libya, 174
Morocco, 188–189
Russia, 249, 251, 253
Saudi Arabia, 256
Suriname, 287
Ukraine, 318
Vietnam, 344
Yemen, 348
See also Arab Spring; specific protests

Provost, Anne Marie, 39–40

P-Square, 210

PT Telekomunikasi Indonesia International (Telin), 306

PTCL (Pakistan Telecommunications Company Ltd.), 221

PTV Sports, 222

Pussy Riot, 252

Putin, Vladamir, 249, 252, 318

Pwar, 196

Pyaw Kyi, 196

Pyra Labs, xxxiv

Qatar, **239–242**

Qatar Telecom, 239

Q-link, xxxviii

Qorqmaymiz, 336

QQ, xxiv, xl

Quakemap.org, 206

Quantum Computer, xxxix

Quito, Ecuador, 72

Qzone, xxiv, xxix, xli

Racism, 272

Radio. See Traditional media

Radio Munna, 23

Raghib, Amine, 190

Rahman, Washiqur, 23

RaincheckPH, 233

Rakuten, xxvi

Rambler&Co, 251

Ramón Torres Psicólogo, 225

Rapid Response Team, 207

Rate, 85

Razak, Najib, 179

Red Nacional de Ciencia, Educación y Tecnología (INTERRED–CETCOL), 55

"Red Patria," 339

Red Social, 61

Red squares, 39

Red Star, 216

Reddit, xv, xxxiii–xxxiv

Refugee crisis, 125, 289–290

Refugees and Repatriations Ministry (Afghanistan), 4

#RefugeesWelcome, 291

Regional Informatics Networks for Africa project, 208

Regulation of Interception of Communications and Provision of Communication-Related Information (RICA) Act, 272

Religion, 108, 147, 255–256

Remittances, 79–80, 351

Renamo, 192

Renewable sources of electricity, 243

Ren-Ren, 52

Renren, xxix

Renzi, Matteo, 150–151

Replay, xxxii

"Report on the Internet Economy," 120

Republic of China (ROC). *See* Taiwan
Rhodes, Cecil John, statue of, 273
#RhodesMustFall, 273
RICA (Regulation of Interception of Communications and Provision of Communication-Related Information) Act, 272
Rightel, 137
RobiAxiata Limited, 23
Roboscan, 243
ROC (Republic of China). *See* Taiwan
Rodman, Dennis, 215
Romania, **243–247**
Roy, Avijit, 23
Royal Decree No. M/17 (2008, Saudi Arabia), 256
Rumors, 192–193, 196–197
Russia, **247–254**
    control of internet by, 251–252, 253, 319
    history, 247–248
    internet penetration, 250
    Moscow Spring, 253
    national security concerns, 249–250
    Putin and, 249
    Ukraine and, 319
    United States compared, 248
"Russian Federation's Doctrine of Information Security," 249

Safety Check, 43–44, 99, 206
Said, Khaled, 76
Sajde KZ, 163
Salam Pax, 139
Salas, José Antonio Rodríguez, 283–284
Saleh, Ali Abdullah, 348–349
Samsung Senegal, 259
Sánchez, Yoani, 62
Sankalpa, 207
Santadar, 29–30
Santiesteban-Prats, Angel, 62
Sanz, Fernando, xxxiii, 13–14
SAPO Timor-Leste, 307
Sarawak Report, 179
SAT-3 (South Atlantic 3) cable, 34
Satellite Solutions Worldwide, 236
Saudi Arabia, **255–258**, 320
Saudi Telecom Company, 255–256
SBC Communications, xl

Scare PewDiePie, 330
Schrems, Maximilian, 143–144
Scottish Rural Development Programme (SRDP), 327
SEACOM cable, 298
Search engines
    Cameroon, 35
    Ecuador, 71
    Germany, 103
    human search engines in Iceland, 124
    Kazakhstan, 162
    North Korea, 216
    South Korea, 277
    Taiwan, 293
    Tunisia, 311–312
    Vietnam, 344
    See also specific search engines by name
Sears, xxxviii
Sedition, 26
Sedition Act (Singapore), 265
SEE Egypt, 77
Self-censorship, 115, 116, 300, 335. *See also* Government control of Internet
Selfies, 231
Senegal, **258–262**
Seneweb, 259
#SenStopEbola, 261
Serfaty, Dan, xxvii
al-Sheikh, Abdul Aziz, 256
Shufersal, 147
al-Shugairi, Ahmed, 257
@ sign, 150
Al Sihir, Kadim, 140
Silbermann, Ben, xxxiii
Silent stands, 76
"Silicon Savanah," 164
"Silicon Wadi," 146
SIM registration, 193, 209, 272, 309–310
Simferopol internet exchange point (IXP), 319
Sina Weibo, xxxiv–xxxv, 52, 53, 295
Singapore, **262–266**
Singapore Airlines, 264
Singapore–Massachusetts Institute of Technology Alliance for Research and Technology (SMART), 264
Singnet, 262

SingTel, 262–266
SixDegrees, xxiii, xl
Skjerdal, Terje, 352–353
Skolkovo hub, 249–250
Skype, xxiv–xxv, 72, 150, 264
Skyrock, 98
Skyrock FM, 98
Sloth rescue, 71–72
SMART, 264
Smart filtering, 135
Smart Government initiative, 321
Smartphones. *See* Mobile device use
Smuggling, 217
Snake game, xvi
Snapchat
    France, 98
    Italy, 150
    overview of, xxxi–xxxii
    Somalia, 269
    United States, 332
Snapdeal, 127
Social activism
    Honduras, 117
    Indonesia, 130, 133
    Kenya, 164–165
    Malaysia, 179
    Morocco, 190
    Namibia, 201
    Peru, 228
    Romania, 244
    Senegal, 261
    Somalia, 269
    South Africa, 272, 273–274
    Syria, 291
    Yemen, 348–349
Social justice, Mapuche peoples, 43, 44
Social Media Lab, 128
Social media officers, 128
Social Media Summit, 2–3
Social media/social networking
    Afghanistan, 2
    Africa, xviii–xix
    Algeria, 6, 7
    Angola, 10
    apps for, xvii–sviii
    Argentina, 12–14
    Australia, 16
    Bangladesh, 23–24
    Botswana, 25

    Brazil, 27–30
    Cameroon, 34–36
    Canada, 38–41
    China, 49
    Colombia, 57
    comparison of, xiv
    Cuba, 60–61
    defined, xiv, xxvi
    Denmark, 66–67
    Ecuador, 71, 72–73
    Egyptian Revolution and, 76
    El Salvador, 81
    Estonia, 84–85
    Ethiopia, 89
    Finland, 94–95
    France, 97, 98
    Georgia, 102
    Germany, 103–104
    Ghana, 109
    government control of, 7
    Greece, 110, 111–112
    Honduras, 130–133
    Hong Kong, 52–53
    human search engines in Iceland, 124
    Hungary, 119
    Iceland, 123, 124–125
    India, 128
    Indonesia, 130
    Iran, 134, 135–137
    Ireland, 142–143
    Israel, 147
    Italy, 149–150
    Japan, 158
    Kazakhstan, 163
    Kenya, 165–166
    Libya, 173–175
    Macau, 54–55
    Malaysia, 177, 178–180
    Mexico, 182–183
    Morocco, 188–189
    Mozambique, 192
    Myanmar, 194–195, 196
    Nepal, 206–207
    North Korea, 216, 217
    overview of, xxiii–xxiv, xxvi–xxx
    Pakistan, 222
    Paraguay, 225
    penetration versus importance of
        access, 66

Peru, 227, 229
Philippines, 231–232
Poland, 235–236
popularity of, xlii, xliii
Qatar, 239–240
range of apps available, xxiii
reasons to use, xiv–xv
Romania, 244–245
Saudi Arabia, 255, 257
Senegal, 259–260
Singapore, 264
social activism with, 117
Somalia, 268–269
South Africa, 270–271, 273–274
South Korea, 275–278
Suriname, 285–287
Syria, 288, 289
Taiwan, 294
Tanzania, 299
Thailand, 301, 302
Timor-Leste, 306–307
Tunisia, 311
Turkey, 313–314
Ukraine, 318
United Arab Emirates, 321
United Kingdom, 325–326
United States, 328–332
Uzbekistan, 335–336
Venezuela, 337, 338–341
Vietnam, 343–344
Yemen, 347–348
Zimbabwe, 351–354
See also Blogs
Social Suriname, 286
Sol, Armando Calderón, 80
Somalia, xvii, **266–270**
#SomeoneTellCNN, 166
Somkid Jatusripitak, 304
#SOSVenezuela, 340–341
So-Net, 156
Sony Entertainment hack, 218
Soon, Tay Eng, 262
South Africa, **270–274**
South Atlantic 3 (SAT-3) cable, 34
South Korea, **274–281**
  app usage, 275–276
  connection speeds, xii, 326
  cybercrime, 279
  demilitarized zone, 214

eating broadcasts, 277–278
  internet penetration, 275
  North Korea media banned, 217, 278
  social media use, 275–278
  websites banned in North Korea, 215
Space race, 248
Spain, **281–284**
Sparks, xxvii–xxviii
Sparrow, Penny, 272
SparrowSMS, 207
Speak-to-tweet, 60, 61
Speigel, Evan, xxxi–xxxii
SpyEye, 7
SRDP (Scottish Rural Development
    Programme), 327
Stache Passions, xxiii
Stackoverflow, 127
Star balloons, 277
Star Trek Dating, xxiii
StarHub, 262–263, 264
Start-up businesses, 84–85, 236–237,
    344–345
"Start-Up Nation," 146
State of the Internet report, xii, 65, 173
States of emergency in Tunisia,
    310–311
Statista survey, 95
StatsMonkey, 35
#staystrong, 180
Streaming media, 68, 210, 277
StumbleUpon, xxxiii
Suess, Randy, xxxvii
Sunday Standard, 26
#sunu2012, 261
Sunu2012, 260–261
#sunucause, 261
Sunucause, 261
SuP Media, 318
SuperWiFi broadband network in
    Jamaica, 153
#SupportYemen, 349
Support Yemen, 349
Suriname, **284–288**
Surveillance, 29, 77. *See also*
    Government control of Internet;
    Self-censorship
Sweden, xii
Swiftkey, 245
Switzerland, xii

#SY, 349
Synergie Media, 188, 189
Syria, **288–291**
"Syria is Calling" Facebook page, 125
Syrian Refugee Support Group, 40–41
Systrom Kevin, xxxi

Tablets. *See* Mobile device use
Tablets in School (TIS), 154
Taiwan, **293–297**
Taliban, 1, 2
Tallinn, Estonia, 86
Tamazight, 7
Tank Commando, 295
Tanzania, **297–300**
Tanzania Communications Commission
    (TCC), 298
Tanzania Communications Regulatory
    Authority (TCRA), 298
Tanzanian Broadcasting Commission
    (TBC), 298
Taringa!, xxxiii, 13–14, 116
Taringa! Creadores, 14
Tasmania. *See* Australia
Tax Committee of the Kazakh Finance
    Ministry, 163
Taxes, 73, 120, 143, 163
Taxi services, 106, 313
TBC (Tanzanian Broadcasting Commis-
    sion), 298
TCC (Tanzania Communications Com-
    mission), 298
TCRA (Tanzania Communications
    Regulatory Authority), 298
Technical Agency for Telecommunica-
    tions (l'Agence Tunisienne des
    Télécommunications), 311
Tele2, 162
Telecom Regulatory Authority of India
    (TRAI), 126, 128
Telecommunications Act (2001, Bangla-
    desh), 21
Telecommunications Regulatory Author-
    ity (TRA), 321–322
Telecommunications Regulatory National
    Agency (Agence Nationale de
    Reglementation des Telecommunica-
    tions), 189
Telefónica, 72

Telegram, 28, 137, 141
Telegraph, 137
Tele-medicine, 26, 35
Telemor (Viettel Global Investment JSC),
    306
Telenor, 119
Telephone and Telephone Board (Bangla-
    desh), 21
Telephony. *See* Voice over internet
    protocol (VoIP)
Telesur, 286
Television. *See* Traditional media
Telin (PT Telekomunikasi Indonesia
    International), 306
Telkom, 271
Telstra Corporation, 17
Tenashar, 264
Tencent, xxiv, xli, xlii–xliii, 49
La Tendedera, 61
Tengrinews.kz, 163
Teoti.com, xxxiii
Tequila, Tila, 264
Terrorism, 98–99, 256. *See also* Daesh
    (Islamic State)
Tetum, 305
Th3professional blog, 190
Thailand, **301–304**
Al Thani, Hamad bin Khalifa, 239
The Source, xxxviii, xxxix
#ThisFlag campaign, 352
They Too ("Ook Zij"), 287
Threat Intelligence Brief, 18
Threema, 104
Throttling, 135
Tiger Leap project, 83
Tigo Communications, 115–116
Tigo Senegal, 259
Time magazine selfies capital article, 231
Time Warner, xl
Timor Telcom, 305–306
Timor-Leste (East Timor), **305–308**
TIS (Tablets in School), 154
Tokyo Olympics, effects on internet of,
    158–159
Tomlinson, Ray, xxxvii
Torres, Ramón, 225
Tourism, 124, 286, 326–327
TRA (Telecommunications Regulatory
    Authority), 321–322

Traditional media
  Georgia, 104
  Ghana, 108
  Iraq, 139–140
  Israel, 144, 147
  Kenya, 165
  Mozambique, 192
  social media use and, 67–68
  Somalia, 267
  Spain, 283
  streaming media and, 68
  Timor-Leste, 307
TRAI (Telecom Regulatory Authority of
  India), 126, 128
Trans-Asia-Europe Fiber Optic Line,
  169, 334
Transparency International survey,
  101–102
Tree removal, 344
Trintex, xxxviii
Trivia Crack, 14
Troll farms, 252
Trubshaw, Roy, xxxviii
Tsai Ing-wen, 296
Tudou, xxx–xxxi
Tuenti, 281–282, 337
Tuenti Local, 282
Tuenti Móvil, 282
Tumblr, 183. *See also* Social media/social
  networking
Tunisia, **308–312**
  Arab Spring, 308, 312
  Bouazizi self-immolation in, 174, 348
  government control of Internet, 308,
    309–310
  language use and issues, 311–312
  states of emergency in, 310–311
  Yemen compared, 348
Tunisian Internet Agency (ATI, Agence
  Tunisienne d'Internet), 311
"Tunisian Jasmine," 241
Tunisie Télécom, 309
Turkey, **313–316**
Tweefs, 165
"Tweeting Pothole," 117
Twitch.tv, 331
Twitter
  Africa, xix
  Algeria, 7

Canada, 38
Colombia, 57
country withheld content tool,
  302–303
Cuba, 60
Daesh (Islamic State) using, 141
Denmark, 66
Ecuador, 72
El Salvador, 81
France, 98
Greece, 112
hashtags on, 211
historic role plays on, 281
Honduras, 116
Hong Kong, 53
Indonesia, 130–131
Instagram advertised on, xxxi
Iran, 136
Islamic State (Daesh) using, 141
Italy, 150
Japan, 158
job recruitment using, 295
Kazakhstan, 163
Kenya, 165–166
length limits, xliii
Libya, 173–174
Malaysia, 179
Maple Spring in Canada, 39–40
Nepal, 204, 206
North Korea, 215, 217
overview of, xxxiv, xli
Pakistan, 222
Paraguay, 226
Peru, 228
Poland, 235
popularity of, xv
Qatar, 240
Russia, 250
Saudi Arabia, 255
Singapore, 264
Somalia, 269
South Africa, 271
Spain, 283–284
Suriname, 285–286
Syria, 289
Taiwan, 295
Thailand, 301, 302–303
Timor-Leste, 307
Tunisia, 311

Twitter (*continued*)
  Turkey, 314
  Ukraine, 318
  United Arab Emirates, 321
  United Kingdom, 325
  United States, 331, 332
  Uzbekistan, 335
  Venezuela, 337, 338, 339, 340
  World Cup Final tweets, 104
  Yemen, 347
  2channel, 157

UAE. *See* United Arab Emirates
UA-IX (Ukrainian Internet Exchange),
    317
Uber, 106
Ubiquitous computing, xiii
Udal'tsov, Sergei, 252
UFO Social, xxiii
Ukraine, 229, 249, **317–320**
Ukrainian Internet Exchange (UA-IX),
    317
Ukrtelecom, 317
UM study, 232
Underwater cable (EMOS-1), 145
"The Unintelligent Facebook Users,"
    231
Union of Soviet Socialist Republics
    (USSR), 247–248
United Arab Emirates, xvii, **320–324**
United Kingdom, **324–327**
United Online, Inc., xli
United States, **328–333**
  ARPA created, xi
  connection speeds, xii
  hate speech, 229
  Russia compared, 248
  social media use, 328–332
El Universo, 73
UPC, 118
Upload speeds, defined, xxii
Upstream filters, 170
Urdu, 221
Urdupoint, 222
Uriminzokkiri, 217–218
U.S. Agency for International Develop-
    ment (USAID), 61
U.S. Federal Communications Commis-
    sion (FCC), 234

U.S. National Security Agency (NSA),
    271–272
Ushahidi, 164–165
USSR (Union of Soviet Socialist Repub-
    lics), 247–248
Utube.uz, 335
Uzbekistan, **333–336**
Uztelecom, 335

Varzesh3.com, 136
Venezuela, **337–342**
Vera, Carlos, 73
Verizon, xlii
Viacom, xlii
Viadeo, xxvii
Vibanda vya simu (calling booths), 298
Viber, xxvi, 195
Video gamers, watching, 330–331
Vietnam, **342–346**
Viettel Global Investment JSC (Telemor),
    306
Vigilantism, Peru, 228–229
Violence in Somalia, 269
Virtual private networks (VPNs), 48
Visafone, 208
Vista (Orange Botswana), 25, 26
Vive Digital (Live Digital), 55–56
VKontakte
  history of, xxix
  Kazakhstan, 163
  Kyrgyzstan, 169
  popularity of, xv
  Russia, 250
  Ukraine, 318
  Uzbekistan, 335
Vodacom, 192, 271
Vodacome, 298
Vodafone, 119
Voice over internet protocol (VoIP)
  Brazil, 28–29
  Jamaica, 154
  Moroccan ban on, 189–190
  overview of, xxiv–xxvi
  Tunisia, 309
  United Arab Emirates, 322
  See also WhatsApp
VoIP. *See* Voice over internet protocol
    (VoIP)
von Meister, William, xxxviii

Voting, internet for, 84
VPN-Hotspot Shield, 136, 137
VPNs (virtual private networks), 48
VR-Zone, 264
VTR, 42

Wade, Abdoulaye, 260–261
Walla.co.il, 147
Wang, Gary, xxx–xxxi
WASC (West African Submarine Cable), 34
Wataniya Telecom Algerie (Ooredoo), 6
"Water Army," 48
Waymu.pk, 223
Waze, 146
Weather apps, 233
Web apps, xvi
Web defacement, 7
Web Pass, 209–210
Webretho, 344–345
Webroot, 18
Website use
    Argentina, 12
    Bangladesh, 22–23
    Chile, 45
    El Salvador, 81
    Ghana, 109
    Greece, 112
    Honduras, 116, 132
    Hungary, 119
    Iceland, 123
    India, 127
    Iran, 134–135, 136
    Iraq, 140
    Ireland, 142
    Israel, 147
    Italy, 149
    Jamaica, 153
    Kazakhstan, 162
    Kenya, 165
    Kyrgyzstan, 168–169
    Libya, 173–174
    Malaysia, 177
    Mexico, 181
    Mongolia, 185–186
    Mozambique, 192
    Myanmar, 196
    Nepal, 205
    Pakistan, 222

Peru, 227
Philippines, 231
Russia, 250–251
Senegal, 260
Spain, 282
Suriname, 285
Taiwan, 293, 294
United States, 329
Uzbekistan, 335
Venezuela, 337, 338
Vietnam, 342
See also specific websites by name
WeChat
    history of, xli
    Hong Kong, 53
    Malaysia, 177
    popularity of, xliii
    Singapore, 264
Weibo, xxxiv–xxxv, 52, 295
Weinreich, Andrew, xl
Weiser, Mark, xiii
Weixin, xxiv
Weizmann Institution, 145
West African Submarine Cable (WASC), 34
Wetruwe Mapuche, 43
WhatsApp
    Brazil, 28–29
    Chile, 44–45
    Daesh using, 141
    Ecuador, 72
    Egypt, 77
    encryption by, xlii
    France, 98
    Germany, 103–104
    Honduras, 116
    Hong Kong, 52
    Iran, 137
    Israel, 147
    Italy, 150
    Jamaica, 154
    Mexico, 182
    Mozambique, 193
    Namibia, 200
    overview of, xxv–xxvi, xli
    purchased by Facebook, xxvi
    Saudi Arabia, 255
    Singapore, 264
    South Africa, 271

WhatsApp (*continued*)
  Syria, 289
  Tanzania, 299
  United Arab Emirates, 322
  United Kingdom, 325
"Where Is Raed?," 139
Whisper, 141
Widowo, Joko, 131
Wiener, Norbert, 248
Wiexin. *See* WeChat
Wi-Fi kiosks, 327
WiFiChileGob, 42
WikiBilim Public Foundation, 162
Wikileaks, 174
Wikipedia, 204, 246
Wikis, 94
Wildlife smuggling, 228
Windows Store, xvii
Wolof, 260
Women, violence against, 14–15
Women Journalists Without Chains, 349
Women's rights, 3–4
WordPress, xxxiv, 183
World Cup Final tweets, 104
World Map of Social Networks, xv
World Wide Web, xxxix
Wretch.cc, 294

Xiaonei, xxix
Xing, xxvii, 105
X-road system, 83

yad2.co.il, 147
Yahoo, xxx, xlii, 136, 222
Yahoo Search, 325
Yammer, 94
Yandex, 162, 250–251
Yanukovych, Viktor, 249
Yemeksepeti.com, 313
#Yemen, 348
Yemen, 173, **347–350**
Yessentayeva, Bayan, 163
Yingluck Shinawatra, 302
Ynet.co.il, 147
Youku, xxxi
Youku Tudou, xxx–xxxi
YouthStream Media Networks, xl

YouTube
  Chile, 42–43
  Denmark, 66
  Egypt, 77
  El Salvador, 81
  France, 98
  Georgia, 102
  Ghana, 108
  Greece, 112
  history of, xxx, xli
  Honduras, 116
  Hungary, 119
  Ireland, 142
  Kenya, 165
  Libya, 173
  Mongolia, 186
  Morocco, 190
  North Korea, 215, 217–218
  notification policy, xliii
  Pakistan, 222, 223
  Poland, 237
  purchased by Google, xli
  Romania, 246
  Saudi Arabia, 255
  South Korea, 277
  Spain, 283
  Suriname, 285
  Thailand, 302
  Turkey, 315
  United States, 330
  Uzbekistan, 335
YY, xxxi

Zalo, 343, 344
Zang, Arthur, 35
Zap.co.il, 147
Zelaya, Manuel, 116–117
Zennström, Niklas, xxv
Zetas drug cartel, 183
ZhZh, 251
Zimbabwe, **351–354**
Zimbra, 72
Zindagi Gulzar Hai (Life Is a Garden), 222–223
ZiyoNet, 335
Zolspot, 353
Zuckerberg, Mark, xxvii, xli
ZunZuneo, 61, 338